SCIENTIFIC AMERICAN | Resource Library

Earth Sciences 3

SCIENTIFIC AMERICAN

Resource Library

READINGS IN THE
Earth Sciences VOLUME 3

OFFPRINTS 875–899

W. H. FREEMAN AND COMPANY   San Francisco

Each of the articles in this volume is available as a separate Offprint. For a complete listing of Offprints available in the Life Sciences, Chemistry, Physics, Technology, Psychology, the Social Sciences, the History and Philosophy of Science, and the Earth Sciences, write to W. H. Freeman and Company, 660 Market Street, San Francisco, California 94104 or to W. H. Freeman and Company, Ltd., 58 Kings Road, Reading, England RG1 3AA.

Printed in the United States of America
Library of Congress Catalog Card Number: 69–15600
International Standard Book Number: 0–7167–0988–0

Organization of the  Resource Library

Subject Series:

The *SCIENTIFIC AMERICAN Resource Library* is a multi-volume compilation of articles selected from the magazine. These are organized under five subject classifications.

> Readings in Earth Sciences (3 volumes)
> Readings in Life Sciences (10 volumes)
> Readings in Physical Sciences and Technology (3 volumes)
> Readings in Psychology (3 volumes)
> Readings in Social Sciences (2 volumes)

Numbering System:

Each article is numbered and the articles in the volumes are arranged in numerical order. The numbers assigned to each series are:

> Earth Sciences, 801–999.
> Life Sciences, 1–199 and 1001–1999.
> Physical Sciences and Technology, 201–399.
> Psychology, 401–599.
> Social Sciences, 601–799.

Topic Index:

This index classifies the Readings in Earth Sciences by topic. Note that articles from other subject series that are relevant to the topic are also listed.

Author Index:

The authors of the articles in all five subject series are given. The numbers after the authors' names are the article numbers, not page numbers.

Scientific American Offprints:

Every article in these volumes is published separately in the SCIENTIFIC AMERICAN Offprint Series and may be purchased in any quantity. Order by article number and title.
The number of each Offprint corresponds to the number on each article in the Resource Library.
A catalog of SCIENTIFIC AMERICAN Offprints may be obtained from the publisher.

Additions to the Series:

New titles are added to the SCIENTIFIC AMERICAN Offprint Series every month and when enough new articles on a subject are available, a new bound volume will be added to the *SCIENTIFIC AMERICAN Resource Library*.

W. H. Freeman and Company
660 Market Street, San Francisco, California 94104
58 Kings Road, Reading, England RG1 3AA

Contents

VOLUME **1**

Topic Index
Author Index

Introduction

VOLUME **2**

Topic Index
Author Index

VOLUME **3**

Topic Index
Author Index

Topic Index

(This index includes not only the articles in Readings in Earth Sciences
but also relevant articles from other subject series.)

Author Index

(The authors of the articles in all five subject series are given here.)

A Abelson, P. H. 101
Adams, E. 1081
Adams, R. M. 606
Adolph, E. F. 1067
Agranoff, B. W. 1077
Alder, B. J. 265
Alexander, A. S. 650
Alfvén, H. 311
Allen, R. D. 182
Allfrey, V. G. 92
Allison, A. C. 1065, 1085
Almond, R. 534
Amerine, M. A. 190
Amoore, J. E. 297
Anderson, D. L. 855, 896
Andrew, R. J. 627
Arditti, J. 1031
Arnon, D. I. 75
Asch, S. E. 450
Asmundson, S. J. 1219
Astin, A. V. 326
Atkinson, R. C. 538
Attneave, F. 540
Axelrod, J. 1015

B Bailey, H. S., Jr. 830
Baker, D. J., Jr. 890
Baker, P. F. 1038
Bales, R. F. 451
Bandura, A. 505
Barghoorn, E. S. 895
Barnea, J. 898
Barnes, V. E. 802
Barnett, S. A. 1060
Barron, F. 432, 483
Bartholomew, G. A. 1120
Bascom, W. 828, 845, 887
Baserga, R. 165
Bass, A. M. 263
Bassett, C. A. L. 1021
Bassham, J. A. 122
Batra, L. R. 1086
Batra, S. W. T. 1086
Beadle, G. W. 1, 1214
Beale, I. L. 535
Bearn, A. G. 150
Beermann, W. 180
Bell, R. H. V. 1228
Bennet, E. L. 541
Bennet-Clark, H. C. 1183
Benson, H. 1242
Benzer S. 120
Benzinger, T. H. 129
Beranek, L. L. 306
Berelson B. 621
Berger, H. 287
Berkowitz, L. 481
Berlyne, D. E. 500
Bernal, J. D. 267
Berns, M. W. 1170
Bernstein, J. 829
Beroza, M. 189
Best, J. B. 149
Bethe, H. A. 201
Bettelheim, B. 433, 439
Biale, J. B. 118
Biddulph, S. & O. 53

Bilaniuk, O. 284
Billingham, R. E. 148
Binford, L. R. 643
Binford, S. R. 643
Birch, H. G. 529
Bitterman, M. E. 490
Blackler, A. W. 94
Bloom, J. L. 339
Blough, D. S. 458
Bodmer, W. F. 1199
Boehm, G. A. W. 258
Boerma, A. H. 1186
Bogert, C. M. 1119
Bolin, B. 1193
Bonner, J. T. 164, 1051, 1145
Bormann, F. H. 1202
Bornstein, P. 1225
Botelho, S. Y. 194
Bower, T. G. R. 502, 539
Bowers, R. 332
Boycott, B. B. 1006
Brachet, J. 90
Brady, J. V. 425
Bragg, L. 325
Braidwood, R. J. 605
Braude, A. I. 177
Braun, A. C. 1024
Brazier, M. A. B. 125
Brett, J. R. 1019
Brindley, G. S. 1089
Britten, R. J. 1173
Broadbent, D. E. 467
Bronowski, J. 291
Brooks, H. 332
Broom, R. 832
Brower, L. P. 1133
Brown, H. 102, 1198
Brown, J. F., Jr. 280
Brown, J. H. 1236
Brown, L. R. 1196
Brues, C. T. 838
Buchhold, T. A. 270
Bullard, E. 880
Bullen, K. E. 804
Bunnell, S., Jr. 483
Burbidge, G. 202, 203, 305
Burbidge, M. 203
Burnet, M. 2, 3, 78, 138
Bustad, L. K. 1045
Buswell, A. M. 262
Butler, R. A. 426
Butler, W. L. 107
Butterfield, H. 607

C Cairns, J. 1030
Calhoun, J. B. 506
Calvin, M. 308
Camhi, J. M. 1231
Caplan, N. S. 638
Carey, N. 4
Carr, A. 1010
Carter, A. P. 629
Cattell, R. 475
Cavalli-Sforza, L. L. 1154, 1199
Ceraso, J. 509
Champagnat, A. 1020

Changeaux, J. 1008
Chapanis, A. 496
Chapman, C. B. 1011
Chess, S. 529
Chew, G. F. 296
Chisolm, J. J., Jr. 1211
Clark, B. F. C. 1092
Clark, J. D. 820
Clark, J. R. 1135
Clarke, C. A. 1126
Clements, J. A. 142
Clevenger, S. 186
Clever, U. 180
Cloud, P. 1192
Cobb, W. C. 109
Cockrill, R. 1088
Coe, M. D. 648
Cohen, J. 427, 489
Colbert, E. H. 806
Cole, L. C. 144
Collier, H. O. J. 132, 169
Comer, J. P. 633
Comroe, J. H., Jr. 1034
Constantinides, P. F. 4
Converse, P. E. 656
Cook, E. 667
Cooper, A. F., Jr. 1237
Cooper, C. F. 1099
Coopersmith, S. 511
Corballis, M. C. 535
Corey, R. B. 31
Crary, A. P. 857
Crawford, B., Jr. 257
Crick, F. H. C. 5, 54, 123, 1052
Crombie, A. C. 184
Crow, J. F. 55
Crowe, J. H. 1237
Cruxent, J. M. 652
Csapo, A. 163
Cuff, F. B., Jr. 204
Curtis, B. C. 1140

D Darrow, K. K. 205
Davenport, H. W. 1240
Davidson, E. H. 1013
Davis, H. M. 206
Davis, K. 645, 659
Dawkins, M. J. R. 1018
Dayhoff, M. O. 1148
De Benedetti, S. 207, 271
de Duve, C. 156
Deering, R. A. 143
Deevey, E. S., Jr. 608, 811, 834, 840, 1195
de Heinzelin, J. 613
Delwiche, C. C. 1194
Denenberg, V. H. 478
Denton, E. 1209
Derjaguin, B. V. 266
DeVore, I. 614
Diamond, M. C. 541
DiCara, L. V. 525
Dickerson, R. E. 1245
Dietz, R. S. 801, 866, 892, 899
Dilger, W. C. 1049
Dirac, P. A. M. 292
Dobzhansky, T. 6, 609

SCIENTIFIC AMERICAN | Resource Library

Earth Sciences 3

SCIENTIFIC AMERICAN December 1968, Vol. 219, No. 6, pp. 60–70

OFFPRINT **875**

SEA-FLOOR SPREADING

by J. R. Heirtzler

Geophysical phenomena ranging from earthquakes to continental drift are being explained by a new theory that gives promise of eventually relating geomagnetism and the earth's internal and orbital dynamics.

Comprehensive new theories that rationalize large numbers of observations and explain major aspects of the physical world are rare in any field of investigation. Such a synthesis may be within reach in geophysics. The past few years have seen the emergence of a new theory concerning systematic movements of the sea floor. It deals with vast and formerly unsuspected forces that churn the interior of the earth and account for the arrangement of ocean basins and land masses as we know them today. The theory is based on a variety of observations and hypotheses concerning the topography of the sea floor and the distribution of its sediments, the occurrence of faults and earthquakes, the internal structure of the earth, its magnetic field and periodic reversals of the field. It neatly supports the developing theory of continental drift. Together these theories have already been successful in explaining many surface features of the earth and providing information on internal earth processes. And it is possible that their full importance has yet to be appreciated—that they point toward a major synthesis relating the internal dynamics of the earth, its magnetic field and the dynamics of its orbital motions.

History of the Theory

The stage was set for the discovery of sea-floor spreading by the long debate over continental drift [see "Continental Drift," by J. Tuzo Wilson, SCIENTIFIC AMERICAN Offprint 868, and "The Confirmation of Continental Drift," by Patrick M. Hurley, Offprint 874]. Evidence from the shape, geological structure and paleontology of various continents and, within the past 20 years, studies of the "paleomagnetism" frozen into volcanic rocks had suggested that the continents

have drifted to their present locations from appreciably different positions in the course of millions of years. Even after the possibility of such drifting began to be recognized, it was not at all clear what forces could have caused great land masses to move over the surface of the globe.

By the late 1950's oceanographers had discovered that a continuous range of undersea mountains twists and branches through the world's oceans, that this ridge is usually found in the middle of the ocean and that earthquakes are associated with it. Marine geologists were aware too of the striking youth of the ocean floor: no bottom samples were ever found to be older than the Cretaceous period, which began some 135 million years ago. About 1960 Harry H. Hess of Princeton University proposed that the ocean floor might be in motion. He suggested a kind of convective movement that forced material from deep in the earth to well up along the axis of the mid-ocean ridges, to spread outward across the ocean floor and to disappear into trenches at the edges of continents. (The hypothesis seemed particularly attractive in the case of the Pacific Ocean, which is bordered by trenches, but it was less satisfactory for other oceans, which lack them.)

At about the same time Ronald G. Mason, Arthur D. Raff and Victor Vacquier of the Scripps Institution of Oceanography discovered that the ocean floor off the west coast of North America exhibited a remarkably regular striped pattern of variations in magnetic intensity [see "The Magnetism of the Ocean Floor," by Arthur D. Raff; SCIENTIFIC AMERICAN, October, 1961]. The pattern suggested great filamentary magnetic bodies, oriented north-south and offset at intervals along distinct lines running approximately at a right angle to the linea-

tions. No structural features that could explain such a pattern had ever been observed. The origin of these unique magnetic bodies remained a mystery for nearly five years. In 1963, after it had been noted that a distinct magnetic body could often be detected at the axis of a

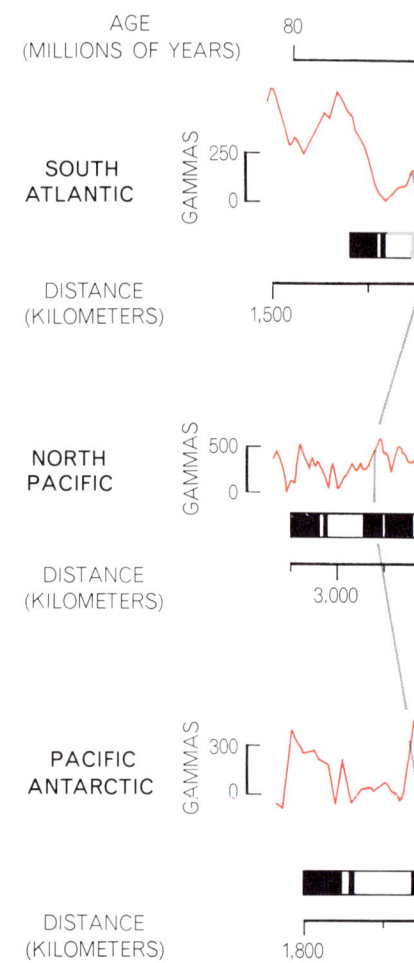

MAGNETIC ANOMALIES (*color*) recorded in all the world's oceans reveal the same succession of magnetic bodies (*black and white bands in strips*) parallel to the mid-ocean ridge. The bodies represent rock that

mid-ocean ridge, F. J. Vine and D. H. Matthews of the University of Cambridge proposed a convincing test of the hypothesis advanced by Hess. It was based on the discovery (which was just then being confirmed in detail) that the earth's magnetic field had reversed direction a number of times in past ages. They reasoned that if molten rock were pushed up along the axis of the mid-ocean ridge, it would become magnetized in the direction of the earth's prevailing magnetic field as it cooled. If the newly cooled material was subsequently pushed out away from the ridge, it would form strips of alternately "normal" and "reversed" magnetism, depending on the polarity of the earth's magnetic field when the rock solidified. A magnetometer at the surface of the ocean should detect these strips as positive or negative anomalies in the earth's smooth field.

Confirmation

As the Vine-Matthews proposal was being published, I was engaged, with colleagues from the Lamont Geological Observatory of Columbia University and the U.S. Naval Oceanographic Office, in a careful magnetic survey of the Reykjanes Ridge, a section of the Mid-Atlantic Ridge south of Iceland that was known to have large magnetic anomalies. We found that the anomalies were linear and symmetrically distributed parallel to the axis of the ridge. This strongly supported the idea of sea-floor spreading from the ridge and the formation of magnetic anomalies, just as Vine and Matthews had suggested. A little later Vine and J. Tuzo Wilson of the University of Toronto pointed out that the recent reversals of the field matched, one for one, part of the extensive pattern of magnetic lineations recorded just off the west coast of North America by Mason, Raff and Vacquier.

By 1965 it was clear to us and others that magnetism could be the key to reconstructing the history of the ocean floor and the movements of the conti-nents. In only three years a great deal has been learned. Indeed, so many different workers have made significant contributions that it is impossible to name them all in a brief review or even always to know who was the first to make a new observation or propose a new model.

Pioneering efforts to measure the earth's magnetic field at sea had begun at Lamont about 20 years ago. Simple and precise instruments were designed to be towed behind ships and in time efficient techniques were devised for recording, storing and interpreting the data to delineate sea-floor structures hidden under layers of sediments. In 1965, when the new importance of magnetic anomalies became apparent, we had a large stock of data from all the oceans of the world and computer techniques with which to process the data. Examining the data in the light of the new hypotheses, we were able to recognize the same sequence of magnetic bodies extending away from the axis of the mid-ocean

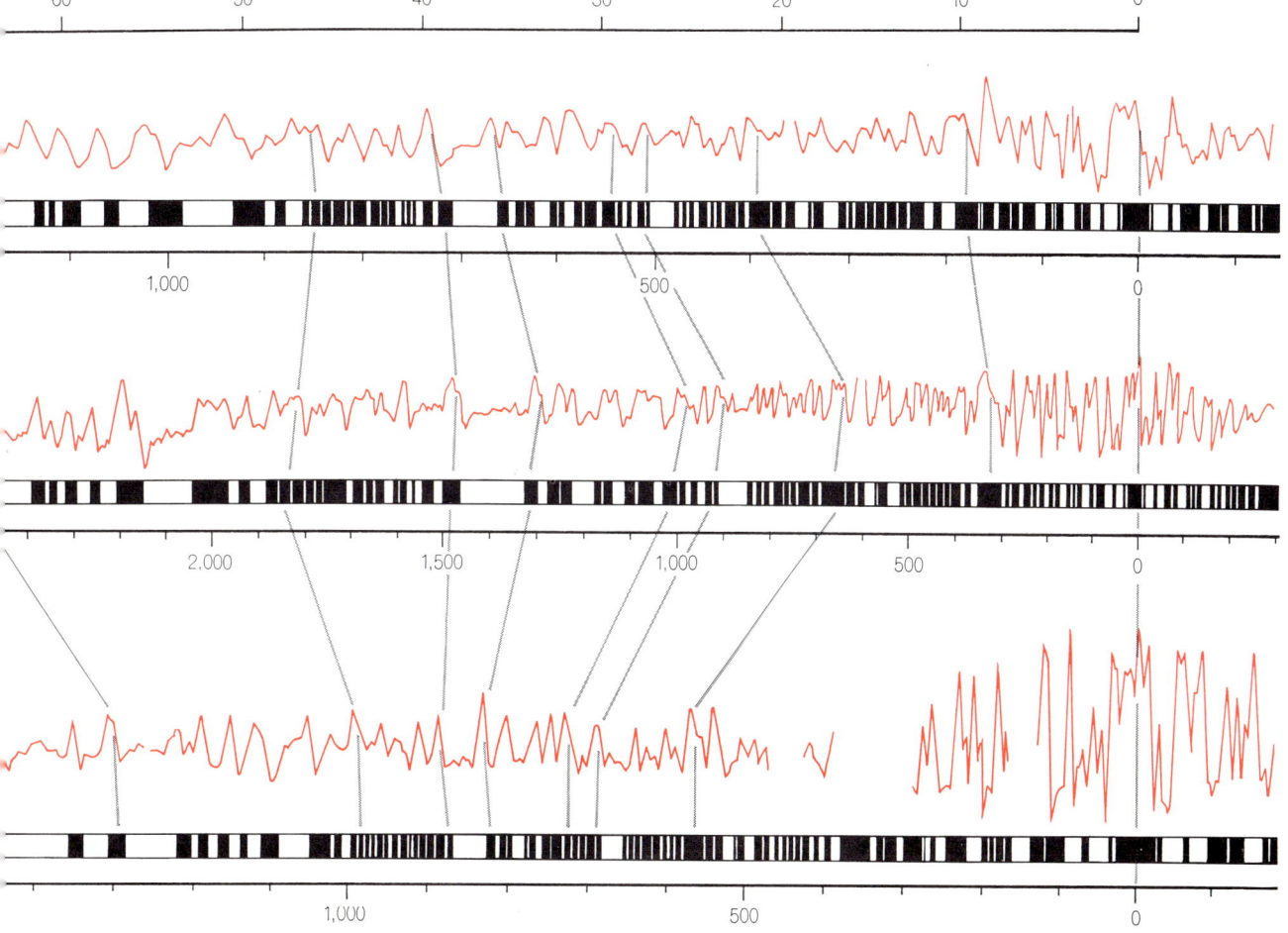

welled up along a "spreading axis" at the ridge during successive periods when the earth's magnetic field was "normal" as it is today or "reversed." The rock was magnetized in the ambient field and later forced out from the axis by subsequent flows. Here three magnetic traces are shown, from three oceans. The anomalies (in gammas, a measure of field intensity) and the magnetic bodies associated with each are spaced differently in each ocean because the spreading rates were different (the rate in the South Atlantic is believed to have been the most constant), but each ocean has the same sequence of 171 reversals extending back 76 million years.

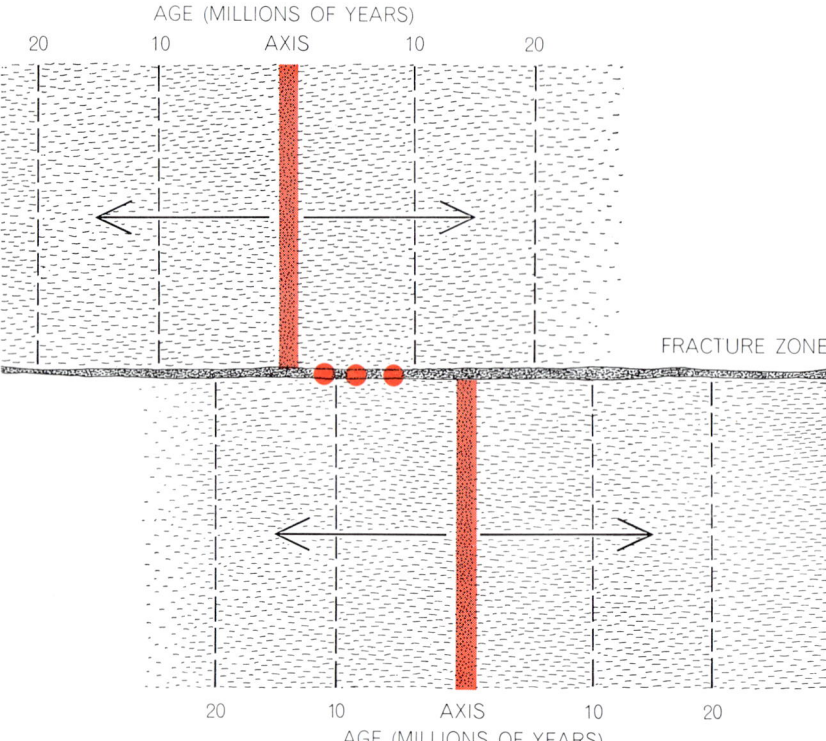

AGE (MILLIONS OF YEARS)

20   10   AXIS   10   20

FRACTURE ZONE

20   10   AXIS   10   20

AGE (MILLIONS OF YEARS)

MOLTEN ROCK wells up from the deep earth along a spreading axis, solidifies and is pushed out (*arrows*) by subsequent upwelling. The axis is offset by a fracture zone. Between two offset axes material on each side of the fracture zone moves in opposite directions and the friction between two blocks of the crust causes shallow earthquakes (*colored disks*).

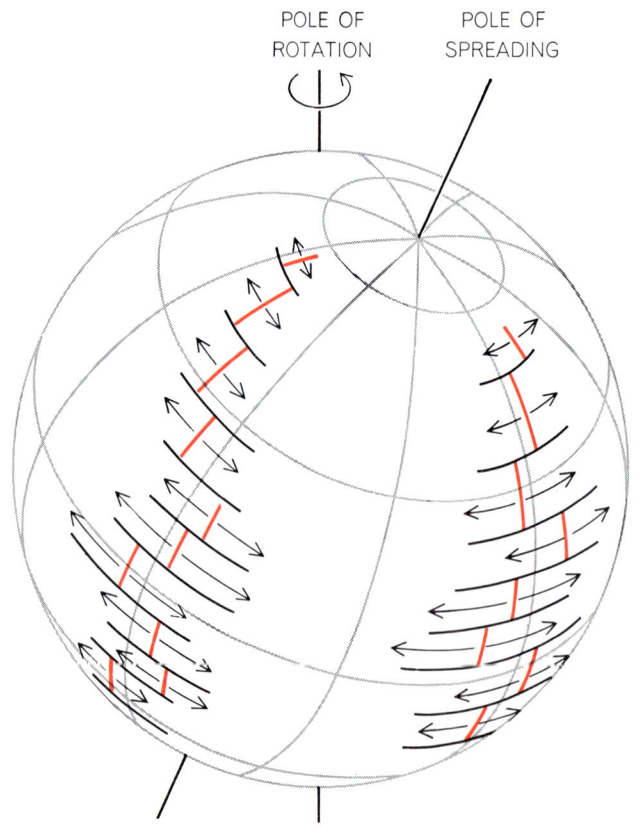

POLE OF
ROTATION

POLE OF
SPREADING

GEOMETRICAL RELATION between ridge axis and fracture zone becomes evident if one conceives of lines of latitude and longitude drawn about a "pole of spreading" rather than the pole of rotation. In each ocean the fracture zones are perpendicular to the spreading axis, and the rate of spreading (*arrows*) varies directly with distance from the equator.

ridge in the South Pacific, South Atlantic and Indian oceans.

Further examination has revealed a worldwide pattern of sea-floor spreading that makes sense of a wide variety of observations. It seems to explain the occurrence of many earthquakes. It establishes a detailed time scale for magnetic reversals and accounts for the direction and rate of continental drift. However, the precise geological events involved in the upwelling along the ridge and the downwelling at the edges of continents are still not understood in detail.

The Worldwide Pattern

At this point we know the main features of the spreading pattern in about half of the world's ocean areas, for the most part those that are adjacent to the better-explored sections of the mid-ocean ridge. The areas where the spreading pattern is unknown either are unexplored or simply resist explanation. For example, many oceans are not wide enough to show the very oldest magnetic bodies, and so not enough magnetic anomalies can be recorded to establish a rate of spreading. Some areas of the ocean floor show no magnetic anomalies. This may be because they are actually not part of the moving floor or because they were created in ancient geologic ages when the earth's field may not have been reversing. There may be places where the spreading is so slow that magnetic bodies crowd one another and confuse the magnetic-anomaly pattern. We have even found places where the axis of spreading is apparently under the continents or does not coincide with the axis of the mid-ocean-ridge system.

The mid-ocean ridge, the axis of the sea-floor spreading, does not meander smoothly from ocean to ocean; instead it is abruptly offset at many points. The offsets, or fracture zones, often extend to great distances on each side of the axis and are usually marked by some irregularity in the topography of the sea floor. Since material wells up at the axis and moves outward on both sides of a fracture zone, it is clear that there will be substantial friction along the zone between offset sections of the axis, where the material is moving in opposite directions [*see top illustration at left*]. Earthquakes, of course, are generated by just such rubbing and scraping between blocks of the earth's crust. Seismologists are now able to pinpoint the location and depth of an earthquake to within a few miles and also to identify the direction of initial motion of the crustal blocks in-

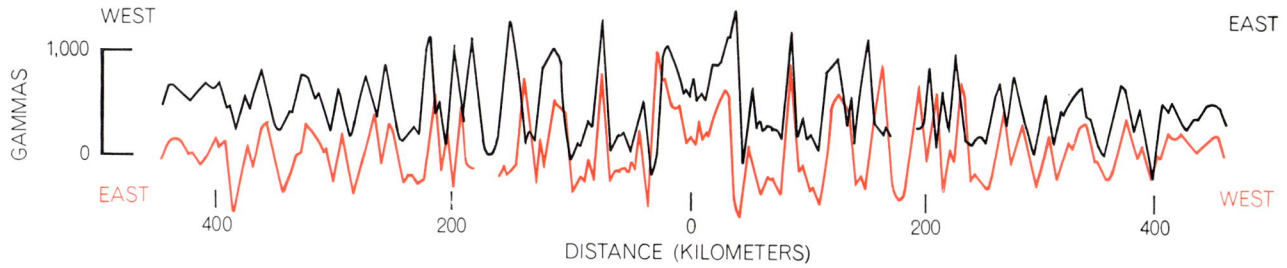

SYMMETRY of the magnetic record reflects the fact that similarly magnetized rock is pushed out on both sides of the axis. The sym-metry is demonstrated by reversing a record covering about 1,000 kilometers (*color*) and superposing it on the original (*black*).

volved. Such measurements confirm the theory of sea-floor spreading: Most earth-quakes along the ridge system occur in fracture zones between offset sections of the spreading axis—and the direction of first motion is just what one would pre-dict on the basis of the sea-floor motion.

Although it does not appear so on a map with the usual Mercator projection, an axis of spreading and its associated fracture zones are at right angles. In the case of the Pacific and South Atlantic oceans the relation is particularly clear and includes even the rate of spreading. If one conceives of latitude and longi-tude lines drawn not about the earth's pole of rotation but about a "spreading pole," then the axis of spreading is par-allel to the new lines of longitude and the fracture zones are perpendicular to the axis and parallel to the lines of lati-tude [*see bottom illustration on opposite page*]. Furthermore, the fastest spread-ing occurs along the equator of this new coordinate system and the rate decreases regularly with distance from the equa-tor—as if giant splits in the floor of the oceans were taking place along the axes. The spreading poles vary for different oceans. For the Pacific and South Atlan-tic they appear to be very close to the earth's magnetic poles: near Greenland and in Antarctica south of Australia. Spreading in the Indian Ocean seems to be related to a different pair of poles, which can be located, with less certainty, in North Africa and in the Pacific north of New Zealand. Present information in-dicates that spreading in the North At-lantic occurs about a third set of poles, and fracture zones in older parts of the northeast Pacific floor seem to have been associated with yet another set of poles. There are indications that at least some of and perhaps all the spreading axes are themselves in motion, so that the loca-tion of their poles must be changing.

Some workers, notably W. Jason Mor-gan of Princeton, have attempted to gen-eralize the motions of the earth's crust still further, suggesting that the crust of

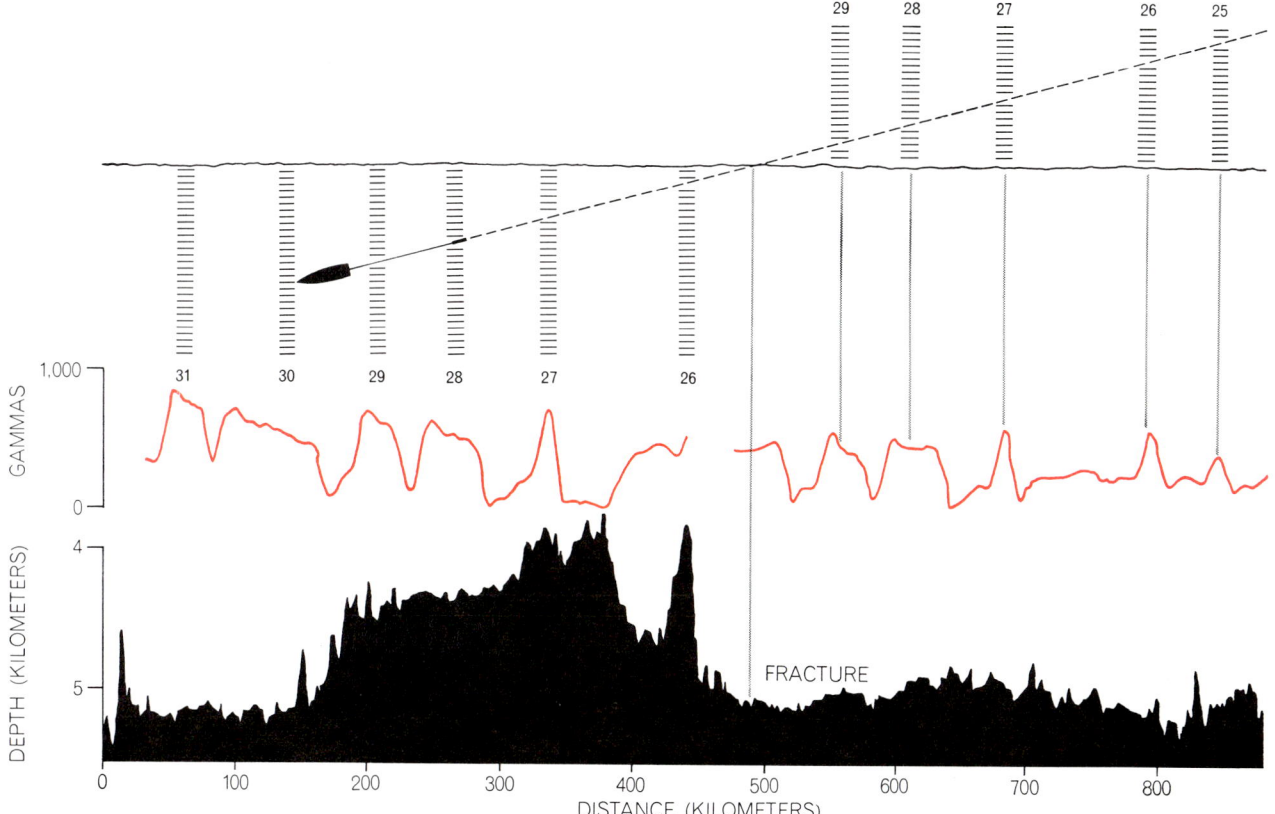

FRACTURE ZONE is identified by magnetic and bottom-profile data. A magnetometer was towed along a course (*broken line at top*) that twice crossed the same magnetic bodies (*Nos. 26–29*), offset along a fracture zone. The bodies caused similar magnetic anomalies, recognizable in the magnetic record (*middle*). Sound-ings revealed a prominent feature near the fracture zone (*bottom*).

the earth is divided into perhaps six rigid plates. These plates grow by the addition of new crust as new material wells up at the spreading axis; at their outer edges they override or are overridden by other plates. The concept of such strong and rigid plates is appealing to the seismolo-gist since, as will be explained later, deep earthquakes seem to be located at the proposed edges of many of them.

The Geomagnetic Time Scale

The magnetized bodies in the ocean floor have provided an amazingly complete history of magnetic-field reversals that now extends back about 76 million years, into the Cretaceous period. We began with a time scale that had been established by workers at Stanford University and the U.S. Geological Survey,

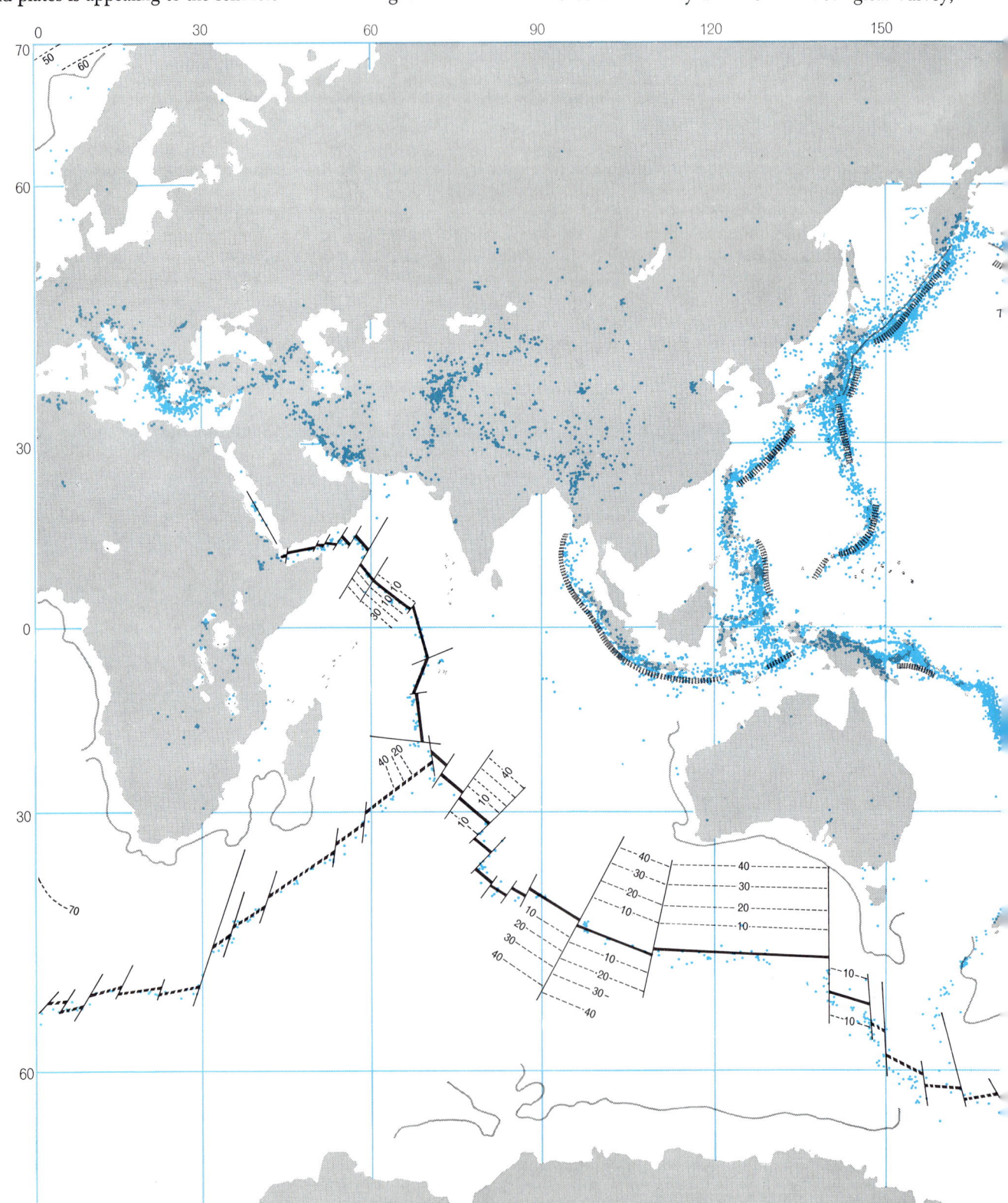

WORLDWIDE PATTERN of sea-floor spreading is evident when magnetic and seismic data are combined. Mid-ocean ridges (*heavy black lines*) are offset by transverse fracture zones (*thin lines*). On the basis of spreading rates determined from magnetic data the author and his colleagues established "isochrons" that give the age of the sea floor in millions of years (*broken thin lines*). The edges

who correlated the magnetism and radioactive-decay dates of rock samples going back some 3.5 million years [see "Reversals of the Earth's Magnetic Field," by Allan Cox, G. Brent Dalrymple and Richard R. Doell; SCIENTIFIC AMERICAN, February, 1967]. By comparing the ages they had assigned to specific reversals of the geomagnetic field with the distance of the corresponding anomalies from the ridge axis, we were able to extend the observations of Vine and Wilson over much of the ocean floor and thus to determine the rate at which the floor had spread in the various oceans.

The rates were found to be different for different sections of the ridge but seemed to have been remarkably steady over the ages in many areas. They vary from about half an inch to a little more

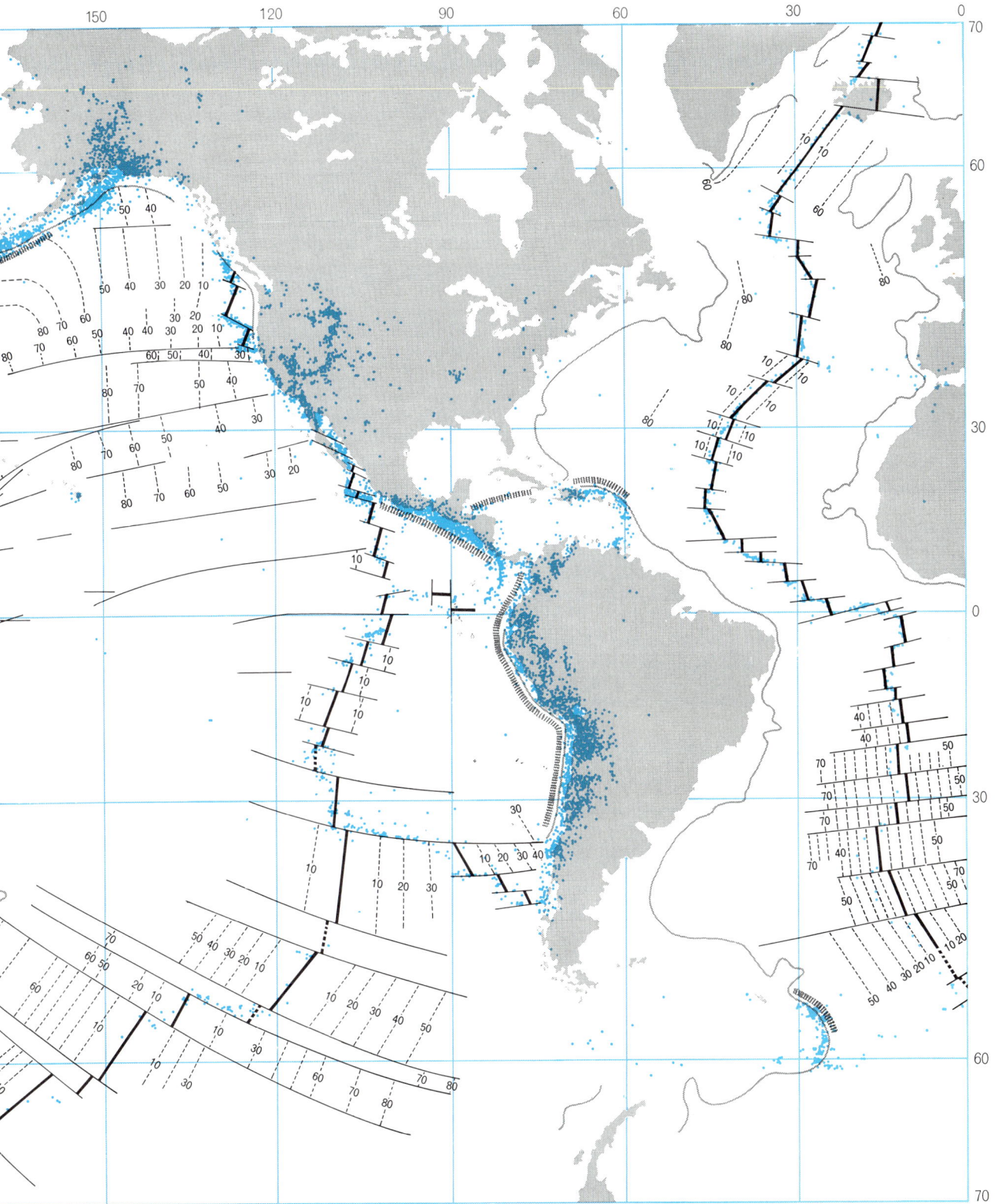

of many continental masses (*gray lines*) are rimmed by deep ocean trenches (*hatching*). When the epicenters of all earthquakes recorded from 1957 to 1967 (plotted by Muawi Barazangi and James Dorman from U.S. Coast and Geodetic Survey data) are superposed (*colored dots*), the vast majority of them fall along mid-ocean ridges or along the trenches, where the moving sea floor turns down.

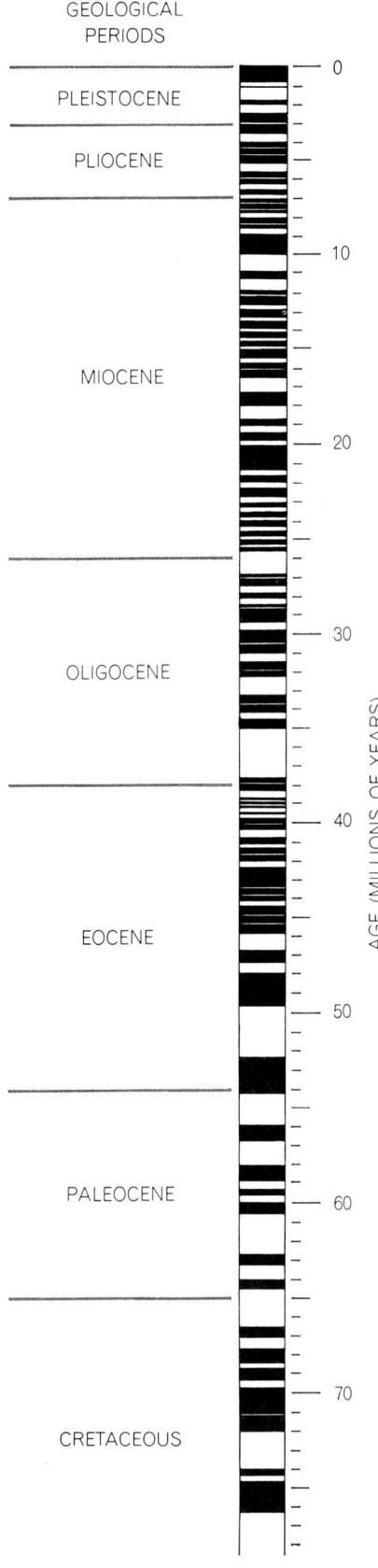

GEOLOGICAL
PERIODS

PLEISTOCENE

PLIOCENE

MIOCENE

OLIGOCENE

EOCENE

PALEOCENE

CRETACEOUS

AGE (MILLIONS OF YEARS)

0

10

20

30

40

50

60

70

CHRONOLOGY of geomagnetic-field reversals was worked out by extrapolation, beginning with the dates that had been established for recent reversals and assuming a constant rate for spreading within each ocean. The dates were confirmed by geological data.

than two inches per year away from the axis of spreading. Although these rates of movement are small by everyday standards, they are large on a geological time scale. Their magnitude came as a surprise to geophysicists even though comparable rates of slippage had been observed along the San Andreas fault, an exposed fracture zone in southern California.

In several areas we saw that there was no obvious hiatus in spreading. We therefore felt free to assume a constant spreading rate and on that basis to assign dates to geomagnetic field reversals extending far beyond the 3.5-million-year scale. This extrapolation of spreading rates may not seem justified at first, but it is supported by consistency from ocean to ocean and by agreement with other geophysical evidence, such as radioactive and paleontological dating of rock samples. Within the probable errors of the methods no discrepancies have been found. We have now identified 171 reversals of the earth's magnetic field over the 76 million years and believe that the dates assigned to each are quite accurate. (Of course, if spreading stopped abruptly all over the world for a certain time, our time scale before the spreading stopped would be too young by that number of years; this is improbable but cannot be entirely ruled out.) On the whole the evidence for the correctness of the time scale is so strong that it can now be used in turn to study variations in spreading rates over the ages in certain disturbed areas.

The average "normal" interval (when the magnetic field was oriented as it is today) works out to 420,000 years and the average "reversed" interval to 480,000 years. The closeness of the two numbers means that the earth is just about as likely to be found in one state of magnetic polarity as in the other. The present era of normal polarity has lasted 700,000 years. Are we due for a change? Only 15 percent of the normal intervals have been longer than that, although some were apparently as long as three million years. The shortest intervals seem to have been less than 50,000 years, but brief instants of geologic time are difficult to confirm by absolute dating methods. This suggests one of the drawbacks of the magnetic scale: simply because it is so detailed, it is unlikely soon to be proved wrong; by the same token, however, it is difficult to use to date a piece of magnetized earth material. Just as one may need a "finder" telescope to orient a high-powered telescope, so one must start by knowing the approximate age of

materials in order to utilize the geomagnetic time scale.

The geomagnetism I have been discussing is frozen into igneous rock of the basaltic type that wells up from deep within the earth. Over much of the ocean floor, of course, this bedrock is not exposed but is covered by varying thicknesses of accumulated sediments. These sediments can also be magnetized, since the tiny particles of which they are composed can become oriented in the earth's field as they settle slowly to the ocean floor. (They are magnetized only one ten-thousandth as strongly as the basalt, however, so that even a thick layer of sediment does not interfere with measurements of anomalies caused by the underlying basalt.) By dropping a hollow cylindrical pipe into the bottom mud one can bring to the surface long "cores" of successive layers of sediment, each constituting an undisturbed record of magnetic-field reversals. Workers at Scripps and Lamont have recently developed sensitive techniques for measuring these weakly magnetized specimens, and the record of geomagnetic reversals has been verified back about 10 million years. The ability to correlate sedimentary layers of the same age in different oceans has proved of immense value to marine geologists, who often lack good paleontological or other indicators of geological horizons. In making detailed comparisons of magnetic reversals and the populations of microscopic animal fossils in such cores, investigators have noted a striking correlation between reversals and major changes in microfaunal species. It has been proposed that such changes are the result of mutations caused by increased exposure to cosmic rays if the earth's protective magnetic field is largely attenuated during a reversal. The evidence for this is not strong, however, and an alternative explanation can be put forward, as I shall suggest later in this article.

Footprints of Continents

The indicated movements of the ocean floor are of the right size and in the right directions to account for continental drift. Topographic and geological evidence had pointed to the probable existence some 200 or 300 million years ago of a single large land mass in the Southern Hemisphere (named Gondwanaland for a key geological province in India) that included Africa, India, South America, Australia, New Zealand and Antarctica, and another mass in the Northern Hemisphere called Laurasia. The positions of the present continents within these land masses were not clear, nor was it possible to trace in detail the sequence of events in their breakup.

The magnetic lineations in the ocean floor serve as "footprints" of the continents, marking their consecutive positions before they reached their present locations. We found that the slow but steady and prolonged rates of motion were sufficient to separate South America from Africa—thus creating the entire South Atlantic Ocean—in about 200 million years and to separate Australia from Antarctica in about 40 million years. As more of the sea floor was dated we could establish more exactly just when the various continents separated and how they moved [see illustration at right]. It is now possible to reconstruct the original positions of the continents and the shallow continental shelves, so that land geologists can go into the field and check the continuity of ancient geological structures that were torn apart when the separations occurred.

Although it is possible to tell when and how the continents pulled away from one another, it is not always entirely clear whether or not—and for how long—a continent may have stood still. The impression is that both Africa and Antarctica have remained fixed with respect to the rotational axis for the last 100 million years, or since Africa split off from the remainder of Gondwanaland. If this is true, then the fact that the spreading axis between South America and Africa remains about midway between them indicates that the axis itself must be moving too.

Source and Sink

Until the theory of sea-floor spreading was advanced the only known sources of deep-earth material at the surface were volcanic eruptions. It now appears, however, that most currently active volcanoes are not at the sites of upwelling along the mid-ocean ridge but rather in areas where the moving sea floor turns down under the continents. Recent volcanic eruptions in the Philippines, Mexico and Guatemala are examples, as is the continuing intermittent activity of volcanoes in western North America, Alaska and Japan. The upwelling that initiates sea-floor spreading must therefore represent some unfamiliar geophysical phenomenon, and investigators have concentrated much attention on the axis of the mid-ocean ridge where it occurs. The axis has several unusual properties: a large heat flux, a concentration of shallow earthquakes, unusual seismic-wave velocities, a lack of sediment and eroded rocks, and the presence of a prominent magnetic anomaly. While most of these unusual conditions extend over a distance of from tens to hundreds of miles on both sides of the crest of the ridge, the magnetic anomaly is sharply localized; by studying the symmetry of the magnetic pattern it is possible to locate the spreading axis to within a few miles. The terrain of the ridge is rugged and it is not possible to associate the spreading axis with any single topographic feature,

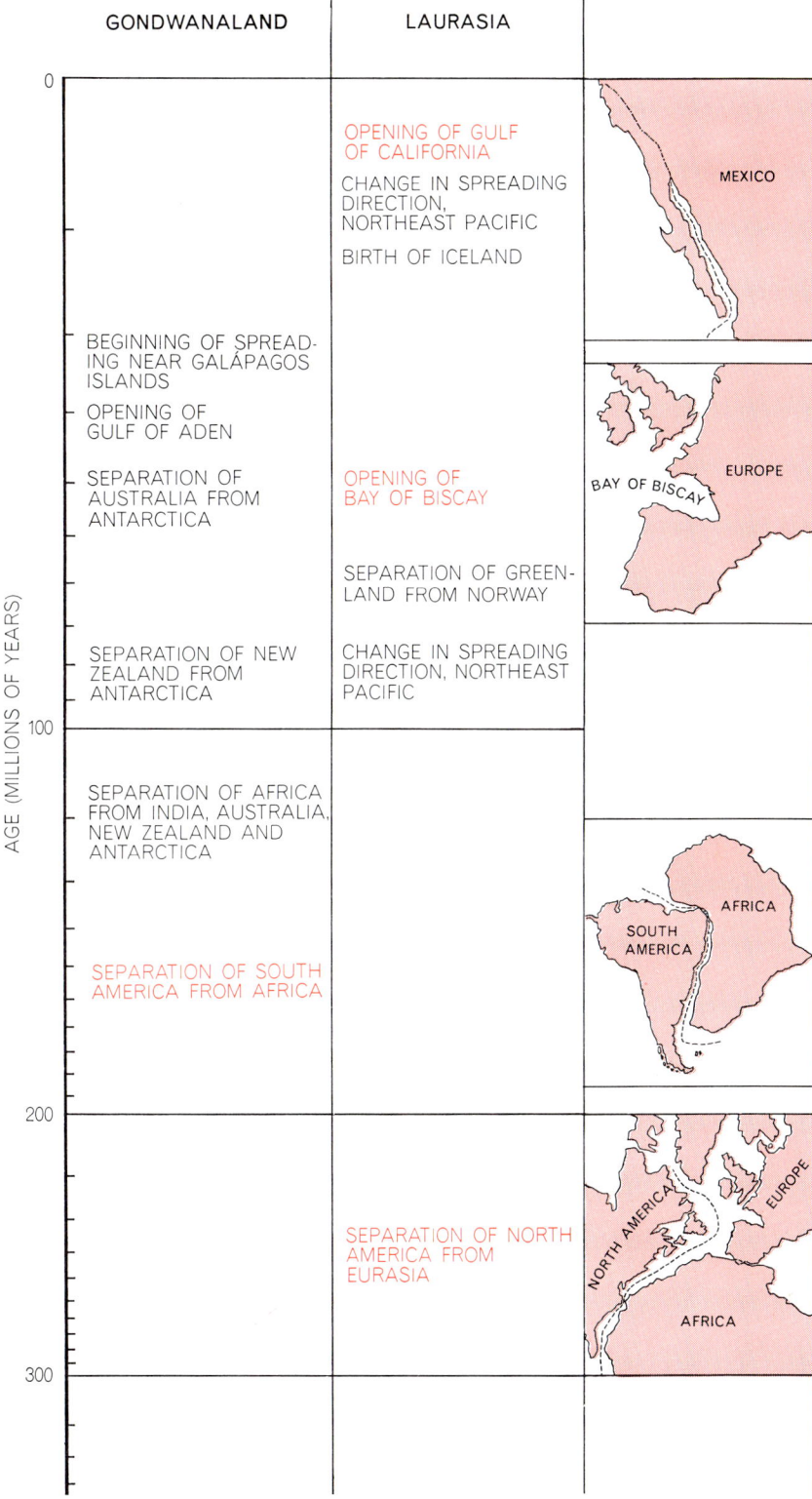

TIME SCALE for continental movements and other changes can now be established, since isochrons show the age of the ocean floor and the direction of spreading at any time.

although one can say that when the ridge contains a median "rift valley," as many do, the axis lies within it.

The magnetized plate is not very thick. Analyses of the observed magnetic anomalies suggest that the thickness is from a half-mile to a few miles; studies of the transmission of seismic waves and the distribution of heat flux, on the other hand, indicate a thickness of a few tens of miles or perhaps 100 miles for the moving plate. The precise linearity of the magnetized bodies shows that the upwelling material was not tumbled about after it became cooled and magnetized; in this it is different from the usual folded rocks seen on land or the irregular flows that surround a volcanic eruption. Attempts to locate the axis with a magnetometer submerged near the sea floor show that there are a number of very magnetic bodies where the axis would be expected.

The evidence suggests to most investigators that the upwelling mechanism is an injection of molten deep-earth material by linear intrusions called dikes. Such bodies have a high probability of being injected along the line of the spreading axis. Each injection may be quite localized rather than greatly elongated along the axis. The new material is hot enough to reheat adjacent rock so that both the new material and the slightly older material nearby is magnetized in the direction of the ambient field before being quenched by the cold seawater. This explanation does not indicate anything precise about the thickness of the moving layer; it does account for the lack of tumbling of magnetized blocks, since a dike would tend to push the older material horizontally away from the ridge.

The deep oceanic trenches found around the periphery of the Pacific are thought to be places where surface material returns to the deep earth; so, very likely, are smaller trenches in the North Atlantic and South Atlantic and the Indian Ocean. In many parts of the world where plates of moving sea floor have been identified the outer edge of the plate has not been located, and so we probably do not know where all the sinks are. In most places the spreading sea floor seems to be turning down under the continents, but in some places it seems to be pushing continents ahead of it or even tearing continents apart; we know of no place where the spreading floor is overriding a continent.

If the epicenters of deep earthquakes are mapped, they are almost all found to be on a plane that starts at the floor of a trench and makes an angle of about 45 degrees with the horizontal [*see top illustration on page 633*]. The slippage of earth material parallel to the plane and extending hundreds of miles below the surface is the cause of the earthquakes. These earthquake planes almost certainly define regions of downwelling; studies of first motion in such areas show that the sea floor moves down with respect to the adjacent continent. A thin crustal layer with a characteristic speed of seismic-wave propagation has been shown to underlie the ocean sediments and turn down at 45 degrees to great depths.

The magnetic evidence for what happens at the trenches is ambiguous, however. The sea-floor magnetic pattern is altered suddenly. Measurements at the Aleutian Trench, for example, show a

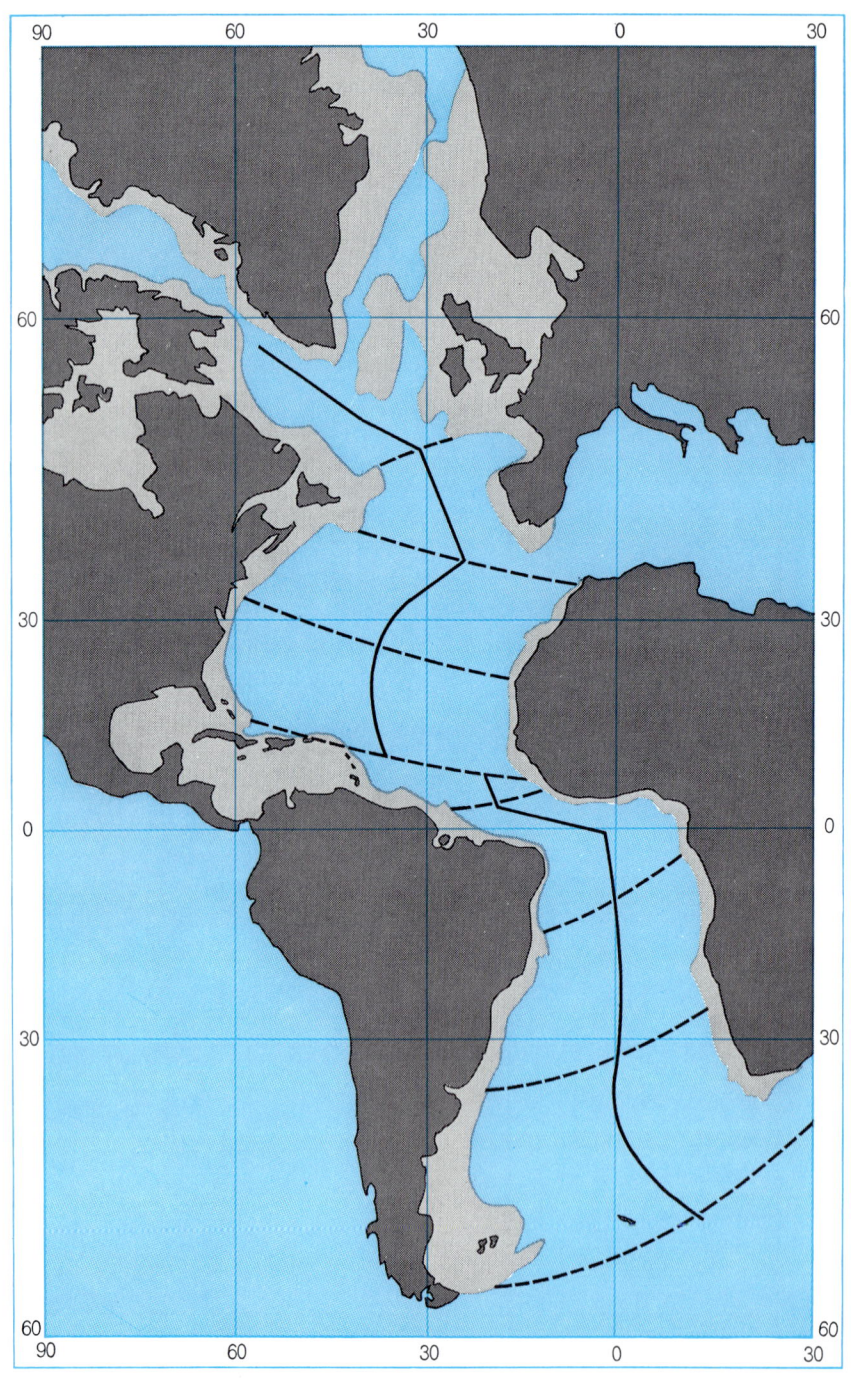

ANCIENT POSITIONS of the continents can be plotted on the basis of the time scale. This map shows the relation of the Americas to Eurasia and Africa 70 million years ago. The continents fitted together generally along the edges of continental masses (*light grey*) rather than of present dry-land areas (*dark gray*). Broken lines trace directions of movement.

sharp discontinuity [*see bottom illustration on this page*]. There is no sign of a magnetic body that should (on the basis of measurements made elsewhere) be located about three kilometers below the trench floor. Its absence might be explained by heating or mechanical deformation, but neither sufficiently high temperatures nor enough seismic activity to cause deformation are indicated so close to the trench floor. Another problem at the Aleutian Trench is that the magnetic bodies seem to be in the wrong sequence. If the trenches are sinks, the older bodies should be in the trench and the younger ones to seaward; the opposite seems to be the case! Moreover, south of the Alaska Peninsula bodies oriented essentially north-south turn and run east-west, parallel to the coast. This complexity is difficult to explain if trenches are sinks. It is one of the most awkward sticking points in the theory.

Speculations

The feeling among many geophysicists that we are on the brink of even more comprehensive theories about the earth stems from the striking geographic or temporal coincidence of certain geophysical phenomena, many having to do with the dynamics of our planet. Existing theories suggest no clear causal relation among these phenomena, and so one hopes for some higher-order synthesis that will establish such a relation.

What are some of these coincidences? The present pole of spreading (for several of the oceans, at least) is near the geomagnetic axis; in Cretaceous times the spreading pole for the North Pacific was near the Cretaceous geomagnetic pole. There have often been significant changes in the microfaunal population of the sea at magnetic reversals. A major meteorite (tektite) fall occurred just at the time of the last reversal [see "Tektites and Geomagnetic Reversals," by Billy P. Glass and Bruce C. Heezen; SCIENTIFIC AMERICAN, July, 1967]. Some authors have speculated recently on a relation between mountain-building activity and magnetic reversals; others see a relation between changes in sea-floor spreading and mountain-building. Mechanisms have been suggested whereby the earth's magnetic field could be generated by convective motion caused in turn by irregularities in the earth's orbit. There has been a revival of a 30-year-old theory that the glacial ages were caused by changes in the tilt of the earth's axis. Finally, there is clear evidence that large earthquakes occur

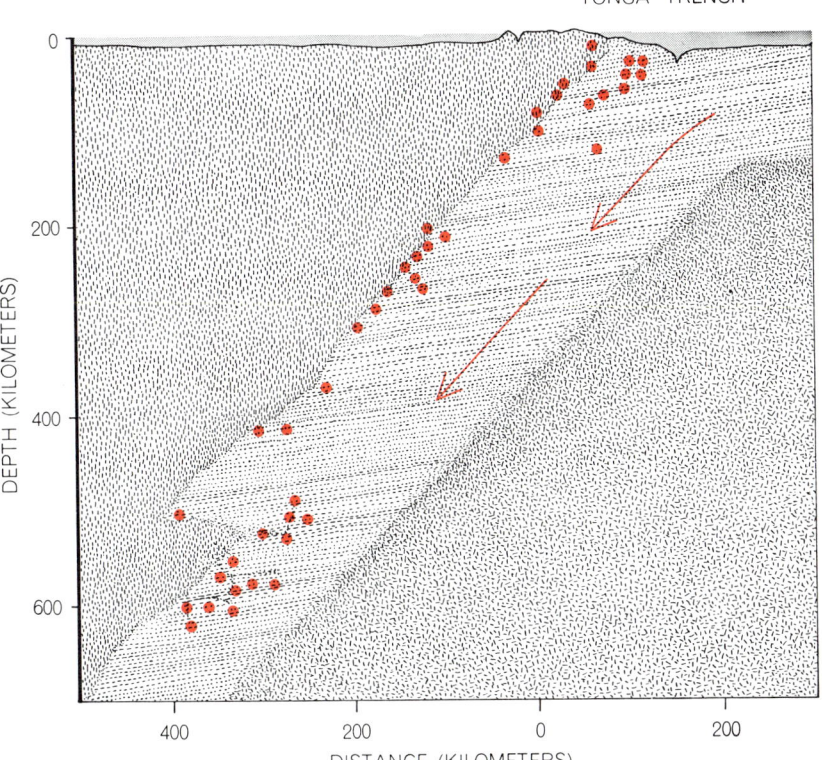

MOVING SEA FLOOR turns down at the trenches rimming the Pacific Ocean. Deep earthquake epicenters line the downward path where floor material moves under land masses. Symbols (*color*) show epicenters recorded at the Tonga Trench in the South Pacific.

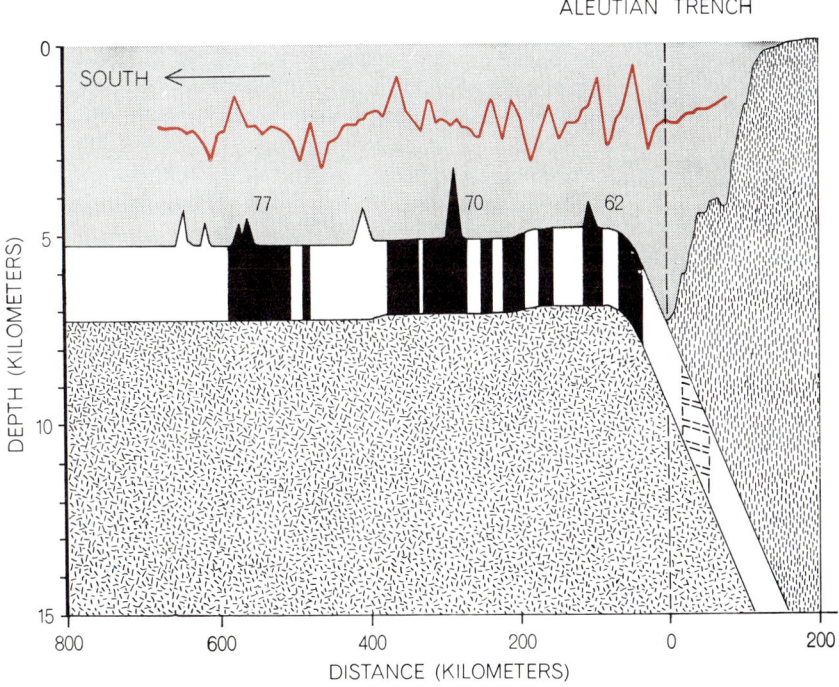

ALEUTIAN TRENCH data fail to fit the model, however. The magnetic record (*color*) shows the expected sequence of normal and reversed bodies (*black and white bands*) approaching the trench, but there is no evidence of the next body, which should appear about three kilometers below the trench (*hatched band*). Moreover, the magnetic bodies seem to be younger (*77 to 62 million years*) as one approaches the trench rather than older.

at about the same time as certain changes in the earth's rotational motion.

This article is not the place for a full evaluation of all these developments. It is interesting to note, however, that a common thread running through these and many other proposals at the frontier of geophysical research is the role played by displacements of the earth's axis of rotation. It seems that rather minor variations can affect to a surprising extent both the climate at the surface of the earth and forces and stresses within the earth.

The intimate relation between the spreading pole and the magnetic pole suggests that the convective motion within the earth and the earth's field may have a common cause. It could be due to the proposed effect of orbital irregularities or to some dynamo action within the earth. Since the direction of convection does not reverse when the field reverses, it is clear that the convective motion does not simply generate the field, nor is it likely that the reversing field could "pump" convection cells. Whatever the driving force is for these two phenomena, it would seem to be related to the motion of the earth. Recently it has been shown that earthquakes of a magnitude of 7.5 or greater on the Richter scale either cause or are caused by changes in the earth's "wobble," a small circular motion described by the north pole of rotation [see "Science and the Citizen," SCIENTIFIC AMERICAN, November, 1968]. Whatever the mechanism of these changes, it is not hard to believe that similar changes in the earth's axial motion in times past could have caused major earthquake and mountain-building activity and could even have caused the magnetic field to flip.

To summarize: Every few months there are changes in the earth's rotational motion that affect sea-floor spreading and cause the earthquakes associated with it. If such a change is large enough, it may even reverse the earth's magnetic field. (Both the presence of a geomagnetic field and the spreading of the sea floor seem to be due to the mere fact that the earth is rotating; it is only the changes in motion that are associated with certain large earthquakes and reversals of the field.) Changes in the microfaunal population of the sea are related to changes in the climate, which are related in turn to variations in the earth's motions.

Such speculation cannot yet be confirmed but neither can it be firmly denied; indeed, it is no more outlandish than theories of sea-floor spreading seemed to be a few years ago.

The Author

J. R. HEIRTZLER is director of the Hudson Laboratories of Columbia University. He was educated as a physicist, receiving his Ph.D. from New York University in 1953, but has long made hobbies of geology and astronomy. After obtaining his doctorate he taught in universities, worked as a nuclear physicist in industry and, from 1960 to 1967, was in charge of a group that was engaged in research on geomagnetism at the Lamont Geological Observatory of Columbia University. He writes that "firsthand experience with the oceans during World War II and a love of travel combined to give me a strong interest in marine geophysics."

Bibliography

DEBATE ABOUT THE EARTH: APPROACH TO GEOPHYSICS THROUGH ANALYSIS OF CONTINENTAL DRIFT. H. Takeuchi, S. Uyeda and H. Kanamori, translated by Keiko Kanamori. Freeman, Cooper & Co., 1964.

SPREADING OF THE OCEAN FLOOR: NEW EVIDENCE. F. J. Vine in Science, Vol. 154, No. 3755, pages 1405–1415; December 16, 1966.

THE HISTORY OF THE EARTH'S CRUST. Edited by Robert A. Phinney. Princeton University Press, 1968.

SEISMOLOGY AND THE NEW GLOBAL TECTONICS. Bryan Isacks, Jack Oliver and Lynn R. Sykes in Journal of Geophysical Research, Vol. 73, No. 18, pages 5855–5899; September 15, 1968.

SCIENTIFIC
AMERICAN   December 1968, Vol. 219, No. 6, pp. 74–82

OFFPRINT **876**

FOG

by Joel N. Myers

A kind of grounded cloud, fog can halt sea, air and highway travel.
When combined with air pollutants, it can be lethal. Ways are known,
however, not only to dissipate fog but also to inhibit its formation.

Fog, once little more than a nuisance except at sea, has become an important hazard to modern man. Its effects on travel are intensified by the speed of the airplane and the automobile. Dense fogs close airports in the U.S. for an average of 115 hours per year, and in 1967 they cost the nation's airlines an estimated $75 million in disrupted schedules as well as inestimable inconvenience to passengers. On present-day turnpikes fog can be disastrous; a single pileup on a fog-shrouded freeway in Los Angeles involved more than 100 vehicles. Above all, fog in combination with air pollution now increasingly afflicts large cities. Its potential was suggested alarmingly by the London smog of December, 1952.

On December 5 a dense fog settled over the city. A strong inversion—warm air lying above cold air—blocked the removal of polluted air by vertical movement, and at the same time the winds were so light that such air was only gradually removed by horizontal movement. Within 24 hours the tons of smoke, dust and chemical fumes given off by the city's furnaces, factories and automobiles turned the fog brown and then black. Two days later the visibility in the city had been reduced to a matter of inches. People fell off wharfs into the Thames and drowned. Others wandered blindly until they died of exposure to the cold. The toxic air afflicted millions of Londoners with smarting eyes, coughing, nausea and diarrhea. It was estimated that the smog killed at least 4,000 persons, mainly from respiratory disorders, and caused permanent injury to tens of thousands.

The principal source of toxicity in most pathological smogs appears to be sulfur dioxide from the smokestacks. The sulfur dioxide is oxidized in the air to sulfur trioxide, which in turn combines with fog droplets to form sulfuric acid. One therefore inhales sulfuric acid with each breath, with resulting acute irritation of the throat and lungs. This seems to have been the chief cause of injury and death in the London smog and in the 1948 smog in Donora, Pa., which killed 20 victims and sickened nearly half of the 14,000 inhabitants.

Strictly speaking, the perennial "smogs" of Los Angeles are misnamed, as they frequently consist of a haze of pollutants that are trapped by an inversion layer of dry air rather than by fog. Polluted air, however, can generate fog and cause it to persist. Once formed, the fog reflects solar radiation back into space, thereby providing an environment favorable to the accumulation of further pollution. In effect fog, itself partly a product of pollution, causes pollution to increase.

Some metropolitan areas in the U.S. suffer from light smogs up to 100 days of the year, and during fall and winter dangerous smog conditions can become frequent. The long-term effects of the urban smogs are believed to include chronic bronchitis, emphysema, asthma and lung cancer. The polluted air corrodes metals, rots wood, causes paint to discolor and flake and may cause extensive damage to vegetation and livestock. In addition to the directly damaging effects, there are indications that the combination of fog and air pollution may be upsetting the delicate balance of man's ecosystem. Smog plays an important role in influencing the earth's gain and loss of radiation, which ultimately determines most of the climatic variables. Smog reduces sunlight, lowers daytime temperature and wind speed, raises humidity and is even suspected of causing a decrease in rainfall. Thus for many reasons—the hazards to travel, the damaging effects

of smog, the modification of climate—fog is a growing problem that will demand increasing research attention in the years ahead.

Fog is simply a cloud on the ground, composed, like any cloud, of tiny droplets of water or, in rare cases, of ice crystals, forming an ice fog. Ice fogs usually occur only in extremely cold climates, because the water droplets in a cloud are so tiny they do not solidify until the air temperature is far below freezing, generally 30 degrees below zero Celsius or lower. The droplets of fogs are nearly spherical; they vary in diameter between two and 50 microns and in concentration between 20 and 500 droplets per cubic centimeter of air. The transparency of a fog depends mainly on the concentration of droplets; the more droplets, the denser the fog. A wet sea fog may contain a gram of water per cubic meter; a very light fog may have as little as .02 gram of water per cubic meter.

Since water is 800 times denser than air, investigators were long puzzled as to why fogs did not quickly disappear through fallout of the water particles to the ground. Even allowing for air resistance, a 10-micron water droplet falls in still air with a velocity of .3 centimeter per second, and a 20-micron droplet falls at 1.3 centimeters per second. To explain the persistence of fogs many early investigators concluded that the droplets must be hollow (that is, bubbles). It turns out, however, that the droplets are fully liquid and do fall at the predictable rate, but in fog-creating conditions they either are buoyed up by rising air currents or are continually replaced by new droplets condensing from the water vapor in the air.

The atmosphere always contains

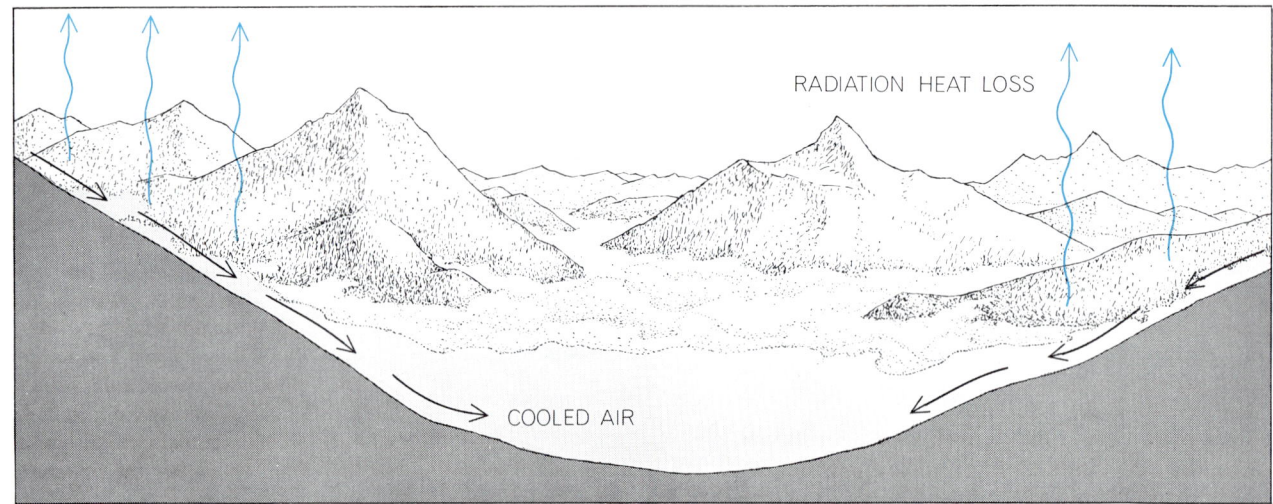

FOG IS FORMED when moist air is cooled; the air then cannot hold as much moisture in the form of water vapor as it can when it is warmer. As the humidity approaches the saturation point tiny water droplets form, obscuring visibility. The diagram indicates how a radiation fog, one of three kinds of "cooling fog," forms. Night cooling has reduced the air's temperature and cause water droplets to appear. The cold, fog-filled air drains downhill and accumulates in low-lying areas (*bottom illustration on page 640*).

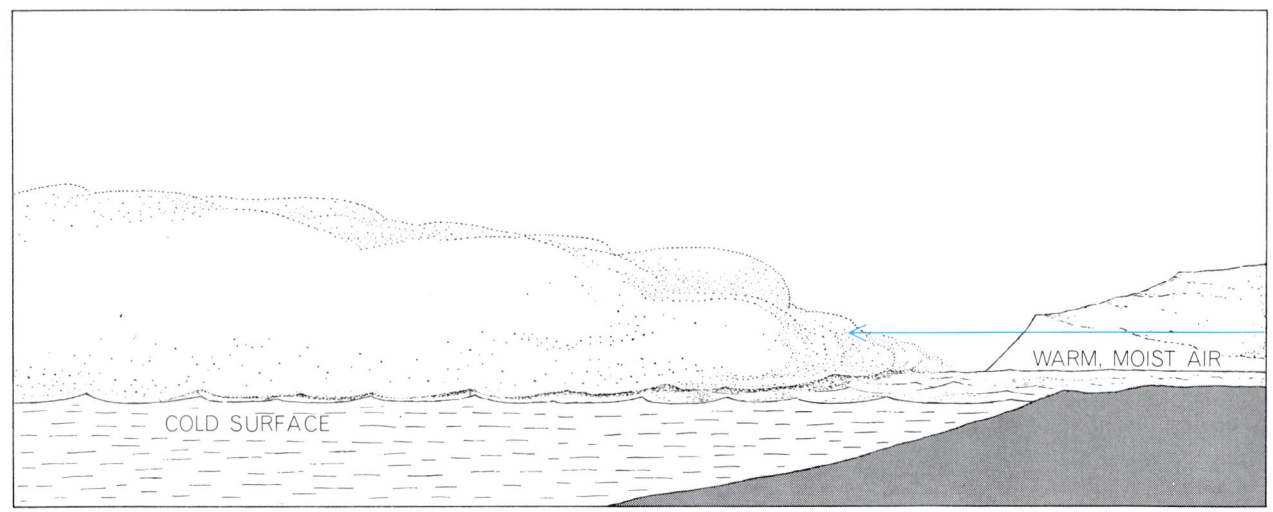

ADVECTION FOG, the second kind of cooling fog, forms when warm, moist air is carried across a cold surface and is cooled to the saturation point. Advection fogs are common at sea where warm air masses come in contact with cold ocean currents (*see top illustration on pages 638–39*) and on land during the cold months when moist tropical air masses are carried across chilled ground.

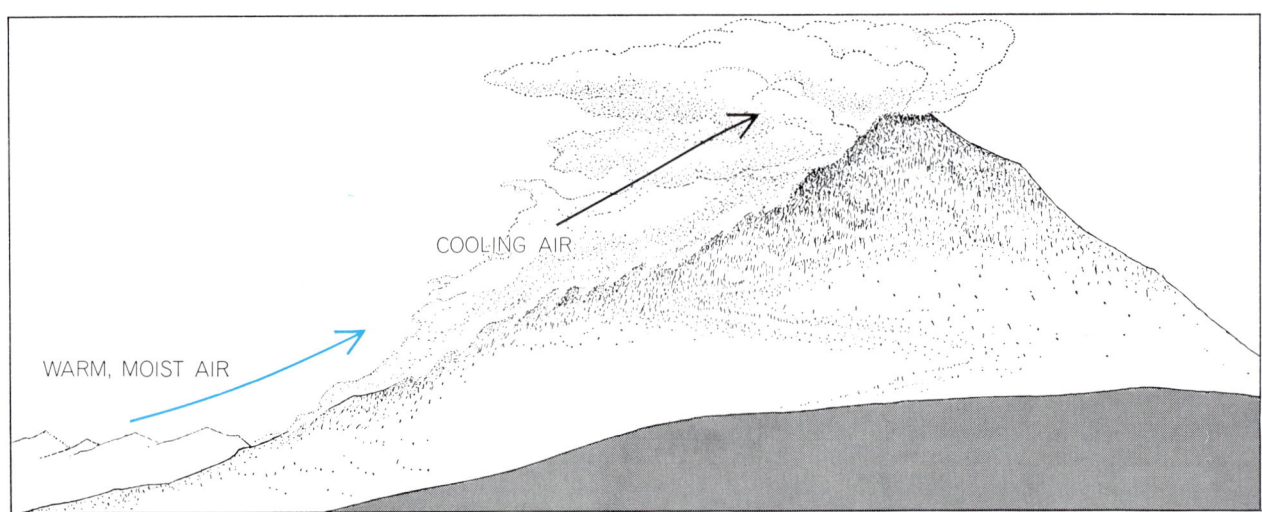

UPSLOPE FOG, the third kind of cooling fog, forms when an air mass is forced upward. In the diagram a topographic barrier forces the air mass to rise. As pressure diminishes the air mass expands, grows cooler and, as in both other cases, soon becomes saturated.

some water vapor, supplied by evaporation from bodies of water, vegetation and other sources. Since the air's capacity for holding water in the form of vapor decreases with falling temperature, even comparatively dry air will reach the saturation point—100 percent relative humidity—when it is cooled sufficiently. At that point the vapor of course begins to condense into liquid water. Fogs often form, however, at humidities well below 100 percent. The droplets condense on tiny particles of dust in the air called condensation nuclei. These are hygroscopic particles, which, because of their affinity for water vapor, initiate condensation at subsaturation humidities—sometimes as low as 65 percent. The nucleus on which the water condenses, which may be a soil particle or a grain of sea salt, a combustion product or cosmic dust, usually dissolves in the droplet. Because the saturation point is lower for solutions than it is for pure water, the droplets of solution tend to condense more water vapor on them and grow in size. A rise in the air's humidity will also enlarge the droplets and will form more of them, thereby thickening a light fog into a dense one.

Given suitable conditions of temperature and humidity, the density of a fog and its microphysical properties will depend on the availability of condensation nuclei and their nature. Fogs become particularly dense near certain industrial plants because of the high concentration of hygroscopic combustion particles in the air. This is not to say that air pollution is generally a primary cause of fog formation, but it does cause fogs to form sooner and persist longer, and it makes them dirtier—hence less transparent—than fogs that develop at higher humidities in relatively clean air.

From a meteorological point of view fogs are classified in several types according to the gross natural processes that generate them. Over land the most common type is the "radiation fog" that arises from nighttime cooling of the earth's surface and the lower atmosphere. As the earth radiates away its heat during the night, fog may form if the air in contact with the cooling ground is moist or, even though it is fairly dry at first, if it is cooled a great deal. Radiation fogs occur most frequently over swampy terrain and in deep, narrow valleys where cold air draining down from the hillsides concentrates in the valley bottom. The likelihood of fog formation depends considerably on the wind speed. A moderate to strong wind, by moving the air about and diluting the cooling effect,

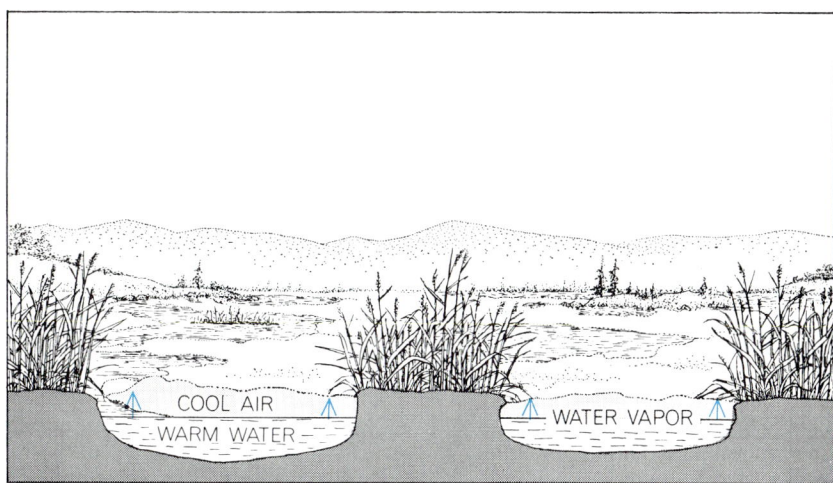

WARM-WATER FOG forms like steam over water that is covered by a much colder air mass. Water vapor from the comparatively warm water rises into the colder air and is rapidly condensed into fog droplets. The fog's intensity depends on the temperature differential.

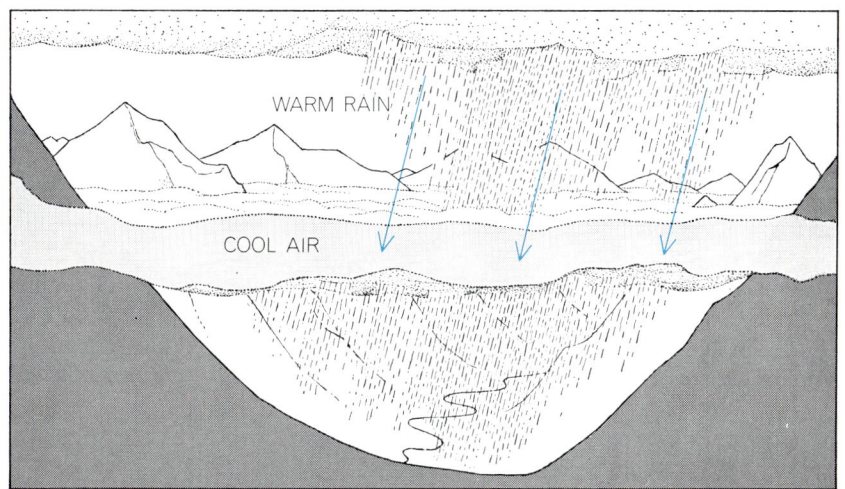

WARM-RAIN FOG is formed when raindrops from higher clouds encounter a layer of cooler air near the ground; evaporating raindrops saturate the cold air layer. As the diagram indicates, the fog looks like a low cloud to an observer looking up from the valley floor.

tends to prevent the formation of fog. If the air is calm, only a thin layer of air next to the ground is much affected by the cooling, so that the condensation may be restricted to dew or a shallow ground fog. The condition most likely to produce an extensive fog is a slight breeze; by generating turbulence near the ground such a breeze may spread out the cooled surface air to form a layer of fog several hundred feet deep. Above this cold layer the air remains warm.

One might suppose that radiation fogs should reach a maximum around dawn, when the air temperature ordinarily is at its daily low point. Actually it has been found that this type of fog sometimes becomes thickest shortly after sunrise. The reason appears to be that the sun's early rays, not yet strong enough to evaporate the fog droplets, generate turbulence that intensifies and thickens the

fog layer. The fog does not begin to dissipate until the sun is high enough to heat the atmosphere, stirring the foggy air so that it mixes with the warm, dry air above. Obviously in pockets that are topographically shielded from the sun the fog will remain longer.

A somewhat different process produces what are called advection fogs. In this case the fog arises from the movement of humid air over a surface that is already cold. Most sea fogs are of this type; indeed, the foggiest places in the world are the areas above cold ocean currents. Advection also commonly plays a part, in combination with nocturnal cooling, in the generation of land fogs. In a different way air moving up a mountainside sometimes produces fog: the air expands and cools as it rises because the atmospheric pressure diminishes with altitude. Advection and upslope fogs can provide additions to the water supply. On some coastal mountains in California, for example, the ground receives more water from fog dripping off vegetation than it does from rain. Residents of some arid regions take advantage of this phenomenon by suspending arrays of nylon threads to extract water from drifting fogs.

Fog is also produced by the familiar steaming process we observe above a hot bath or a heated kettle or on a hot roof or parking lot after a summer shower; the vapor rising from the warm water quickly condenses into steam in the cooler air above. In this way the evaporation from a body of water on an unseasonably cold night may generate a shallow fog (up to perhaps 50 feet). Warm rain falling through cool air also can give rise to a steamlike fog; it is a common cause of the fogging in of airfields during rainy weather.

The principal natural agents of fog dispersal are sunshine and brisk wind, but a dense fog has built-in resistance to the sun. Because fog is an excellent reflector of sunlight, only 20 to 40 percent of the impinging solar energy penetrates it to warm the ground and the foggy air, and only part of that heat is available to evaporate the fog droplets. Consequently a dense fog strongly resists dissipation

by the sun. Moreover, the resistance is increased when the fog becomes a thick smog.

To what extent has our era of industrialism intensified the fog problem? This is very difficult to determine, because the meteorological records are incomplete and the phenomenon itself is highly variable. The incidence and intensity of fogs vary widely with time and season in any given place and from one place to another, even within a few miles. Furthermore, weather-observation practices have changed considerably over the past century: the observations are much more frequent than they used to be, cover many more locations and are complicated by changes both in the location of observatories and in the observers' nomenclature. The definition of "dense" fog, for example, has been revised repeatedly by the U.S. Weather Bureau. Nevertheless, although reliable comparisons cannot be made in detail, a survey of past records indicates some general trends. It appears that the incidence, duration and probably the density of fogs vary directly with the amount of industrial activity and air pollution in the area involved.

Perhaps the most reliable set of data for a single location is the one for the

CLASSIC EXAMPLE of an advection fog is the light, low-lying bank seen obscuring

city of Prague in Czechoslovakia, where consistent observations of fogs have been made for the past century and a half. The records show that in the period since 1881 Prague has had nearly twice as many fogs as in the preceding 80-year period. In general it appears

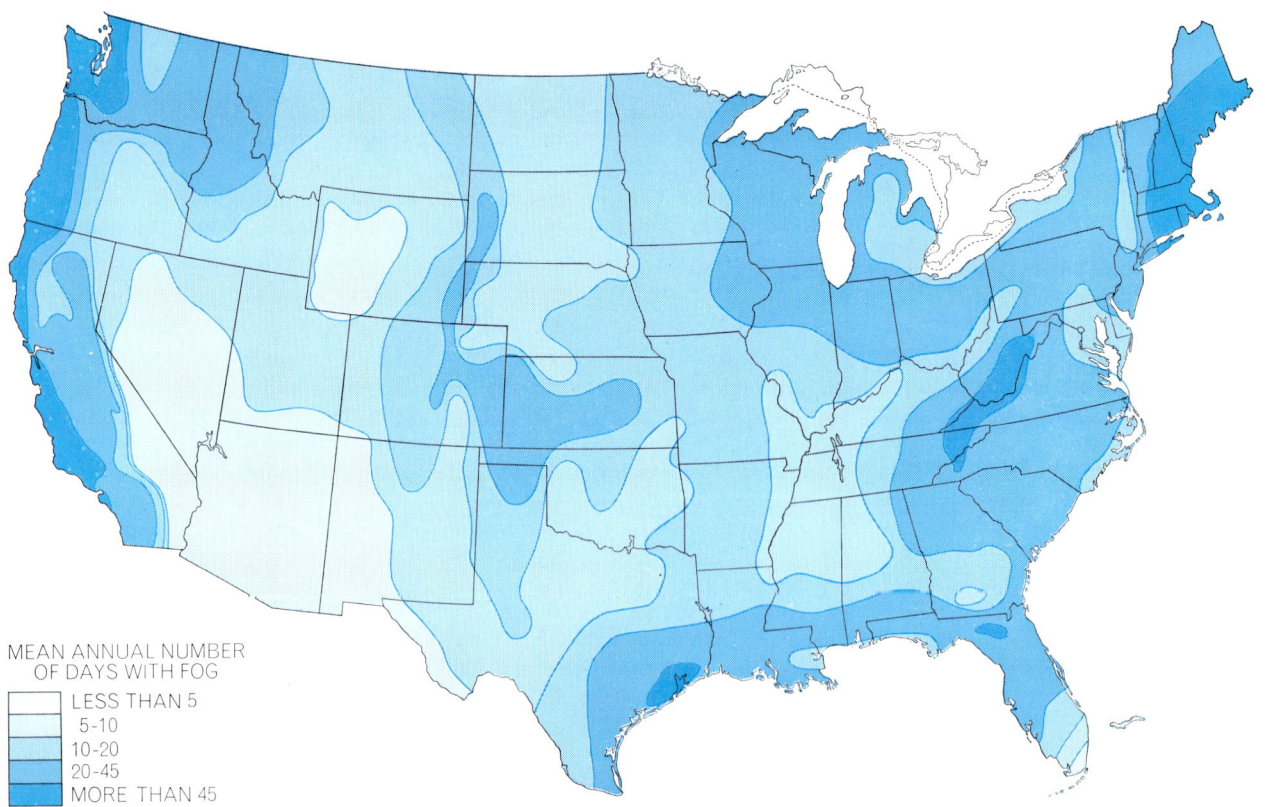

MEAN ANNUAL NUMBER
OF DAYS WITH FOG

LESS THAN 5
5-10
10-20
20-45
MORE THAN 45

DISTRIBUTION OF FOGS in the continental U.S. is shown by area shading that indicates the days of dense fog per year reported by 251 weather stations from 1900 to 1960. The least foggy parts of the mainland are the desert areas of Arizona, California and Nevada; foggiest are the Pacific and New England coasts. Appalachian hills often have rain fogs; the valleys, radiation fogs.

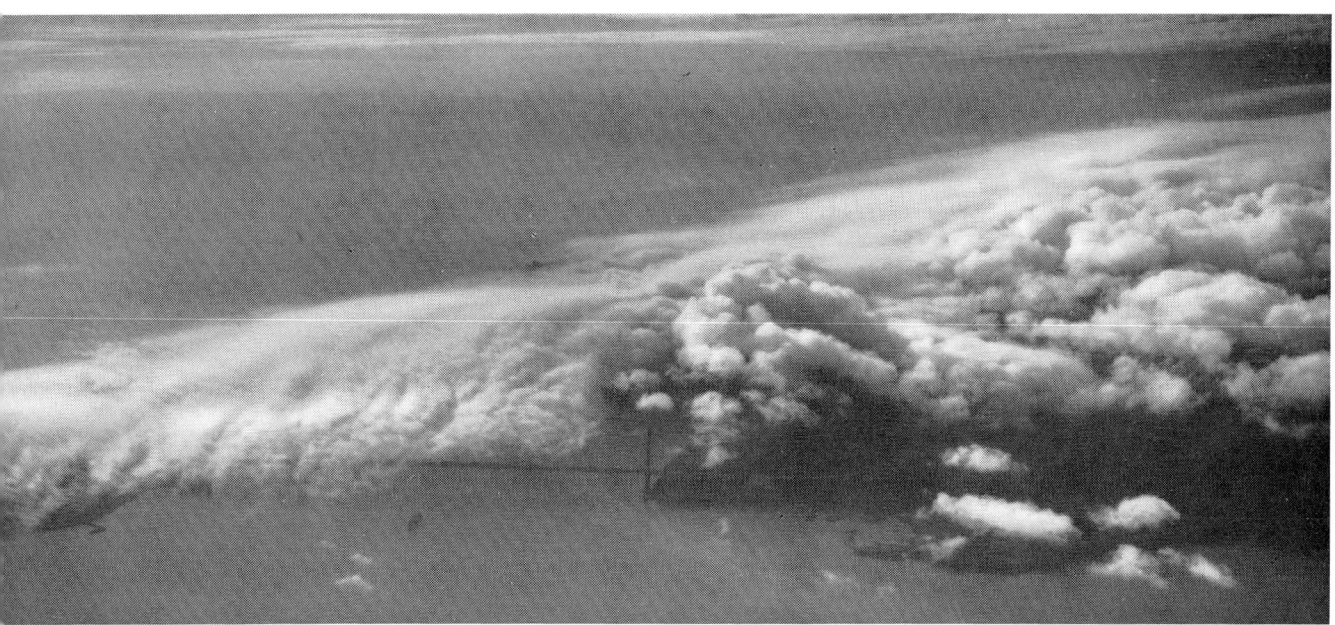

the Golden Gate Bridge at the entrance to San Francisco Bay in the aerial photograph. When a warm, moist air mass passes over the cold waters of the California Current, it is cooled to the saturation point and some of its water vapor is condensed into fog droplets.

that fog tends to be more frequent in and near cities than in unpolluted rural areas. The available data do not indicate, however, if fogs tend to be denser in cities than they are elsewhere. It is conceivable that the relatively warm microclimate of large cities (due to the solar heating of their streets and buildings and the urban artificial heat) may prevent the strong nocturnal cooling that is usually a prerequisite for dense fogs. If that is so, perhaps the places where dense fogs occur most frequently are highly industrialized small cities and suburban communities that lie downwind of metropolises.

A fog recently studied by Charles L. Hosler, Jr., a meteorologist at Pennsylvania State University, may serve as an illustration of the problem. The fog is seen shortly after sunrise in Bald Eagle Valley in central Pennsylvania on days when little or no fog is observed elsewhere in the state. The valley contains large industrial plants that daily discharge billions of hygroscopic particles and tons of water vapor into the atmosphere. Although fog formation is aided by the overnight drainage of cold air into Bald Eagle Valley, the pollution is a major contributing factor.

What can be done about the growing fog problem? The phenomenon has become a matter of active concern for many interests—government, industry, the airlines, the military—and it is being attacked by investigators in a wide range of disciplines. Clearly one of the prime needs is a vigorous assault on the complex problem of air pollution. Nor is pollution confined exclusively to the atmosphere. A form of pollution that is responsible for some fogs and that has received little attention is the "thermal pollution" of natural water resulting from industrial practices. It gives rise to the formation of fogs by increasing the normal rate of water evaporation.

Enormous quantities of heated water are discharged into our streams and lakes by industrial operations that use water for various cooling purposes. The principal offenders are steel mills, paper factories, sewage-treatment plants, certain chemical and manufacturing plants and particularly electric-power generating stations. Power plants based on nuclear fuel, to which many utilities are now turning, use enormous quantities of water for cooling. The power industry estimates that by 1980 it will be using one-fifth of the total free water runoff in the U.S. for cooling. Thus, ironically, nuclear plants, which are counted on to reduce air pollution by replacing plants that use fossil fuels, may become a major factor in spawning fog and thereby trapping air pollution in the form of smog.

Some industrial plants, notably in the power industry, have taken to getting rid of their heated water by evaporating it in massive cooling towers instead of dumping it into bodies of water. Unfortunately this method is expensive and may still produce fogs. Some towers evaporate hundreds of cubic meters of water per hour, and the tremendous flux of vapor, if trapped under an inversion on a windless night, could generate and maintain a dense fog over a large city. Federal aviation officials believe a cooling tower recently built three miles north of the Morgantown Municipal Airport in West Virginia is responsible for local fogs that have begun to trouble the area. And it appears that fog plumes from other cooling towers may have contributed to automobile accidents on highways near them.

Cooling towers might be turned into assets rather than liabilities in farming areas. The fog they produce could prevent heat from escaping from the soil and thus protect cold-sensitive crops against frost. Some farmers in localities of frequent fogs report that fog blankets extend the frost-free growing season and thereby increase their crop yields. Russian meteorologists have successfully used artificial fogs to protect vineyards from frost. Perhaps power companies should be encouraged to build their new plants in rural regions where farming could benefit from such frost protection.

Among the various possibilities for dealing with the fog problem the approach that has been explored most actively up to now is the idea of dissipating fog by some artificial means. Over the past several decades hundreds of schemes for doing this have been proposed and many have been tried, but so far no universally practicable method has been found. What are the prospects that an effective and not too expensive technique could be developed?

The most direct method would be

ICE FOG (*above*) fills the Nenana River valley near Fairbanks, Alaska. Ice fogs consist mainly of ice crystals that form in air cooled to 30 degrees below zero Celsius or lower. They may appear because a temperature drop has produced nearly 100 percent humidity or because the air is saturated by the addition of water vapor.

RADIATION FOG (*below*) spills out of Bald Eagle Valley throu a mountain gap near Lock Haven, Pa. Radiation fogs are caused the night cooling of the earth's surface, which reduces the air te perature and raises the humidity to near-saturation. In Bald Ea Valley industrial wastes contribute to the fog-forming process

simply to blow the fog away, using an artificial wind. This tactic is actually employed around some settlements in the Arctic to disperse ice fogs. In the frigid atmosphere the water vapor emitted from human settlements (indeed, even the exhalations of a reindeer herd) can easily saturate the air and produce a local fog; at a temperature of 30 degrees or more below zero C. the fog consists mainly of ice crystals. Giant fans have been used successfully to free settlements of such fogs. Obviously, however, the method is applicable only on a small scale and in special situations.

Another attack on fog, based on the principle of evaporating the droplets by one means or another, has been applied in several interesting ways. The best-known of these is the FIDO method (Fog Investigation and Dispersal Operation), in which fuel-oil fires have been used to burn off (that is, evaporate) fog on airfields. The British resorted to this technique on military fields in World War II and successfully cleared them for more than 2,000 takeoffs and landings that could not have been undertaken otherwise. The method has important drawbacks, however. It is expensive, requiring hundreds of dollars' worth of fuel to clear a jet runway for 10 to 15 minutes; it creates a fire hazard for planes landing on the field, and it cannot dissipate all dense fogs. Moreover, the smoke and moisture released by the combustion of oil hinder evaporation of the fog. It has been proposed that this problem might be obviated by using electricity, jet-engine exhausts or anthracite coal to provide the heat, but the high cost would still remain a major objection. Drying the air with chemicals rather than heat has been effective in clearing fog in some cases; here again, however, the procedure is too expensive for wide use, and the chemicals employed tend to be corrosive.

Among the various principles of fog dispersal that have been tested, the most promising seems to be the injection of a catalyst or some other agent that will cause the droplets to coalesce and thus grow large enough to fall quickly to the ground. This type of attack has proved its worth in fogs consisting of supercooled droplets. By seeding the fog with particles of a very cold substance such as dry ice or liquid propane, one can cause some of the fog droplets to freeze. Water vapor in the air then condenses onto these ice crystals; the resultant drying of the air turns additional fog droplets back into vapor. The vapor, in turn, acceler-

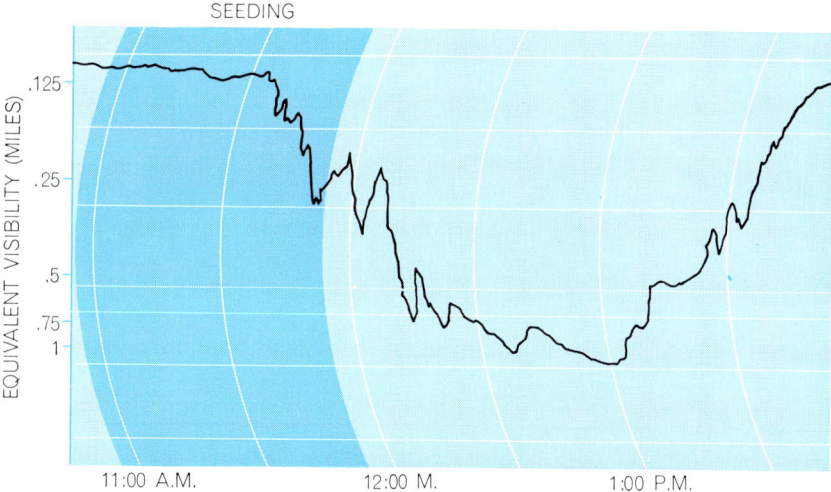

COLD-SEEDING TEST at the airport in Medford, Ore., began shortly before 11:00 A.M.; visibility, as measured by a transmissometer, was less than an eighth of a mile. Thirty minutes after the first seeding run visibility began to improve. By 12:15 P.M. it exceeded the half-mile minimum required at the airport and remained above minimum for over an hour.

ates the growth of the ice crystals, which fall to the ground as they enlarge. United Air Lines has been seeding supercooled fogs at fields in the Pacific Northwest and in Alaska for several years and has found the method to be about 80 percent successful in dissipating such fogs. The airline estimates that this investment in fog control has repaid it fivefold by maintaining the regularity of flight schedules. The method has also worked well in other areas where it is applicable. It is effective, however, only for supercooled fogs, which account for only about 5 percent of all the fogs in the U.S. Temperate Zone.

Several ideas for dispersing warm fogs by droplet coalescence are currently being explored. The Air Transport Association is sponsoring a series of tests of a chemical mixture (composition undisclosed) that is said to make droplets combine by an electrical attraction effect. It is reported to have achieved some success in dissipating radiation fogs at Sacramento, Calif., during calm or very light winds. The ability of the method to disperse moving fogs, however, remains to be demonstrated. In any attempt to control fog, the wind condition is crucial. When the air is calm, clearing it of fog presents a comparatively uncomplicated problem because the volume of air that needs to be treated is limited. On the other hand, a breeze can quickly refog a space that has been cleared of fog; when moderate or strong winds are blowing, it is almost impossible to maintain a clearing.

Another scheme that has been proposed for fog dispersal involves dropping

carefully controlled doses of salt particles into the fog. The theory is that when these hygroscopic particles deliquesce into solution droplets, they will gain water and grow at the expense of natural fog droplets because the humidity is higher with respect to the solution than it is with respect to natural water. It is hoped that the collection of the water into fewer and larger drops will produce a rapid improvement in visibility and that the fall of the large drops to the ground will maintain the improvement for some time after seeding has ended.

For effective progress toward the economical control or modification of fogs we shall have to learn a great deal more about the basic structure of fog and the chemical, physical and electrical properties of the fog droplets. Intensive studies are going forward on various types of natural fogs and on artificial fogs produced in the laboratory. The results of these investigations will be used to test mathematical models that describe fogs in terms of such quantities as temperature, humidity, wind, condensation nuclei, concentration of droplets and amount of liquid water. It should then be possible to obtain insight into the mechanisms and energy exchanges involved in the formation and maintenance of fogs and allow meteorologists to determine which kind of fog will respond to which dispersal method.

The artificial dissipation of fog will be rather costly in any case. Much might be done to prevent the formation of fogs in the first place. For example, spreading a chemical film over swamps and

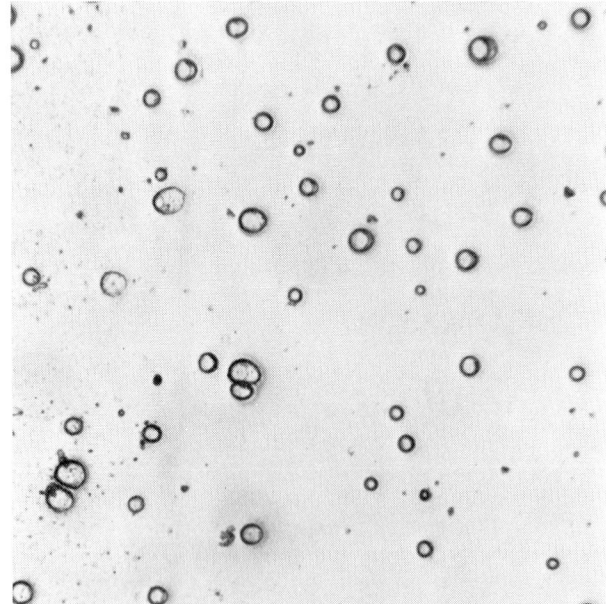

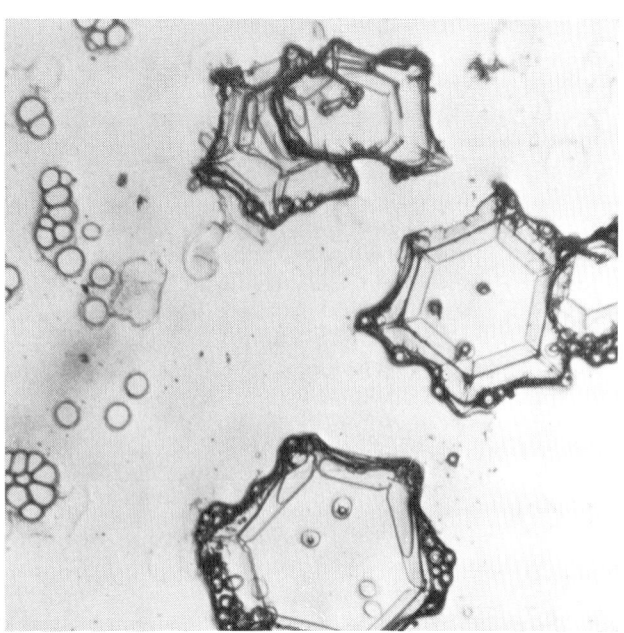

WATER DROPLETS that comprise a fog vary in size from about two microns to as much as 50 microns in diameter; their average size is 20 microns. The droplets in the photomicrograph are from a sample of supercooled fog prepared for a "cold seeding" study.

ICE CRYSTALS are formed in a sample of supercooled fog by seeding with propane. The crystals grow as water vapor is deposited on them. Reduced humidity makes the fog droplets evaporate, clearing the air and furnishing more water vapor for crystal growth.

FOG CLEARANCE is achieved by dropping dry ice into a bank of supercooled fog overlying Elmendorf Air Force Base in Anchorage, Alaska. Seeding with dry ice initiated the growth of ice crystals, de-priving the air of the liquid water that had made it foggy. Only 5 percent of the fogs in those parts of the U.S. that lie within the Temperate Zone are supercooled and dispersable by cold-seeding.

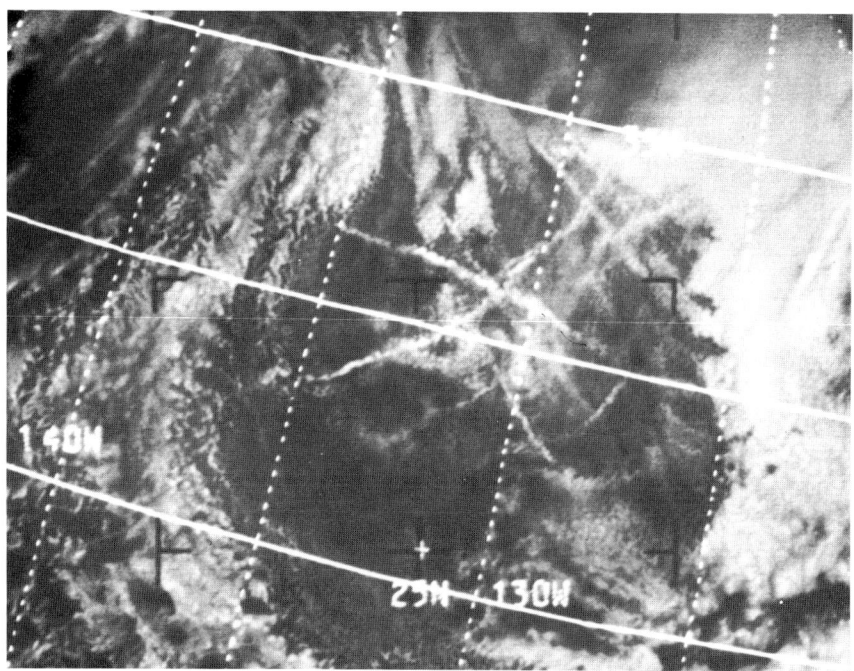

POLLUTION FOGS over the Pacific Ocean off the coast of California are visible in a weather satellite photograph as white trails in the area between 25 and 35 degrees north latitude and 125 and 135 degrees west longitude. Each trail is a narrow fogbank formed in response to the discharge of hygroscopic particles into the atmosphere from the funnels of ships at sea.

more careful attention to the selection of proper locations for activities that generate fog and for those that may be troubled by fog.

Factories giving rise to air pollution, for example, should be located at sites where the topography and prevailing winds favor effective dispersal of the pollutants. Plants that must dispose of heated water should not be built near densely populated or well-traveled areas. On the other hand, in the selection of sites for airports, highways, sports stadiums, golf courses and so forth consideration should be given to finding locations where the geography, topography, soil characteristics and other features would tend to minimize the formation and persistence of fogs. For example, ideally an airport should be situated (1) upwind of any nearby source of air pollution, (2) on a plateau that stands high enough to shed cold air into a valley but low enough to avoid hilltop immersion in clouds, (3) away from rivers, lakes or marshy ground and (4) in full, unobstructed exposure to the sun.

By the application of meteorological knowledge to planning and by the development of further methods for fog prevention and dispersal, we may eventually be able to deal with the growing fog and smog problem. The success of these efforts, however, will hinge on the achievement of effective control over the pollution of the air and waters.

lakes in the vicinity of sensitive areas such as airports and highways might considerably reduce evaporation and thus reduce the frequency and intensity of fogs in those areas. There are indications that in places where shallow radiation fogs are common the fogs could be prevented from spreading by planting vegetation thickly around the area of origin. It would be desirable to prohibit pollution-generating factories from operating at times when the air is calm or an inversion is present. And the fog menace could be diminished greatly by giving

The Author

JOEL N. MYERS is an instructor in meteorology at Pennsylvania State University, where he was graduated in 1961, received a master's degree in 1963 and is now working on his doctorate. He writes: "At the age of seven I became interested in the weather and began keeping a daily record of the local conditions. At the age of 14 I was appointed a cooperative weather observer by the U.S. Weather Bureau." In addition to his teaching he does a daily broadcast on weather that is carried by two educational television stations. He remarks that the program is somewhat unusual "in that it serves the double role of providing an in-depth analysis and forecast of the weather while attempting to educate the viewers about basic meteorological concepts." Myers is interested in applying specialized weather forecasts to industry and has developed a consulting service used by most of the major ski areas in the northeastern U.S. A headline on a newspaper account of his service read, "Slopes' Top Aide an Avid Nonskier."

Bibliography

FOG. Joseph J. George in *Compendium of Meteorology*, edited by Thomas F. Malone. American Meteorological Society, 1951.

WEATHER ANALYSIS AND FORECASTING, VOL. II: WEATHER AND WEATHER SYSTEMS. Sverre Petterssen. McGraw-Hill Book Company, Inc., 1956.

CLOUD PHYSICS AND CLOUD SEEDING. Louis J. Battan. Anchor Books, Doubleday & Company, Inc., 1962.

THE UNCLEAN SKY: A METEOROLOGIST LOOKS AT AIR POLLUTION. Louis J. Battan. Anchor Books, Doubleday & Company, Inc., 1966.

AIR POLLUTION. 1968. R. S. Scorer. Pergamon Press, 1968.

SCIENTIFIC AMERICAN March 1969, Vol. 220, No. 3, pp. 54–64

OFFPRINT **877**

CONTINENTAL DRIFT AND EVOLUTION

by Björn Kurtén

The breakup of ancient supercontinents would have had major effects on the evolution of living organisms. Does it explain the difference in the diversification of reptiles and mammals?

The history of life on the earth, as it is revealed in the fossil record, is characterized by intervals in which organisms of one type multiplied and diversified with extraordinary exuberance. One such interval is the age of reptiles, which lasted 200 million years and gave rise to some 20 reptilian orders, or major groups of reptiles. The age of reptiles was followed by our own age of mammals, which has lasted for 65 million years and has given rise to some 30 mammalian orders.

The difference between the number of reptilian orders and the number of mammalian ones is intriguing. How is it that the mammals diversified into half again as many orders as the reptiles in a third of the time? The answer may lie in the concept of continental drift, which has recently attracted so much attention from geologists and geophysicists [see "The Confirmation of Continental Drift," by Patrick M. Hurley; SCIENTIFIC AMERICAN Offprint 874]. It now seems that for most of the age of reptiles the continents were collected in two supercontinents, one in the Northern Hemisphere and one in the Southern. Early in the age of mammals the two supercontinents had apparently broken up into the continents of today but the present connections between some of the continents had not yet formed. Clearly such events would have had a profound effect on the evolution of living organisms.

The world of living organisms is a world of specialists. Each animal or plant has its special ecological role. Among the mammals of North America, for instance, there are grass-eating prairie animals such as the pronghorn antelope, browsing woodland animals such as the deer, flesh-eating animals specializing in large game, such as the mountain lion, or in small game, such as the fox, and so on. Each order of mammals

comprises a number of species related to one another by common descent, sharing the same broad kind of specialization and having a certain physical resemblance to one another. The order Carnivora, for example, consists of a number of related forms (weasels, bears, dogs, cats, hyenas and so on), most of which are flesh-eaters. There are a few exceptions (the aardwolf is an insect-eating hyena and the giant panda lives on bamboo shoots), but these are recognized as late specializations.

Radiation and Convergence

In spite of being highly diverse, all the orders of mammals have a common origin. They arose from a single ancestral species that lived at some unknown time in the Mesozoic era, which is roughly synonymous with the age of reptiles. The American paleontologist Henry Fairfield Osborn named the evolution of such a diversified host from a single ancestral type "adaptive radiation." By adapting to different ways of life—walking, climbing, swimming, flying, planteating, flesh-eating and so on—the descendant forms come to diverge more and more from one another. Adaptive radiation is not restricted to mammals; in fact we can trace the process within every major division of the plant and animal kingdoms.

The opposite phenomenon, in which stocks that were originally very different gradually come to resemble one another through adaptation to the same kind of life, is termed convergence. This too seems to be quite common among mammals. There is a tendency to duplication—indeed multiplication—of orders performing the same function. Perhaps the most remarkable instance is found among the mammals that have specialized in large-scale predation on termites

and ants in the Tropics. This ecological niche is filled in South America by the ant bear *Myrmecophaga* and its related forms, all belonging to the order Edentata. In Asia and Africa the same role is played by mammals of the order Pholidota: the pangolins, or scaly anteaters. In Africa a third order has established itself in this business: the Tubulidentata, or aardvarks. Finally, in Australia there is the spiny anteater, which is in the order Monotremata. Thus we have members of four different orders living the same kind of life.

One can cite many other examples. There are, for instance, several living and extinct orders of hoofed herbivores. There are two living orders (the Rodentia, or rodents, and the Lagomorpha, or rabbits and hares) whose chisel-like incisor teeth are specialized for gnawing. Some extinct orders specialized in the same way, and an early primate, an iceage ungulate and a living marsupial have also intruded into the "rodent niche" [see top illustration on page 649]. This kind of duplication, or near-duplication, is an essential ingredient in the richness of the mammalian life that unfolded during the Cenozoic era, or the age of mammals. Of the 30 or so orders of land-dwelling mammals that appeared during this period almost two-thirds are still extant.

The Reptiles of the Cretaceous

The 65 million years of the Cenozoic are divided into two periods: the long Tertiary and the brief Quaternary, which includes the present [see chart on page 646]. The 200-million-year age of reptiles embraces the three periods of the Mesozoic era (the Triassic, the Jurassic and the Cretaceous) and the final period (the Permian) of the preceding era. It is instructive to compare the number

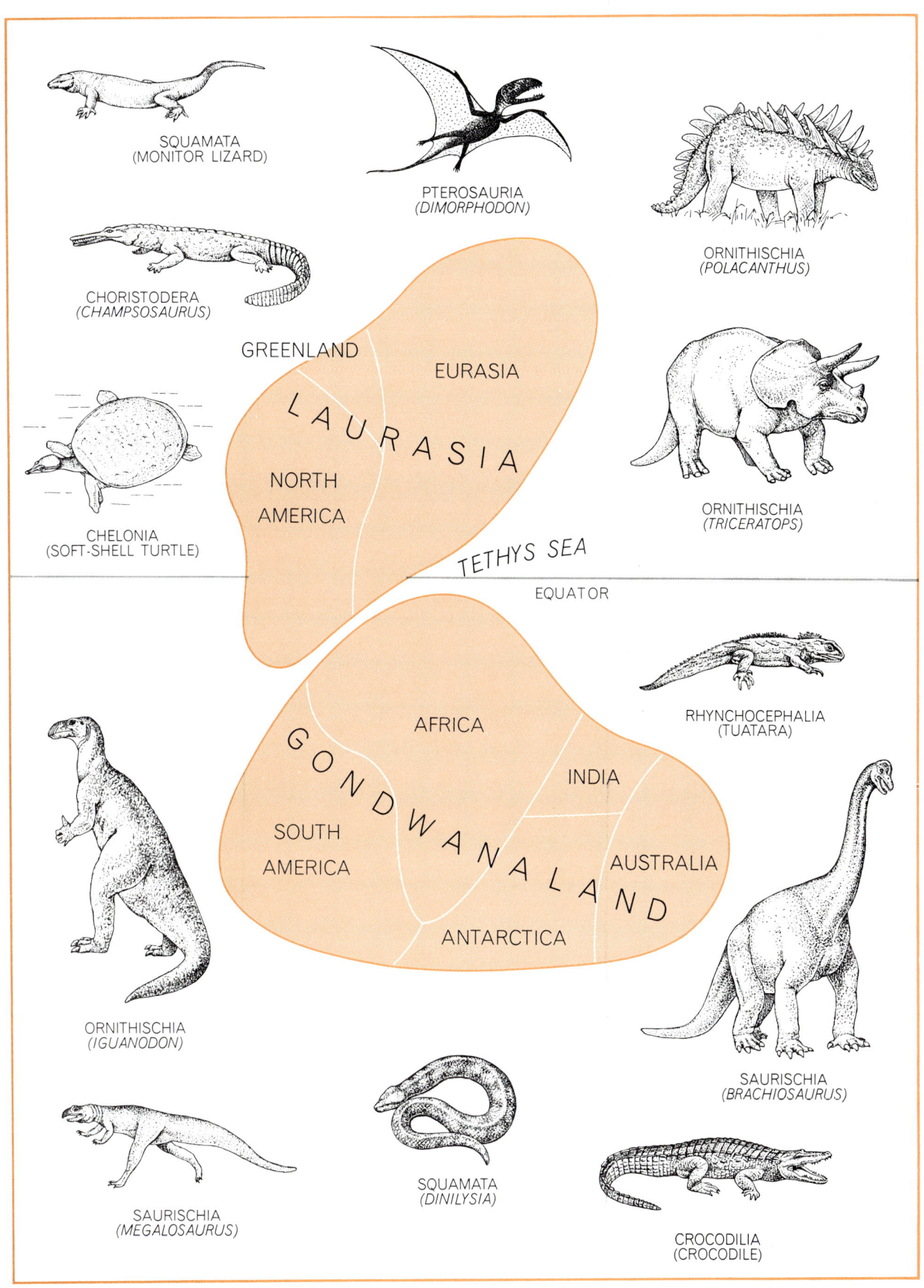

SQUAMATA
(MONITOR LIZARD)

PTEROSAURIA
(*DIMORPHODON*)

ORNITHISCHIA
(*POLACANTHUS*)

CHORISTODERA
(*CHAMPSOSAURUS*)

CHELONIA
(SOFT-SHELL TURTLE)

GREENLAND

EURASIA

L A U R A S I A

NORTH
AMERICA

ORNITHISCHIA
(*TRICERATOPS*)

TETHYS SEA

EQUATOR

AFRICA

RHYNCHOCEPHALIA
(TUATARA)

G O N D W A N A L A N D

INDIA

SOUTH
AMERICA

AUSTRALIA

ANTARCTICA

ORNITHISCHIA
(*IGUANODON*)

SAURISCHIA
(*BRACHIOSAURUS*)

SAURISCHIA
(*MEGALOSAURUS*)

SQUAMATA
(*DINILYSIA*)

CROCODILIA
(CROCODILE)

TWO SUPERCONTINENTS of the Mesozoic era were Laurasia in the north and Gondwanaland in the south. The 12 major types of reptiles, represented by typical species, are those whose fossil remains are found in Cretaceous formations. Most of the orders inhabited both supercontinents; migrations were probably by way of a land bridge in the west, where the Tethys Sea was narrowest.

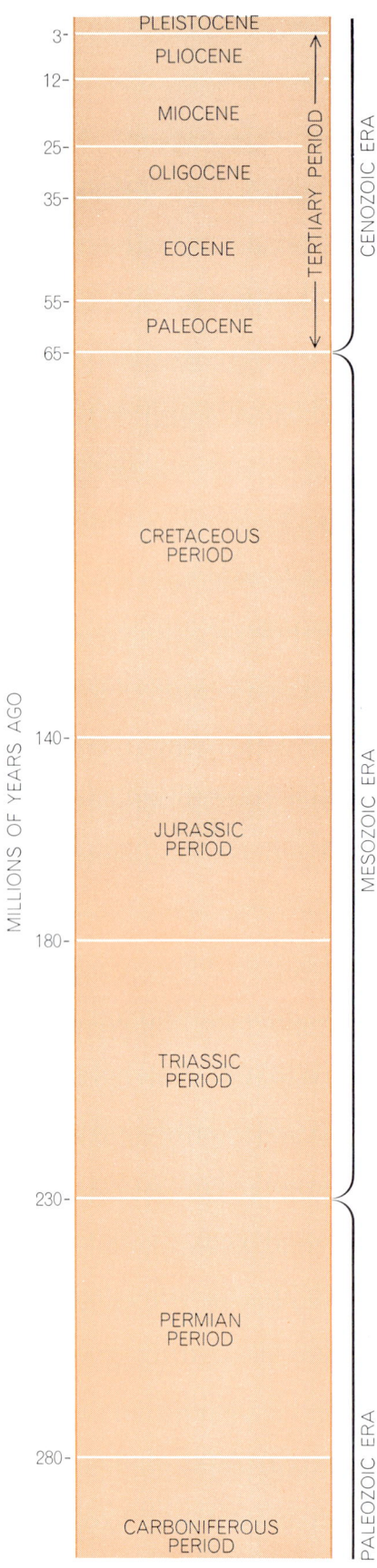

PLEISTOCENE

PLIOCENE

MIOCENE

OLIGOCENE

EOCENE

PALEOCENE

TERTIARY PERIOD

CENOZOIC ERA

CRETACEOUS PERIOD

JURASSIC PERIOD

MESOZOIC ERA

TRIASSIC PERIOD

PERMIAN PERIOD

CARBONIFEROUS PERIOD

PALEOZOIC ERA

MILLIONS OF YEARS AGO

3 —
12 —
25 —
35 —
55 —
65 —
140 —
180 —
230 —
280 —

SIX PERIODS of earth history were occupied by the age of reptiles and the age of mammals. The reptiles' rise began 280 million years ago, in the final period of the Paleozoic era. Mammals replaced reptiles as dominant land animals 65 million years ago.

of reptilian orders that flourished during some Mesozoic interval about as long as the Cenozoic era with the number of mammalian orders in the Cenozoic. The Cretaceous period is a good candidate. Some 75 million years in duration, it is only slightly longer than the age of mammals. Moreover, the Cretaceous was the culmination of reptilian life and its fossil record on most continents is good. In the Cretaceous the following orders of land reptiles were extant:

Order Crocodilia: crocodiles, alligators and the like. Their ecological role was amphibious predation; their size, medium to large.

Order Saurischia: saurischian dinosaurs. These were of two basic types: bipedal upland predators (Theropoda) and very large amphibious herbivores (Sauropoda).

Order Ornithischia: ornithischian dinosaurs. Here there were three basic types: bipedal herbivores (Ornithopoda), heavily armored quadrupedal herbivores (Stegosauria and Ankylosauria) and horned herbivores (Ceratopsia).

Order Pterosauria: flying reptiles.

Order Chelonia: turtles and tortoises.

Order Squamata: The two basic types were lizards (Lacertilia) and snakes (Serpentes). Both had the same principal ecological role: small to medium-sized predator.

Order Choristodera (or suborder in the order Eosuchia): champsosaurs. These were amphibious predators.

One or two other reptilian orders may be represented by rare forms. Even if we include them, we get only eight or nine orders of land reptiles in Cretaceous times. One could maintain that an order of reptiles ranks somewhat higher than an order of mammals; some reptilian orders include two or even three basic adaptive types. Even if these types are kept separate, however, the total rises only to 12 or 13. Furthermore, there seems to be only one clear-cut case of ecological duplication: both the crocodilians and the champsosaurs are sizable amphibious predators. (The turtles cannot be considered duplicates of the armored dinosaurs. For one thing, they were very much smaller.) A total of somewhere between seven and 13 orders over a period of 75 million years seems a sluggish record compared with the mammalian achievement of perhaps 30 orders in 65 million years. What light can paleogeography shed on this matter?

The Mesozoic Continents

The two supercontinents of the age of reptiles have been named Laurasia (after

Laurentian and Eurasia) and Gondwanaland (after a characteristic geological formation, the Gondwana). Between them lay the Tethys Sea (named for the wife of Oceanus in Greek myth, who was mother of the seas). Laurasia, the northern supercontinent, consisted of what would later be North America, Greenland and Eurasia north of the Alps and the Himalayas. Gondwanaland, the southern one, consisted of the future South America, Africa, India, Australia and Antarctica. The supercontinents may have begun to split up as early as the Triassic period, but the rifts between them did not become effective barriers to the movement of land animals until well into the Cretaceous, when the age of reptiles was nearing its end.

When the mammals began to diversify in the late Cretaceous and early Tertiary, the separation of the continents appears to have been at an extreme. The old ties were sundered and no new ones had formed. The land areas were further fragmented by a high sea level; the waters flooded the continental margins and formed great inland seas, some of which completely partitioned the continents. For example, South America was cut in two by water in the region that later became the Amazon basin, and Eurasia was split by the joining of the Tethys Sea and the Arctic Ocean. In these circumstances each chip of former supercontinent became the nucleus for an adaptive radiation of its own, each fostering a local version of a balanced fauna. There were at least eight such nuclei at the beginning of the age of mammals. Obviously such a situation is quite different from the one in the age of reptiles, when there were only two separate land masses.

Where the Reptiles Originated

The fossil record contains certain clues to some of the reptilian orders' probable areas of origin. The immense distance in time and the utterly different geography, however, make definite inferences hazardous. Let us see what can be said about the orders of Cretaceous reptiles (most of which, of course, arose long before the Cretaceous):

Crocodilia. The earliest fossil crocodilians appear in Middle Triassic formations in a Gondwanaland continent (South America). The first crocodilians in Laurasia are found in Upper Triassic formations. Thus a Gondwanaland origin is suggested.

Saurischia. The first of these dinosaurs appear on both supercontinents in the Middle Triassic, but they are more

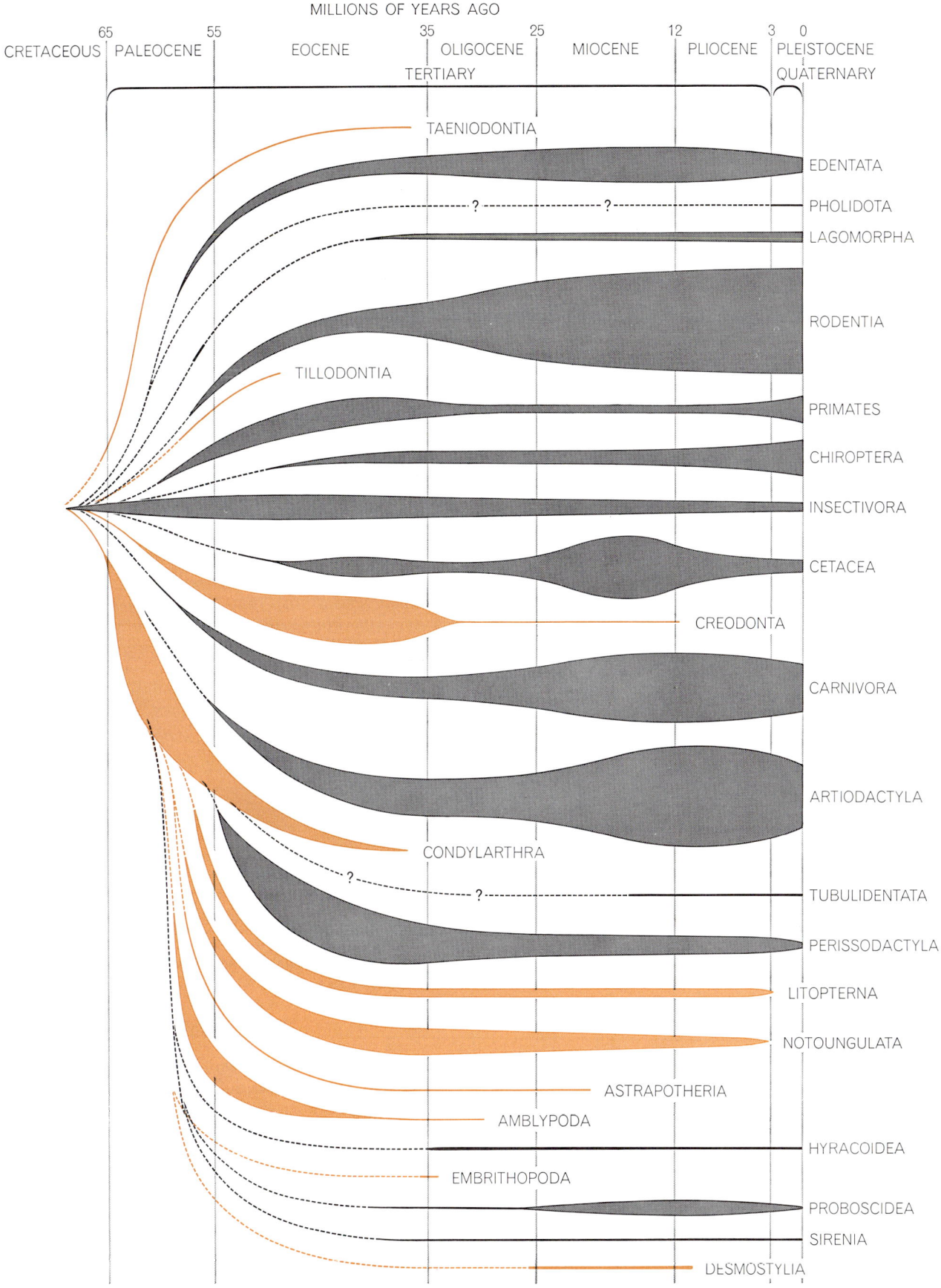

MILLIONS OF YEARS AGO

CRETACEOUS | PALEOCENE | EOCENE | OLIGOCENE | MIOCENE | PLIOCENE | PLEISTOCENE

65   55   35   25   12   3   0

TERTIARY   QUATERNARY

TAENIODONTIA
EDENTATA
PHOLIDOTA
LAGOMORPHA
RODENTIA
TILLODONTIA
PRIMATES
CHIROPTERA
INSECTIVORA
CETACEA
CREODONTA
CARNIVORA
ARTIODACTYLA
CONDYLARTHRA
TUBULIDENTATA
PERISSODACTYLA
LITOPTERNA
NOTOUNGULATA
ASTRAPOTHERIA
AMBLYPODA
HYRACOIDEA
EMBRITHOPODA
PROBOSCIDEA
SIRENIA
DESMOSTYLIA

ADAPTIVE RADIATION of the mammals has been traced from its starting point late in the Mesozoic era by Alfred S. Romer of Harvard University. Records for 25 extinct and extant orders of placental mammals are shown here. The lines increase and decrease in width in proportion to the abundance of each order. Extinct orders are shown in color; broken lines mean that no fossil record exists during the indicated interval and question marks imply doubt about the suggested ancestral relation between some orders.

varied in the south. A Gondwanaland origin is very tentatively suggested.

Ornithischia. These dinosaurs appear in the Upper Triassic of South Africa (Gondwanaland) and invade Laurasia somewhat later. A Gondwanaland origin is indicated.

Pterosauria. The oldest fossils of flying reptiles come from the early Jurassic of Europe. They represent highly specialized forms, however, and their antecedents are unknown. No conclusion seems possible.

Chelonia. Turtles are found in Triassic formations in Laurasia. None are found in Gondwanaland before Cretaceous times. This suggests a Laurasian origin. On the other hand, a possible forerunner of turtles appears in the Permian of South Africa. If the Permian form was in fact ancestral, a Gondwanaland origin would be indicated. In any case, the order's main center of evolution certainly lay in the northern supercontinent.

Squamata. Early lizards are found in the late Triassic of the north, which may suggest a Laurasian origin. Unfortunately the lizards in question are aberrant gliding animals. They must have had a long history, of which we know nothing at present.

Choristodera. The crocodile-like champsosaurs are found only in North America and Europe, and so presumably originated in Laurasia.

The indications are, then, that three orders of reptiles—the crocodilians and the two orders of dinosaurs—may have originated in Gondwanaland. Three others—the turtles, the lizards and snakes and the champsosaurs—may have originated in Laurasia. The total number of basic adaptive types in the Gondwanaland group is six; the Laurasia group has four. The Gondwanaland radiation may well have been slightly richer than the Laurasian because it seems that the southern supercontinent was somewhat larger and had a slightly more varied climate. Laurasian climates seem to have been tropical to temperate. Southern parts of Gondwanaland were heavily glaciated late in the era preceding the Mesozoic, and its northern shores (facing the Tethys Sea) had a fully tropical climate.

Although some groups of reptiles, such as the champsosaurs, were confined to one or another of the supercontinents, most of the reptilian orders sooner or later spread into both of them. This means that there must have been ways for land animals to cross the Tethys Sea. The Tethys was narrow in the west and wide to the east. Presumably whatever land connection there was—a true land bridge or island stepping-stones—was located in the western part of the sea. In any case, migration along such routes meant that there was little local differentiation among the reptiles of the Mesozoic era. It was over an essentially uniform reptilian world that the sun finally set at the end of the age of reptiles.

Early Mammals of Laurasia

The conditions of mammalian evolution were radically different. In early and middle Cretaceous times the connections between continents were evidently close enough for primitive mammals to spread into all corners of the habitable world. As the continents drifted farther apart, however, populations of these primitive forms were gradually isolated from one another. This was particularly the case, as we shall see, with the mammals that inhabited the daughter continents of Gondwanaland. Among the Laurasian continents North America was drifting away from Europe, but at the beginning of the age of mammals the distance was not great and there is good evidence that some land connection remained well into the early Tertiary. North American and European mammals were practically identical as late as early Eocene times. Furthermore, throughout the Cenozoic era there was a connection between Alaska and Siberia, at least intermittently, across the Bering Strait. On the other hand, the inland sea extending from the Tethys to the Arctic Ocean formed a complete barrier to direct migration between Europe and Asia in the early Tertiary. Migrations could take place only by way of North America.

In this way the three daughter continents of ancient Laurasia formed three semi-isolated nuclear areas. Many orders of mammals arose in these Laurasian nuclei, among them seven orders that are now extinct but that covered a wide spectrum of specialized types, including primitive hoofed herbivores, carnivores, insectivores and gnawers. The orders of mammals that seem to have arisen in the northern daughter continents and that are extant today are:

Insectivora: moles, hedgehogs, shrews and the like. The earliest fossil insectivores are found in the late Cretaceous of North America and Asia.

Chiroptera: bats. The earliest-known bat comes from the early Eocene of North America. At a slightly later date bats were also common in Europe.

Primates: prosimians (for example, tarsiers and lemurs), monkeys, apes, man. Early primates have recently been found in the late Cretaceous of North America. In the early Tertiary they are common in Europe as well.

Carnivora: cats, dogs, bears, weasels and the like. The first true carnivores appear in the Paleocene of North America.

Perissodactyla: horses, tapirs and other odd-toed ungulates. The earliest forms appear at the beginning of the Eocene in the Northern Hemisphere.

Artiodactyla: cattle, deer, pigs and other even-toed ungulates. Like the odd-toed ungulates, they appear in the early Eocene of the Northern Hemisphere.

Rodentia: rats, mice, squirrels, beavers and the like. The first rodents appear in the Paleocene of North America.

Lagomorpha: hares and rabbits. This order makes its first appearance in the Eocene of the Northern Hemisphere.

Pholidota: pangolins. The earliest come from Europe in the middle Tertiary.

The fact that a given order of mammals is found in older fossil deposits in North America than in Europe or Asia does not necessarily mean that the order arose in the New World. It may simply reflect the fact that we know much more about the early mammals of North America than we do about those of Eurasia. All we can really say is that a total of 16 extant or extinct orders of mammals probably arose in the Northern Hemisphere.

Early Mammals of South America

The fragmentation of Gondwanaland seems to have started earlier than that of Laurasia. The rifting certainly had a much more radical effect. Looking at South America first, we note that at the beginning of the Tertiary this continent was tenuously connected to North America but that for the rest of the period it was completely isolated. The evidence for the tenuous early linkage is the presence in the early Tertiary beds of North America of mammalian fossils representing two predominantly South American orders: the Edentata (the order that includes today's ant bears, sloths and armadillos) and the Notoungulata (an order of extinct hoofed herbivores).

Four other orders of mammals are exclusively South American: the Paucituberculata (opossum rats and other small South American marsupials), the Pyrotheria (extinct elephant-like animals), the Litopterna (extinct hoofed herbivores, including some forms resembling

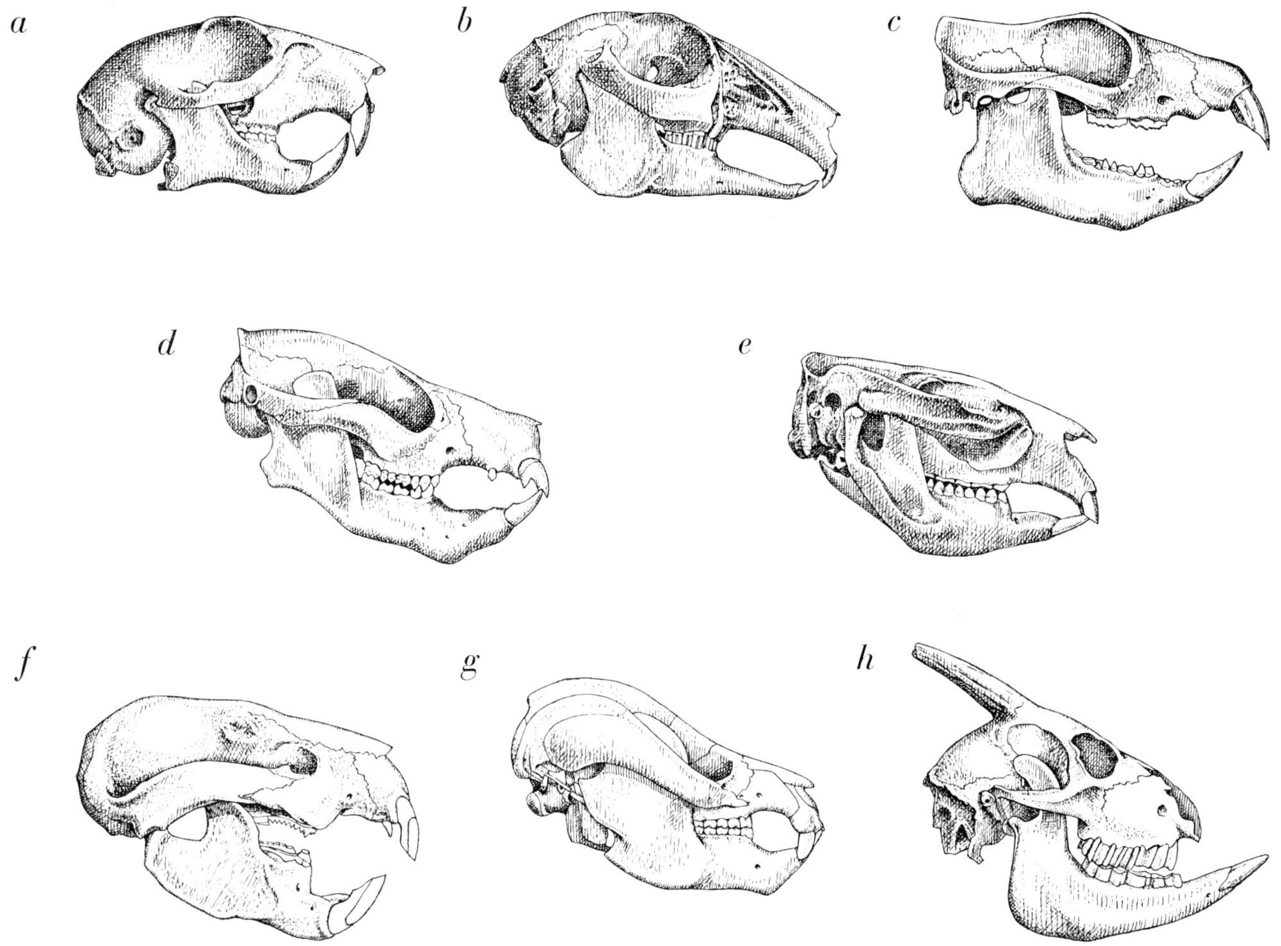

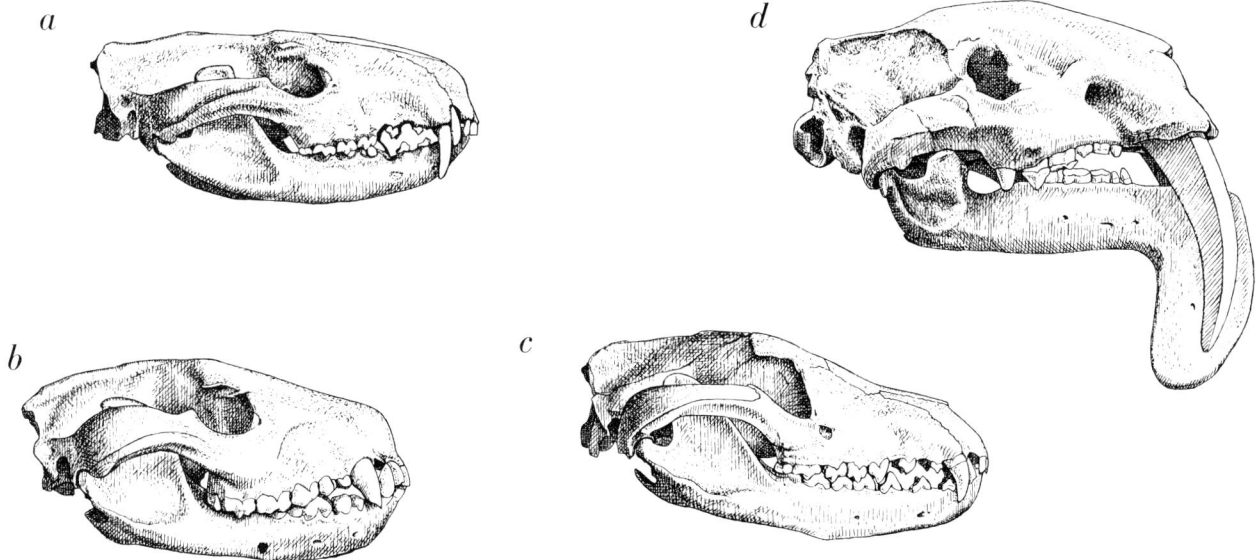

a *b* *c*

d *e*

f *g* *h*

CHISEL-LIKE INCISORS, specialized for gnawing, appear in animals belonging to several extinct and extant orders in addition to the rodents, represented by a squirrel (*a*), and the lagomorphs, represented by a hare (*b*), which are today's main specialists in this ecological role. Representatives of other orders with chisel-like incisor teeth are an early tillodont, *Trogosus* (*c*), an early primate, *Plesiadapis* (*d*), a living marsupial, the wombat (*e*), one of the extinct multituberculate mammals, *Taeniolabis* (*f*), a mammal-like reptile of the Triassic, *Bienotherium* (*g*), and a Pleistocene cave goat, *Myotragus* (*h*), whose incisor teeth are in the lower jaw only.

a *d*

b *c*

CARNIVOROUS MARSUPIALS, living and extinct, fill an ecological niche more commonly occupied by the placental carnivores today. Illustrated are the skulls of two living forms, the Australian "cat," *Dasyurus* (*a*), and the Tasmanian devil, *Sarcophilus* (*b*). The Tasmanian "wolf," *Thylacinus* (*c*), has not been seen for many years and may be extinct. A tiger-sized predator of South America, *Thylacosmilus* (*d*) became extinct in Pliocene times, long before the placental sabertooth of the Pleistocene, *Smilodon*, appeared.

horses and camels) and the Astrapotheria (extinct large hoofed herbivores of very peculiar appearance). Thus a total of six orders, extinct or extant, probably originated in South America. Still another order, perhaps of even more ancient origin, is the Marsupicarnivora. The order is so widely distributed, with species found in South America, North America, Europe and Australia, that its place of origin is quite uncertain. It includes, in addition to the extinct marsupial carnivores of South America, the opossums of the New World and the native "cats" and "wolves" of the Australian area.

The most important barrier isolating South America from North America in the Tertiary period was the Bolívar Trench. This arm of the sea cut across the extreme northwest corner of the continent. In the late Tertiary the bottom of the Bolívar Trench was lifted above sea level and became a mountainous land area. A similar arm of sea, to which I have already referred, extended across the continent in the region that is now the Amazon basin. This further enhanced the isolation of the southern part of South America.

Africa's role as a center of adaptive radiation is problematical because practically nothing is known of its native mammals before the end of the Eocene. We do know, however, that much of the continent was flooded by marginal seas, and that in the early Tertiary, Africa was cut up into two or three large islands. Still, there must have been a land route to Eurasia even in the Eocene; some of the African mammals of the following epoch (the Oligocene) are clearly immigrants from the north or northeast. Nonetheless, the majority of African mammals are of local origin. They include the following orders:

Proboscidea: the mastodons and elephants.

Hyracoidea: the conies and their extinct relatives.

Embrithopoda: an extinct order of very large mammals.

Tubulidentata: the aardvarks.

In addition the order Sirenia, consisting of the aquatic dugongs and manatees, is evidently related to the Proboscidea and hence presumably also originated in Africa. The same may be true of another order of aquatic mammals,

the extinct Desmostylia, which also seems to be related to the elephants. The one snag in this interpretation is that desmostylian fossils are found only in the North Pacific, which seems rather a long way from Africa. Nonetheless, once they were waterborne, early desmostylians might have crossed the Atlantic, which was then only a narrow sea, navigated the Bolívar Trench and, rather like Cortes (but stouter), found themselves in the Pacific.

Early Mammals of Africa

Thus there are certainly four, and possibly six, mammalian orders for which an African origin can be postulated. Here it should be noted that Africa had an impressive array of primates in the Oligocene. This suggests that the order Primates had a comparatively long history in Africa before that time. Even though the order as such does not have its roots in Africa, it is possible that the higher primates—the Old World monkeys, the apes and the ancestors of man—may have originated there. Most of the fossil primates found in the Oligocene

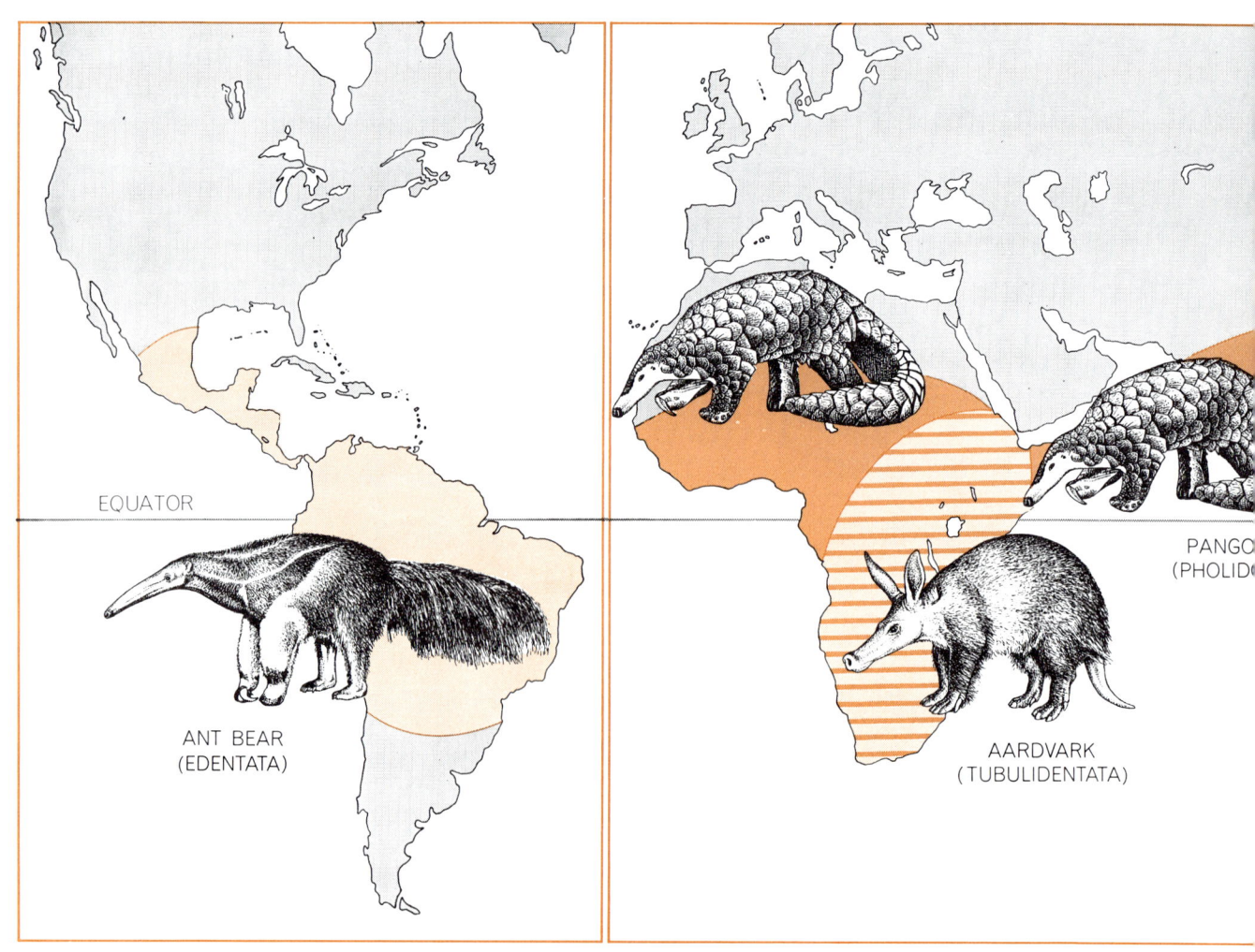

EQUATOR

ANT BEAR
(EDENTATA)

AARDVARK
(TUBULIDENTATA)

PANGO
(PHOLID

formations of Africa are primitive apes or monkeys, but there is at least one form (*Propliopithecus*) whose dentition looks like a miniature blueprint of a set of human teeth.

The Rest of Gondwanaland

We know little or nothing of the zoogeographic roles played by India and Antarctica in the early Tertiary. Mammalian fossils from the early Tertiary are also absent in Australia. It may be assumed, however, that the orders of mammals now limited to Australia probably originated there. These include two orders of marsupials: the Peramelina, comprised of several bandicoot genera, and the Diprotodonta, in which are found the kangaroos, wombats, phalangers and a number of extinct forms. In addition the order Monotremata, a very primitive group of mammals that includes the spiny anteater and the platypus, is likely to be of Australian origin. This gives us a total of three orders probably founded in Australia.

Summing up, we find that the three Laurasian continents produced a total of 16 orders of mammals, an average of five or six orders per continent. As for Gondwanaland, South America produced six orders, Africa four to six and Australia three. The fact that Australia is a small continent probably accounts for the lower number of orders founded there. Otherwise the distribution—the average of five or six orders per subdivision—is remarkably uniform for both the Laurasian and Gondwanaland supercontinents. The mammalian record should be compared with the data on Cretaceous reptiles, which show that the two supercontinents produced a total of 12 or 13 orders (or adaptively distinct suborders). A regularity is suggested, as if a single nucleus of radiation would tend in a given time to produce and support a given amount of basic zoological variation.

As the Tertiary period continued new land connections were gradually formed, replacing those sundered when the old supercontinents broke up. Africa made its landfall with Eurasia in the Oligocene and Miocene epochs. Laurasian orders of mammals spread into Africa and crowded out some of the local forms, but at the same time some African mammals (notably the mastodons and elephants) went forth to conquer almost the entire world. In the Western Hemisphere the draining and uplifting of the Bolívar Trench was followed by intense intermigration and competition among the mammals of the two Americas. In the process much of the typical South American mammal population was exterminated, but a few forms pressed successfully into North America to become part of the continent's spectacular ice-age wildlife.

India, a fragment of Gondwanaland that finally became part of Asia, must have made a contribution to the land fauna of that continent but just what it was cannot be said at present. Of all the drifting Noah's arks of mammalian evolution only two—Antarctica and Australia—persist in isolation to this day. The unknown mammals of Antarctica have long been extinct, killed by the ice that engulfed their world. Australia is therefore the only island continent that still retains much of its pristine mammalian fauna. [*see illustration on next page*].

If the fragmentation of the continents at the beginning of the age of mammals promoted variety, the amalgamation in the latter half of the age of mammals has promoted efficiency by means of a large-scale test of the survival of the fittest. There is a concomitant loss of variety; 13 orders of land mammals have become extinct in the course of the Cenozoic. Most of the extinct orders are island-continent productions, which suggests that a system of semi-isolated provinces, such as the daughter continents of Laurasia, tends to produce a more efficient brood than the completely isolated nuclei of the Southern Hemisphere. Not all the Gondwanaland orders were inferior, however; the edentates were moderately successful and the proboscidians spectacularly so.

As far as land mammals are concerned, the world's major zoogeographic provinces are at present four in number: the Holarctic-Indian, which consists of North America and Eurasia and also northern Africa; the Neotropical, made up of Central America and South America; the Ethiopian, consisting of Africa south of the Sahara, and the Australian. This represents a reduction from seven provinces with about 30 orders of mammals to four provinces with about 18 orders. The reduction in variety is proportional to the reduction in the number of provinces.

In conclusion it is interesting to note that we ourselves, as a subgroup within the order Primates, probably owe our origin to a radiation within one of Gondwanaland's island continents. I have noted that an Oligocene primate of Africa may have been close to the line of human evolution. By Miocene times there were definite hominids in Africa, identified by various authorities as members of the genus *Ramapithecus* or the genus *Kenyapithecus*. Apparently these early hominids spread into Asia and Europe toward the end of the Miocene. The cycle of continental fragmentation and amalgamation thus seems to have played an important part in the origin of man as well as of the other land mammals.

PINY ANTEATER
MONOTREMATA)

FOUR ANT-EATING MAMMALS have become adapted to the same kind of life although each is a member of a different mammalian order. Their similar appearance provides an example of an evolutionary process known as convergence. The ant bears of the New World Tropics are in the order Edentata. The aardvark of Africa is the only species in the order Tubulidentata. Pangolins, found both in Asia and in Africa, are members of the order Pholidota. The spiny anteater of Australia, a very primitive mammal, is in the order Monotremata.

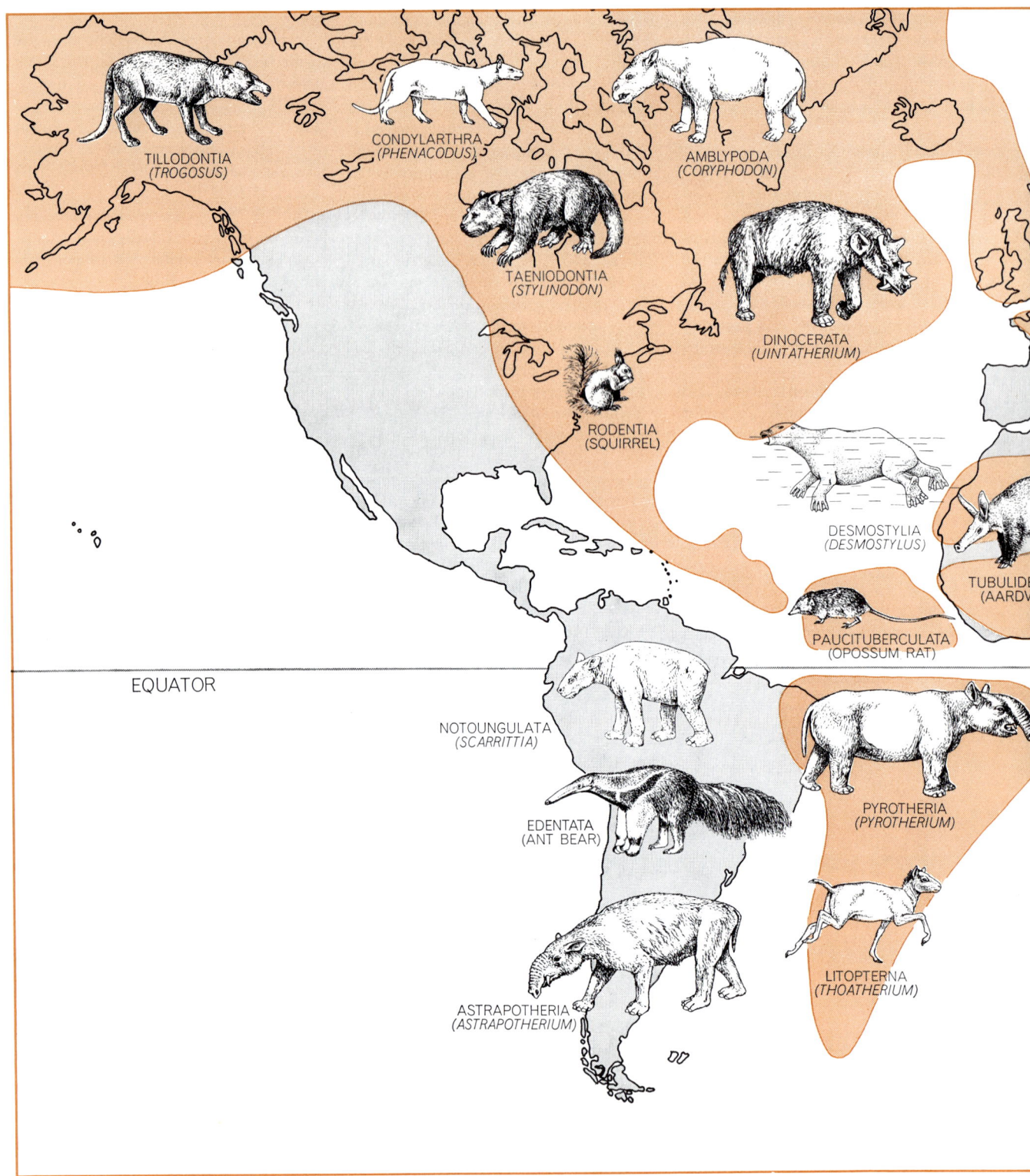

CONTINENTAL DRIFT affected the evolution of the mammals by fragmenting the two supercontinents early in the Cenozoic era. In the north, Europe and Asia, although separated by a sea, remained connected with North America during part of the era. The

CHIROPTERA
(LITTLE BROWN BAT)

PHOLIDOTA (PANGOLIN)

CREODONTA
(*HYAENODON*)

CARNIVORA
(WOLF)

PERISSODACTYLA
(BLACK RHINOCEROS)

PRIMATES
(RINGTAILED LEMUR)

ARTIODACTYLA
(GRANT'S GAZELLE)

LAGOMORPHA
(HARE)

MULTITUBERCULATA
(*MESODMA*)

INSECTIVORA
(WHITE-TOOTHED SHREW)

DINOCERITHOPODA
(*DINOITHERIUM*)

SIRENIA
(MANATEE)

HYRACOIDEA
(ROCK CONEY)

MONOTREMATA
(SPINY ANTEATER)

DIPROTODONTA
(KANGAROO)

PROBOSCIDEA
(AFRICAN ELEPHANT)

PERAMELINA
(LONG-NOSED BANDICOOT)

free migration that resulted prevents certainty regarding the place of origin of many orders of mammals that evolved in the north.

The far wider rifting of Gondwanaland allowed the evolution of unique groups of mammals in South America, Africa and Australia.

The Author

BJÖRN KURTÉN is a lecturer in paleontology at the University of Helsinki, where he obtained his Ph.D. in 1954. He has written a number of articles dealing with such subjects as fossil carnivores, dating of early man, late Tertiary and Quaternary stratigraphy, and aspects of population dynamics, evolutionary theory and paleobiogeography. His recent books include *Pleistocene Mammals of Europe* and *The Age of the Dinosaurs*.

Bibliography

VERTEBRATE PALEONTOLOGY. Alfred Sherwood Romer. The University of Chicago Press, 1966.

THE AGE OF THE DINOSAURS. Björn Kurtén. World University Library, 1968.

SCIENTIFIC AMERICAN September 1963, Vol. 209, No. 3, pp. 92–108

OFFPRINT **878**

WATER

by Roger Revelle

Men need water to drink and for many other purposes, but by far the largest amount of water they have available must go to agriculture. Again the basic need in the proper utilization of water is education.

*Did you ever hear of Sweet Betsy
    from Pike
Who crossed the wide prairie with her
    lover Ike?
The alkali desert was burning and bare
And Ike got disgusted with everything
    there.
They reached California with sand
    in their eye,
Saying, "Good-by Pike County, we'll
    stay till we die."*

This bleary and partly unprintable ballad of the 1850's marks the time when most Americans first became aware of the problems of water in national development. In northern Europe, where most of their ancestors had lived, there had always been plenty of water; in the eastern U.S., where they had learned to farm, abundant rain supplied all the water needs of their crops. But when the pioneers crossed the Missouri River, they came to an arid country where water was more precious than land: its presence meant life, its absence death.

Today water problems are part of the national consciousness, and most Americans are aware that the future development of their country is intimately related to the wise use of water resources. The same obviously holds true for the less developed countries. The water problems of the U.S. and the poorer countries are fundamentally similar, but they also differ in significant ways.

Water is both the most abundant and the most important substance with which man deals. The quantities of water required for his different uses vary over a wide range. The amount of drinking water needed each year by human beings and domestic animals is of the order of 10 tons per ton of living tissue. Industrial water requirements for washing, cooling and the circulation of materials range from one to two tons per ton of product in the manufacture of brick to 250 tons per ton of paper and 600 tons per ton of nitrate fertilizer. Even the largest of these quantities is small compared with the amounts of water needed in agriculture. To grow a ton of sugar or corn under irrigation about 1,000 tons of water must be "consumed," that is, changed by soil evaporation and plant transpiration from liquid to vapor. Wheat, rice and cotton fiber respectively require about 1,500, 4,000 and 10,000 tons of water per ton of crop.

When we think of water and its uses, we are concerned with the volume of flow through the hydrologic cycle; hence the most meaningful measurements are in terms of volume per unit time: acre-feet per year, gallons per day, cubic feet per second. An acre-foot is 325,872 gallons, the amount of water required to cover an acre of land to a depth of a foot. Eleven hundred acre-feet a year is approximately equal to a million gallons a day, or 1.5 cubic feet per second. A million gallons a day fills the needs of 5,000 to 10,000 people in a city; 1,100 acre-feet a year is enough to irrigate 250 to 300 acres of farmland.

The total amount of rain and snow falling on the earth each year is about 380 billion acre-feet: 300 billion on the ocean and 80 billion on the land. Over the ocean 9 per cent more water evaporates than falls back as rain. This is balanced by an equal excess of precipitation over evaporation on land; consequently the volume of water carried to the sea by glaciers, rivers and coastal springs is close to 27 billion acre-feet per year. About 13 billion acre-feet is carried by 68 major river systems from a drainage area of 14 billion acres. Somewhat less than half the runoff of liquid water from the land to the ocean is carried by thousands of small rivers flowing across coastal plains or islands; the area drained is about 11 billion acres, but part of this is desert with virtually no runoff.

Eight billion acres on the continents drain into inland seas, lakes or playas. This includes most of the earth's six billion acres of desert and also such relatively well-watered areas as the basins of the Volga, Ural, Amu Darya and Syr Darya rivers, which transport several hundred million acre-feet of water each year into the Caspian and Aral seas. The remainder of the land surface, about four billion acres, is covered by glaciers.

Even agriculture, man's principal consumer of water, takes little of the available supply. A billion acre-feet per year —less than 4 per cent of the total river flow—is used to irrigate 310 million acres of land, or about 1 per cent of the land area of the earth. Roughly 10 billion acre-feet of rainfall and snowfall is evaporated and transpired each year from the remaining three billion acres of the earth's cultivated lands and thus helps to grow mankind's food and fiber. Most river waters flow to the sea almost unused by man, and more than half of the water evaporating from the continents—particularly that part of the evaporation taking place in the wet rain forests and semihumid savannas of the Tropics—plays little part in human life.

Although it is not usually reckoned as such in economic statistics, water can be considered a raw material. In the U.S. the production of raw materials has a minor role in the total economy, and water costs are small even when compared with those of other raw materials. The cost of all the water used by U.S. householders, industry and agriculture is around $5 billion a year: only 1 per cent of the gross national product.

ATMOSPHERE   ANNUAL PRECIPITATION **100**

NONIRRIGATED LAN

FORESTS AND BROWSE VEGETATION **16**

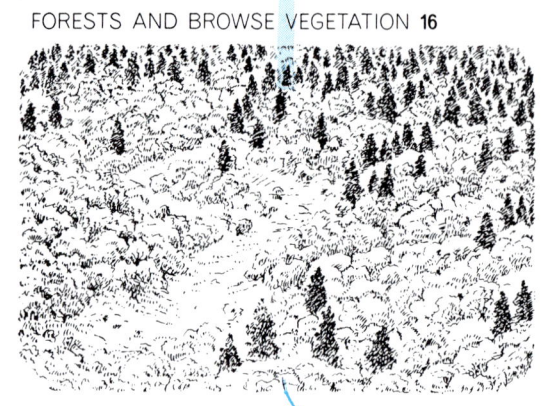

FARM CROP AND PASTURE **23**

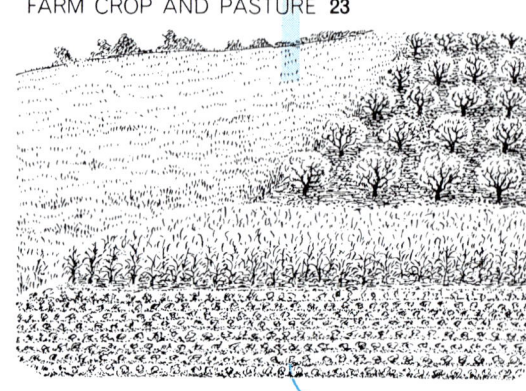

IRRIGATION

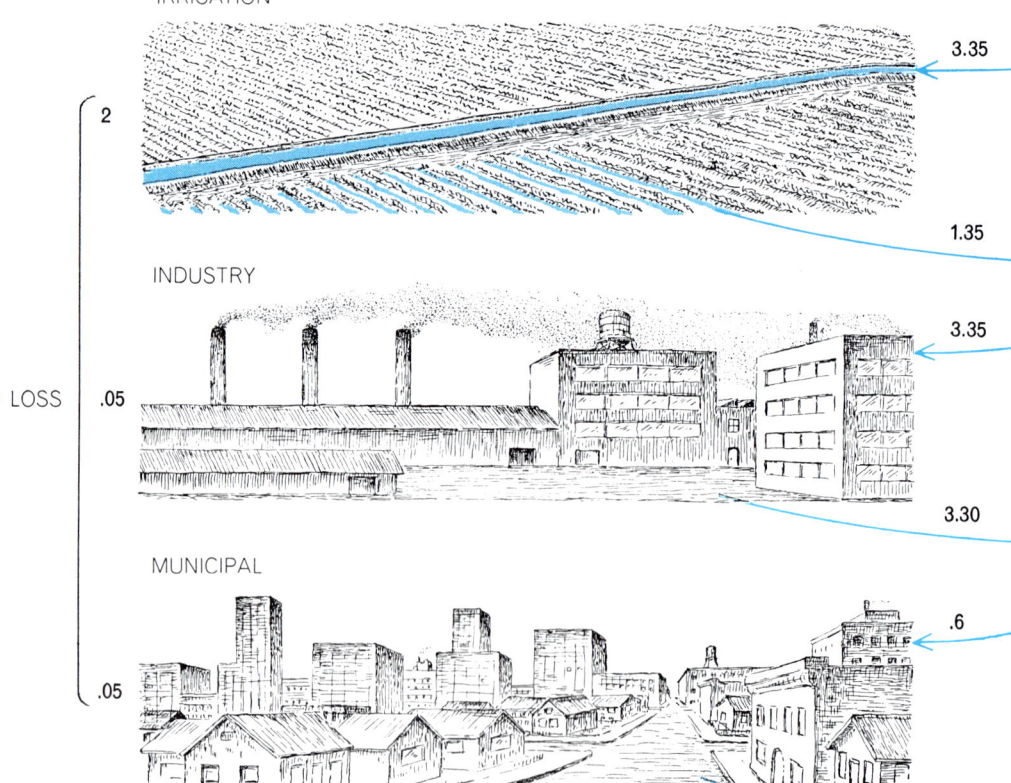

3.35

1.35

INDUSTRY

3.35

CONSUMED LOSS

2

.05

3.30

MUNICIPAL

.6

.05

.55

OCEAN RESERVOIR

HYDROLOGIC CYCLE for the U.S. shows the fraction of annual precipitation used in a highly developed nation. Twenty-nine per cent of the rainfall arrives at the oceans (*bottom*) via stream flow; 71 per cent falls on various types of nonirrigated land, returning

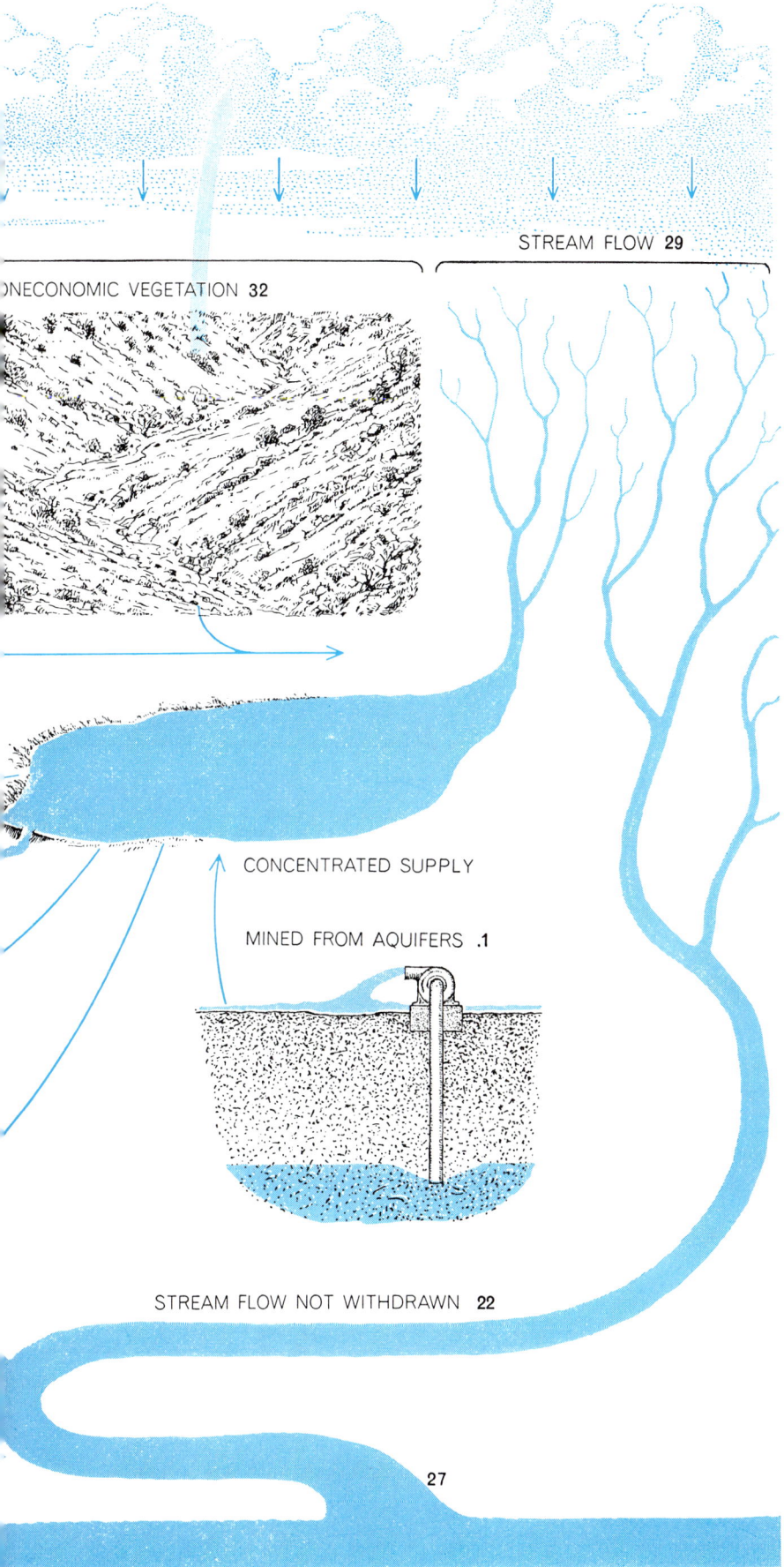

STREAM FLOW **29**

NECONOMIC VEGETATION **32**

CONCENTRATED SUPPLY

MINED FROM AQUIFERS **.1**

STREAM FLOW NOT WITHDRAWN **22**

27

directly to the atmosphere (top) by transpiration and evaporation. Water withdrawn for irrigation, industry and municipal use is shown at left to constitute only 7.3 per cent of total.

The less developed countries, where raw materials are a major component of the economy, cannot afford water prices that would be acceptable in the U.S.

In the U.S. water costs $10 to $20 an acre-foot, compared with wholesale prices of $22,000 an acre-foot for petroleum, $100,000 an acre-foot for milk and $1 million an acre-foot (not counting taxes) for bourbon whiskey. The largest tanker ever built can hold less than $1,000 worth of water. Yet Americans use so much water—about 1,700 gallons a day per capita—that capital costs for water development are comparable to other kinds of investment. Although the water diverted from streams and pumped from the ground is equivalent to only about 7 per cent of the rain and snow falling on the U.S., this is still an enormous quantity: 200 times more than the weight of any other material used except air. The annual capital expenditure for water structures in the U.S. —dams, community and industrial water works, sewage-treatment plants, pipelines and drains, irrigation canals, river-control structures and hydroelectric works—is about $10 billion.

One of the most critical water problems of the U.S. is represented by the vast water-short region of the Southwest and the high Western plains. In some parts of the Southwest water stored underground is being mined at an alarmingly high rate, and new sources must soon be found to supply even the present population. The average annual supply of controllable water in the entire region is 76 million acre-feet. If agriculture continued to develop at the present rate, 98 to 131 million acre-feet would be required by the year 2000. Provided that the neighboring water-surplus regions could be persuaded to share their abundance, this deficit could be met by long-distance transportation of 22 to 55 million acre-feet per year. But the annual cost would be $2 billion to $4 billion, or $60 to $100 per acre-foot of water, including amortization of capital costs of $30 billion to $70 billion. The cost per acre-foot would be too high for most agriculture, although not too high for municipal, industrial and recreational needs.

Nathaniel Wollman of the University of New Mexico and his colleagues have shown that the average value added to the economy of the Southwest through the use of water in irrigation is only $44 to $51 an acre-foot, whereas the value gained from recreational uses could be about $250 an acre-foot and from industrial uses $3,000 to $4,000 an acre-foot. Because the quantities of water consumed by city-dwellers and their industries are much less than those in agriculture, the arid Western states would not require such a vast increase in future supply if they shifted from a predominantly agricultural to a predominantly industrial economic base.

The value of water in the water-short regions of the U.S. that are in a phase of rapid economic development increases more rapidly than the cost. Even high-cost water is a small burden on the gross product of a predominantly industrial and urban economy, and high water costs are only a small economic disadvantage. This is easily overcome if other conditions, such as climate, happen to be propitious.

Throughout the country favorable benefit-to-cost ratios can usually be attained from relatively high-cost multi-purpose water developments for city residents, industry, irrigation agriculture, the oxidation and dispersal of municipal and industrial wastes, the generation of hydroelectric power, pollution control, fish and wildlife conservation, navigation, recreation and flood control.

In the less developed countries water development by itself does not produce much added value for the present economy. Municipal and industrial water re-

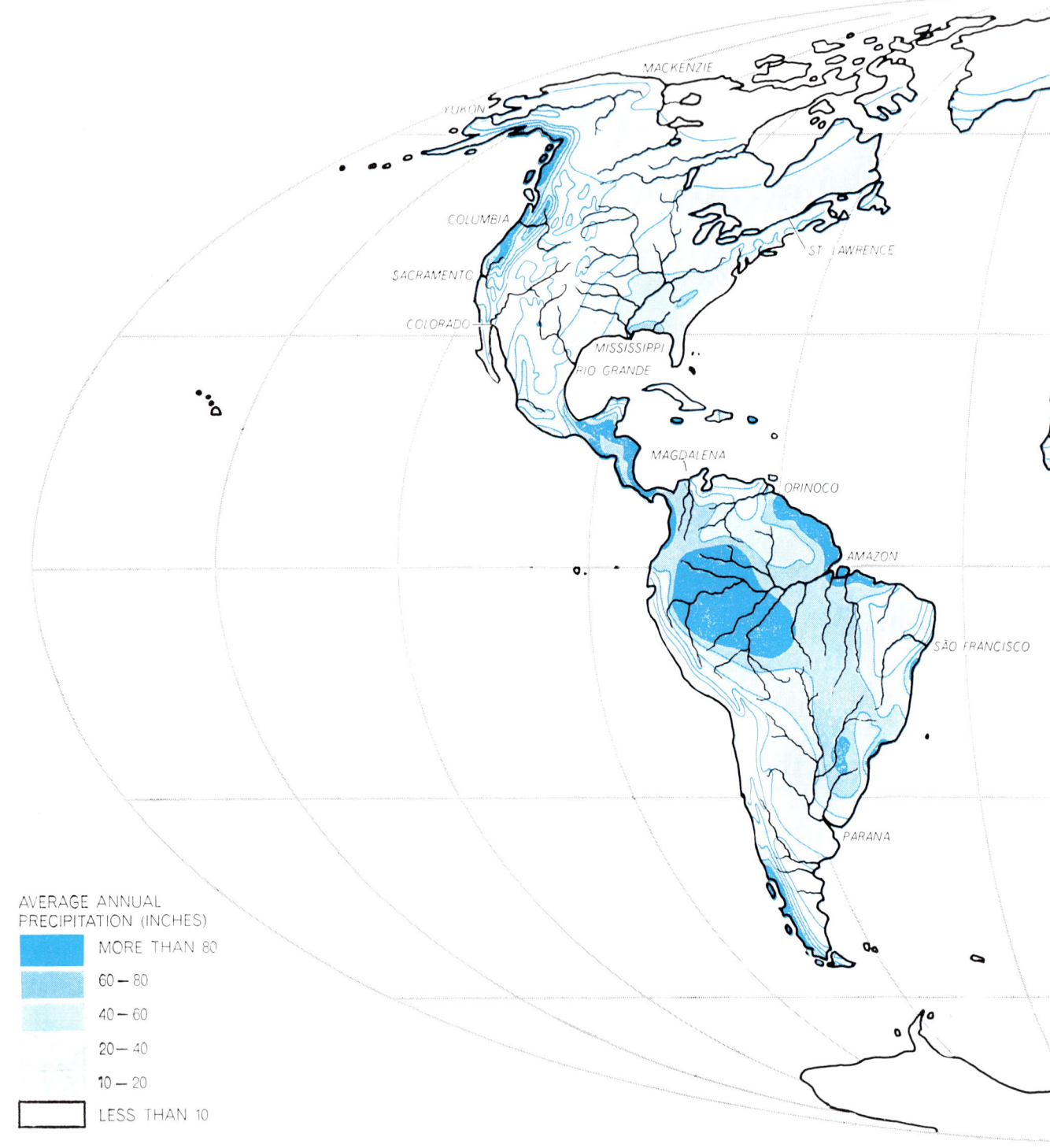

AVERAGE ANNUAL
PRECIPITATION (INCHES)

MORE THAN 80

60 — 80

40 — 60

20 — 40

10 — 20

LESS THAN 10

MAJOR RIVERS carry water to the oceans at rate of more than 11 billion acre-feet every year, but even in the U.S. less than a quarter of this flow is diverted by man for his own purposes. Key at lower left distinguishes areas by amount of precipitation they

quirements are much smaller than they are in the U.S., and the immediate water needs are chiefly for agriculture, which calls for about the same amount of water in any warm region. Most of these countries have a low-yielding subsistence agriculture that brings in very little cash per acre-foot of water, and their farmers can afford to pay only a few dollars per acre-foot. Development of water resources must be accompanied

by other measures to raise agricultural yields per acre-foot and per man-hour, and in general to increase the economic value of water.

One means of coping with water problems in both the U.S. and the less developed countries is to improve the present rather low efficiency of water use. Here much could be done by effective research. For example, about half the water provided for irrigation is lost in trans-

port, and less than half the water that reaches the fields is utilized by plants.

New mulching methods are already being applied to reduce evaporation from soil surfaces, thereby making more water available for transpiration by the plants. Through research on the physiology of water uptake and transport in plants, and on plant genetics, transpiration could probably be lowered without a proportional reduction in growth. De-

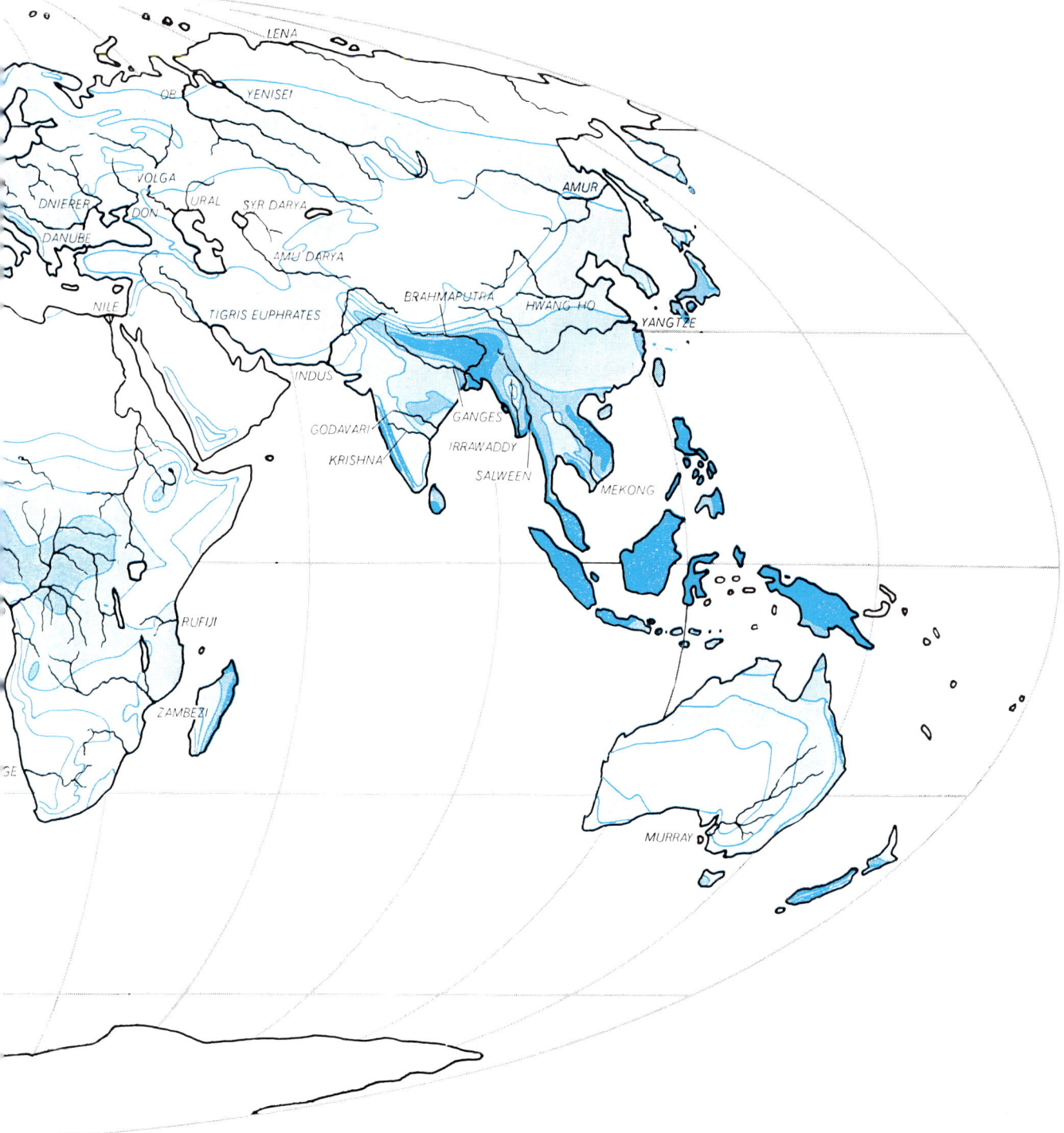

receive each year. Several regions where the average annual precipitation is less than 10 inches (*white*) can be seen to lie close to large rivers. Plans for overcoming the aridity of Egypt and central Russia call for giant dams on the Nile and the Ob.

velopment of salt-tolerant crops would reduce the amounts of irrigation water needed to maintain low salt concentrations in the solution around the plant roots. The loss of water by seepage from irrigation canals and percolation from fields would be lowered by the development of better linings for canals and better irrigation practices. Losses from canals would also be reduced if we could learn how to control useless water-loving plants that suck water through the canal banks and transpire it to the air.

In arid regions the runoff from a large area must be concentrated to provide water for a relatively small fraction of the land, and techniques are needed to increase the proportion of total precipitation that can be concentrated. Development of such techniques requires re-search on means of increasing the runoff from mountain areas (for example, by reducing evaporation from snow fields and modifying the plant cover in order to reduce transpiration) and on methods for accelerating the rate of recharge of valley aquifers.

Finally, water problems could be dealt with by steps that—in contrast to those seeking to make better use of existing supplies—sought to increase the total volume of fresh water. Here research moves on two fronts: attempts to modify precipitation patterns by exerting control over weather and climate, and development of more economical methods of converting sea water or brackish water to fresh water. The ability to control weather and climate, even to a small degree, would be of the greatest im-portance to human beings everywhere. Whether or not a measure of control can be obtained will remain uncertain until we understand the natural proc-esses in the atmosphere much better than we do now. As for desalination, this could be accomplished more economical-ly than at present if the amount of energy required to separate water and salt could be reduced or the cost of energy lowered. Research on the properties of water, salt solutions, surfaces and membranes is fundamental to the desalination prob-lem. So is research aimed at lowering energy costs.

We know too little to be able to make more than a rough appraisal of the po-tentialities of water-resources develop-ment for agriculture in the less devel-oped countries. The modern technology

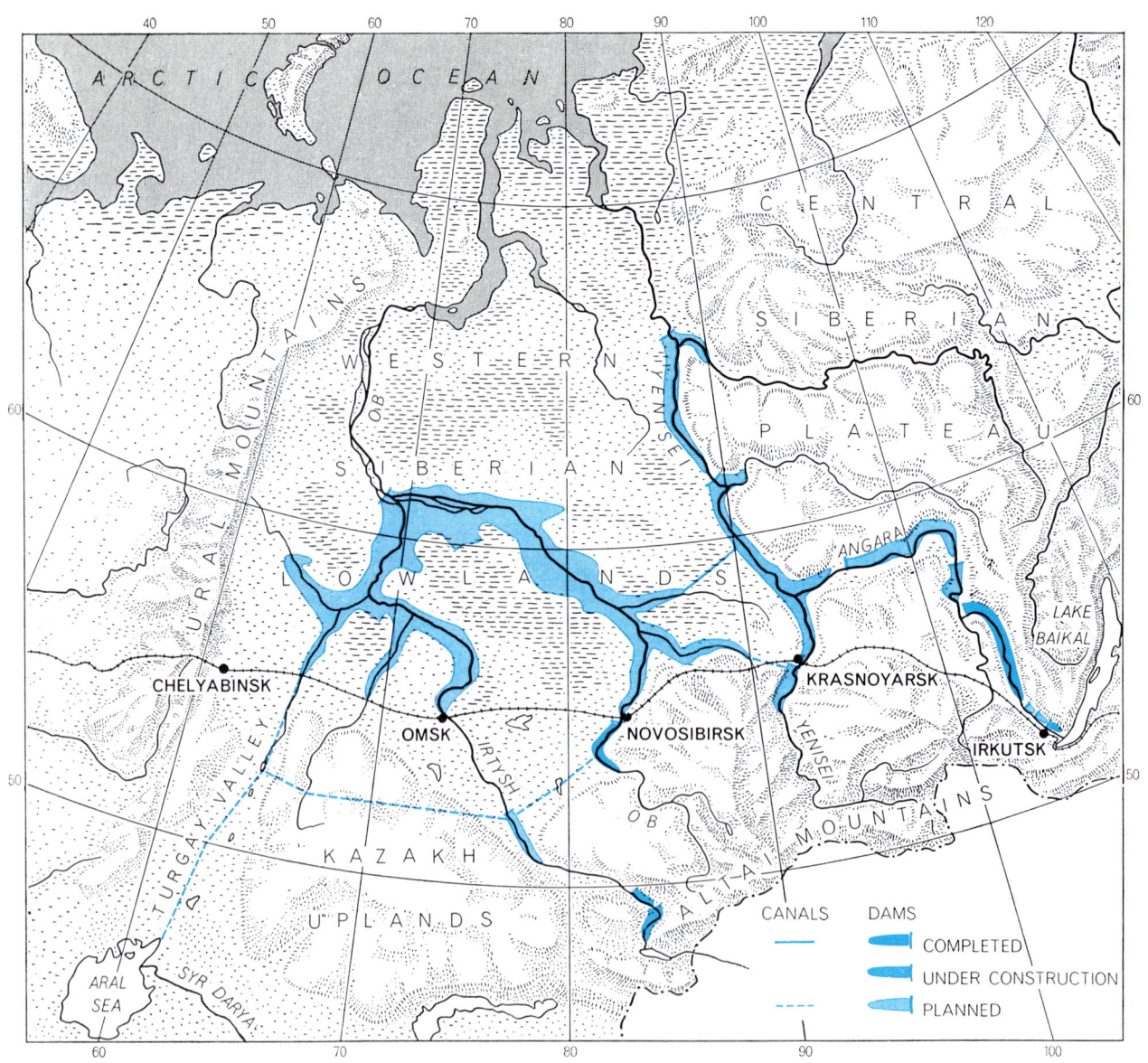

OB-YENISEI PROJECT calls for huge dam on the Ob, creating an inland sea five-sixths the size of Italy. Canal would link Ob and Yenisei rivers so that 12.5 per cent of the water now flowing unused to the Arctic would irrigate the central Soviet steppes.

of irrigation engineering, drainage, sanitation and agricultural practice is quite different from that which determined patterns of land and water use in the past. At the same time technology is almost completely lacking for expanding productive agriculture in the areas of most abundant water and almost unused land: the humid Tropics. Our concern should be not only to find ways of increasing total production in order to feed and clothe the world's expanding human population but also to raise production per farm worker, that is, to raise living standards. A world-wide strategy for development of land and water will require a careful analysis of existing knowledge, region by region, together with field surveys and experimental research in each region by experienced and imaginative specialists.

In humid areas agriculture is limited only by the extent of good land; in arid lands water is the absolute limiting factor. Unless climates can be modified or sea water can be cheaply converted and economically transported, the area of arable land in the arid zone will always exceed the available water. At present, however, neither surface nor underground waters are fully utilized, either for double-cropping in presently cultivated lands or for bringing new land under cultivation.

In addition to improving the utilization of water and increasing agricultural yields other problems that contributed to the destruction of desert civilizations in the past must still be overcome in arid land development. Among them is the fact that the spreading of water over large areas provides a fertile ground for human diseases, such as malaria and bilharzia, and for plant pests. Egyptian records show an average of one plague every 11 years. Uncontrollable malaria might well have been the cause of the mysterious disappearance of the great civilization of the cities Mohenjo-Daro and · Harappa, which flourished 4,500 years ago in the Indus valley of Pakistan.

Soil drainage in a nearly level flood plain is very difficult and is usually neglected, with the result that the water table comes close to the surface and drowns the roots of most crop plants. Water rises through the soil by capillary action and evaporates, leaving an accumulation of salt that poisons the plants. The related disasters of waterlogging and salinity may have caused the ruin of the Babylonian civilization in the valley of the Tigris-Euphrates, and they are a frightening menace today in West Pakistan.

Another threat is the conflict between

WATERLOGGED FIELDS near Sargodha in Pakistan reflect leaky canal system and inadequate drainage. Cultivated plots can be seen under water in center of photograph. When it evaporates, the water will deter renewed cultivation by leaving salts in topsoil.

SALINE FIELDS stand out against darker cultivated land in this aerial photograph. The related problems of waterlogging and salt accumulation in the soil have made five million acres of West Pakistan's irrigated farmland either impossible or unprofitable to cultivate.

the sedentary farmers of the plain and nomadic herdsmen. The present-day Powindahs of West Pakistan remind us of this ancient conflict. In our own West the feuds between cattlemen and farmers are still a vivid memory.

In considering the possibilities of agricultural development in the world's arid lands one thinks first of the famous rivers that have played so large a role in human history: among them the Nile, the Indus and its tributaries and the Tigris-Euphrates.

For thousands of years the Egyptians carried out irrigation by allowing the Nile waters during flood stage to spread in ponded basins broadly over the delta and the valley. When the flood subsided, the basin banks were cut and the ponded water flowed back to the river. The Nile and the sun were said to be the prime farmers of Egypt. It was thought that the river's silt, deposited during the annual flood, fertilized the soil. Sun-drying and -cracking, during the fallow season before the flood, deeply furrowed the soil and killed off weeds and micro-organisms, making plowing unnecessary. The flood arrived in July, reached its height in September and subsided quickly. The fields were sown in early winter with wheat, barley, beans, onions, flax and clover. Summer crops were grown only on the river levees and in areas that contained a shallow water table, where water could be lifted by hand from the river banks or from wells. High floods left the basins pestilential morasses that brought plagues and epidemics. Low floods brought famine.

During the past 140 years this ancient system has been transformed. In 1820 Egypt had reached a nadir, with a population of only 2.5 million and with three million cultivated acres. This date marked the beginning of perennial canal irrigation and widespread planting of summer crops, including cotton, corn, rice and sugar cane as well as the traditional winter crops. Low dams called barrages were built across the river; the water backed up behind these structures was diverted through large new canals that flowed the year round. By 1955–1956 the cultivated area had increased to 5.7 million acres and the intensity of cultivation to 177 per cent; that is, more than 10 million acres of crops were harvested. Salinity and waterlogging became serious menaces in the early part

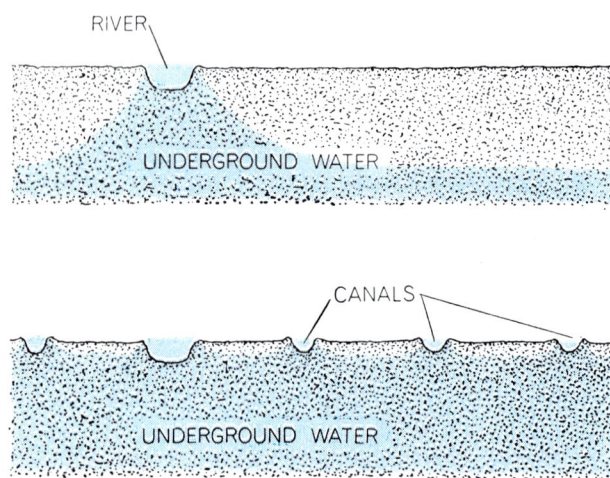

LACK OF DRAINAGE has caused the underground water table to rise disastrously in parts of Pakistan. Before construction of leaky canals, water approached ground level only near rivers. Now water table in many areas is high enough to drown crop roots.

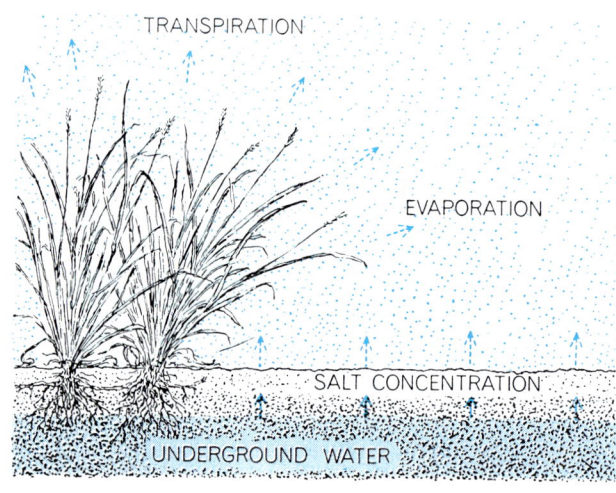

SALT ACCUMULATES on topsoil in two ways. Underground water rises by capillary action, lifting dissolved salts that will be left behind after evaporation. If farmer uses thin layer of water to irrigate topsoil, it will evaporate before percolating down.

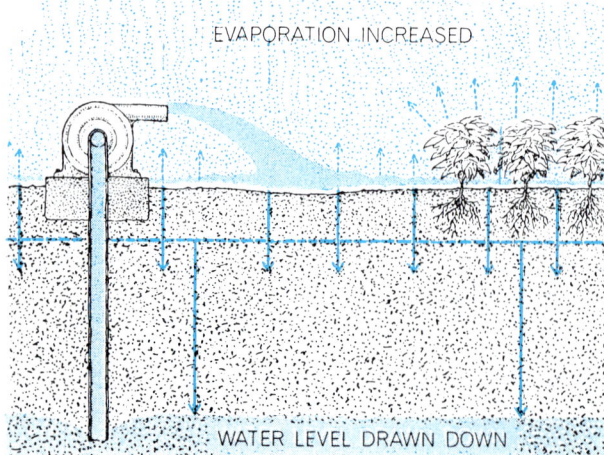

VERTICAL DRAINAGE might solve related problems of salinity and waterlogging. Pumped through cased well from underground table, as illustrated here, enough water could reach surface for salts to percolate down beneath topsoil before full evaporation occurs.

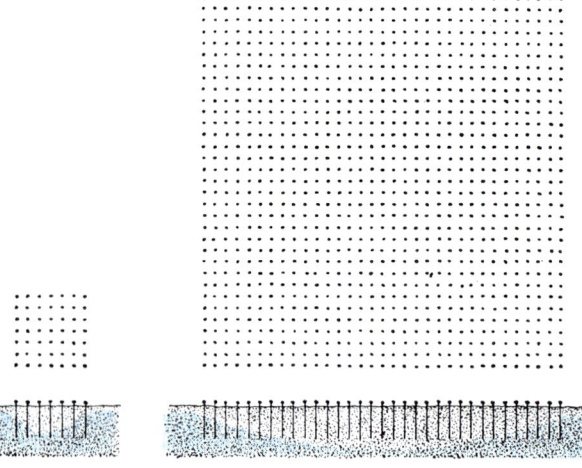

LARGE-SCALE PROJECT will be required to lower the water table. Drainage of a field of one million acres (right) could negate seepage from adjacent land. But pumping out a small field (left) would have no appreciable effect on underground water level.

of this century, but they have been fairly well controlled by an extensive drainage system. Chemical fertilizers are used in large amounts and crop yields per acre are high, even though the Nile silt no longer settles on the fields but is deposited back of the barrages. Sufficient food is grown to feed the present population of 27 million. From the standpoint of crop yields per acre, although not per man, Egypt is a developed country.

The average annual flow of the Nile is 72 million acre-feet, but it is occasionally as high as 105 million acre-feet or as low as 36 million. If all the average flow could be utilized, it would be enough to irrigate 12 to 15 million acres on a year-round basis. At present the area of irrigated land in Egypt is less than half that. During the flood season much of the water flows to the sea unused, and during the rest of the year a shortage of surface and underground water limits the size of the cultivated area.

Now the construction of the Aswan High Dam promises to bring the river under complete control. The dam will have a storage capacity of 105 million acre-feet, equal to the highest annual flow during the past century. There will no longer be a Nile flood; the tamed river will become simply a huge feeder canal for irrigation. With the average of 55 million acre-feet per year available to Egypt (17 million acre-feet from the reservoir is allocated to the Sudan), it will be possible to increase the cultivated area in the delta and the valley floor by 2.2 million acres, or nearly 40 per cent, and to convert .7 million acres from flood to perennial irrigation. Hydroelectric power generation of more than a million kilowatts will make power available for pump drainage, which may increase crop production by 20 per cent, and for the manufacture of chemical fertilizers. The electric power will also be used to lift water to the desert margins of the valley, where it is hoped that an additional one to two million acres can be brought under the plow. If all these benefits can be realized, total agricultural production in Egypt can be increased by 90 per cent, enough to feed almost twice the present population and at the same time provide crops for export.

W hen Alexander the Great pushed his tired armies eastward some 2,300 years ago, they came at last to an old desert civilization on the banks of the mightiest river they had ever seen. The Aryans, who had preceded Alexander by 1,000 years, did not give the river a

name; they called it simply the Indus, which was their word for river, and they named the subcontinent they had invaded "India": the land of the river.

The Indus and its five tributaries of the Punjab, together with the flat plain through which they flow, are one of the major natural resources of the earth. In the Punjab and Sind regions of West Pakistan 30 million persons dwell on the plain; 23 million make their living from farming it. They produce most of the food and fiber that feed and clothe nearly 50 million people.

The rivers carry more than twice the flow of the Nile. Half this water is diverted into a highly developed system of irrigation canals and is used to irrigate some 23 million acres—by far the largest single irrigated region on earth. Underneath the northern part of the plain lies a huge reservoir of fresh ground water, equal in volume to 10 times the annual flow of the rivers.

In spite of the great potentialities of the plain, the fact is that poverty and hunger, not well-fed prosperity, are today the common lot of the people of West Pakistan. These afflictions are nowhere more desperately evident than in the farming villages of the countryside. In a country of farmers food must be imported to provide the most meager of diets; the gap between food production and the number of mouths to be fed is widening.

The problem of agriculture in West Pakistan is both a physical and a human one. It is a problem of land, water and people and of the interactions among them. One of its aspects is the waterlogging and salt accumulation in the soil, caused by poor drainage in the vast, nearly flat plain, that are slowly destroying the fertility of much of the irrigated land. The area of canal-irrigated and cultivated land already seriously damaged by waterlogging and salinity is close to five million acres, or about 18 per cent of the gross sown area. Three other difficulties also beset agriculture: shortage of irrigation water, problems of land tenure and poor farming practices.

Although crops can be grown throughout the year, and both a winter and a summer growing season are traditional, the irrigation canals lose so much water by seepage that the amount carried to the fields is sufficient to irrigate only about half the land during each season. Even so, the crops are inadequately irrigated, particularly in summer. Much of the cropped area receives insufficient water to prevent salt accumulation.

Many of the farmers are sharecropping tenants who have little incentive to

increase production. Nearly all of them struggle with small and widely separated plots that multiply the difficulties of efficient use of irrigation water and farm animals and gravely inhibit change in traditional practices.

In West Pakistan we have the wasteful paradox of a great and modern irrigation system pouring its water onto lands cultivated as they were in the Middle Ages. Plowing is done by a wooden plow of ancient design, pulled by undernourished bullocks. Unselected seeds are sown broadcast. Pakistan uses only a hundredth as much fertilizer per acre as Egypt.

Careful investigation shows that in most of the Punjab the problems of waterlogging and salinity could be cured, and at the same time adequate water could be supplied to the crops, by sinking fields of large wells to pump the underground water and spread it on the cultivated lands. Part of the pumped water would be carried off by evaporation and transpiration and part would percolate back into the ground, in the process washing the salt out of the soil.

If the well fields are too small in area, lateral infiltration of underground water from the surrounding land will be large compared with the rate at which the pumped water can evaporate, and the process of dewatering will be retarded or completely inhibited. For this and other reasons each Punjab project area should be about a million acres in size.

Removal of salt and provision of additional water are necessary, but by no means sufficient, measures to raise agriculture in West Pakistan from its desperate poverty. Equally essential are chemical fertilizers, higher-yielding seeds, pest control, credit and marketing facilities, and above all incentives and knowledge to adopt better farming practices. The job cannot be done all at once; it is necessary to concentrate on project areas of manageable size. Initial capital costs for a million-acre project in the Punjab would be of the order of $55 million, including costs of wells and electrification, nitrogen-fertilizer plants, pest-control facilities and filling of administrative, educational and research pipelines.

In the Sind region initial capital costs would be considerably higher, probably between $130 million and $165 million per million acres. That is largely because the underground water in most of the Sind is too salty to be used for irrigation, and drainage is therefore a more difficult matter than in the Punjab.

After a few years the minimum net increase in crop value in each million-acre project in the Punjab could be $55 million to $60 million a year, equal to the capital costs and to twice the present gross production, excluding livestock. In the Sind the net increase, including livestock, could probably be at least equal to the present output.

The same interrelated problems of water, land and people that afflict the Indus plain also exist in the valley of the Tigris-Euphrates, but on a much smaller scale. Salty soil is found over large areas; because of waterlogging it is possible to cultivate only about a third of the seven million acres of irrigated land each year. The remainder is left fallow and unirrigated to dry out the subsoil and to build up a little soil nitrogen. Great damage was done long ago when the ancient canal systems were destroyed and the land was depopulated by waves of nomadic invaders. But the nomads merely hastened the salt accumulation and waterlogging that were the seeds of destruction. These had begun centuries earlier as a result of inadequate drainage and inability to control floods.

If the flow of the Tigris-Euphrates could be fully utilized, through combined development of surface and ground water, and if the soils were adequately leached and drained, the irrigated area cultivated each year could be increased to 10 to 12 million acres. If greater water usage were combined with perennial cropping, better farming practices and the application of chemical fertilizers, total agricultural production could be raised at least fivefold.

The largest opportunities for expansion of the area of irrigated arid and semiarid lands exist in the U.S.S.R. Between 1950 and 1960, 15 million acres in the neighborhood of the Black and Caspian seas were provided with irrigation water from the Volga, Dnieper, Amu Darya and Syr Darya rivers. The total flow of these rivers is more than 300 million acre-feet, sufficient, under the cold-winter and warm-summer climate of the steppes, to supply all the water needed to irrigate 70 to 100 million acres.

Because of the relatively advanced economic level of the country, large multipurpose water developments in the U.S.S.R. are economically feasible, and a high percentage of the capital invested goes for power, transportation, industrial water supplies and flood control.

Soviet engineers have outlined a plan to build an immense dam on the Ob River, creating an inland sea five-sixths the size of Italy, and to dig a canal connecting the Yenisei with the Ob above the dam [see illustration on page 660].

The impounded waters would be transported through a giant system of canals, rivers and lakes to the Aral Sea and thence by canal to the Caspian Sea. Several hundred million acre-feet of water that now goes to waste each year in the Arctic Ocean would be conserved. This water would be used to irrigate 50 million acres of crop lands and a somewhat larger area of pasture in arid western Siberia and Kazakhstan. Accompanying hydroelectric power installations would have a capacity of more than 70 million kilowatts. Major storage, irrigation and hydroelectric works are also under construction or planned in the northern Caucasus and in the Azerbaijan, Georgian and Armenian Soviet Socialist republics. These will bring additional tens of millions of acres under irrigation.

In some parts of the arid zone both surface and ground water are so scarce that it is difficult to see how irrigation agriculture can be developed to support the rapidly expanding population. In the Maghreb countries of North Africa—Tunisia, Algeria and Morocco—there is probably not enough water in the region north of the Sahara to irrigate more than 3.5 million acres of land, yet the combined population of these three countries is already 26 million (equal to Egypt's) and will double in 20 to 25 years. Elaborate systems of dry farming have been developed in the Maghreb; for example, the planting of olive trees far apart in light, sandy soils that catch and hold the nighttime dew. With this technique it has been possible to grow olive and other fruit trees on more than a million acres in Tunisia. In the long run it may be necessary to employ most of the available water in the Maghreb countries for industrial purposes, because these can provide a tenfold to hundredfold higher marginal value for water than agriculture can.

A new possibility for water development has recently been opened, however. During the past few years evidence has been obtained that large areas in the Sahara may be underlain by an enormous lake of fresh water. In some places the water-bearing sands are 3,000 feet thick, and they appear to extend for at least 500 miles south of the Atlas Mountains and perhaps eastward into Tunisia and Libya. If this evidence is correct, the amount of useful water may be very large indeed—of the order of 100 billion acre-feet, sufficient to irrigate many millions of acres for centuries.

In general the possibilities of expanding the area of irrigated land in the arid zone outside the U.S.S.R. are not large

when measured in numbers of acres. But crop yields under irrigation in the arid lands are high and assured if all the factors of agricultural production are properly applied. In fact, irrigation agriculture in arid regions can be successful only if it is intensive and high-yielding; it is costly to construct and maintain drainage systems that will keep the water table from rising too close to the surface, and to provide enough water on each acre to leach the salts out of the soil. In hot, arid lands some kinds of irrigation agriculture can be so productive that very expensive irrigation water, such as could be produced by sea-water desalination, may soon become economical.

Much greater possibilities (and also greater difficulties) exist for agricultural expansion in the regions of savanna climate, which are characterized by an annual cycle of heavy rainfall during one season, followed by drought the remainder of the year, and by warm weather at all seasons. In Africa, for example, many millions of what are now barren acres could be brought under irrigated cultivation, provided that interested farmers could be found, in the neighborhood of the great bend of the Niger River in former French West Africa, in the basin of the Rufiji River of Tanganyika and near Lake Kyoga in Uganda. Similarly, in the area extending from India east through Burma, Thailand and Vietnam to the northern Philippines, air temperature and solar radiation are suitable for year-round crop growth, and water and land are the limiting factors [see "The Mekong River Plan," by Gilbert F. White; SCIENTIFIC AMERICAN, April, 1963].

In the lower basin of the Ganges and Brahmaputra rivers, comprising East Pakistan and the Indian states of Bengal, Bihar and Assam, some 140 million people live on 70 million cultivated acres. The basic resources of soil and water are grossly underutilized in this land

MULTIPURPOSE DAM at Watts Bar, Tenn., appears in the aerial photograph on the opposite page. The deep lake at left is formed where the dam halts the Tennessee River. Spilled back to river depth beyond the dam, the water regains a slate-gray hue. The turbulent area at lower right is caused by the flow of water through the square-capped hydroelectric generators. At bottom center is a power-distribution station. At top is the lock that enables shipping to pass the dam. The ribs in the dam are spillways, which here are closed. They can adjust the lake level to help control flooding.

of ancient civilization, extreme present poverty and strong population pressure. Each year the rivers carry about a billion acre-feet to the Bay of Bengal, and in the process they flood most of the countryside. Yet only one crop is grown a year. The land is left idle half the year because of the shortage of water and there is a lack of useful occupation for the people six to eight months of the year. Agricultural practices are adjusted to the rhythm of the monsoon.

The opportunities for increasing production are enormous in this region of land shortage and overabundant water. Through surface and underground storage of a portion of the flood waters, water could be provided for three crops each year over more than half of the cultivable area in the alluvial plain, and a considerable additional area could receive sufficient water for two crops. An assured year-round water supply would also provide favorable conditions for intensive use of fertilizers, higher-yielding plant varieties and better farming practices, which could result in a tripling of yields per crop and per acre for cereals, pulses (the edible seeds of leguminous plants such as peas and beans) and oilseeds.

A well-fed livestock industry could be developed in addition to improvements in field crops, and a balanced diet, instead of the present completely inadequate one, could be provided for twice the present population. Expansion of agricultural production here, based on irrigation, would raise few basic problems of land and settlement, but it would require a reorientation of thinking regarding patterns of land and water use. Because of the enormous volumes of water involved and the flatness of the alluvial basin, the cost of water storage and distribution and of flood control and drainage would be high, but the returns through increased farm and livestock production could be several times higher than the cost. The yields per worker must also be increased, however, and a large degree of industrialization accomplished if the project is to finance itself.

Development of water resources is not an end in itself. The investment can be justified only if it leads to higher agricultural or industrial production, or in other ways to an increase of human well-being. To gain these objectives water development must be accompanied by other actions needed to use the water effectively. This is well illustrated in agriculture. One of the basic principles of agricultural science is the principle of interaction: the concurrent use of all the factors of production on the same parcel of land, which will give a much larger harvest than if these factors are used separately on different parcels. Adequate water and water at the right time are essential if seeds of a particular crop variety planted in a given soil are to yield a good crop. But a much larger crop is possible if seeds of a higher-yield variety are planted. This potential increase in the harvest will be realized, however, only if the soil contains sufficient plant nutrients. Usually nitrogen fertilizers and phosphate fertilizers must be added in large amounts to provide the maximum yield. Increased soil fertility will be drained off by weeds unless these are rigorously controlled, and an eager host of insect pests and plant diseases will fight to share the crop with the farmer unless he can combat them with pest-control measures. Improved seed varieties planted without adequate water, abundant fertilizer and rigorous pest control may not do even as well as the traditionally planted varieties. The potentialities for double- or triple-cropping in a perennial irrigation system cannot be achieved if the farmers do not have tractors and efficient tools to enable them to prepare their fields in the short interval between harvest and planting.

To meet the cost of new irrigation systems the farmer must produce much more per acre-foot of water than he has in the past, and this can be done only if all the factors of production are made available to him and if he is taught how to use them effectively. The human, educational, social and institutional problems of bringing the necessary knowledge to millions of farmers are immense. The task of remaking methods of production that are intimately tied to ways of living and of overcoming institutional and political resistance to change is more difficult than any of the engineering problems. Illiteracy, malnutrition and disease; poverty so harsh that the farmer does not dare risk innovation because failure will mean starvation; small and fragmented farm holdings; land-rental and taxation systems that destroy incentive; extreme difficulties in obtaining a farm loan promptly at a reasonable interest rate; poor marketing and storage systems; administrative inefficiency and corruption; the shortage of trained teachers and farm advisers; inadequate government services for agricultural research, education and extension and for control of water-borne diseases—all must be overcome if investments in water resources in the developing countries are to produce really beneficial results.

The Author

ROGER REVELLE is University Dean of Research at the University of California and director of that university's Scripps Institution of Oceanography. Revelle began his long association with the Scripps Institution in 1931, two years after acquiring an A.B. in geology from Pomona College. He received a Ph.D. from Scripps in 1936 and was professor of oceanography there in 1951, when he became the first alumnus of that institution to be appointed its director. During World War II Revelle served as a commander in the U.S. Navy and immediately after the war joined the Office of Naval Research as head of the Geophysics Branch. In 1946 he organized the oceanographic expedition associated with the atomic bomb test in Bikini Lagoon, measuring the diffusion of radioactive waters and their effects on marine organisms. During the early 1950's he led several other expeditions to the central and southern Pacific, developing new methods for measuring the flow of heat out through the floor of the ocean. He served as president of the first International Oceanographic Congress held by the United Nations in 1959, and in 1961 he became the first man to hold the post of science adviser to the Secretary of the Interior. Revelle is currently president of the Committee on Oceanographic Research of the International Council of Scientific Unions and a member of the U.S. Commission to the UNESCO Office of Oceanography.

Bibliography

DESIGN OF WATER RESOURCES SYSTEMS. Arthur Mass, Maynard Hufschmidt, Robert Dorfman and Harold Thomas. Harvard University Press, 1962.

A HISTORY OF LAND USE IN ARID REGIONS: ARID ZONE RESEARCH, XVII, edited by L. Dudley Stamp. UNESCO, 1961.

MISSION TO THE INDUS. Roger Revelle in *New Scientist,* Vol. 17, No. 326, pages 340–342; February, 1963.

POSSIBILITIES OF INCREASING WORLD FOOD PRODUCTION: BASIC STUDY No. 10. Food and Agriculture Organization of the UN, 1963.

THE VALUE OF WATER IN ALTERNATIVE USES. Nathaniel Wollman. University of New Mexico Press, 1962.

WATER FACTS FOR THE NATION'S FUTURE. W. B. Langbein and W. G. Hoyt. The Ronald Press Company, 1959.

WATER RESOURCES: A REPORT TO THE COMMITTEE ON NATURAL RESOURCES. PUBLICATION 1000-B. National Academy of Sciences—National Research Council, 1962.

SCIENTIFIC
AMERICAN September 1969, Vol. 221, No. 3, pp. 54–65
OFFPRINT **879**

MAN AND THE SEA

by Roger Revelle

Introducing an entire issue devoted to this topic, with special
reference to man's advancing utilization of the ocean and his
new understanding of how the present oceans were created.

I grew up not far from the ocean and
have lived most of my adult life near
the seashore. The ocean holds me in
an enduring spell. Part of the spell comes
from mystery—the fourfold mystery of
the shoreline, the surface, the horizon
and the timeless motion of the sea. The
thin, moving line between land and water on an open cost presents a nearly
unpenetrable wall. The ships and fishing boats I watch from my living-room
window exist in a separate world, as remote as another planet. Below the surface there is a multitude of living things,
darting and watching, living and dying;
theirs is an alien world I cannot see
and can hardly imagine. At the horizon,
where my line of sight touches the edge
of the great globe itself, I watch ships
slowly disappear, first the hulls and then
the tall masts bound on voyages to unknown ports 10,000 miles away. From
beyond the horizon come the waves that
break rhythmically on the beach, sounding now loud, now soft, as they did long
before I was born and as they will in the
far future. The restless, ever changing
ocean is timeless on the scale of my life,
and this also is a mystery.

Part of the ocean spell comes from the
interplay of light and reflection between
the sea and the sky, the track of sunlight
on the water and the pale or rosy colors
of clouds. It was these that fascinated
the greatest of all English painters,
J. M. W. Turner, and inspired part of
Debussy's tone poem "La Mer." In order
to observe and study the infinite variety

of color patterns over the ocean, Turner
found a house by the sea in east Kent.
He allowed the natives to think he was
an eccentric sea captain named Puggy
Booth who even in retirement could not
stop looking at the sea.

The ocean has an impact on all our
senses: the unique sea smell, the crash-
ing sound of breakers, the glitter of
waves dancing under the sun and the
moon, the feel of spindrift blowing
across one's face, the salty, bitter taste
of the water. Yet the spell of the ocean
is more than mystery and sensory delight. Part of it must come from outside
the senses, from half-forgotten memories
and images beyond imagining, deep below the surface of consciousness.

Being an oceanographer is not quite
the same as being a professional sailor.
Oceanographers have the best of two
worlds—both the sea and the land. Yet
many of them like many sailors, find it
extraordinarily satisfying to be far from
the nearest coast on one of the small, oily
and uncomfortable ships of their trade,
even in the midst of a vicious storm, let
alone on one of those wonderful days in
the Tropics when the sea and the air are
smiling and calm. I think the chief reason is that on shipboard both the past
and the future disappear. Little can be
done to remedy the mistakes of yester-
day; no planning for tomorrow can reck-
on with the unpredictability of ships and
the sea. To live in the present is the es-
sence of being a seaman.

The work of an oceanographer, how-

ever, is inextricably related to time. To
understand the present ocean he must
reconstruct its history, and to test and
use his understanding he needs to be
able to predict—both what he will find
by new observations and future events
in the sea.

Over the past two decades there has
been a marvelous increase in our
understanding of the geological history
of the ocean. This has come about chiefly
through wide-ranging exploration of the
earth under the sea by investigators of
many countries, using a variety of new
instruments and powerful new methods.
I believe that in times to come these 20
years when men gained a new level of
understanding of their planetary home
will be thought of as one of the great
ages of exploration.

In the wealth of new knowledge there
is no discovery more paradoxical than
this: the ocean floor is younger than the
ocean. Below the ancient waters the
sediments and the underlying rocks are
constantly renewed. An almost continuous ridge 40,000 miles long and several
hundred miles wide, rifted along much
of its length by deep valleys and broken
by numerous fractures, lies in the central part of the oceans—a structure on
the same planetary scale as the oceans
and continents themselves. Magnetic,
thermal and seismic observations show
that near the summit of the ridge new
rock wells up from the mantle below
and slowly moves outward across the

ocean basins. Along some margins of the ocean the spreading ocean floor moves the adacent continents: elsewhere the floor plunges downward in deep trenches and disappears into the earth's interior [see "the Deep-Ocean Floor," by H. W. Menard; Scientific American Offprint 883].

The simplicity and grandeur of this hypothesis of sea-floor spreading has caught the imagination of earth scientists throughout the world. As a result,

as Sir Edward Bullard says in the article following this one: "We are in the middle of a rejuvenating process in geology comparable to the one that physics experienced in the 1890's and to the one that is now in progress in molecular biology."

In the same brief period, in pursuit of both scientific and economic or military objectives, man has acquired a new freedom of operation in the third dimension of the ocean. Below the forbidding mystery of the surface, developing tech-

nology gives access to greater and greater depths and to the resources awaiting discovery and exploitation there. Offshore wells now supply a fifth of the world's oil and gas. The value of their output is already equal to half the product of the fisheries and to a fourth of the value of the services rendered by the freighters and tankers that ply the sealanes. Anticipation of still greater yields has brought an ominous extension of claims of national sovereignty [see "The

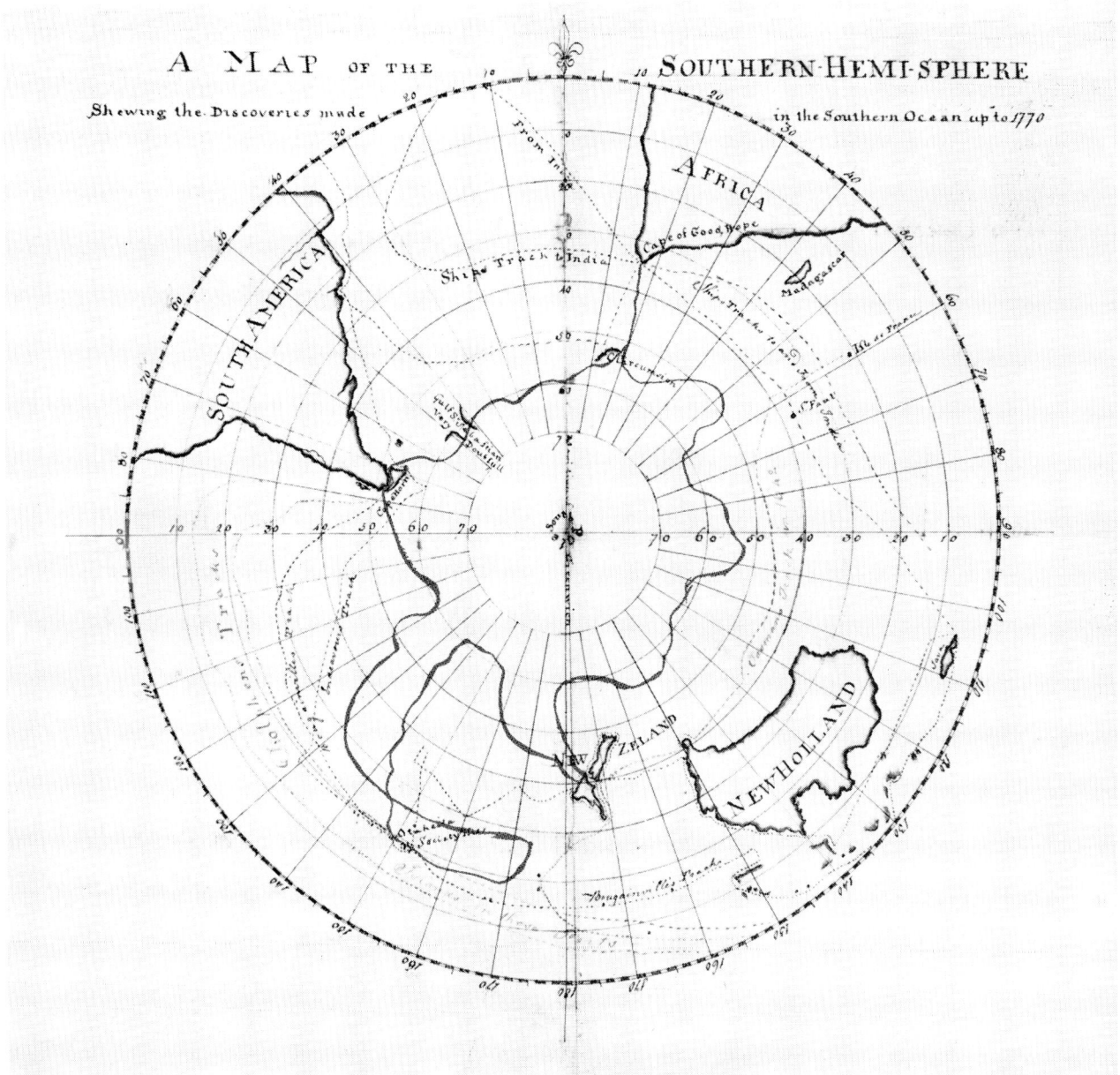

EXPLORATION OF THE OCEAN is symbolized by this map of the Southern Hemisphere drawn by Captain James Cook and enclosed in his memorandum to Lord Sandwich dated 6 February 1772. The strong continuous line (yellow in the original) indicates the route Cook proposed for his second circumnavigation of the globe. The tracks of earlier navigators are shown by dotted lines. One of these lines (labeled "Endeavour's Track") shows the route followed by Cook in his first great voyage of discovery, which lasted from 1768 to 1771. On these two voyages Cook proved the

nonexistence of Terra Australis, a huge imaginary continent that for hundreds of years had been thought to cut off the Southern Ocean beyond about 50 degrees south latitude. Although he encountered many icebergs near the southern limits of his track and deduced that they must have broken off from very large glaciers on a land mass beyond the Antarctic Circle, Cook never glimpsed Antarctica itself except in his imagination. The original of this map, which measures about 12 inches on a side, was photographed at the Public Library of New South Wales in Sydney, Australia.

Ocean and Man," by Warren S. Wooster; SCIENTIFIC AMERICAN Offprint 888]. Beyond the narrow limits of the territorial seas, established by the range of cannon in earlier days, the 1958 Geneva Convention on the Continental Shelf gives a coastal state the exclusive right to exploit the seabed out to a depth of 200 meters or "beyond that limit, to where the depth of the superjacent waters admits of the exploitation of the natural resources of the said area."

Some international lawyers have interpreted this provision to mean that as deep-water technology advances the coastal states will be able to extend their jurisdiction out to the midpoint of the ocean basins. The long-run consequences of such a division of the ocean into national territories are appalling to contemplate. They would constitute a *reductio ad absurdum* of the concept of nation-states.

Even today the ocean both divides and links the nations of mankind. In the past the fate of many peoples was shaped by the sea. The list of these peoples includes the Phoenicians, who are said to have been the first seamen to dare to sail at night, guided by the North Star; the sea kings of Crete; the Athenians; the Norsemen of the Middle Ages and their nearly antipodal contemporaries the Polynesians; the Genoans and Venetians of the late Middle Ages; the Portuguese, the Spaniards and the Dutch during the Renaissance, and the English from the 16th to the 20th centuries.

All the sea peoples shared several characteristics. In the beginning their populations were small and their lands were poor. If they were to prosper, it had to be by commerce and trade, or by finding new lands to conquer and colonize. All were courageous, ingenious, rapacious and ruthless. All were filled with a spirit of curiosity and a drive for discovery, and this was probably their unique quality. Although some of them built great empires, none lasted in any real sense for more than a few hundred years. Their home resources were inadequate to maintain the burdens they had assumed, and the overseas outposts were without roots and unstable.

The energy, will and creative power of the sea peoples are symbolized in the design of their ships. Consider the marvelous wooden ships of the Vikings, as represented by the ship found so remarkably intact in 1880 in the burial mound at Gokstad in Norway. In contrast to the static or soaring grandeur of the structures of stone built by the nations that are rooted in the land, the

Gokstad ship is mobile and light. She is only 78 feet long, about the length of Columbus' *Niña;* her draft is no more than three feet and her planks are less than an inch thick. Her seaworthiness was proved for modern skeptics when an exact replica crossed the Atlantic under oars and sail in 1893. The carved ornamentation of the prows of many of the Viking ships somewhat resembles the adornment of old stave churches that still stand in Norway, but it has a rhythmic, moving pattern that leaps forward toward the unknown. Its delicate sophistication reminds us that on some of their long voyages the Vikings sailed down

the Volga and across the Caspian Sea to Persia, returning home laden with exotic treasures and new ideas. On another expedition they carved a runic rhyme on a marble lion at Apollo's sacred island of Delos.

The lines of the best sailing ships were always functional, designed to give the combination of speed, sea-keeping ability, cargo and passenger space and, if necessary, protection against enemies that was needed for a particular trade. Speed depended not only on the ship's resistance to the water and on its effective sail area but also on the hull form and the rigging that would allow laying

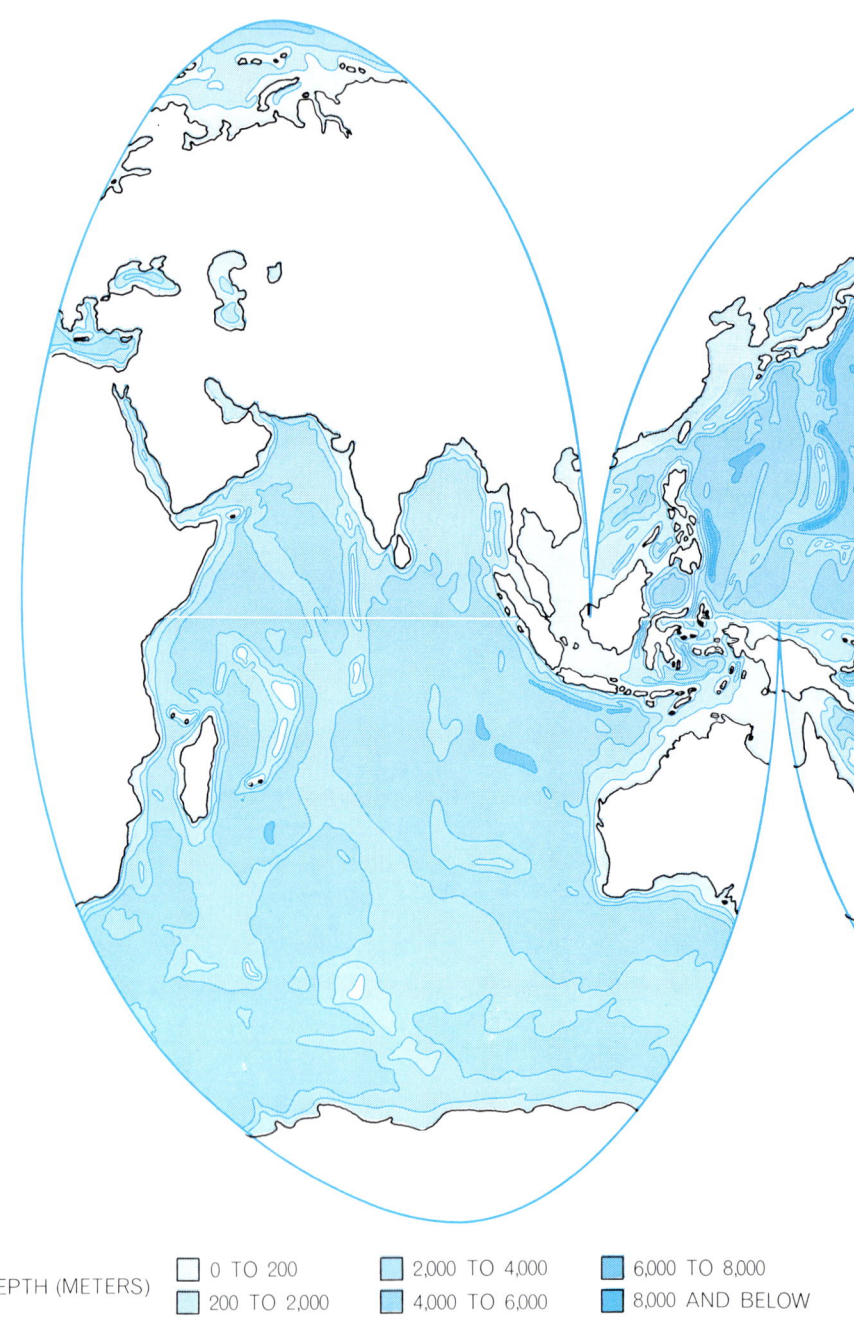

DEPTH (METERS)   ☐ 0 TO 200   ☐ 2,000 TO 4,000   ☐ 6,000 TO 8,000
☐ 200 TO 2,000   ☐ 4,000 TO 6,000   ☐ 8,000 AND BELOW

VASTNESS OF THE WORLD OCEAN is emphasized in this map, which is referred to by cartographers as a homolographic equal-area projection. The sea covers some 140 million

a course as close as possible to the wind, and on the length at the waterline, which determined the velocity of the waves made by the ship. The variety of ways in which the different sea peoples solved these problems, in accordance with their particular needs and available materials, is a constant delight to the amateur of sailing ships. When these diverse vessels were under full sail in a fresh breeze, how beautiful they must have been!

Fortunately the beauty of sailing ships under way has always had wide popular appeal, and numerous realistic paintings and drawings still exist, although almost all the ships themselves have long since disappeared. The splendor of the sea in those days is mirrored in the slim grace of a modern high-masted sailing yacht, sailing full and by or running free under a spinnaker; this must be one of the finest sights ever seen by a sailorman. Yet even a tramp freighter or an oceanographic vessel, when viewed from a distance, has a certain dignity.

Although ships are always feminine and many of them, from *Queen Elizabeth 2* to Italian fishing boats, have been named after women, the ocean world is preeminently a masculine one. Sailors talk a good deal about the lovely, pliant girls they are going to visit in the next port, and mermaids, sea nymphs and sirens have filled their fantasies, but they usually take a dim view of women on shipboard. Because the ocean is a cruel mistress she demands the masculine virtues of courage and strength rather than the feminine ones of compassion and sensitivity. From a human standpoint, therefore, the ocean is only a half-world, and this fact has profoundly affected men's relationships to the sea.

To sense the character of life at sea under sail one must listen to the chanteys sailors sang to help haul in heavy lines and the "forecastle songs" that

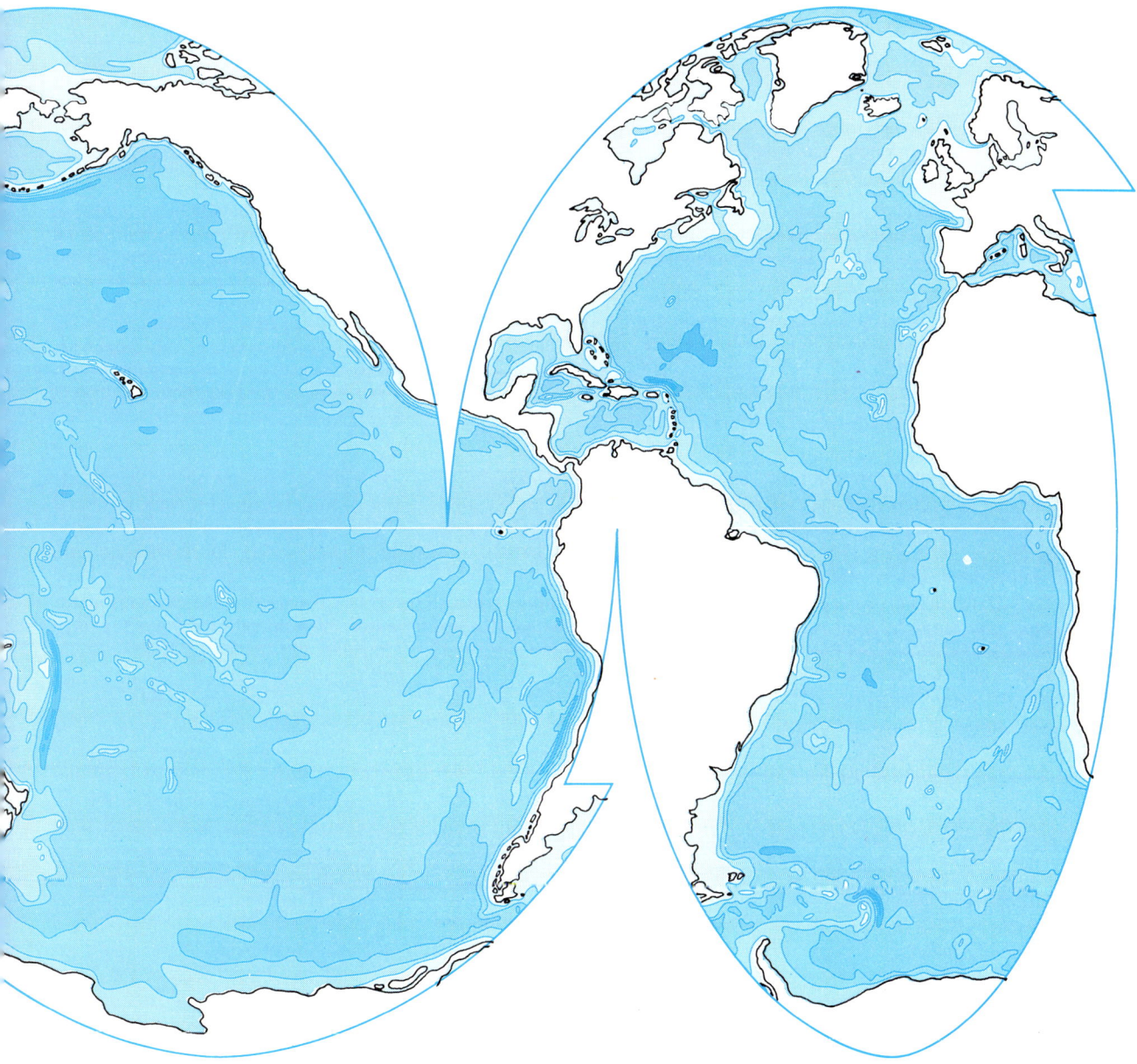

square miles, or 71 percent of the earth's surface, leaving only about 29 percent for the continental masses and lesser islands. The average depth of the sea is two and a half miles. Hence the total volume of water in the ocean is roughly 350 million cubic miles.

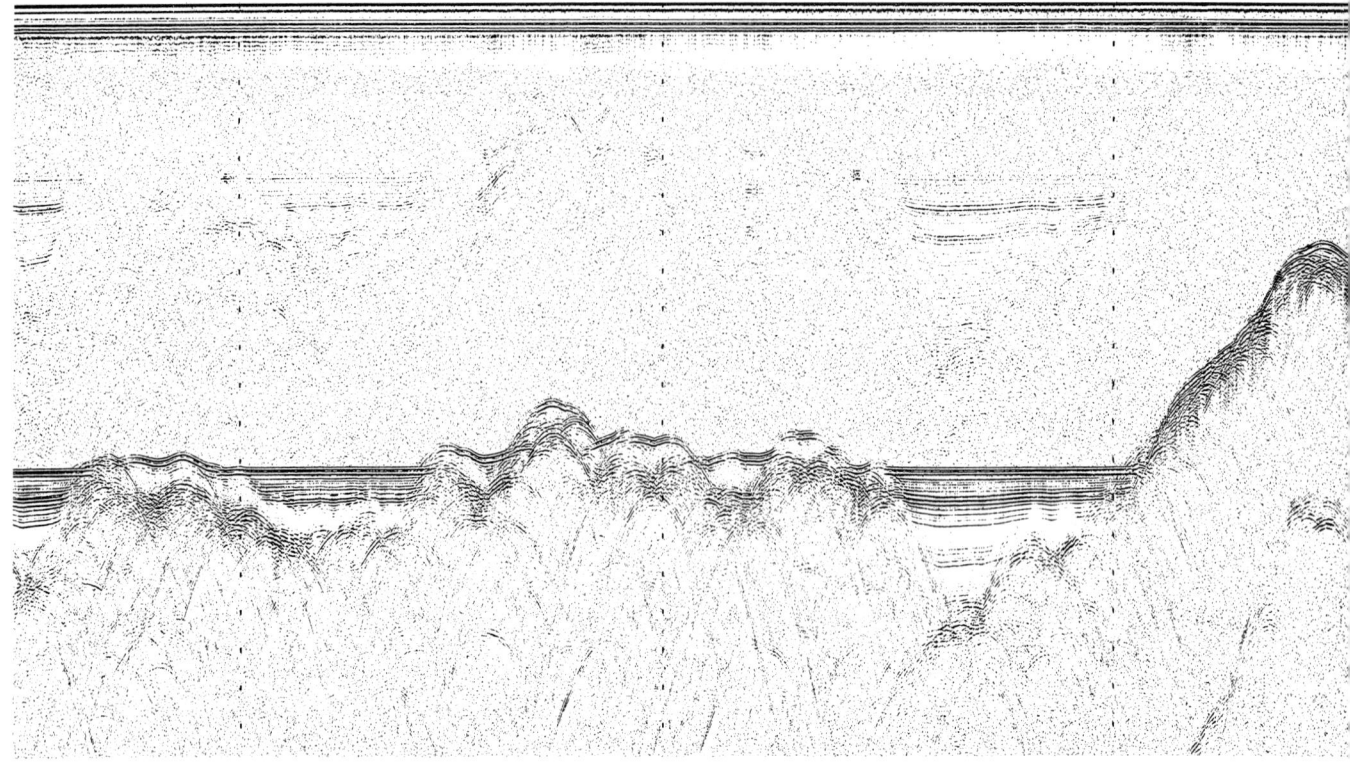

SEISMIC PROFILE on the following six pages shows a continuous cross-sectional view of the geological structure of the deep-ocean floor in the vicinity of the mid-Atlantic rise between South America and Africa. The profile was obtained by means of a seismic recording system carried on board the Teledyne Exploration Company's research vessel *Stranger* during a west-east traverse between Trinidad, B.W.I., and Monrovia, Liberia, in June, 1967. The Teledyne "super sparker" system consists of several banks of high-voltage capacitors, which discharge into the seawater at intervals of a few seconds, generating enough steam bubbles to simulate a gaseous explosion. The resulting acoustical impulse is transmitted through the water and into the sea floor, where seismic reflections are produced by geological structures as deep as two miles below the sea floor. The reflected signals return to the surface of the ocean, where they are detected by an array of hydrophones trailed behind the ship and recorded on continuously moving paper tape. The vertical-to-horizontal-scale ratio of the resulting profile is 8 : 1. The segment of the profile shown on this page shows a portion of

whiled away the dogwatch hours in the evening. In these days of disciplined labor workmen toil in concentrated silence. Seamen on the old sailing ships howled and yelled and sang at their work. As Richard Henry Dana wrote in *Two Years before the Mast*, "A song is as necessary to sailors as a drum and fife to soldiers. They can't pull in time or pull with a will without it." There were the short drag chanteys:

*Haul on the bowline, our bully ship's
  a-rolling,
Haul on the bowline, the bowline haul!*

Halyard chanteys were for hoisting the sail and catting the anchor:

*O blow the man down, bullies,
Blow the man down!
To me way—aye, blow the man down...
Give me some time to
Blow the man down!*

Capstan chanteys were for steady pushing on the capstan bars when the anchor was being weighed or the ship warped:

*Oh say was you ever in Rio Grande?
Way, you Rio!...
Sing fare ye well, my pretty young girls,
For we're bound to the Rio Grande.*

Forecastle songs were often ballads, as familiar on the land as on the sea. Among them is "The High Barbaree," which tells how a British ship sank a pirate:

*Blow high, blow low, 'cause slow
  sailed we,
Look ahead, look astern, look
  to windward and to lee
Cruising down along the coast
  of the High Barbaree.*

The relationship between nations and the sea has often been described in terms of national sea power. The American naval theorist Alfred Thayer Mahan, in his classic work *The Influence of Sea Power upon History, 1660–1783*, held that the sea power of a state rested on three things: the production of manufactured goods in the home country for overseas trade, commerce across the seas in its own ships, and colonies or dependent states on the farther shore to provide safe ports and goods for exchange, usually raw materials to be brought back to the home country for fabrication into finished products. The function of a navy was to advance all three elements of sea power. By dominating the sea the navy protected the homeland from invasion and ensured that wars would be fought on other people's territory, while production proceeded unimpeded at home. The navy guarded the nation's merchant ships and drove the ships of other nations off the ocean. It helped in the acquisition of colonies and ensured their subsequent docility by transporting and supplying troops and administrators and by preventing other states from interfering in these activities.

The Battle of Salamis is a famous example of naval protection of the home country. Greece was saved from Persian conquest by the "wooden walls" of Athens, and Xerxes flogged the sea that had betrayed him. The subsequent organization at Athens of the Delian league (through which many of the Greek islands and maritime cities at first contributed to a common treasury on the sacred

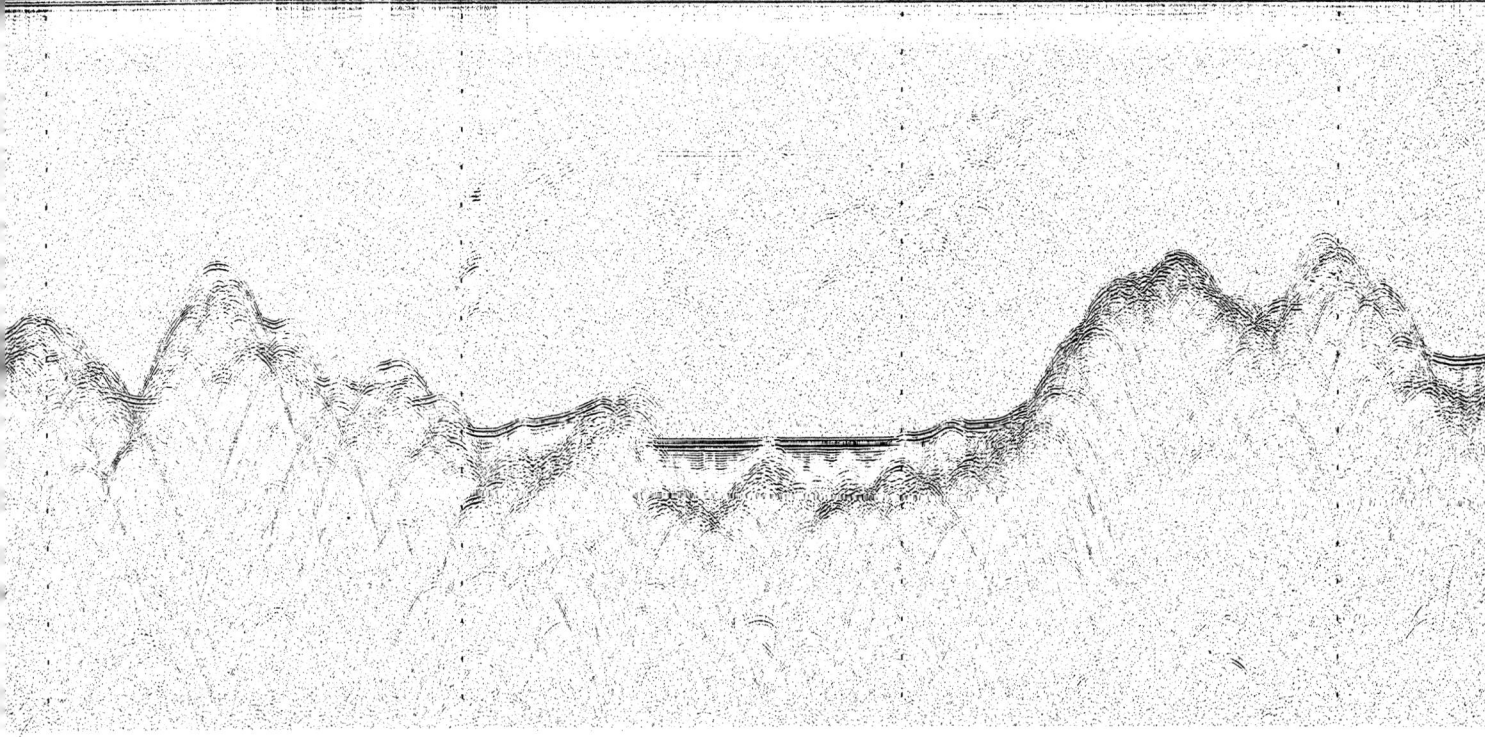

the western abyssal plain near the base of the mid-Atlantic rise. In this region the sediments are comparatively flat and uniform; they are estimated to be from 1,000 to 1,500 meters thick. Although the basement complex underlying the sedimentary layers is quite irregular, within the sediments themselves there is little evidence of recent faulting. The water depth at this point is about 5,300 meters. At the beginning of the mid-Atlantic rise (*pages 673–75*) the crystalline basement structures rise abruptly above the level of the mean sea floor. In this region sediments exist only in small basins over the basement complex. As one approaches the crest of the mid-Atlantic rise (*pages 676–77*) there is no longer evidence of sediments even in the basins. Individual peaks and ridges rise as much as 1,700 meters above the mean sea floor near the crest. The nature of the seismic reflections in this region suggests recent as well as ancient tectonic activity. The entire section of profile reproduced on these six pages covers a distance of about 240 miles centered approximately 1,000 miles east of Trinidad. The mid-Atlantic rise is roughly 150 miles wide at this latitude.

island of Delos and were later forced to submit to Athenian rule) illustrates the use of a navy to acquire and control overseas dependencies. At the same time Athens greatly expanded her commerce throughout the Mediterranean world, thus developing the second of Mahan's three elements of sea power, and for a few decades she enjoyed a golden age.

The tragedy of the Carthaginian brothers Hannibal and Hasdrubal was chosen by Mahan to illustrate the protective role of sea power. At the time of the Second Punic War between Rome and Carthage, although the Mediterranean Sea had not yet become the Roman *mare nostrum*, Roman fleets had effective control of the Adriatic, Tyrrhenian and Sardinian seas, of the eastern coast of Spain as far south as the mouth of the Ebro and of the northern and eastern coast of Sicily as far down as Syracuse. Hannibal's fleet was not strong enough to enable him to invade Italy by sea. From Carthage through his supply base in southern Spain he launched his forces overland across Gaul and the Alps. On this march he lost 33,000 of the 60,000 veteran soldiers with whom he started.

In spite of these losses Hannibal managed to score a series of brilliant military successes in Italy, but his army slowly wasted away because the Roman control of the sea denied him replacements. His ally Philip of Macedon was unable to land any soldiers in Italy, and Carthage could provide only occasional support by sea. Finally Hasdrubal managed to cross the Pyrenees at their extreme western end and to press on through Gaul and over the Alps in an attempt to reinforce his brother. Roman sea power countered this move by transporting 11,-000 men in ships from the Roman base in Spain back to Italy. These reinforcements, together with the Roman troops positioned between Hannibal and Hasdrubal, fell in overwhelming numbers on the younger brother at the river Metaurus. Hannibal's first news of the battle came when his brother's head was thrown into his camp. He cried out in despair that Rome would now be mistress of the world.

The importance of production in the home country for the development of sea power is shown in a negative way by the later history of Portugal. This small nation was the first in Europe to commit her fortunes to the sea and to trade with distant lands, but as Mahan writes: "The mines of Brazil were the ruin of Portugal.... All manufactures fell into insane contempt; ere long the English supplied the Portuguese not only with clothes but with all merchandise, all commodities, even to salt, fish and grain. After their gold, the Portuguese abandoned their very soil. The vineyards of Oporto were finally bought by the English with Brazilian gold, which had only passed through Portugal to be spread throughout England. In fifty years the Brazilian mines yielded $500 million worth of gold to Portugal but at the end of that time Portugal had only $25 million left."

The Spanish galleons that plied between Mexico and the Philippines furnish a similarly negative illustration. In the annals of the conquest of the "perils and hazards of the sea" the story of these galleons is a remarkable one. Every year for 250 years—from 1565 to 1815—the Spanish governors in Mexico sent out well-armed ships from Acapulco across the longest stretch of water in the world, carrying Mexican silver to the colony of

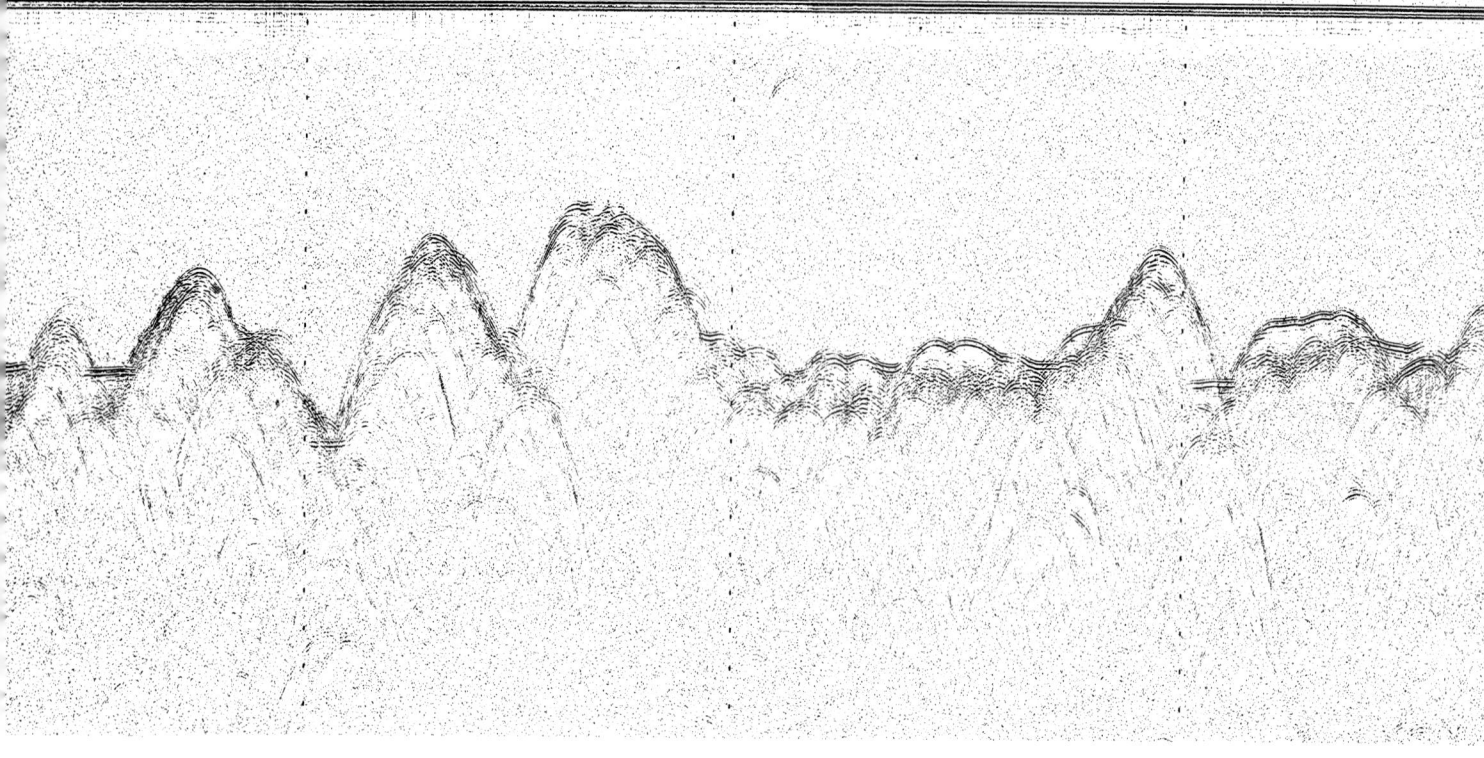

Manila in the Philippines. There the money was exchanged for Chinese silks, porcelains, ladies' combs and other luxury goods, and the galleons laden with these commodities sailed eastward again to Mexico. Throughout the 250 years only four galleons were captured, one in 1587 by a gentleman privateer, Thomas Cavendish, and three in the 18th century during the worldwide wars in which Britain won her empire by achieving primacy at sea.

Because the trade was primarily in consumer goods and in silver, however, the Manila colony never became much more than an entrepôt for the transshipment of Chinese luxuries. It developed no industrial or commercial life of its own, and neither Mexico nor Spain gained long-run benefits to their economies. Anglo-Saxon historians are fond of belittling the maritime skills of Spain. As the history of the Manila galleons shows, however, the Spaniards lacked neither seamanship nor naval prowess. Their failure resulted from a lack of interest and skill in manufacturing and trade.

Throughout the 17th and 18th centuries England pursued all the elements of sea power with a ruthless aggressiveness and self-confident arrogance that is hard to believe today. Even Charles II, who depended on subsidies from Louis XIV, told the Sun King when a dispute arose about the command of the combined French and English fleets against the Dutch Republic, "It is the custom of the English to command at sea." The

wars of the 18th century impoverished France because of the enormous cost of her armies, while England got rich through overseas trade, supported by manufacturing at home and protected by her mastery of the seas.

In the long periods of maritime peace imposed by the supremacy of Roman and English sea power, the sea became a medium for unifying civilization. As Sir Kenneth Clark has observed, if you had visited the public square of any Mediterranean town in the first century A.D.—in Greece, Italy, France, Asia Minor or North Africa—you would hardly have known where you were, any more than you would in a modern airport. Thus in the 18th and 19th centuries did English sea power protect the development of a new civilization around a larger ocean—the Atlantic civilization, which we are still building today.

What meaning does sea power hold for the modern world? Mahan's concepts seem outmoded in an age when superpowers can destroy each other by pushing the right buttons, when colonies have disappeared and when the North Vietnamese can laugh in the teeth of the most expensive navy the world has ever seen.

A new kind of sea power, which Mahan would scarcely recognize, now holds the center of the stage. The growing vulnerability of land-based intercontinental ballistic missiles under the impact of new weapons developments [see "Military Technology and National Se-

curity," by H. F. York; SCIENTIFIC AMERICAN, August, 1969] makes mutual deterrence in the balance of nuclear terror hinge increasingly on the atomic-powered submarines, armed with ballistic missiles, with which the U.S.S.R. and the U.S. confront each other under the seas. In part because of the effectiveness of this deterrent, the traditional uses of sea power appear to have become obsolete.

For the foreseeable future a general war in Europe, which would require the U.S. Navy to protect vast ocean shipments of men and materials, seems the unlikeliest of possibilities. Navies, if they are to be for other purposes, will be employed only in small wars, of which unhappily there seems to be an abundance in this bitter second half of the 20th century, and to maintain the peace, wherever possible, along the coastlines of the world. These would both be major tasks but they require a radical rethinking of our inherited concepts of sea power.

The ocean first became one ocean in the consciousness of man during the age of discovery. Man's venture into the unknown was first carried past the point of no return by the Portuguese. The Infante Dom Henrique, who is called Prince Henry the Navigator by Anglo-Saxon writers, was the epitome of the spirit of this age. He made his headquarters at Sagres on Cape St. Vincent, the southwestern promontory of Europe. Here he built a kind of academy for seafaring in unknown oceans; he collected all the charts and sailing directions he

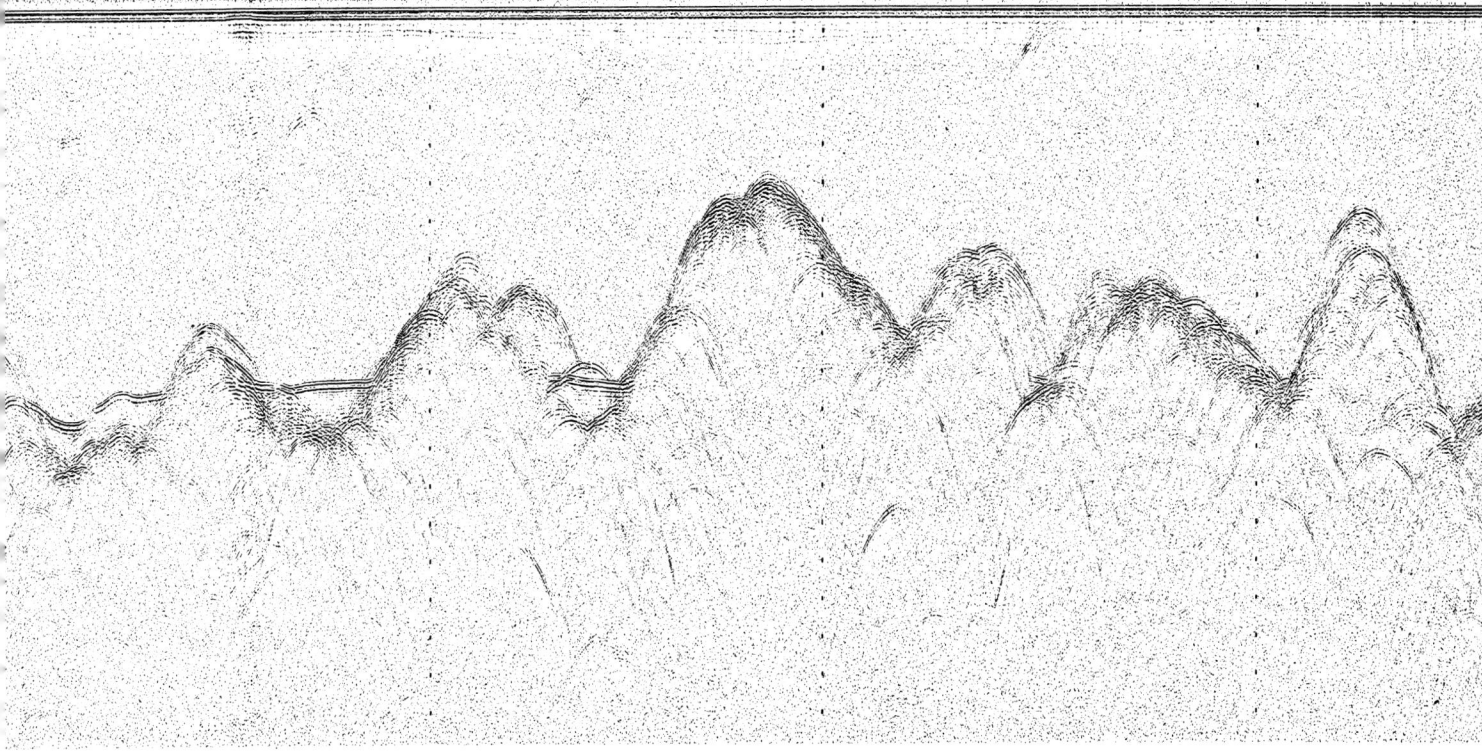

could find, and he employed mathematicians and geographers to decipher them and to make new ones. The most daring and competent master mariners from Italy, Spain and Portugal entered his service. He sent them out each year farther and farther into the western ocean and down the coast of Africa. Searching for the mythical Isles of St. Brendan (a seagoing Irish saint whose largely imaginary sea stories were popular throughout the Middle Ages), his captains found the Azores.

Prince Henry's great objective, however, was to send his ships down the coast of Africa beyond Cape Bojador, and if possible around the southern point of that continent on a new route to the Indies. This was a fearful challenge. Beyond the cape, according to ancient lore, lay the impassable Tropical Zone, with water temperatures that reached the boiling point at the Equator. Equally serious was the undoubted fact that both the wind and the current were dead ahead whenever the Portuguese caravels tried to lay a course for their return voyage home. Cape Bojador was finally rounded in 1434, and by the year of Prince Henry's death (1460) the Portuguese had reached within 10 degrees of the Equator. Then they swept rapidly onward; only 28 years later Bartholomeu Diaz sailed to the southernmost point of Africa, which he called the Cape of Storms. The Portuguese king renamed it the Cape of Good Hope, in joyful anticipation of Vasco da Gama's shortly to be accomplished voyage to Calicut.

For 100 years after Prince Henry's death Portugal, which was called Lusitania by the Romans, merited Camoëns' splendid boast in the "Lusiads," the epic poem that celebrates Vasco da Gama's expedition:

Behold her seated here, both head and key
Of Europe all, the Lusitanian queen;
Where endeth land and where beginneth sea;
Where Phoebus goeth down in ocean green.

The path had been laid to the treasures of the East and to an empire whose bedraggled and unhappy remnants still exist.

Prince Henry compels our admiration for his vision and leadership. His world of faith and terrors was so different from our own, however, that we have difficulty visualizing it. In contrast Captain James Cook is almost a contemporary. He would have felt at home on a modern oceanographic expedition into the deep Pacific and would have approved its objectives.

Cook's journals are models of modest and clear writing. Like all good marine surveyors he also made painstakingly accurate sketches and maps of the islands and shorelines he visited. He had the eye of an artist, seeing many things other men did not see, and the curiosity and powers of logical thought of a scientist.

Of Cook's two great discoveries, one was negative. This was the nonexistence of Terra Australis, a huge imaginary continent that for hundreds of years had been thought to cut off the Southern Ocean beyond about 50 degrees south latitude. Between 1768 and 1775 Cook twice sailed around the world in his small ships *Endeavour, Resolution* and *Adventure,* crisscrossing many times the "roaring forties" and the "howling fifties" and finding no land. He noticed, however, that the icebergs he encountered near the southern limits of his track were free of salt. He deduced that they must have broken off from huge glaciers on a land mass beyond the Antarctic Circle. In these high southern latitudes Cook also noted the presence of numerous albatrosses and stormy petrels. This seemed independent proof of a large land mass to the south, where these birds could nest. Thus in his mind's eye Cook saw Antarctica, although that continent was not actually discovered for another 50 years.

Cook's discovery of the Hawaiian Islands, the largest of the Polynesians and the islands farthest from any continent, was an accident. Cook found them on his way to Alaska and the Arctic Ocean in search of a northwest passage between Europe and Asia. He was as astonished as we are today that the Spanish Manila galleons had never found them.

It has been said that Cook's epitaph is the map of the Pacific, almost as we now know it. If you are planning a cruise to the South Sea Islands, many of the

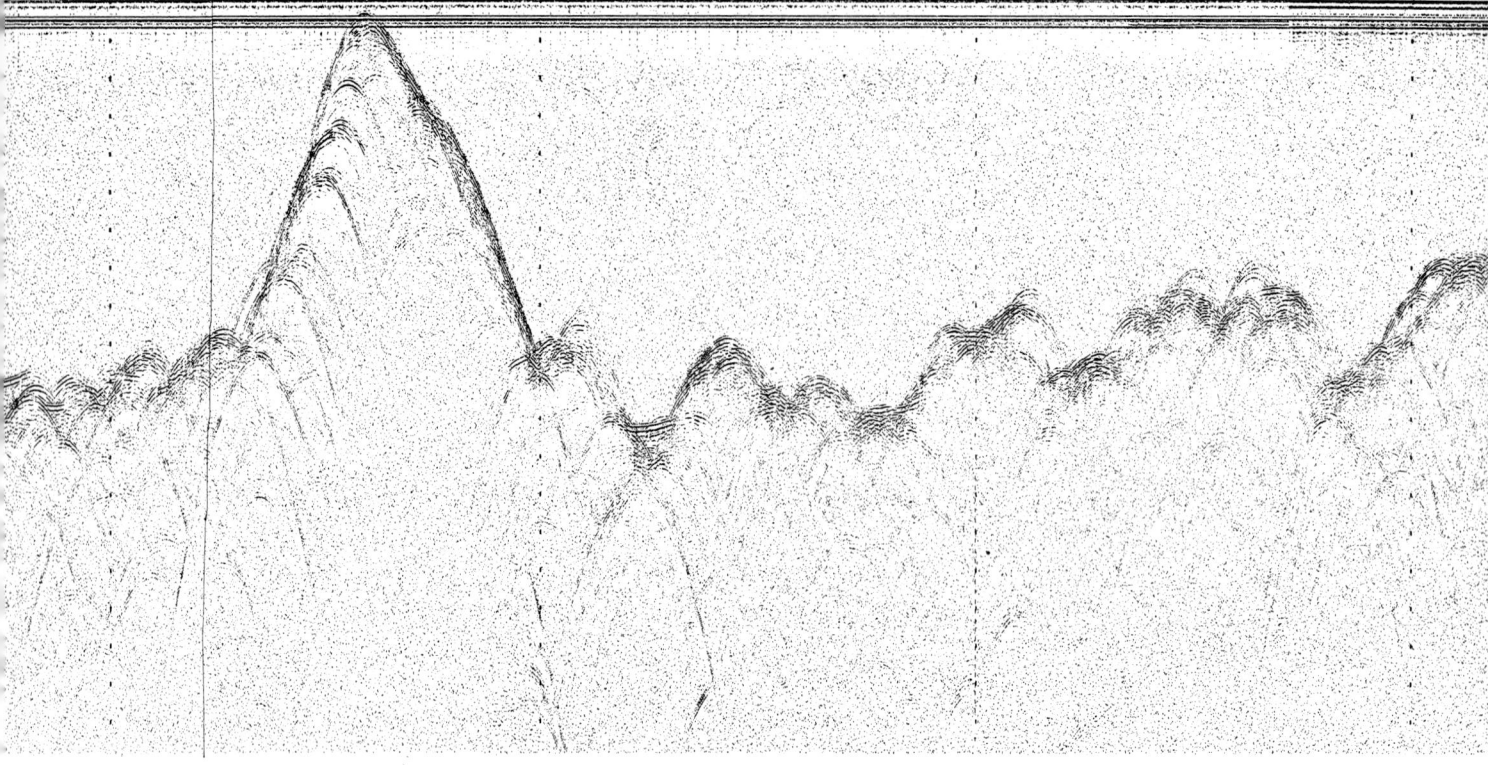

charts you will obtain from the Naval Oceanographic Office will bear this notation: "From an Admiralty Survey in 1769" (or some other date from 1768 to 1777). The survey was made by Cook.

Cook mapped Palmerston Atoll, a lonely speck halfway between Samoa and Tahiti in the archipelago named after him, in 1774. He made a note of strong currents around the atoll. On the Capricorn Expedition of the Scripps Institution of Oceanography we had the great satisfaction of proving that he had made one of his rare mistakes. Without benefit of echo-sounding gear, Cook had evidently become confused in his sextant fixes on the low, almost indistinguishable landmarks around the atoll. As a result his survey had failed to close by about two miles. He concluded that this was due to currents that retarded the ship. Actually the atoll is two miles longer than he thought it was, and the currents do not exist.

From a scientific point of view there are many properties of the oceans and of seawater that combine to determine the relationships between man and the sea. The most obvious are the oceans' vast area, great depth and huge volume. Seawater is the most abundant substance accessible to man, and all the continents are islands in the midst of the world-girdling sea. The periphery, or shoreline, of the oceans is several times longer than it would need to be if the waters were contained in a simple circular basin. This long shoreline gives many peoples access

to the sea and enables them to enjoy its benefits. More important, the shoreline allows relatively easy access to vast areas of the land. In the words of Mahan, "the powers ruling the sea are always near any country whose ports are open to their ships."

Because of the complex shape of the boundaries between the land and the sea, certain small parts of the oceans have an importance that far outweighs their area. These are the narrow seas, the straits and passages between the land that lie athwart the well-traveled trade routes of nations. They are the best-known places in the sea, as is easily seen by reciting some of their names: Bosporus and Dardanelles, Kattegat and Skagerrak, Gibraltar and Bab el Mandeb, Florida and Torres, Tsushima and Malacca, the St. George Channel and the Strait of Dover. To these must be added the narrow isthmuses that join the continents, Panama and Suez.

The fluid character of water on our planet is the miracle that makes life possible, but it also means that the oceans fill all the low places of the earth. Because of this geographical fact the oceans are the ultimate receptacle of the wastes of the land, including the wastes that are produced in ever increasing amounts by human beings and their industries.

The relatively high density and low viscosity of seawater are the essential qualities that make the sea surface a broad and easily traveled highway. In technical terms the water provides a

high lift-drag ratio to the ships that float on it; consequently large ships and heavy cargoes can be moved fairly rapidly across the ocean with comparatively little motive power. Maritime commerce would be impossible if seawater were as viscous as molasses, and the fuel required to carry the huge cargoes of modern ships would be prohibitively expensive if the water were as light as air. At the same time the combination of low viscosity and high density gives rise to the principal hazard of the sea, the giant wind-waves caused by storms that crush small ships and fiercely attack coastal structures. If the water were much more viscous, the wind could not build up high, steep waves, and if it were much lighter, the wave force would be insignificant.

The high surface tension of seawater, although not of fundamental importance, is a matter of considerable convenience to human beings because it means that the water does not stick to surfaces with which it comes in contact but runs off easily and leaves them relatively dry. Seagoing would be a messy business if the oceans consisted of oil. The high density and fluidity of the water create a serious difficulty when men attempt to lower themselves or their equipment deep below the surface: the enormous hydrostatic pressure at great depths, which crushes all but the strongest vessels, forms air pockets in submarine cables and produces high stresses in equipment made of materials with different compressibilities.

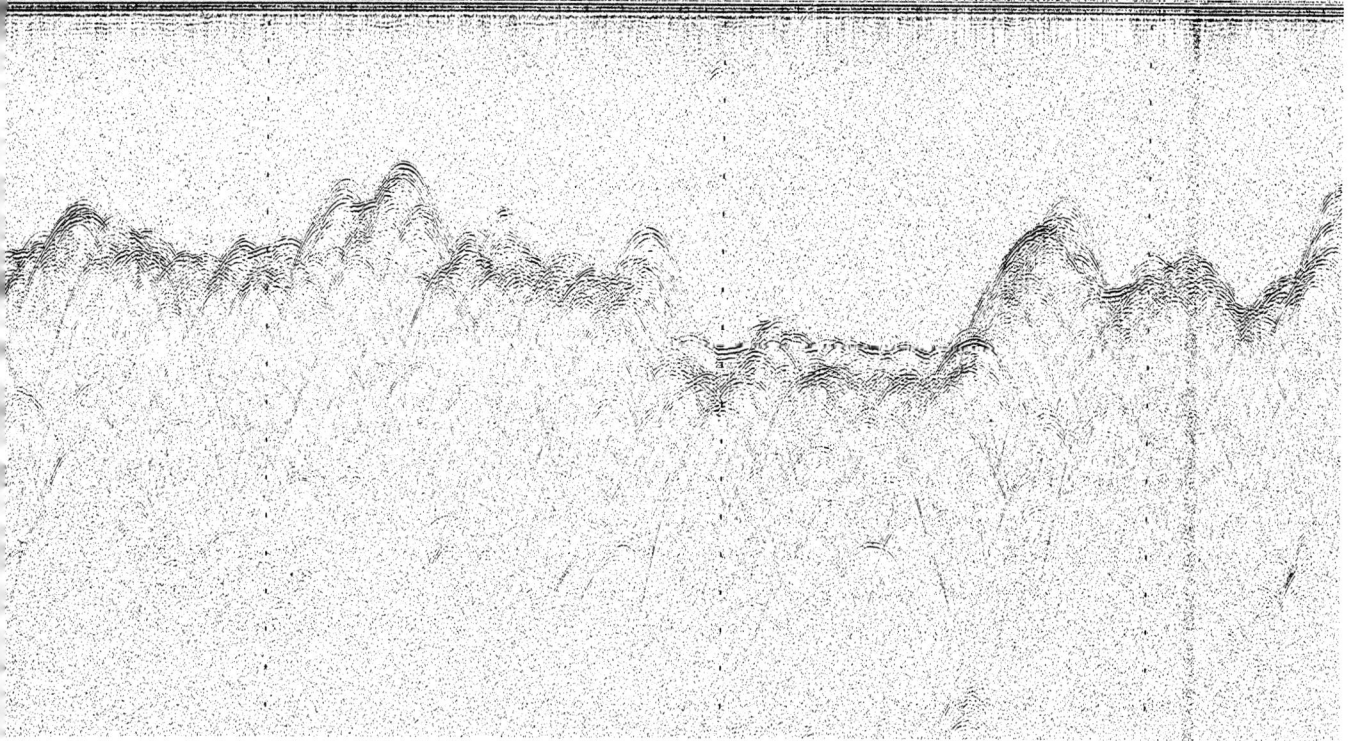

From the human standpoint the saltiness of seawater is its most undesirable quality. Because of its high salt content men can neither drink it nor use it to water their crops. The ionized salts make the water a good conductor of electricity, with the result that the ocean can be penetrated for only a short distance by radio waves, and electrolytic reactions between dissimilar metals or compounds proceed very rapidly. The salts plus dissolved oxygen cause seawater to be a highly corrosive substance for most man-made objects. When immersed in the sea, the works of man soon suffer a sea change into something that may be rich and strange but is usually useless for human purposes. This quality of seawater makes research in materials a strategic salient in the progress of ocean technology [see "Technology and the Ocean," by Willard Bascom; SCIENTIFIC AMERICAN Offprint 887].

Seawater is much less transparent to visible light than air is, but it is more transparent than most other substances. This intermediate transparency, combined with a high heat capacity and a high heat of vaporization, makes the ocean the regulator in the great thermal engine of the ocean and the atmosphere [see "The Atmosphere and the Ocean," by R. W. Stewart; SCIENTIFIC AMERICAN Offprint 881]. Most of the energy of sunlight passes virtually unimpeded throughout the atmosphere into the ocean, where it is absorbed and transformed. Nearly a third of all solar energy reaching the earth's surface goes

to evaporate seawater. The thermal inertia of the sea, the circulation of the water and the geographic distribution of ocean and land profoundly influence our planet's weather and climate.

The moderate transparency and high density of seawater, and the great depth of the ocean, give life in the sea a very different character from that on land [see "The Nature of Oceanic Life," by John D. Isaacs; SCIENTIFIC AMERICAN Offprint 884]. Photosynthesis can take place only in the waters near the surface where bright sunlight penetrates, and consequently marine plants in the open ocean cannot support themselves on solid ground. Most marine plants have solved this problem by being extremely small, so that they have a large surface-to-volume ratio and sink or rise slowly through the water. To maintain their small size they have a short life-span, measured in hours or days rather than in the seasons or centuries that characterize land plants. Consequently, although the total production of organic matter in the most fertile regions of the ocean is higher than that in fertile land areas, the biomass of marine plants at any one time is small.

The animals of the sea have been forced to adjust themselves to this regime of the plants, with the result that the food web in the sea is much more complex than on land, and the animals that can be utilized by man make up only a small part of the organic production of the ocean. The plants themselves

are present in such low concentration that they cannot be economically harvested. The potential harvest of human food from the sea is still far from being realized by human beings. Yet, as S. J. Holt shows [see "The Food Resources of the Ocean," SCIENTIFIC AMERICAN Offprint 886], it will never be possible for men to obtain more than a fraction of their food requirements from the ocean.

From the biologist's point of view one of the most important characteristics of the ocean is its great age, even on a geologic time scale. Time has been available for the evolution of many different forms of life. As a result living things in the sea present an incredible diversity. Some 200,000 species have already been identified and new ones are found on every oceanographic expedition. Some of these are "living fossils" from the ancient past and others have evolved recently.

In former times men were chiefly concerned with the surface of the sea, together with the near-surface waters and shallow seabed that were accessible to fishermen. As our technology advances, however, our ability to penetrate and use the entire huge volume of the ocean and to explore and exploit the seabed, even to the greatest depths, is rapidly increasing. The shape of the sea floor, the properties of the deep waters and the distribution of deep-sea resources have thus become of much greater interest than previously. As K. O. Emery and Edward Wenk, Jr., set forth at greater length [see "The Continental Shelves," Offprint

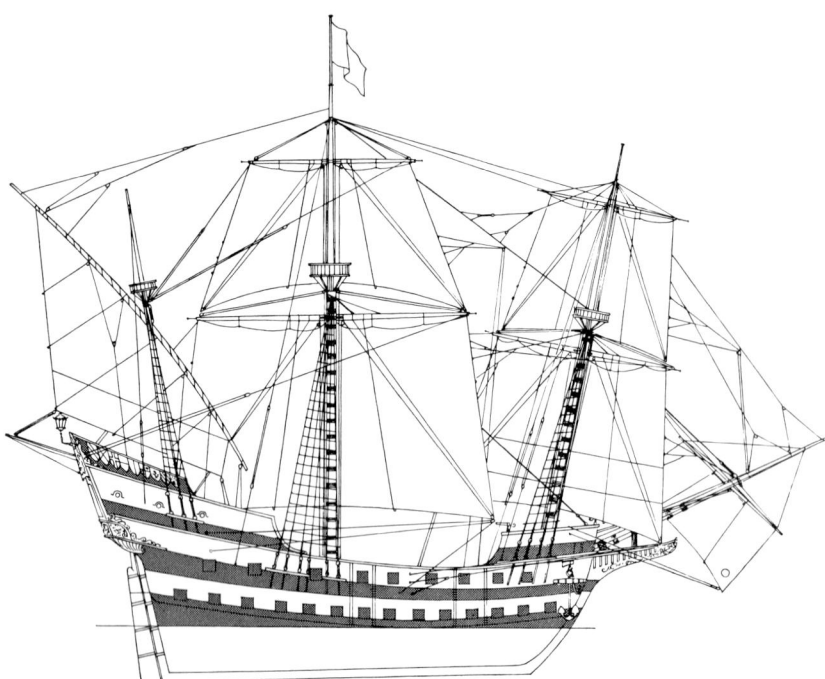

MANILA GALLEON was a remarkably seaworthy ship that for 250 years (from 1565 to 1815) was used to make round-trip voyages between the Spanish colonies in Mexico and the Philippines. Curiously the Spaniards never discovered the Hawaiian Islands, although they sailed both north and south of them. Hawaii was finally discovered by Cook in 1778.

882, and "The Physical Resources of the Ocean," SCIENTIFIC AMERICAN Offprint 885], the margins of the oceans surrounding the continents containing the continental shelves and slopes and the deeper continental rise, are probably the principal location for the most important mineral resources underneath the sea: petroleum and natural gas. The living resources seem to exist almost entirely in the upper 1,500 meters, and like the fossil fuels they are most concentrated over the continental shelves and slopes.

Submarine vehicles for both military and peaceful purposes are a spectacular element of the new marine technology. From a military point of view the great advantage of submarines rests on the low transparency of seawater for visible light and radio waves. A deeply submerged submarine can neither be seen visually nor be detected by radar. The submarine itself would be equally blind were it not for one of the most remarkable qualities of the ocean: its capacity to transmit low-frequency sounds over great distances.

The explosion of a one-pound block of dynamite in the air can be heard for about half a mile. Such an explosion at mid-depths in the ocean can be heard for many thousands of miles. The ocean is far from being a perfect acoustical transmitter, however. It is full of background noises made by animals and volcanoes; the surface and the bottom scatter and absorb sound, variations in temperature and density bend and distort sound waves in complicated ways, and the ocean rings with echoes like a badly designed auditorium. To elucidate and overcome these imperfections in sound propagation is one of the major problems of marine technology.

We may confidently expect that the ocean will be far more important to mankind in the future than it has ever been in the past. Recognition of the new opportunities in the ocean has brought a widening realization that the organization of human society into national states, which works, however imperfectly on the land, is not well suited to the optimum utilization of the sea. A new kind of regime is needed in which the interests of all states are protected but those of humanity as a whole are paramount. The possible nature of such a regime is now being widely discussed in the United Nations and other international forums. It is too early to forecast what may emerge. Nevertheless, it can be hoped that agreement will be reached on a set of principles somewhat as follows:

1. The resources of the high seas and the underlying seabed, outside those areas adjacent to the coasts in which the coastal states exercise certain exclusive rights, are the common heritage of mankind and shall be used and conserved in the common interest of all men. All countries shall participate in an equitable manner in the benefits gained from these rescources. Individual states shall not appropriate any part of the high seas or of the underlying seabed.

2. The areas adjacent to the coasts in which coastal states exercise certain exclusive rights shall be as small as feasible, and the outer limits of these areas shall be fixed by a definite depth or distance from shore or by a formula involving both depth and distance. This outer boundary, together with the nature of the exclusive rights to be exercised, should be determined by international agreement as soon as possible.

3. Internationally coordinated action must be taken to prevent pollution of the ocean, including control of pollutants coming from the land or the air, such as pesticides, radioactive substances, poisonous chemicals and sewage; from ships, submarines or other equipment used at sea, and from exploitation of marine resources, for example exploration, production, storage and transportation of oil and gas.

4. The freedom of scientific research in the ocean shall be kept inviolate. The exclusive rights granted to the coastal states shall not include the right to interfere with scientific research, provided that the coastal state is given prior notification of the plan to conduct the research, has full opportunity to participate in it and has access to all the data obtained and samples collected, and provided that the research does not deleteriously affect marine resources or other uses of the sea.

5. Greatly intensified international cooperation and coordination in all peaceful uses of the ocean is needed to encourage and advance beneficial exploitation of marine resources and technological developments for this purpose, to ensure the conservation and rational use of resources, and to minimize interference among uses of different resources and among different uses of the same resource.

6. The military uses of the ocean floor shall be as limited as practicable; in particular, no nuclear weapons shall be planted on it.

Agreement on these principles and on creation of an international regime to secure them depends largely, although by no means entirely, on the U.S. It is to be hoped that our government and people will be farsighted enough to see that their own long-range interests be in a generous approach to the new age of the oceans.

The Author

ROGER REVELLE is Richard Saltonstall Professor of Population Policy in the Faculty of Public Health at Harvard University and director of the Harvard Center for Population Studies. For many years before going to Harvard in 1964 he was director of the Scripps Institution of Oceanography, which he had joined in 1931 as a research assistant. His work in oceanography has won wide recognition; in 1963 the National Academy of Sciences gave him its Agassiz Medal for "outstanding achievement in oceanography." Revelle has also been active in many other areas: he headed the group that developed a plan for land and water development of the Indus River basin; he was a member of the U.S. delegation to the first Atoms for Peace Conference; he has been a member of several panels of the President's Science Advisory Committee; he was U.S. delegate to the General Assembly of the International Council of Scientific Unions in 1968, and he is vice-president of the American Academy of Arts and Sciences. Revelle was graduated from Pomona College in 1929 and began his career as a teaching assistant there. He received his Ph.D. from the University of California in 1936. At the Scripps Institution he became professor of oceanography in 1948 and director in 1951. From 1958 to 1961 he was also director of the La Jolla campus of the University of California. Between 1961 and 1963 he took a leave of absence to serve as the first science adviser to the Secretary of the Interior. He returned to the University of California in 1963 as University Dean of Research for all campuses. Revelle College, the first of the new colleges established in the University of California at San Diego, was named in his honor by the regents of the university in 1965.

Bibliography

THE INFLUENCE OF SEA POWER UPON HISTORY, 1660–1783. Captain A. T. Mahan. Little, Brown, and Company, 1890.

THE MANILA GALLEON: ILLUSTRATED WITH MAPS. William Lytle Schurz. E. P. Dutton & Company, Inc., 1939.

THE EXPLORATIONS OF CAPTAIN JAMES COOK IN THE PACIFIC AS TOLD BY SELECTIONS OF HIS OWN JOURNALS 1768–1779. Edited by A. Grenfell Price. The Limited Editions Club, 1957.

SCIENTIFIC
AMERICAN   September 1969, Vol. 221, No. 3, pp. 66–75   OFFPRINT 880

THE ORIGIN OF THE OCEANS

by Sir Edward Bullard

In recent years it has become increasingly apparent that the floor
of the deep ocean is remarkably young. It is growing outward from
mid-ocean ridges, pushing most of the continents apart as it does.

The earth is uniquely favored among the planets: it has rain, rivers and seas. The large planets (Jupiter, Saturn, Uranus and Neptune) have only a small solid core, presumably overlain by gases liquefied by pressure; they are also surrounded by enormous atmospheres. The inner planets are more like the earth. Mercury, however, has practically no atmosphere and the side of the planet facing the sun is hot enough to melt lead. Venus has a thick atmosphere containing little water and a surface that, according to recent measurements, may be even hotter than the surface of Mercury. Mars and the moon appear to show us their primeval surfaces, affected only by craters formed by the impact of meteorites, and perhaps by volcanoes. Only on the earth has the repetition of erosion and sedimentation —"the colossal hour glass of rock destruction and rock formation"—run its course cycle after cycle and produced the diverse surface that we see. The mountains are raised and then worn away by falling and running water; the debris is carried onto the lowlands and then out to the ocean. Geologically speaking, the process is rapid. The great plateau of Africa is reduced by a foot in a few thousand years, and in a few million years it will be near sea level, like the Precambrian rocks of Canada and Finland. All trace of the original surface of the earth has been removed, but as far back as one can see there is evidence in rounded, water-worn pebbles for the existence of running water and therefore, presumably, of an ocean and of dry land.

The obvious things that no one comments on are often the most remarkable; one of them is the constancy of the total volume of water through the ages. The level of the sea, of course, has varied from time to time. During the ice ages, when much water was locked up in ice sheets on the continents, the level of the sea was lower than it is at present, and the continental shelves of Europe and North America were laid bare. Often the sea has advanced over the coastal plains, but never has it covered all the land or even most of it. The mechanism of this equilibrium is unknown; it might have been expected that water would be expelled gradually from the interior of the earth and that the seas would grow steadily larger, or that water would be dissociated into hydrogen and oxygen in the upper atmosphere and that the hydrogen would escape, leading to a gradual drying up of the seas. These things either do not happen or they balance each other.

The mystery is deepened by the almost complete loss of neon from the earth; in the sun and the stars neon is only a little rarer than oxygen. The neon was presumably lost when the earth was built up from dust and solid grains because neon normally does not form compounds, but if that is so, why was the water not lost too? Water has a molecular weight of 18, which is less than the atomic weight of neon, and thus should escape more easily. It looks as if the water must have been tied up in compounds, perhaps hydrated silicates, until the earth had formed and the neon had escaped. Water must then have been released as a liquid sometime during the first billion years of the earth's history, for which we have no geological record. The planet Mercury and the moon would have been too small to retain water after it was released. Mars seems to have been able to retain a trace, not enough to make oceans but enough to be detectable by spectroscopy.

These speculations about the early history of the earth are open to many doubts. The evidence is almost non-existent, and all one can say is, "It might have been that...." The great increase in understanding of the present state and recent history of the ocean basins that we have gained in the past 20 years is something quite different. For the first time the geology of the oceans has been studied with energy and resources commensurate with the tremendous task. It turns out that the main processes of geology can be understood only when the oceans have been studied; no amount of effort on land could have told us what we now know. The study of marine geology has unlocked the history of the oceans, and it seems likely to make intelligible the history of the continents as well. We are in the middle of a rejuvenating process in geology comparable to the one that physics experienced in the 1890's and to the one that is now in progress in molecular biology.

The critical step was the realization that the oceans are quite different from the continents. The mountains of the oceans are nothing like the Alps or the Rockies, which are largely built from folded sediments. There is a world-encircling mountain range—the mid-ocean ridge—on the sea bottom, but it is built entirely of igneous rocks, of basalts that have emerged from the interior of the earth. Although the undersea mountains have a covering of sediments in

RED SEA and the Gulf of Aden represent two of the newest seaways created by the worldwide spreading of the ocean floor. In this photograph, taken at an altitude of 390 miles from the spacecraft *Gemini 11* in September, 1966, the Red Sea separates Ethiopia (*at left*) from the Arabian peninsula (*at right*). The Gulf of Aden lies between the southern shore of Arabia and Somalia. The excellent fit between the drifting land masses is shown in the illustrations on page 683.

many places, they are not made of sediments, they are not folded and they have not been compressed.

A cracklike valley runs along the crest of the mid-ocean ridge for most of its length, and it is here that new ocean floor is being formed today [see illustration on next two pages]. From a study of the numerous earthquakes along this crack it is clear that the two sides are moving apart and that the crack would continually widen if it were not being filled with material from below. As the rocks on the two sides move away and new rock solidifies in the crack, the events are recorded by a kind of geological tape recorder: the newly solidified rock is magnetized in the direction of the earth's magnetic field. For at least the past 10,000 years, and possibly for as long as 700,000 years, the north magnetic pole has been close to its present location, so that the magnetic field is to the north and downward in the Northern Hemisphere, and to the north and upward in the Southern Hemisphere. As the cracking and the spreading of the ocean floor go on, a strip of magnetized rock is produced. Then one day, or rather in the course of several thousand years, the earth's field reverses, the next effusion of lava is magnetized in the reverse direction and a strip of reversely magnetized rocks is built up between the two split halves of the earlier strip. The reversals succeed one another at widely varying intervals; sometimes the change comes after 50,000 years, often there is no change for a million years and occasionally, as during the Permian period, there is no reversal for 20 million years. The sequence of reversals and the progress of spreading is recorded in all the oceans by the magnetization of the rocks of the ocean floor. The message can be read by a magnetometer towed behind a ship.

We now have enough examples of these magnetic messages to leave no doubt about what is happening. It is a

PROBABLE ARRANGEMENT of continents before the formation of the Atlantic Ocean was determined by the author with the aid of a computer. The fit was made not at the present coastlines but at the true edge of each continent, the line where the continental shelf (dark brown) slopes down steeply to the sea floor. Overlapping land and shelf areas are reddish orange; gaps where the continental edges do not quite meet are dark blue. At present the entire western Atlantic is moving as one great plate carrying both North America and South America with it. At an earlier period the two continents must have moved independently.

truly remarkable fact that the results of magnetic surveys in the South Pacific can be explained—indeed predicted—from the sequence of reversals of the direction of the earth's magnetic field known from magnetic and age measurements, made quite independently on lavas in California, Africa and elsewhere. The only adjustable factor in the calculation is the rate of spreading. Such worldwide theoretical ideas and such detailed agreement between calculation and theory are rare in geology, where theories are usually qualitative, local and of little predictive value.

The speed of spreading on each side of a mid-ocean ridge varies from less than a centimeter per year to as much as eight centimeters. The fastest rate is the one from the East Pacific Rise and the slowest rates are those from the Mid-Atlantic Ridge and from the Carlsberg Ridge of the northwest Indian Ocean. The rate of production of new terrestrial crust at the central valley of a ridge is the sum of the rates of spreading on the two sides. Since the rates on the two sides are commonly almost equal, this sum is twice the rate on each side and may be as much as 16 centimeters (six inches) per year. Such rates are, geologically speaking, fast. At 16 centimeters per year the entire floor of the Pacific Ocean, which is about 15,000 kilometers (10,000 miles) wide, could be produced in 100 million years.

When the mid-ocean ridges are examined in more detail, they are found not to be continuous but to be cut into sections by "fracture zones" [see top illustration on page 686]. A study of the earthquakes on these fracture zones shows that the separate pieces of ridge crest on the two sides of a fracture zone are not moving apart, as might seem likely on first consideration. The two pieces of ridge remain fixed with respect to each other while on each side a plate of the crust moves away as a rigid body; such a fracture is called a transform fault. The earthquakes occur only on the piece of the fracture zone between the two ridge crests; there is no relative motion along the parts outside this section.

If two rigid plates on a sphere are spreading out on each side of a ridge that is crossed by fracture zones, the relative motion of the two plates must be a rotation around some point, termed the pole of spreading. The "axis of spreading," around which the rotation takes place, passes through this pole and the center of the earth. The existence of a pole of spreading and an axis of spreading is geometrically necessary, as was shown by Leonhard Euler in the

RUPTURE OF MIDDLE EAST is being caused by the widening of the Red Sea and the Gulf of Aden. Some 20 million years ago the Arabian peninsula was joined to Africa, as evidenced by the remarkable fit between shorelines (see illustration below). The area within the *Gemini 11* photograph on page 681 is shown by the broken lines.

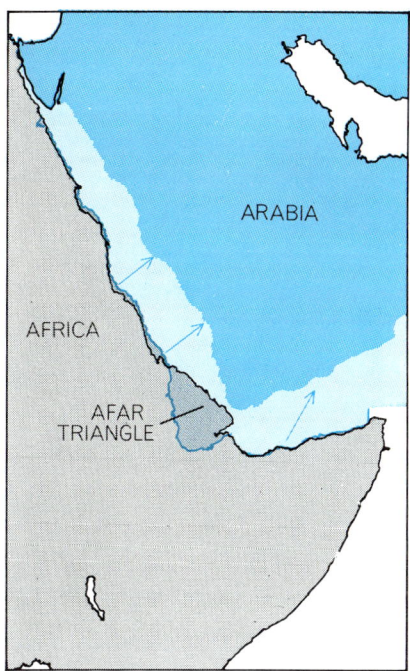

FIT OF SHORELINES of Arabia and Africa works out most successfully if the African coast (*black*) is left intact and if the Arabian coast (*color*) is superposed in two separate sections. In the reconstruction a corner of Arabia overlaps the "Afar triangle" in northern Ethiopia, an area that now has some of the characteristics of an ocean floor.

18th century. If the only motion on the fracture zones is the sliding of the two plates past each other, then the fracture zones must lie along circles of latitude with respect to the pole of spreading, and the rates of spreading at any point on the ridge must be proportional to the perpendicular distance from the point to the axis of spreading [*see bottom illustration on page 686*].

All of this is well verified for the spreading that is going on today. The rates of spreading can be obtained from the magnetic patterns and the dates of the reversals. The poles of spreading can be found from the directions of the fracture zones and checked by the direction of earthquake motions. It turns out that the ridge axes and the magnetic pattern are usually almost at right angles to the fracture zones. This is not a geometrical

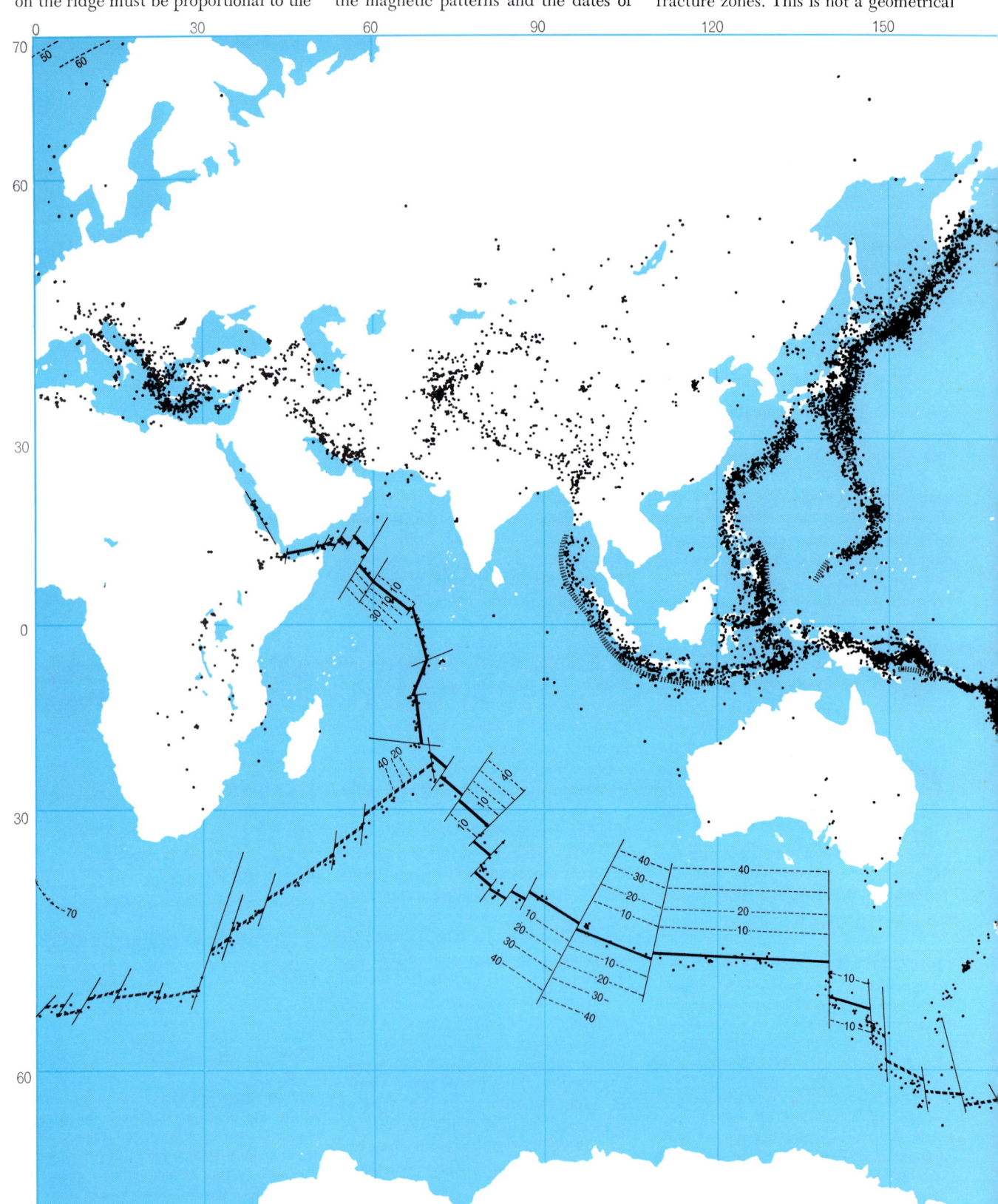

OCEANIC GEOLOGY has turned out to be much simpler than the geology of the continents. New ocean bottom is continuously being extruded along the crest of a worldwide system of ridges (*thick black lines*). The present position of material extruded at intervals of 10 million years, as determined by magnetic studies, appears as broken lines parallel to the ridge system, which is offset by fracture

necessity, but when it does happen it means that the lines of the ridge axes and of the magnetic pattern must, if they are extrapolated, go through the pole of spreading. If the ridge consists of a number of offset sections at right angles to the fracture zones, the axes of these sections will converge on the pole of spreading. It is one of the surprises of the work at sea that this rather simple geometry embraces so large a part of the facts. It seems that marine geology is truly simpler than continental geology and that this is not merely an illusion based on our lesser knowledge of the oceans.

The regularity of the magnetic pattern suggests that the ocean floor can move as a rigid plate over areas several thousand kilometers across. The thickness of the rigid moving plate is quite uncertain,

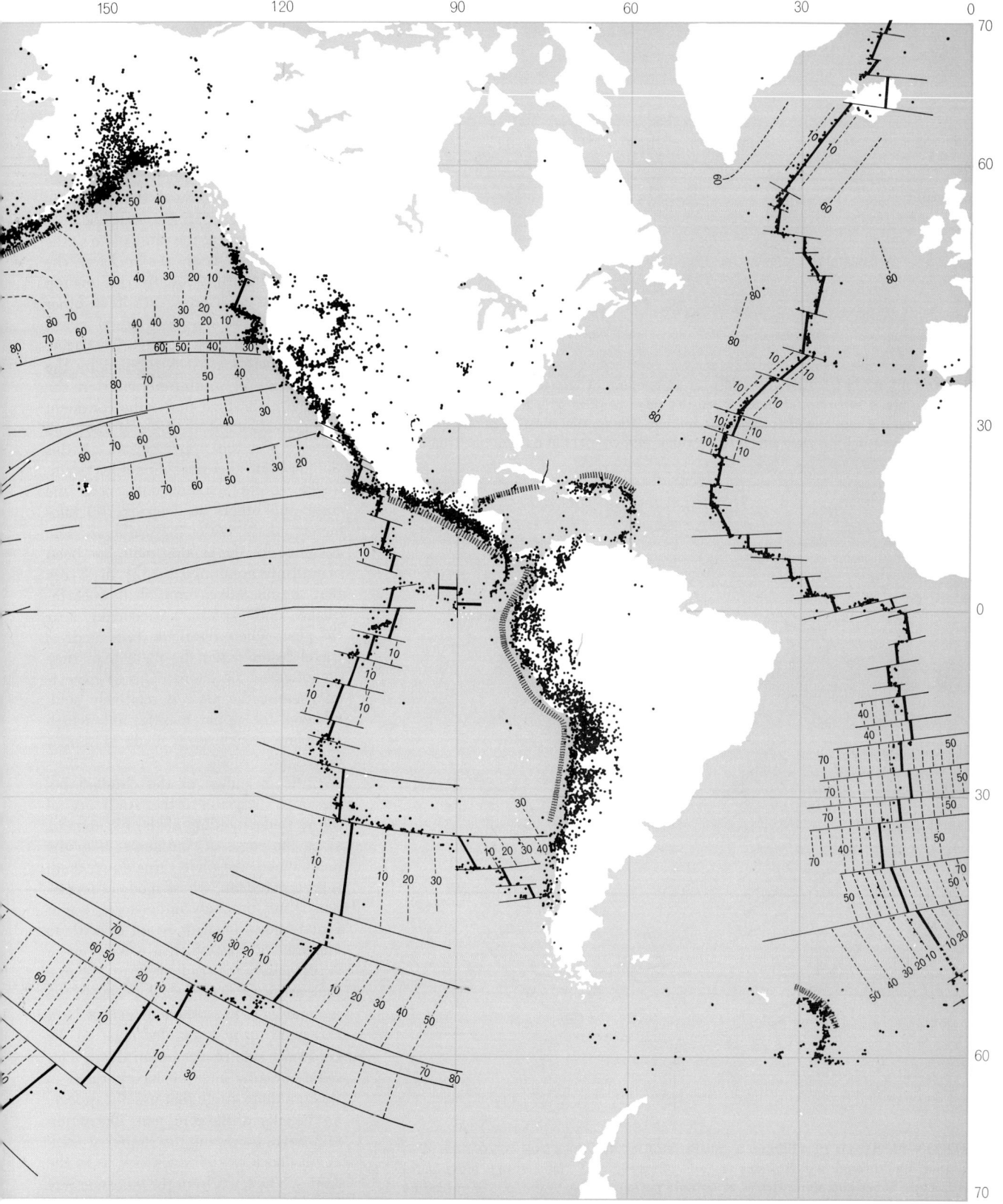

zones (*thin black lines*). Earthquakes (*black dots*) occur along the crests of ridges, on parts of the fracture zone and along deep trenches. These trenches, where the ocean floor dips steeply, are represented by hatched bands. At the maximum estimated rate of sea-floor spreading, about 16 centimeters a year, the entire floor of the Pacific Ocean could be created in perhaps 100 million years.

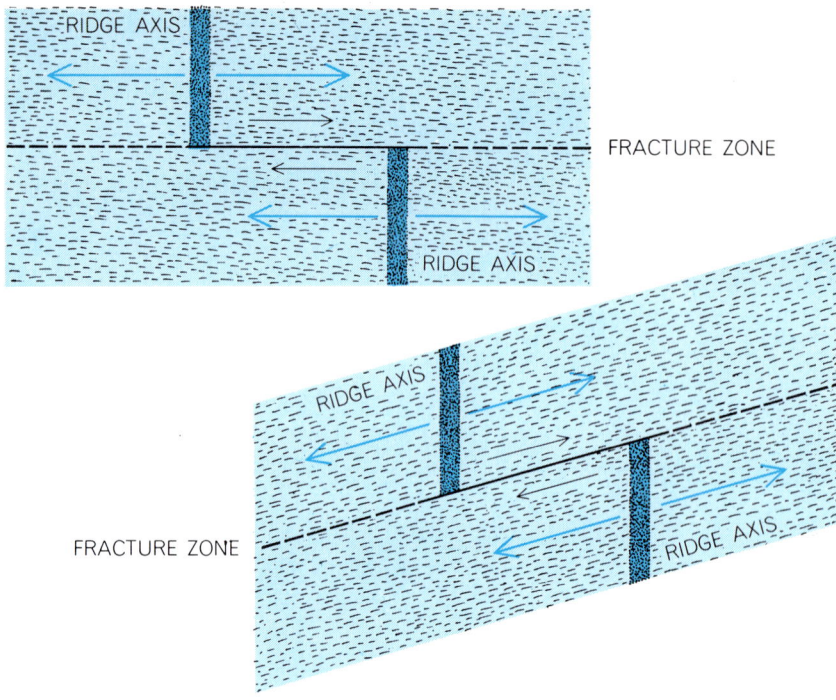

MOTION AT AXIS OF RIDGE consists of an opening of an axial crack (*vertical bands*) where two plates separate (*arrows*). Often the ridge is offset by a fracture zone, making a transform fault where one plate slips past another. The motion must be parallel to the fracture zone. It is usually at right angles to the ridge (*upper left*) but need not be (*lower right*).

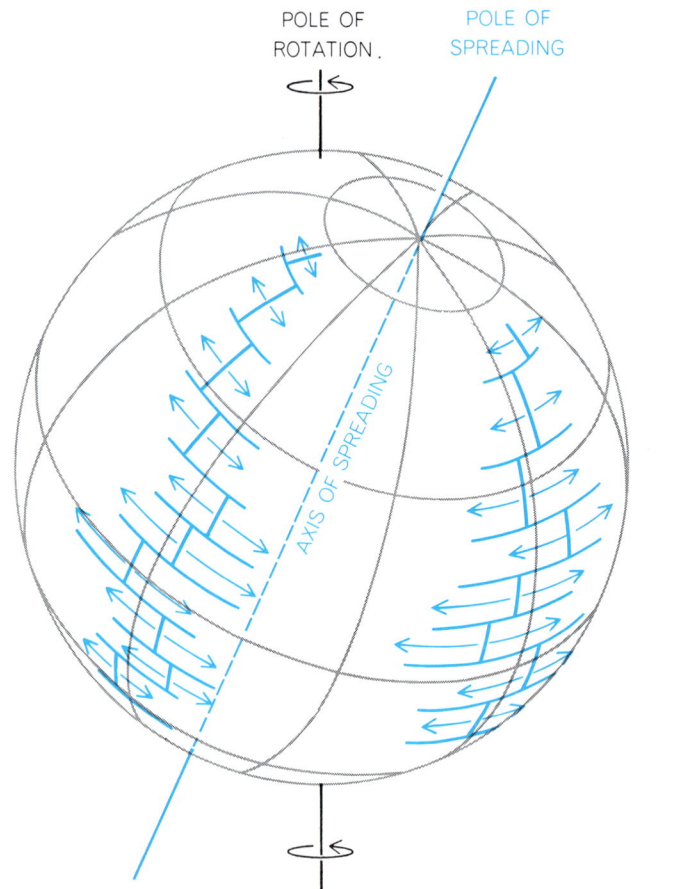

MOTION OF RIGID PLATES on a sphere requires that the plates rotate around a "pole of spreading" through which passes an "axis of spreading." Plates always move parallel to the fracture zones and along circles of latitude perpendicular to the axis of spreading. The rate of spreading is slowest near the pole of spreading and fastest 90 degrees away from it. The spreading pole can be quite remote from the sphere's pole of rotation.

but a value between 70 and 100 kilometers seems likely. If this is so, the greater part of the plate will be made of the same material as the upper part of the earth's mantle—probably of peridotite, a rock largely composed of olivine, a silicate of magnesium and iron, $(Mg, Fe)_2$-SiO_4. The basaltic rocks of the oceanic crust will form the upper five kilometers or so of the plate, with a veneer of sediments on top.

What happens at the boundary of an ocean and a continent? Sometimes, as in the South Atlantic, nothing happens; there are no earthquakes, no distortion, nothing to indicate relative motion between the sea floor and the bordering continent. The continent can then be regarded as part of the same plate as the adjacent ocean floor; the rocks of the continental crust evidently ride on top of the plate and move with it. In other places there is another kind of coast, what Eduard Suess called "a Pacific coast." Such a coast is typified by the Pacific coast of South America. Here the oceanic plate dives under the continent and goes down at an angle of about 45 degrees. On the upper surface of this sloping plate there are numerous earthquakes—quite shallow ones near the coast and others as deep as 700 kilometers inland, under the continent. The evidence for the sinking plate has been beautifully confirmed by the discovery that seismic waves from shallow earthquakes and explosions, occurring near the place where the plate starts its dive, travel faster down the plate than they do in other directions. This is expected because the plate is relatively cold, whereas the upper mantle, into which the plate is sinking, is made of similar material but is hot.

Little is known of the detailed behavior of the plate; further study is vital for an understanding of the phenomena along the edges of continents. Near the point where the plate turns down there is an ocean deep, whose mode of formation is not precisely understood, but if a plate goes down, it is not difficult to imagine ways in which it could leave a depression in the sea floor. It is probable that, as the plate goes down, some of the sediment on its surface is scraped off and piled up in a jumbled mass on the landward side of the ocean deep. This sediment may later be incorporated in the mountain range that usually appears on the edge of the continent. The mountain range bordering the continent commonly has a row of volcanoes, as in the Andes. The lavas from the volcanoes are frequently composed of andesites, which are different from the lavas of the mid-ocean ridges in that they contain more silica. It may reasonably be supposed that they are formed by the partial melting of the descending plate at a depth of about 150 kilometers. The first material to melt will contain more silica than the remaining material; it is also possible that the melted material is contaminated by granite as it rises to the surface through the continental rocks.

In many places the sinking plate goes down under a chain of islands and not under the continent itself. This happens in the Aleutians, to the south of Indonesia, off the islands of the Tonga group, in the Caribbean and in many other places. The volcanoes are then on the islands and the deep earthquakes occur under the almost enclosed sea behind the chain or arc of islands, as they do in the Sea of Japan, the Sea of Okhotsk and the Java Sea.

The destruction of oceanic crust explains one of the great paradoxes of geology. There have always been oceans, but the present oceans contain no sediment more than 150 million years old and very little sediment older than 80 million years. The explanation is that the older sediments have been carried away with the plates and are either piled up at the edge of a continent or are carried down with a sinking plate and lost in the mantle.

The picture is simple: the greater part of the earth's surface is divided into six plates [see *illustration on next two pages*]. These plates move as rigid bodies, new material for them being produced from the upper mantle by lava emerging from the crack along the crest of a mid-ocean ridge. Plates are destroyed at the oceanic trenches by plunging into the mantle, where ultimately they are mixed again with the material whence they came. The scheme is not yet established in all its details. Perhaps the greatest uncertainty is in the section of the ridge running south of South Africa; it is not clear how much of this is truly a ridge and a source of new crust and how much is a series of transform faults with only tangential motion. It is also uncertain whether the American and Eurasian plates meet in Alaska or in Siberia. It appears certain, however, that they do not meet along the Bering Strait.

A close look at the system of ridges, fracture zones, trenches and earthquakes reveals many other features of great interest, which can only be mentioned here. The Red Sea and the Gulf of Aden appear to be embryo oceans [see *illustration on page 681*]. Their floors are truly oceanic, with no continental rocks; along their axes one can find offset lengths of crack joined by fracture zones, and magnetic surveys show the worldwide magnetic pattern but only the most recent parts of it. These seas are being formed by the movement of Africa and Arabia away from each other. A detailed study of the geology, the topography and the present motion suggests that the separation started 20 million years ago in the Miocene period and that it is still continuing. If this is so, there must have been a sliding movement along the Jordan rift valley, with the area to the east having moved about 100 kilometers northward with respect to the western portion. There must also have been an opening of the East African rift valley by 65 kilometers or so.

The first of these displacements is well established by geological comparisons between the two sides of the valley, and it should be possible to verify the second. The reassembly of the pieces requires that the southwest corner of Arabia overlap the "Afar triangle" in northern Ethiopia [see *bottom illustration on page 683*]. This area should therefore be part of the embryo ocean. The fact that it is dry land presented a substantial puzzle, but recently it has been shown that the oceanic magnetic pattern extends over the area; it is the only land area in the world where this is known to happen. It seems likely that the Afar triangle is in some sense oceanic. The results of gravity surveys, seismic measurements and drilling will be awaited with interest. On this picture Arabia and the area to the north comprise a small plate separate from the African and Asian plates. The northern boundary of this small plate may be in the mountains of Iran and Turkey, where motion is proceeding today.

A number of other small plates are known. There is one between the Pacific coast of Canada and the ridge off Vancouver Island; it is probable that this is being crumpled at the coast rather than diving under the continent. Farther south the plate and the ridge from which it spread may have been overrun by the westward motion of North America. The ridge appears again in the Gulf of California, which is similar in many ways to the Red Sea and the Gulf of Aden. From the mouth of the Gulf of California the ridge runs southward and is joined by an east-west ridge running through the Galápagos Islands. The sea floor bounded by the two ridges and the trench off Central America seems to constitute a separate small plate.

For the past four million years we can date the lavas on land with enough

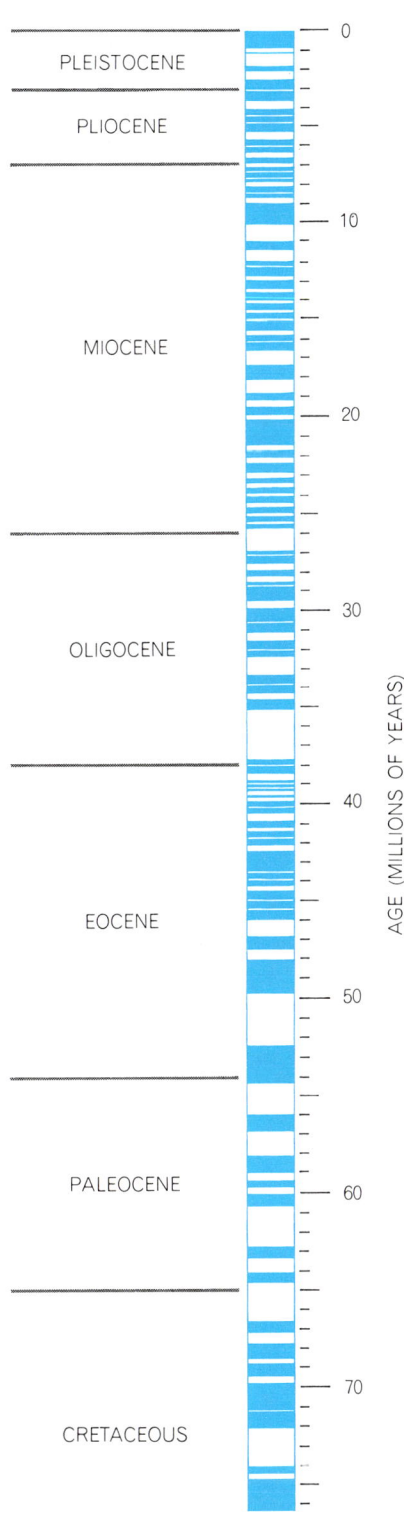

GEOLOGICAL PERIODS

PLEISTOCENE

PLIOCENE

MIOCENE

OLIGOCENE

EOCENE

PALEOCENE

CRETACEOUS

0

10

20

30

40

50

60

70

AGE (MILLIONS OF YEARS)

REVERSALS of the earth's magnetic field can be traced back more than 70 million years using magnetic patterns observed on the sea floor. The timetable of reversals for the most recent four million years was obtained by dating reversals in lava flows on land. Extrapolations beyond that assume that the sea floor spread at a constant rate. Colored bars show periods when the direction of the magnetic field was as it is now.

accuracy to give a timetable of magnetic reversals that can be correlated with the magnetic pattern on the sea bottom. For this period the rates of spreading from the ridges have remained constant. Further back we have a long series of reversals recorded in the ocean floor, but we cannot date them by comparison with lavas on land because the accuracy of the dates is insufficient to put the lavas in order. A rough guess can be made of the time since the oldest part of the magnetic pattern was formed by assuming that the rates have always been what they are today. This yields about 70 million years in the eastern Pacific and the South Atlantic. In fact the spacings of the older magnetic lineations are not in a constant proportion in the different oceans. The rates of spreading must therefore vary with time when long periods are considered. Directions of motion have also changed during this

period, as can be seen from the departure of the older parts of some of the Pacific fracture zones from circles of latitude around the present pole of spreading. A change of direction is also shown by the accurate geometrical and geochronological fit that can be made between South America and Africa [*see illustration, page 682*]. A rotation around the present pole of spreading will not bring the continents together; it is therefore likely that in the early stages of the separation motion was around a point farther to the south.

The ideas of the development of the earth's surface by plate formation, plate motion and plate destruction can be checked with some rigor by drilling. If they are correct, drilling at any point should show sediments of all ages from the present to the time at which this part of the plate was in the central valley of

the ridge. Under these sediments there should be lavas of about the same age as the lowest sediments. From preliminary reports of the drilling by the JOIDES project (a joint enterprise of five American universities) it seems that this expectation has been brilliantly verified and that the rate of spreading has been roughly constant for 70 million years in the South Atlantic. Such studies are of great importance because they will give firm dates for the entire magnetic pattern and provide a detailed chronology for all parts of the ocean floor.

The process of consumption of oceanic crust at the edge of a continent may proceed for tens of millions of years, but if the plate that is being consumed carries a continental fragment, then the consumption must stop when the fragment reaches the trench and collides with the continent beyond it. Because

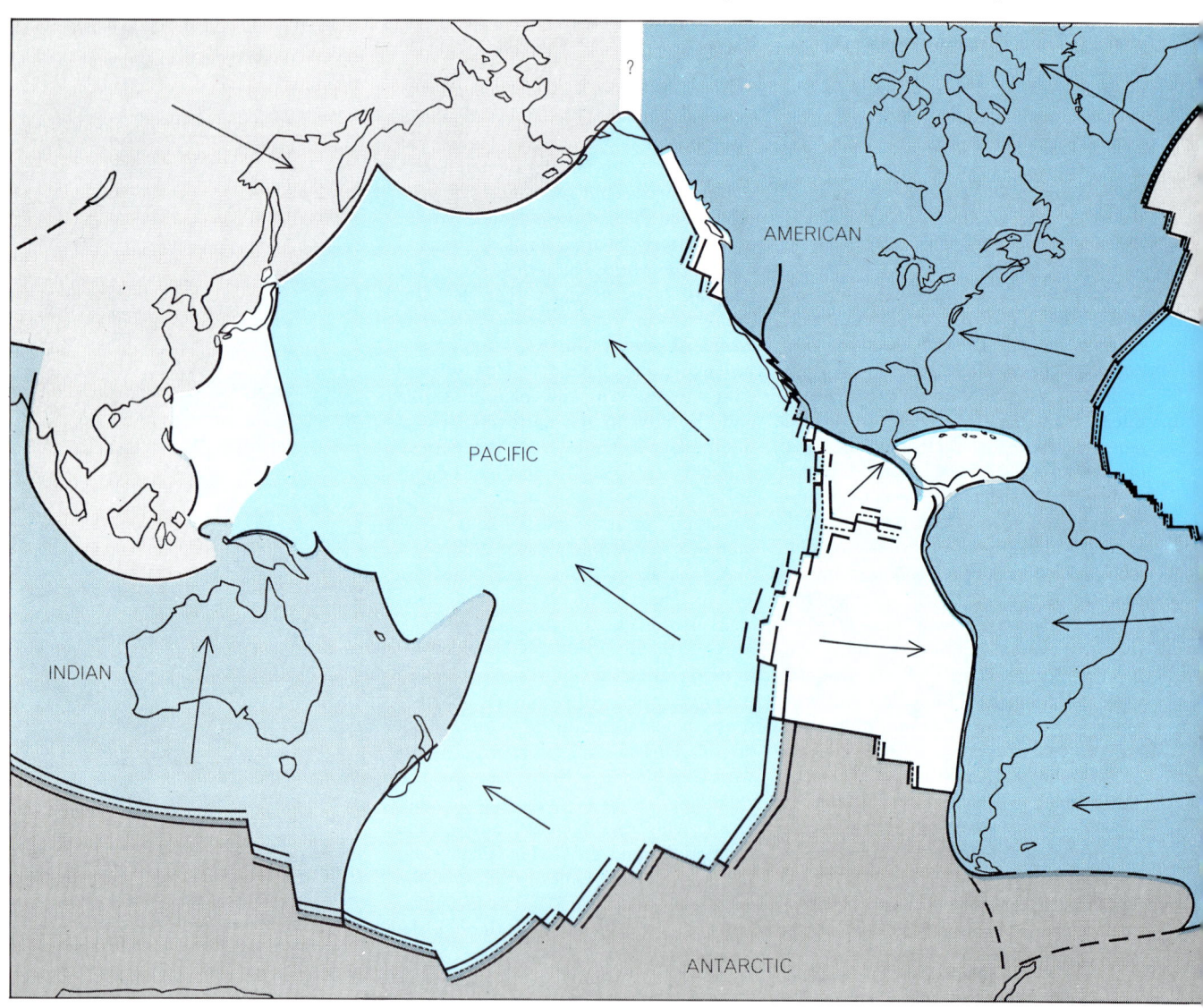

SIX MAJOR PLATES are sufficient to account for the pattern of continental drift inferred to be taking place today. In this model the African plate is assumed to be stationary. Arrows show the direction of motion of the five other large plates, which are generally bounded by ridges or trenches. Several smaller plates, unnamed, also appear. In certain areas, particularly at the junction of

the fragment consists of relatively light rocks it cannot be forced under a continent. The clearest example is the collision of India with what was once the southern margin of Asia. Paleomagnetic work shows that India has been moving northward for the past 100 million years. If it is attached to the plate that is spreading northward and eastward from the Carlsberg Ridge (which runs down the Indian Ocean halfway between Africa and India), then the motion is continuing today. This motion may be the cause of the earthquakes of the Himalayas, and it may also be connected with the formation of the mountains and of the deep sediment-filled trough to the south of them. The exact place where the joint occurs is far from clear and needs study by those with a detailed geological knowledge of northern India.

It seems unlikely that all the continents were collected in a single block for

the American and Eurasian plates and in the region south of Africa, it is hard to say just where the boundaries of the plates lie.

4,000 million years and then broke apart and started their wanderings during the past 100 million years. It is more likely that the processes we see today have always been in action and that all through geologic time there have been moving plates carrying continents. We must expect continents to have split many times and formed new oceans and sometimes to have collided and been welded together. We are only at the beginning of the study of pre-Tertiary events; anything that can be said is speculation and is to be taken only as an indication of where to look.

It is virtually certain that the Atlantic did not exist 150 million years ago. Long before that, in the Lower Paleozoic, 650 to 400 million years ago, there was an older ocean in which the sediments now in the Caledonian-Hercynian-Appalachian mountains of Europe and North America were laid down. Perhaps this ocean was closed long before the present Atlantic opened and separated the Appalachian Mountains of eastern North America from their continuation in northwestern Europe.

The Urals, if they are not unique among mountain ranges, are at least exceptional in being situated in the middle of a continent. There is some paleomagnetic evidence that Siberia is a mosaic of fragments that were not originally contiguous; perhaps the Urals were once near the borders of an ocean that divided Siberia from western Russia. Similarly, it is desirable to ask where the ocean was when the Rockies were being formed. A large part of California is moving rapidly northward, and the entire continent has overrun an ocean ridge; clearly the early Tertiary geography must have been very different from that of the present. Such questions are for the future and require that the ideas of moving plates be applied by those with a detailed knowledge of the various areas.

A history of the oceans does not necessarily require an account of the mechanism behind the observed phenomena. Indeed, no very satisfactory account can be given. The traditional view, put forward by Arthur Holmes and Felix A. Vening-Meinesz, supposes that the upper mantle behaves as a liquid when it is subjected to small forces for long periods and that differences in temperature under oceans and continents are sufficient to produce convection cells in the mantle—with rising currents under the mid-ocean ridges and sinking ones under the continents. These hypothetical cells would carry the plates along as on

a conveyor belt and would provide the forces needed to produce the split along the ridge. This view may be correct; it has the advantage that the currents are driven by temperature differences that themselves depend on the position of the continents. Such a back-coupling can produce complicated and varying motions.

On the other hand, the theory is implausible in that convection does not normally happen along lines. It certainly does not happen along lines broken by frequent offsets, as the ridge is. Also it is difficult to see how the theory applies to the plate between the Mid-Atlantic Ridge and the ridge in the Indian Ocean. This plate is growing on both sides, and since there is no intermediate trench the two ridges must be moving apart. It would be odd if the rising convection currents kept exact pace with them. An alternative theory is that the sinking part of the plate, which is denser than the hotter surrounding mantle, pulls the rest of the plate after it. Again it is difficult to see how this applies to the ridge in the South Atlantic, where neither the African nor the American plate has a sinking part.

Another possibility is that the sinking plate cools the neighboring mantle and produces convection currents that move the plates. This last theory is attractive because it gives some hope of explaining the almost enclosed seas, such as the Sea of Japan. These seas have a typical oceanic floor except that the floor is overlain by several kilometers of sediment. Their floors have probably been sinking for long periods. It seems possible that a sinking current of cooled mantle material on the upper side of the plate might be the cause of such deep basins. The enclosed seas are an important feature of the earth's surface and urgently require explanation; in addition to the seas that are developing at present behind island arcs there are a number of older ones of possibly similar origin, such as the Gulf of Mexico, the Black Sea and perhaps the North Sea.

The ideas set out in this attempt at a history of the ocean have developed in the past 10 years. What we have is a sketch of the outlines of a history; a mass of detail needs to be filled in and many major features are quite uncertain. Nonetheless, there is a stage in the development of a theory when it is most attractive to study and easiest to explain, that is while it is still simple and successful and before too many details and difficulties have been uncovered. This is the interesting stage at which plate theory now stands.

The Author

SIR EDWARD BULLARD is professor of geophysics at the University of Cambridge, which he entered as an undergraduate and where he has spent most of his career. He began his professional work there in 1931 as a demonstrator in geodesy. He was an experimental officer with the British Navy during World War II and then returned to Cambridge as a reader in experimental geophysics. In 1948–1949 he was professor of physics at the University of Toronto, and from 1950 to 1955 he was director of the National Physical Laboratory in the United Kingdom. Returning to Cambridge in 1956, he became successively assistant director of research for the university, reader in geophysics and professor of geophysics. He was elected a Fellow of the Royal Society in 1941 and was knighted in 1953.

Bibliography

THE HISTORY OF THE EARTH'S CRUST. Edited by Robert A. Phinney. Princeton University Press, 1968.

SEA-FLOOR SPREADING AND CONTINENTAL DRIFT. Xavier Le Pichon in *Journal of Geophysical Research*, Vol. 73, No. 12, pages 3661–3697; June 15, 1968.

SEISMOLOGY AND THE NEW GLOBAL TECTONICS. Bryan Isacks, Jack Oliver and Lynn R. Sykes in *Journal of Geophysical Research*, Vol. 73, No. 18, pages 5855–5899; September 15, 1968.

REVERSALS OF THE EARTH'S MAGNETIC FIELD. Sir Edward Bullard in *Philosophical Transactions of the Royal Society of London: Series A, Mathematical and Physical Sciences*, Vol. 263, No. 1143, pages 481–524; December 12, 1968.

SCIENTIFIC
AMERICAN September 1969, Vol. 221, No. 3, pp. 76–86   OFFPRINT **881**

THE ATMOSPHERE AND THE OCEAN

by R. W. Stewart

The two are inextricably linked. The ocean's circulation is driven
by wind and by density differences that largely depend on the air.
The atmospheric heat engine, in turn, is largely driven by the sea.

The atmosphere drives the great ocean circulations and strongly affects the properties of seawater; to a large extent the atmosphere in turn owes its nature to and derives its energy from the ocean. Indeed, there are few phenomena of physical oceanography that are not somehow dominated by the atmosphere, and there are few atmospheric phenomena for which the ocean is unimportant. It is therefore hard to know where to start a discussion of the interactions of the atmosphere and the ocean, since in a way everything depends on everything else. One must break into this circle somewhere, and arbitrarily I shall begin by considering some of the effects of wind on ocean water.

When wind blows over water, it exerts a force on the surface in the direction of the wind. The mechanism by which it does so is rather complex and is far from being completely understood, but that it does it is beyond dispute. The ocean's response to this force is immensely complicated by a number of factors. The fact that the earth is rotating is of overriding importance. The presence of continental barriers across the natural directions of flow of the ocean complicates matters further. Finally there is the fact that water is a fluid, not a solid.

To simplify the picture somewhat, let us start by looking at what would happen to a slab of material resting on the surface of the earth. Let us further assume that the slab can move without friction. Consider the result of a sharp, brief impulse that sets the slab moving, say, due north [see top illustration, page 694]. Looked at by an observer on a rotating earth, any moving object is subject to a "Coriolis acceleration" directed exactly at a right angle to its motion. The magnitude of the acceleration increases with both the speed of the object's motion and the vertical component of the earth's rotation, and in the Northern Hemisphere it is directed to the right of the motion. An acceleration at right angles to the velocity is just what is required to cause motion in a circle, and in the illustration the center of the circle is due east of the original position of the slab. A circular motion of this kind is called an inertial oscillation, and something of this nature may sometimes happen in the ocean, since inertial oscillations are frequently found when careful observations are made with current meters.

An inertial oscillation requires exactly half a pendulum day for a full circle. (A pendulum day is the time required for a complete revolution of a Foucault pendulum. Like the Coriolis effect, it depends on the vertical component of the earth's rotation and therefore varies with latitude, being just under 24 hours at the poles and increasing to several days close to the Equator. To be precise, it is one sidereal—or star time—day divided by the sine of the latitude.) If there were a small amount of friction, the slab would gradually spiral to the center of the circle. Pushing it toward the north thus causes it to end up displaced to the east [see bottom illustration on page 694]. More generally, in the Northern Hemisphere a particle is moved to the right of the direction in which it is impelled, and in the Southern Hemisphere it is moved to the left.

Let us turn to what happens to our frictionless slab if, instead of giving it a short impulse, we give it a steady thrust. Again assume that the force is toward the north [see upper map on page 695]. Under the influence of this force the slab accelerates toward the north, but as soon as it starts to move it comes under the influence of the Coriolis effect and its motion is deflected (in the Northern Hemisphere) to the right—to the east. As long as the slab has at least some component of velocity toward the north the force will continue to add energy to it and its speed will continue to increase. After a quarter of a pendulum day, however, it will be moving due east. In this position the applied force (which is to the north) is pushing at a right angle to the velocity (east), opposing the influence of the Coriolis effect, which is now trying to turn the slab toward the south.

If there has been no loss of energy because of friction, the slab is moving fast enough so that the Coriolis effect dominates, and it turns toward the south. Now there is a component of velocity opposing the applied force, which acts as a brake and takes energy from the motion. At the end of half a pendulum day the process has gone far enough to bring the slab to a full stop, at which point it is directly east of its starting point. If the force continues, it will again accelerate toward the north and the entire process is repeated, so that the slab performs a series of these looping (cycloidal) motions, each loop taking half a pendulum day to execute. Overall, then, a steady force on a frictionless body resting on a rotating earth causes it to move at right angles to the direction of the force. What is happening is that the force is balanced—on the average—by the Coriolis effect.

Now let us look at the situation when there is a certain amount of friction between the slab and the underlying surface [see lower map on page 695]. Any frictional drag reduces the speed attained by the slab, reducing the Coriolis effect until it is no longer entirely able to overcome the driving force. As a result if the force is toward the north, the slab will move in a more or less north-

Wind force: 4   Wind speed: 5½   Wave period: 5   Wave height: 1

Wind force: 5   Wind speed: 11½   Wave period: 6   Wave height: 2

Wind force: 6   Wind speed: 13   Wave period: 7   Wave height: 3

Wind force: 8   Wind speed: 18   Wave period: 6   Wave height: 5

Wind force: 9   Wind speed: 21   Wave period: 9   Wave height: 8

Wind force: 10   Wind speed: 27   Wave period: 9   Wave height: 7

easterly direction—more northerly if the friction is large, more easterly if it is small.

A body of water acts much like a set of such slabs, one on top of the other [see illustration on page 696]. Each slab is able to move largely independently of the others except for the frictional forces among them. If the top slab is pushed by the wind, it will, in the Northern Hemisphere, move in a direction somewhat to the right of the wind. It will exert a frictional force on the second slab down, which will then be set in motion in a direction still farther to the right. At each successive stage the force is somewhat reduced, so that not only does the direction change but also the speed is a bit less. A succession of such effects produces velocities for which the direction spirals as the depth increases. It is known as the Ekman spiral, after the pioneering Swedish oceanographer V. Walfrid Ekman, who first discussed it soon after the beginning of the century. At a certain depth both the current and the frictional forces associated with it become negligibly small. The entire layer above that depth, in which friction is important, is termed the Ekman layer. Since there is negligible friction between the Ekman layer and the water lying under it, the Ekman layer as a whole behaves like the frictionless slab discussed above: its average velocity must be at a right angle to the wind.

The frictional mechanism, which involves turbulence, has proved to be extraordinarily difficult to study either theoretically or through observations, and surprisingly little is known about it. The surface flow does appear to be somewhat to the right of the wind. Primitive theoretical calculations predict that its direction should be 45 degrees from the

EFFECT OF WIND on the surface of the sea is shown in a series of photographs made by the Meteorological Service of Canada. Much of the wind's momentum goes into generating waves rather than directly into making currents. The change in the surface as the wind increases is primarily a change in scale, except for the effect of surface tension: the waves break up more, making more whitecaps. For each photograph the wind force is given according to the Beaufort scale; the wind speed is given in meters per second, the wave period in seconds and the wave height in meters. (In the final photograph the waves are only about half as large as they might become if the force-10 wind, which had blown for less than nine hours, were to continue to blow.)

wind, but this theory is certainly inapplicable in detail. More complicated theoretical models have been attempted, but since almost nothing is known of the nature of turbulence in the presence of a free surface these models rest on weak ground. An educated guess, supported by rather flimsy observational evidence, suggests that the angle is much smaller, perhaps nearer to 10 degrees. All that seems fairly certain is that the average flow in the Ekman layer must be at a right angle to the wind and that there must be some kind of spiral in the current directions. We also believe the bottom of the Ekman layer lies 100 meters or so deep, within a factor of two or three. Of the details of the spiral, and of the turbulent mechanisms that determine its nature, we know very little indeed.

This Ekman-layer flow has some important fairly direct effects in several parts of the world. For example, along the coasts of California and Peru the presence of coastal mountains tends to deflect the low-level winds so that they blow parallel to the coast. Typically, in each case, they blow toward the Equator, and so the average Ekman flow—to the right off California and to the left off Peru—is offshore. As the surface water is swept away deeper water wells up to replace it. The upwelling water is significantly colder than the sun-warmed surface waters, somewhat to the discomfort of swimmers (and, since it is also well fertilized compared with the surface water, to the advantage of fishermen and birds).

The total amount of flow in the directly driven Ekman layer rarely exceeds a couple of tons per second across each meter of surface. That represents a substantial flow of water, but it is much less than the flow in major ocean currents. These are driven in a different way—also by the wind, but indirectly. To see how this works let us take a look at the North Atlantic [lower illustration on page 697]. The winds over this ocean, although they vary a good deal from time to time, have a most persistent characteristic: near 45 degrees north latitude or thereabouts the westerlies blow strongly from west to east, and at about 15 degrees the northeast trades blow, with a marked east-to-west component. The induced Ekman flow is to the right in each case, so that in both cases the water is pushed toward the region known as the Sargasso Sea, with its center at 30 degrees north. This "gathering together of the waters" leads not so much to a piling up (the surface level is only about a meter higher at

the center than at the edges) as a pushing down.

(If it were not for the continental boundaries, the piling up would be much more important. Because water tends to seek a level, the piled-up water would push north above 30 degrees and south below; the pushing force, like any other force in the Northern Hemisphere, would cause a flow to its right, so that in the northern part of the ocean a strong eastward flow would develop and in the southern part a strong westward one. On the earth as it now exists, however, these east-west flows are blocked by the continents; only in the Southern Ocean, around the Antarctic Continent, is such a flow somewhat free. In the absence of the continents the oceans, like the atmosphere, would be dominated by east-west motion. As it is, only a residue of such motion is possible, and it is the pushing down rather than the piling up of water that is important.)

The downward thrust of the surface waters presses down on the layers of water underneath [see illustration on page 698]. For practical purposes water is incompressible, so that pushing it down from the top forces it out at the sides. It must be remembered that this body of underlying water is rotating with the earth. As it is squeezed out laterally its radius of gyration, and therefore its moment of inertia, increase, and so its rate of rotation must slow. If it slows, however, the rotation no longer "fits" the rotation of the underlying earth. There are two possible consequences: either the water can rotate with respect to the earth or it can move to a different latitude where its newly acquired rotation *will* fit. It usually does the latter. Hence a body of water whose rotation has been slowed by being squashed vertically will usually move toward the Equator, where the vertical component of the earth's rotation is smaller; on the other hand, a body whose rotation has been speeded by being bulged up to replace water that has been swept away from the surface will usually move toward the poles.

In the band of water a couple of thousand miles wide along latitude 30 degrees this indirectly wind-driven flow moves water toward the Equator. Of course the regions of the ocean closer to the poles do not become empty of water; somewhere there must be a return flow. The returning water must also attain a rotation that fits the rotation of the underlying earth. If it flows north, it must gain counterclockwise rotation (or lose clockwise rotation). It does this by running in a strong current on the westward side of the ocean, changing its rotation

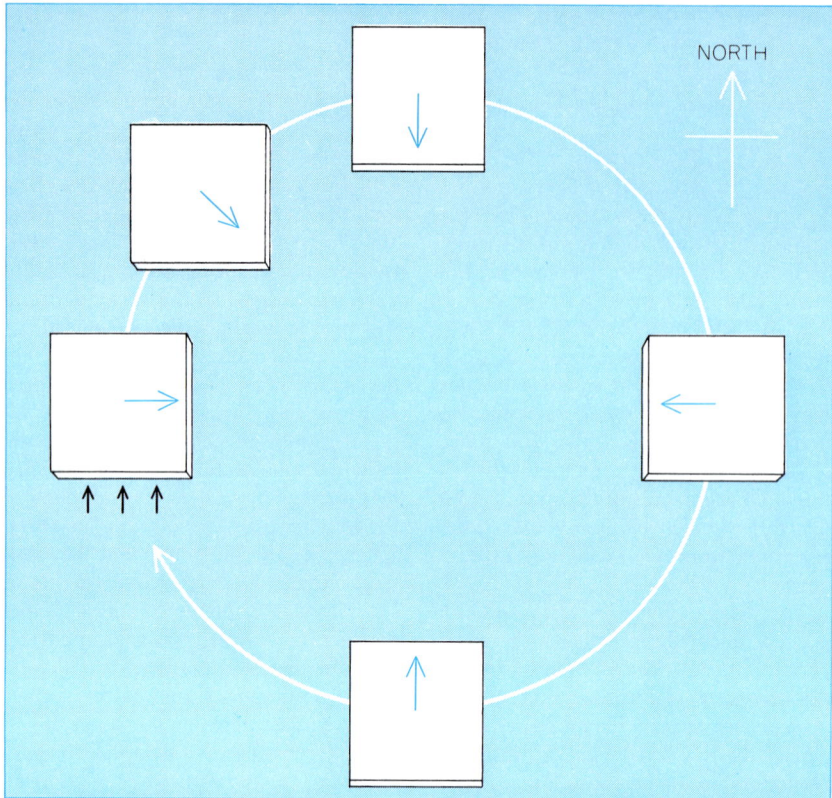

CORIOLIS ACCELERATION, caused by the earth's rotation, affects any object moving on the earth. It is directed at a right angle to the direction of motion (to the right in the Northern Hemisphere). If a frictionless slab is set in motion toward the north by a single impulse (*black arrows*), the Coriolis effect (*colored arrows*) moves the slab in a circle.

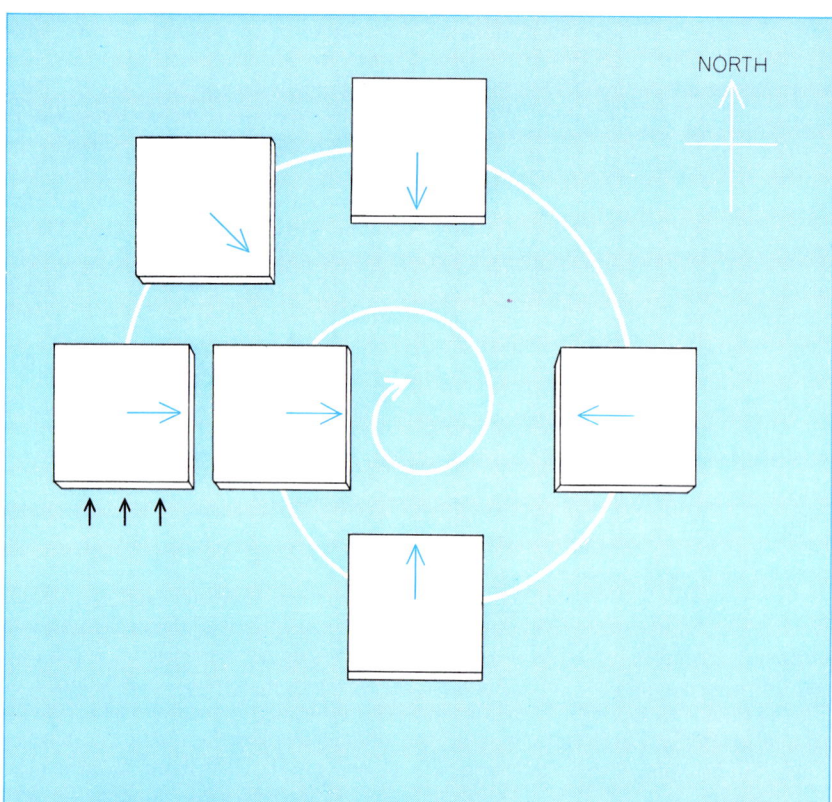

PRESENCE OF FRICTION causes the slab to slow down, spiraling in toward the center of the circle in the top illustration. A push to the north causes a spiral to the east.

by "rubbing its left shoulder" against the shore. The Gulf Stream is such a current; it is the return flow of water that was squeezed south by the wind-driven convergence of surface waters throughout the entire central North Atlantic. Most great ocean currents seem to be indirectly driven in this way.

It is worth noting that these return currents must be on the western side of the oceans (that is, off the eastern coasts of the land) in both hemispheres and regardless of whether the flow is northward or southward. The reason is that the earth's angular velocity of rotation is maximum counterclockwise to an observer looking down at the North Pole and maximum clockwise at the South Pole. Any south-flowing return current in either hemisphere must gain clockwise rotation (or lose counterclockwise rotation) if it is to fit when it arrives. It gains this rotation by friction on its right side, and so it must keep to the right—that is, to the west—of the ocean. On the other hand, a north-flowing return current must keep to the left—again the west!

This description of the general wind-driven circulation accords reasonably well with observations of the long-term characteristics of the ocean circulation. What happens on a shorter term, in response to changes of the atmospheric circulation and the wind-force pattern that results? The characteristic time constant of the Ekman layer is half a pendulum day, and there is every reason to believe this layer adjusts itself within a day or so to changes in the wind field. The indirectly driven flow is much harder to deal with. Its time constant is of the order of years, and we have no clear understanding of how it adjusts; the indirectly driven circulation may still be responding, in ways that are not clear, for years after an atmospheric change.

So far the discussion has been qualitative. To make it quantitative we need to know two things: the nature of the wind over the ocean at each time and place and the amount of force the wind exerts on the surface. Meteorologists are getting better at the first question, although there are some important gaps in our detailed information, notably in the Southern Ocean and in the South Pacific.

Investigation of the second problem, that of the quantitative relation between the wind flow and the force on the surface, is becoming a scientific discipline in its own right. Turbulent flow over a boundary is a complex phenomenon for which there is no really complete theory even in simple laboratory cases. Nevertheless, a great deal of experimental data has been collected on flows over solid surfaces, both in the laboratory and in nature, so that from an engineering point of view at least the situation is fairly well understood. The force exerted on a surface varies with the roughness of that surface and approximately with the square of the wind speed at some fixed height above it. A wind of 10 meters per second (about 20 knots, or 22 miles per hour) measured at a height of 10 meters will produce a force of some 30 tons per square kilometer on a field of mown grass or of about 70 tons per square kilometer on a ripe wheat field. On a really smooth surface such as glass the force is only about 10 tons per square kilometer.

When the wind blows over water, the whole thing is much more complicated. The roughness of the water is not a given characteristic of the surface but depends on the wind itself. Not only that, the elements that constitute the roughness—the waves—themselves move more or less in the direction of the wind. Recent evidence indicates that a large portion of the momentum transferred from the air into the water goes into waves rather than directly into making currents in the water; only as the waves break, or otherwise lose energy, does their momentum

become available to generate currents or produce Ekman layers. Waves carry a substantial amount of both energy and momentum (typically about as much as is carried by the wind in a layer about one wavelength thick), and so the wave-generation process is far from negligible. So far we have no theory that accounts in detail for what we observe.

A violently wavy surface belies its appearance by acting, as far as the wind is concerned, as though it were very smooth. At 10 meters per second, recent measurements seem to agree, the force on the surface is quite a lot less than the force over mown grass and scarcely more than it is over glass; some observations in light winds of two or three meters per second indicate that the force on the wavy surface is less than it is on a surface as smooth as glass. In some way the motion of the waves seems to modify the airflow so that air slips over the surface even more freely than it would without the waves. This seems not to be the case at higher wind speeds, above about five meters per second, but the force remains strikingly low compared with that over other natural surfaces.

One serious deficiency is the fact that there are no direct observations at all in those important cases in which the wind speed is greater than about 12 meters per second and has had time and

fetch (the distance over water) enough to raise substantial waves. (A wind of even 20 meters per second can raise waves eight or 10 meters high—as high as a three-story building. Making observations under such circumstances with the delicate instruments required is such a formidable task that it is little wonder none have been reported.) Some indirect studies have been made by measuring how water piles up against the shore when driven by the wind, but there are many difficulties and uncertainties in the interpretation of such measurements. Such as they are, they indicate that the apparent roughness of the surface increases somewhat under high-wind conditions, so that the force on the surface increases rather more rapidly than as the square of the wind speed.

Assuming that the force increases at least as the square of the wind speed, it is evident that high-wind conditions produce effects far more important than their frequency of occurrence would suggest. Five hours of 60-knot storm winds will put more momentum into the water than a week of 10-knot breezes. If it should be shown that for high winds the force on the surface increases appreciably more rapidly than as the square of the wind speed, then the transfer of momentum to the ocean will turn out to be dominated by what happens during

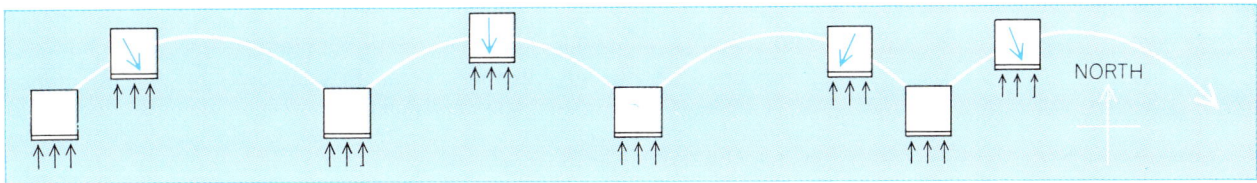

STEADY PUSH (*black arrows*), rather than a single impulse, is balanced, in the absence of friction, by the Coriolis effect (*colored arrows*), causing a series of loops. A steady force on a frictionless slab makes it move at a right angle to the direction of the force.

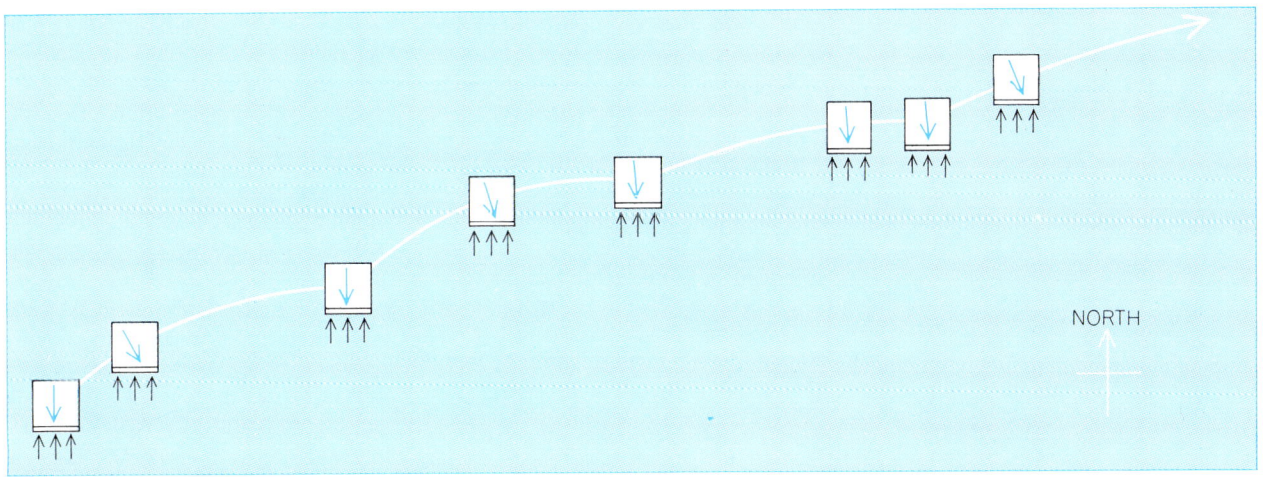

FRICTIONAL DRAG reduces the speed of the slab and thus of the Coriolis effect, which can no longer balance the driving force, and the amplitude of the loops is damped out gradually. A force toward the north therefore moves the slab toward the northeast.

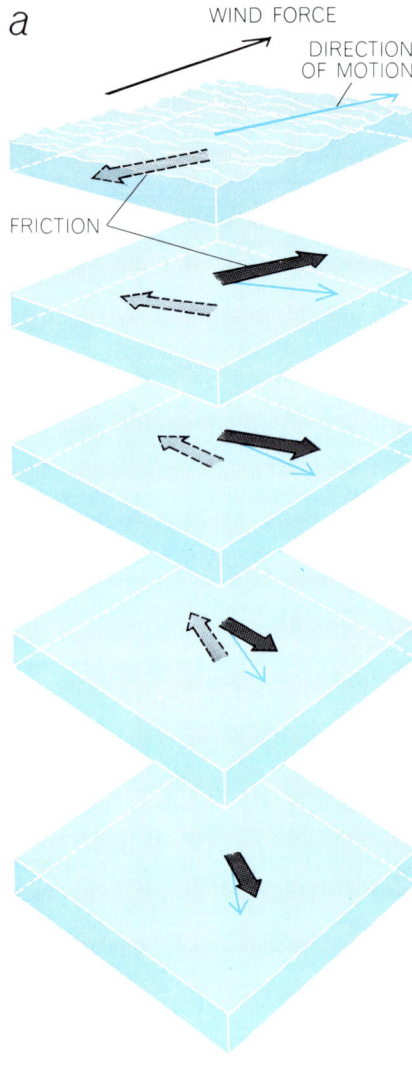

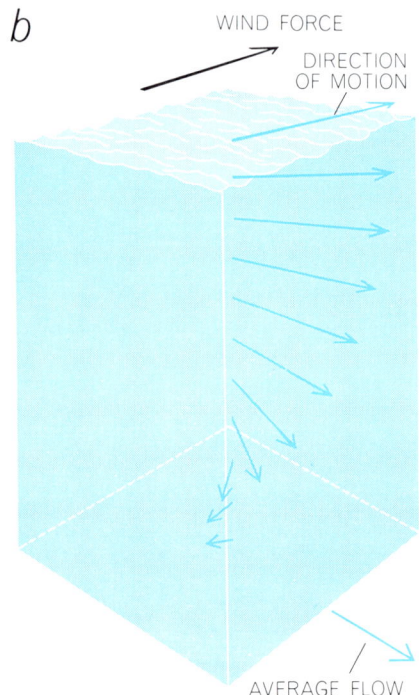

BODY OF WATER can be thought of as a set of slabs (*a*), the top one driven by the wind and each driving the one below it by friction. At each stage the speed of flow is reduced and (in the Northern Hemisphere) directed more to the right. This "Ekman spiral" persists until friction becomes negligible. The "Ekman layer" in which this takes place (*b*) behaves like the frictionless slabs in the preceding illustrations. Its average flow is at right angle to wind driving it.

the occasional storm rather than by the long-term average winds.

It is tempting to try to infer high-wind behavior from what we know about lower wind speeds. Certainly the shapes of wavy surfaces appear nearly the same notwithstanding the size of the waves—as long as one disregards waves less than about five centimeters long, which are strongly affected by surface tension. Yet, curious as it may seem, the only thing that makes one wind-driven wave field different in any fundamental way from another is surface tension, even though it directly affects only these very short waves. Indeed, surface tension is the basis of the entire Beaufort wind scale, which depends on the number and nature of whitecaps; only the fact that the surface tension is better able to hold the surface together at low wind speeds than at high speeds enables us to see a qualitative difference in the nature of the sea surface at different wind speeds [*see illustration on page 692*]. Otherwise the waves would look just the same except for a difference in scale. If we were sure we could ignore surface-tension effects, then we could calculate the force the wind would exert at high wind speeds on the basis of data obtained at lower speeds, but one should be extremely cautious about such calculations, at least until some confirming measurements are available.

Whereas the ocean seems primarily to be driven by surface forces, the atmosphere is a heat engine that makes use of heat received from the sun to develop the mechanical energy of its motion. Any heat engine functions by accepting thermal energy at a comparatively high temperature, discharging some of this thermal energy at a lower temperature and transforming the rest into mechanical energy. The atmosphere does this by absorbing energy at or near its base and radiating it away from much cooler high levels. A substantial proportion of the required heating from below comes from the ocean.

This energy comes in two forms. If cooler air blows over warmer water, there is a direct heat flow into the air. What is usually more important, though, is the evaporation of water from the surface into the air. Evaporation causes cooling, that is, it removes heat, in this case from the surface of the water. When the moisture-laden air is carried to a high altitude, where expansion under reduced atmospheric pressure causes it to cool, the water vapor may recondense into water droplets and the heat that was given up by the surface of the water is transferred to the air. If the cloud that is formed evaporates again, as it sometimes does, the atmosphere gains no net thermal energy. If the water falls to the surface as rain or snow, however, then there has been a net gain and it is available to drive the atmosphere. Typically the heat gained by the atmosphere through this evaporation-condensation process is considerably more than the heat gained by direct thermal transfer through the surface.

Virtually everywhere on the surface of the ocean, averaged over a year, the ocean is a net source of heat to the atmosphere. In some areas the effect is much more marked than in others. For example, some of the most important return currents, such as the Gulf Stream in the western Atlantic and the Kuroshio Current in the western Pacific off Japan, contain very warm water and move so rapidly that the water has not cooled even when it arrives far north of the tropical and subtropical regions where it gained its high temperature. At these northern latitudes the characteristic wind direction is from the west, off the continent. In winter, when the continents are cold, air blowing from them onto this abnormally warm water receives great quantities of heat, both by direct thermal transfer and in the form of water vapor.

The transfer of heat and water vapor depends on a disequilibrium at the interface of the water and the air. Within a millimeter or so of the water the air temperature is not much different from that of the surface water, and the air is nearly saturated with water vapor. The small differences are nevertheless crucial, and the lack of equilibrium is maintained by the mixing of air near the surface with air at higher levels, which is typically appreciably cooler and lower in water-vapor content. The mixing mechanism is a turbulent one, the turbulence gaining its energy from the wind. The higher the wind speed is, the more vigorous the turbulence is and therefore the higher the rates of heat and moisture transfer are. These rates tend to increase linearly with the wind speed, but even less is known about the details of this phenomenon than about the wind force on water. One source of complication is the fact that, as I mentioned above, the wind-to-water transfer of momentum is effected partly by wave-generation mechanisms. When the wind makes waves, it must transfer not only momentum but also important amounts of energy—energy that is not available to provide the turbulence needed to produce the mixing that would effect the transfer of heat and water vapor.

At fairly high wind speeds another phenomenon arises that may be of considerable importance. I mentioned that when surface tension is no longer able to hold the water surface together at high wind speeds, spray droplets blow off the top of the waves. Some of these drops fall back to the surface, but others evaporate and in doing so supply water vapor to the air. They have another important role: The tiny residues of salt that are left over when the droplets of seawater evaporate are small enough and light enough to be carried upward by the turbulent air. They act as nuclei on which condensation may take place, and so they play a role in returning to the atmosphere the heat that is lost in the evaporation process.

The ocean's great effect on climate is illustrated by a comparison of the temperature ranges in three Canadian cities, all at about the same latitude but with very different climates [see top illustration at right]. Victoria is a port on the southern tip of Vancouver Island, on the eastern shore of the Pacific Ocean; Winnipeg is in the middle of the North American land mass; St. John's is on the island of Newfoundland, jutting into the western Atlantic. The most striking climatic difference among the three is the enormous temperature range at Winnipeg compared with the two coastal cities. The range at St. John's, although much less, is still greater than at Victoria, probably because at St. John's the air usually blows from the direction of the continent and the effect of the water is somewhat less dominant than at Victoria, which typically receives its air directly from the ocean. St. John's is colder than Victoria because it is surrounded by cold water of the Labrador Current.

The influence of the ocean is associated with its enormous thermal capacity. Every day, on the average, the earth absorbs from the sun and reradiates into space enough heat to raise the temperature of the entire atmosphere nearly two degrees Celsius (three degrees Fahrenheit). Yet the thermal ca-

	VICTORIA	WINNIPEG	ST. JOHN'S
MEAN JULY MAXIMUM	68	80.1	68.9
MEAN JANUARY MINIMUM	35.6	−8.1	18.5

MODERATING EFFECT of the ocean on climate is illustrated by a comparison of the temperature range (in degrees Fahrenheit) at three Canadian cities. The range between minimums and maximums is much greater at Winnipeg than at coastal Victoria or St. John's.

pacity of the atmosphere is equivalent to that of only the top three meters of the ocean, or only a few percent of the 100 meters or so of ocean water that is heated in summer and cooled in winter. (The great bulk of ocean water, more than 95 percent of it, is so deep that surface heating does not penetrate, and its temperature is independent of season.) If the ocean lost its entire heat supply for a day but continued to give up heat in a normal way, the temperature of the upper 100 meters would drop by only about a tenth of a degree.

Compared with the land, the ocean heats slowly in summer and cools slowly in winter, so that its temperature is much less variable. Moreover, because air has so much less thermal capacity, when it blows over water it tends to come to the water temperature rather than vice versa. For these reasons maritime climates are much more equable than continental ones.

Although the ocean affects the atmo-

sphere's temperature more than the atmosphere affects the ocean's, the ocean is cooled when it gives up heat to the atmosphere. The density of ocean water is controlled by two factors, temperature and salinity, and evaporative cooling tends to make the water denser by affecting both factors: it lowers the temperature and, since evaporation removes water but comparatively little salt, it also increases the salinity. If surface water becomes denser than the water underlying it, vigorous vertical convective mixing sets in. In a few places in the ocean the cooling at the surface can be so intense that the water will sink and mix to great depths, sometimes right to the bottom. Such occurrences are rare both in space and in time, but once cold water has reached great depths it is heated from above very slowly, and so it tends to stay deep for a long time with little change in temperature; there is some evidence of water that has remained cold and deep in the ocean for more than

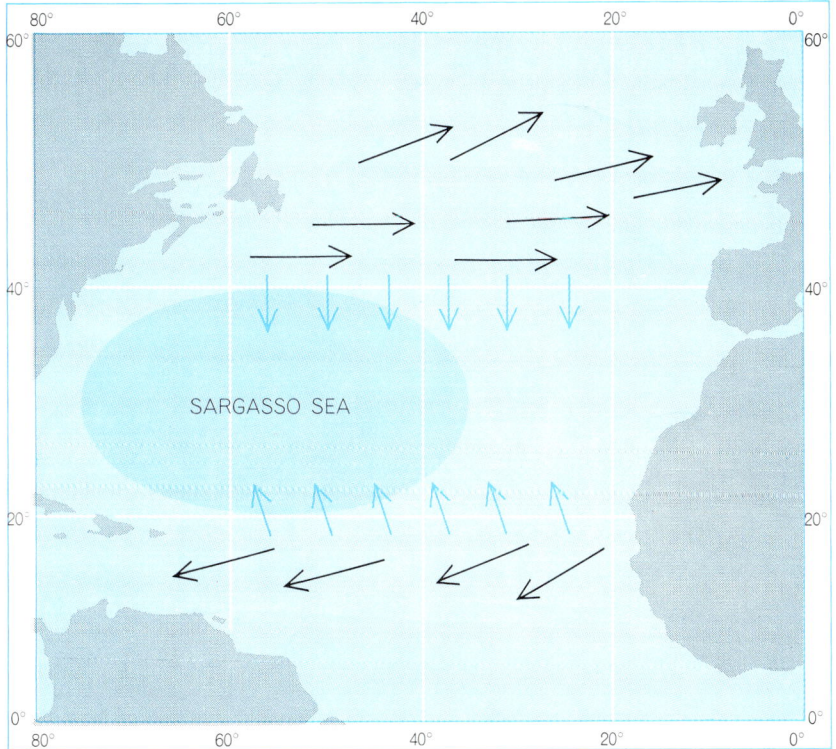

PREVAILING-WIND DIRECTIONS (black arrows) and the resulting Ekman-layer flows (colored arrows) in the North Atlantic drive water into the region of the Sargasso Sea.

1,000 years. With this length of residence not much of the heavy, cold water needs to be produced every year for it to constitute, as it does, the bulk of the ocean water.

The sinking of water cooled at the surface is one aspect of another important feature of the ocean: the flow induced by differences in density, which is to say the flow induced principally by temperature and salt content. This thermohaline circulation of the ocean is in addition to the wind-driven circulation discussed earlier.

In its thermohaline aspects the ocean itself acts as a heat engine, although it is far less efficient than the atmosphere. Roughly speaking, the ocean can be divided into two layers: a rather thin upper one whose density is comparatively low because it is warmed by the sun, and a thick lower one, a fraction of a percent denser and composed of water only a few degrees above the freezing point that has flowed in from those few areas where it is occasionally created. Somewhere—either distributed over the ocean or perhaps only locally near the shore and in other special places—there is mixing between these layers. The mixing is of such a nature that the cold deep water is mixed into the warm upper water rather than the other way around,

that is, the cold water is added to the warm from the bottom [see upper illustration, page 700]. Once the water is in the upper layer its motion is largely governed by the wind-driven circulation, although density differences still play a role. In one way or another some of this surface water arrives at a location and time at which it is cooled sufficiently to sink again and thus complete the circulation.

This picture can be rounded out by consideration of the effects of the earth's rotation, which are in some ways quite surprising. The deep water that mixes into the upper layer must have a net upward motion. (The motion is far too

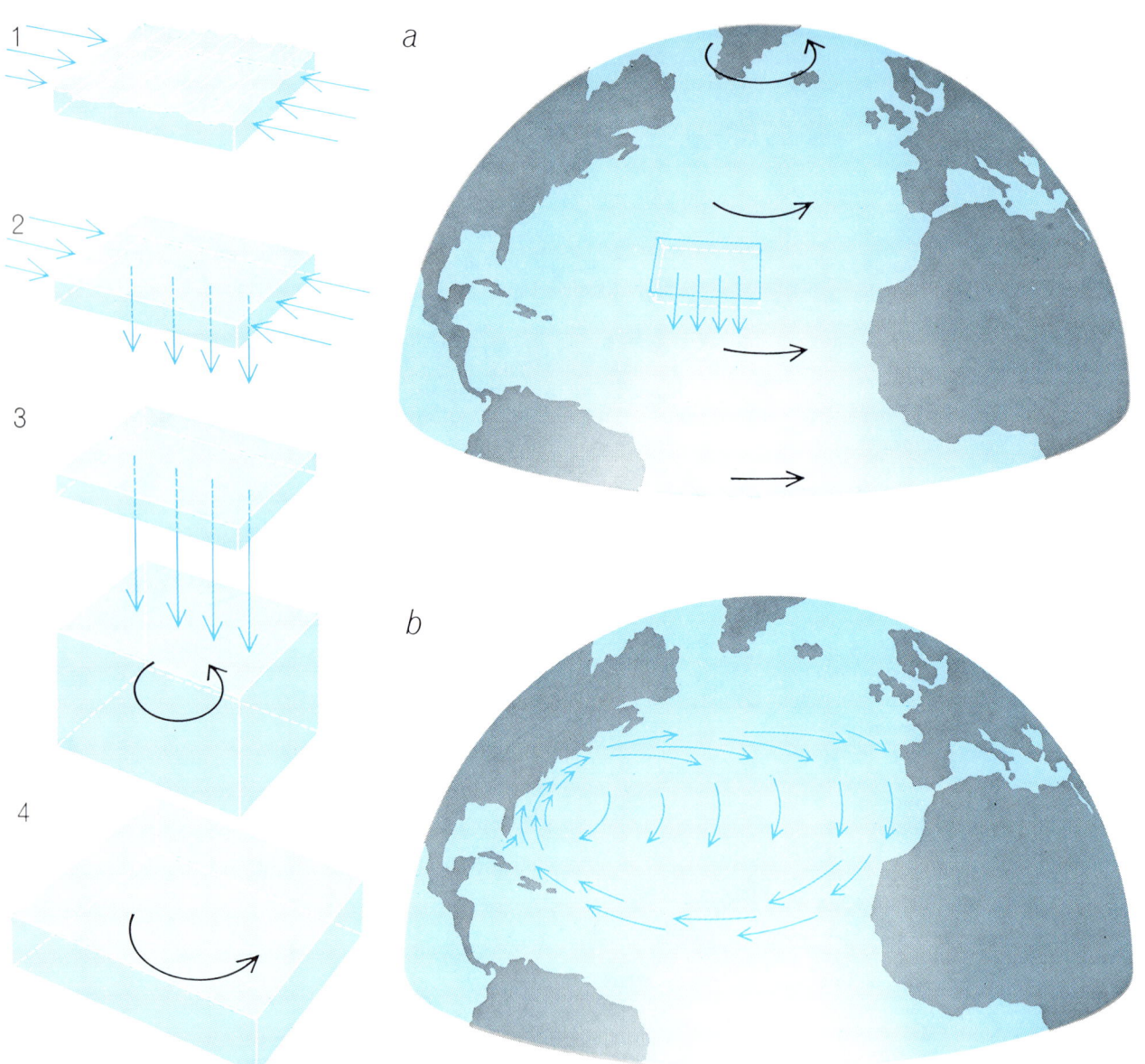

MAJOR CURRENTS are generated by a mechanism involving the Ekman-layer flow and the earth's rotation. The Ekman-layer inflow shown in the preceding illustration (1) produces a downflow (2) that presses on the underlying water (3), squeezing it outward (4) and thus reducing its rate of rotation (curved black arrow). There is a rate of rotation appropriate to each latitude, and when the rotation of a body of water is reduced, it must move (colored arrows) toward the Equator until its new rotation "fits" (a). For this reason there is a general movement of water from the mid-latitudes toward the Equator (b). That water must be replaced, and the water replacing it must have the proper rotation. This is accomplished by a return flow that runs along the western shore of the ocean, changing its rotation by "rubbing its shoulder" against the coast, as the Gulf Stream does in the Atlantic Ocean.

OCEAN AND ATMOSPHERE, the two thin fluid films in which life is sustained and whose nature and motion determine the environment, dominate this color photograph of the watery hemisphere of the earth. The picture was made on January 21, 1968, by a spin-scan camera on NASA's Applications Technology Satellite 3, in synchronous orbit 22,300 miles above the Pacific Ocean. The camera experiment was proposed and developed by Verner E. Suomi and Robert J. Parent of the University of Wisconsin's Space Science and Engineering Center. As the camera spins it scans a 2.2-mile-wide strip across the earth, then steps down in latitude and scans another strip; in about 25 minutes a 2,400-strip picture is completed. What the camera transmits to the earth is an electrical signal representing the amount of green, red and blue light in each successive picture element, and from these signals a color negative is built up at the receiving station. Such photographs yield information on the interrelation of atmospheric and oceanic phenomena. In this picture, for example, the convective pattern over the southeastern Pacific indicates that air heated by the sea is rising.

small to measure, only a few meters per year, but we infer its existence indirectly.) To make possible this upward flow there must be a compensating lateral inflow. Remember that on the rotating earth this lateral inflow results in an increase in speed of rotation, and so for it to continue to fit the rotation of the underlying earth the water must move toward the nearest pole; it must flow away from the equatorial regions. Yet

the source of this cold deep water is at high latitudes! How does it get near the Equator to supply the demand?

The answer is similar to the one for the wind-driven circulation: The cold water must flow in a western boundary current, in order to gain the proper rotation as it moves [see lower illustration on this page]. There is some direct evidence of the inferred concentrated western boundary current in the North At-

lantic, and there are hints of it in the South Pacific, but most of the rest is based on inference. There seems to be no source of cold deep water in the North Pacific, so that the deep water there must come from Antarctic regions.

We have seen that the atmosphere drives the ocean and that heat supplied from the ocean is largely instrumental in releasing energy for the atmo-

POLE ←   EQUATOR →

THERMOHALINE CIRCULATION, the flow induced by density rather than wind action, begins with the creation of dense, cold water that sinks to great depths. Under certain conditions this deep water mixes upward into the warm surface layer (color) as shown here. As it moves up, this water increases its rotation and so it must move generally from the Equator toward the two poles.

COLD DEEP WATER must flow in western boundary currents in order to arrive at the Equator and thus be able to move poleward as it mixes upward. Details of deep circulation are still almost unknown and the chart is intended only to suggest its approximate directions. There is some evidence of such boundary currents in the North Atlantic and there are some hints in the South Pacific.

INFRARED IMAGERY delineates the temperature structure of bodies of water and is used to study currents and wave patterns. This image of the shoulder of the Gulf Stream is from the Antisubmarine Warfare Environmental Prediction Services Project of the Naval Oceanographic Office. It was made by an airborne scanner at low altitude and shows several hundred yards of the boundary between the warm current and cooler water off Cape Hatteras. The range is from about 13 to 21 degrees Celsius, with the warm water darker.

dramatic events of the great Pleistocene glaciations.

There are any number of theories for these events and, since experts disagree, it is incumbent on the rest of us to refrain from dogmatic statements. Nevertheless, it does not seem impossible that the ocean-atmosphere system has a number of more or less stable configurations. That is, there may be a number of different patterns in which the atmosphere can drive the ocean in such a way that the ocean releases heat to the atmosphere in the right quantity and at the right places to allow the pattern to continue. Of course the atmosphere is extremely turbulent, so that its equilibrium is constantly being disturbed. If the system is stable, then forces must come into play that tend to restore conditions after each such disturbance. If there are a number of different stable patterns, however, it is possible that a particularly large disturbance might tip the system from one stable condition to another

One can imagine a gambler's die lying on the floor of a truck running over a rough road; the die is stable on any of its six faces, so that in spite of bouncing and vibration the same face usually remains up—until a particularly big bump jars it so that it lands with a different face up, whereupon it is stable in its new position. It seems not at all impossible that the ocean-atmosphere system behaves something like this. Perhaps in recent years we have been bouncing along with, say, a four showing. Perhaps 200 years ago the die flipped over to three for a moment, then flipped back to four. It could one day jounce over to a snake eye and bring a new ice age!

sphere. There is a great deal of feedback between the two systems. The atmospheric patterns determine the oceanic flows, which in turn influence where—and how much—heat is released to the atmosphere. Further, the atmospheric flow systems determine how much cloud cover there will be over certain parts of the ocean and therefore how much—and where—the ocean will be heated. The system is not a particularly stable one. Every locality has its abnormally cold or mild winters and its abnormally wet or dry summers. The persistence of such anomalies over several months almost certainly involves the ocean, because the characteristic time constants of purely atmospheric phenomena are simply too short. Longer-term climatological variations such as the "little ice age" that lasted for about 40 years near the beginning of the 19th century are even more likely to have involved changes in the ocean's circulation. And then there are the more

The Author

R. W. STEWART is professor in the department of physics and Institute of Oceanography at the University of British Columbia, chairman of the Physical Oceanographic Commission of the International Association of the Physical Sciences of the Ocean and a Fellow of the Royal Society of Canada. He was graduated from Queen's University in Ontario in 1945 with a degree in engineering physics and took his master's degree there two years later. He received his Ph.D. in 1952 from the University of Cambridge, where he also played lacrosse for the university. From 1955 to 1960, when he was appointed to his present position, he worked at the Pacific Naval Laboratory of the Canadian Defence Research Board, mostly on underwater sound propagation. Stewart writes that he has had "a continuing interest in turbulence, which is probably the only connecting thread through my scientific career," and that he has "also been interested in the teaching of physics at the university level."

Bibliography

AN INTRODUCTION TO PHYSICAL OCEANOGRAPHY. William S. von Arx. Addison-Wesley Publishing Company, Inc., 1962.

THE GULF STREAM: A PHYSICAL AND DYNAMICAL DESCRIPTION. Henry Stommel. University of California Press and Cambridge University Press, 1965.

THE INFLUENCE OF FRICTION ON INERTIAL MODELS OF OCEANIC CIRCULATION. R. W. Stewart in Studies on Oceanography, edited by Kozo Yoshida. University of Washington Press, 1965.

ENCYCLOPEDIA OF OCEANOGRAPHY. Edited by Rhodes W. Fairbridge. Reinhold Company, 1966.

DESCRIPTIVE PHYSICAL OCEANOGRAPHY. G. L. Pickard. Pergamon Press, 1968.

SCIENTIFIC
AMERICAN September 1969, Vol. 221, No. 3, pp. 106–122

OFFPRINT **882**

THE CONTINENTAL SHELVES

by K. O. Emery

The shallow regions adjacent to the continents are equal in extent
to 18 percent of the earth's total land area. They are alternately
exposed and drowned as the continental glaciers advance and retreat.

The continental shelves were the first part of the sea floor that was studied by man, chiefly as an aid to navigation and fishing. Perhaps the earliest recorded observation was one made by Herodotus about 450 B.C. "The nature of the land of Egypt is such," he wrote, "that when a ship is approaching it and is yet one day's sail from the shore, if a man try the sounding, he will bring up mud even at a depth of 11 fathoms." A more recent example is found in the diary of a 19th-century seaman: "An old captain once told me to take a cast of the lead at 4 a.m. We were bound to Hull from the Baltic. He came on deck before breakfast and on showing him the arming of the lead, which consisted of sand and small pebbles, I was surprised to see him take a small pebble and put it in his mouth. He tried to break it with his teeth. I was very curious and asked him why he did so. He told me that the small pebbles were called Yorkshire beans, and if you could break them you were toward the westward of the Dogger Bank; if you could not, you were toward the eastward."

Fishing success often depends on knowledge of the kind of bottom frequented by particular fish and on the avoidance of rocky areas that can catch and tear nets. As a result governmental agencies routinely chart bottom topography and materials to aid the fishing industry, but the successful fisherman generally keeps much additional information to himself. Similarly, the production of oil and gas from the continental shelves during the past two decades has led governmental and international agencies to make broad geological surveys, which help to guide the oil industry to the areas of greatest economic promise. The oil companies make studies that are much more detailed and so expensive that the results are considered proprietary, at least until after exploitation rights are secured.

During World War II submarines took a large toll of the ships that crossed the continental shelves, mainly at the approaches of ports. The effectiveness of acoustical detection equipment on submarine hunting ships was much increased by a knowledge of bottom materials and their effects: long ranges over sand, short ones over mud, confusing echoes over rock or coral. Accordingly charts of bottom sediments were compiled by the American and German navies for many areas of the world. This problem had not arisen earlier because both submarines and the search gear of surface ships were too primitive during World War I, and it may not be important in the future owing to the greatly increased sophistication of submarines and to the different role they may play in any future war.

The conflict between disseminating information and keeping it secret is about what one would expect in an environment that is both economically and militarily important. The recent political interest in the sovereignty of the ocean is also to be expected, considering the way the economic potential of the shelves has often been exaggerated in recent years. Thus it is not surprising that there have been a number of proposals to redefine the continental shelf so as to extend it seaward to whatever distance and to whatever depth are necessary to give a nation access to the resources presumably lying or hidden there. In 1953, before the world developed its present large appetite for seafood and minerals, an international commission defined the continental shelf, shelf edge and continental borderland as: "The zone around the continent extending from the low-water line to the depth at which there is a marked increase of slope to greater depth. Where this increase occurs the term shelf edge is appropriate. Conventionally the edge is taken at 100 fathoms (or 200 meters), but instances are known where the increase of slope occurs at more than 200 or less than 65 fathoms. When the zone below the low-water line is highly irregular and includes depths well in excess of those typical of continental shelves, the term continental borderland is appropriate."

Somewhat similar, but shorter, definitions are presented by most textbooks of geology. Where a depth limit is given it is 100 fathoms, an inheritance from the time when navigational charts had only three depth contours: 10, 100 and 1,000 fathoms. On a global basis the edge of the continental shelf ranges in depth from 20 to 550 meters, with an average of 133 meters; the shelf ranges in width from zero to 1,500 kilometers, with an average of 78 kilometers [see illustration on next two pages].

CONTINENTAL SHELF in the Atlantic Ocean off Cape Hatteras is delineated by puffy clouds that show where cold surface water on the eastern edge of the shelf meets warmer surface water. The boundary is near the edge of the shelf, which at this point averages about 120 meters in depth. The picture also shows turbid water moving from Pamlico Sound into the ocean, where it is carried northward by a fringe of the Gulf Stream that lies atop the shelf. The photograph was taken from *Apollo 9* on March 12 of this year, at an altitude of 134 statute miles; the view extends 175 miles in the north-south direction. The astronauts on this mission were James A. McDivitt, David R. Scott and Russell L. Schweikart.

The continental shelves underlie only 7.5 percent of the total area of the oceans, but they are equal to 18 percent of the earth's total land area. A geological understanding of this huge region requires a knowledge of its topography, sediments, rocks and geologic structure. For nearly all shelves there is some information about topography; for perhaps a fourth of them something can be said about the surface sediments, but the rocks and geologic structure are known for less than 10 percent. Detailed knowledge is far less available. The best-known large areas are the ones off the U.S., eastern Canada, western Europe and Japan—in short, the shelves next to countries where scientific knowledge is well developed and freely disseminated. Smaller areas of knowledge are found where oil companies have worked (such as parts of northern South America, parts of Australia and the Persian Gulf) and where oceanographic institutions have conducted repeated operations (northwestern Alaska, the Gulf of California, northwestern Africa, the shelf off Argentina, the Red Sea and the Yellow Sea). Recently some developing countries have effectively closed their shelves to foreign scientific studies; these areas are fated to remain unknown and unexploited for the foreseeable future.

The information that is most costly and most difficult to obtain concerns the underlying rocks and the structural geology of the continental shelves. This information is essential for understanding the origin and most of the history of the shelves. Data about the surface topography and sediments, which are readily accessible, tell only the late history.

Samples of bedrock have been dredged from the surface of many shelves, mainly from the top of small projecting hills and the sides and heads of submarine canyons that incise the shelves. Additional rock samples have come from the top of the adjacent continental slopes. Care must be taken in deciding whether the rock samples are from outcrops, whether they are loose pieces that were deposited by ancient streams or glaciers, or whether they were rafted to their present location by ice, kelp, marine animals or man. The decision is usually based on the size of the piece, on the presence or absence of fresh fractures, on the similarity of lithologic types within a given dredging area or between adjacent dredging areas, and on the amount of tension of the dredge cable. It is helpful to have submarine photographs, which may reveal rock outcrops in the dredging area. Rock also can be sampled by coring: by dropping or forcing a heavily weighted pipe into the bottom. This method can show the dip of the strata, and it can sample rock that is covered by sediments if the sediment is thin or if it is first removed by hydraulic jetting. A better but more expensive method of rock sampling is provided by well-drilling methods. Many holes have been drilled for geological information in shelf areas of structural interest; they provide good information on the sequence and depth of strata, the date of original deposition and geologic structure. In addition several thousand oil wells have been sunk into shelves, but most have been drilled in abnormal geologic structures such as salt domes and folds.

Geophysical methods provide excellent, although indirect, data from which geological cross sections can be constructed. These methods include seismic reflection and refraction, as well as measurements of geomagnetism and gravity. Each method has its advantages, but the most generally successful one is seismic reflection. In practice a ship traverses the shelf and produces a loud acoustical signal in the water at intervals of a few seconds. The chief source of sound energy a few years ago was a chemical explosive, usually dynamite; other sources are now preferred because they are

CONTINENTAL SHELVES underlie about 7.5 percent of the total ocean; all together they occupy an area roughly equal to that of Europe and South America combined, or some 10 million square miles. The shelf is defined as the zone around a continent extending from

cheaper, easier to trigger accurately and greatly reduce the danger both to the operators on the ship and to the fish in the sea. These newer energy sources include electric spark, compressed air and propane gas. Although part of the sound energy is reflected from the sea floor, much of it enters the bottom to be reflected upward from various layers of rock under the bottom. The reflected energy is received by hydrophones trailed behind the ship, the signal is amplified, filtered and recorded on continuously moving paper tapes. This method, termed continuous seismic-reflection profiling, is rapid and can yield information from depths of several kilometers under the sea floor, making it possible to construct geological cross sections. When the interpretations are supplemented by dredging or drilling, they provide the best information now available about the structure of continental shelves.

When existing geological and geophysical information is assembled on a worldwide basis, it shows that continental shelves can be classified into two main types by composition: those that are underlain by sedimentary strata and those underlain by igneous and metamorphic rocks. A large majority of the world's continental shelves mark the top surface of long, thick prisms of sedimentary strata [see *illustration on next two pages*]. Many of the prisms are held in position against the continents by long, narrow fault blocks. Such is true of almost the entire perimeter of the Pacific Ocean, where tectonic activity has also produced deep trenches that are parallel to the base of the continental slope. In some areas, such as the West Coast of the U.S., a single geologic dam is known to have extended for thousands of kilometers along the coast. Locally part of the dam rises above sea level to form the granitic Farallon Islands that lie immediately off San Francisco. These rocks are some 100 million years old, but they were thrust up to form the dam only about 25 million years ago. Elsewhere, as in the Yellow Sea of Asia, half a dozen such fault dams or fold dams

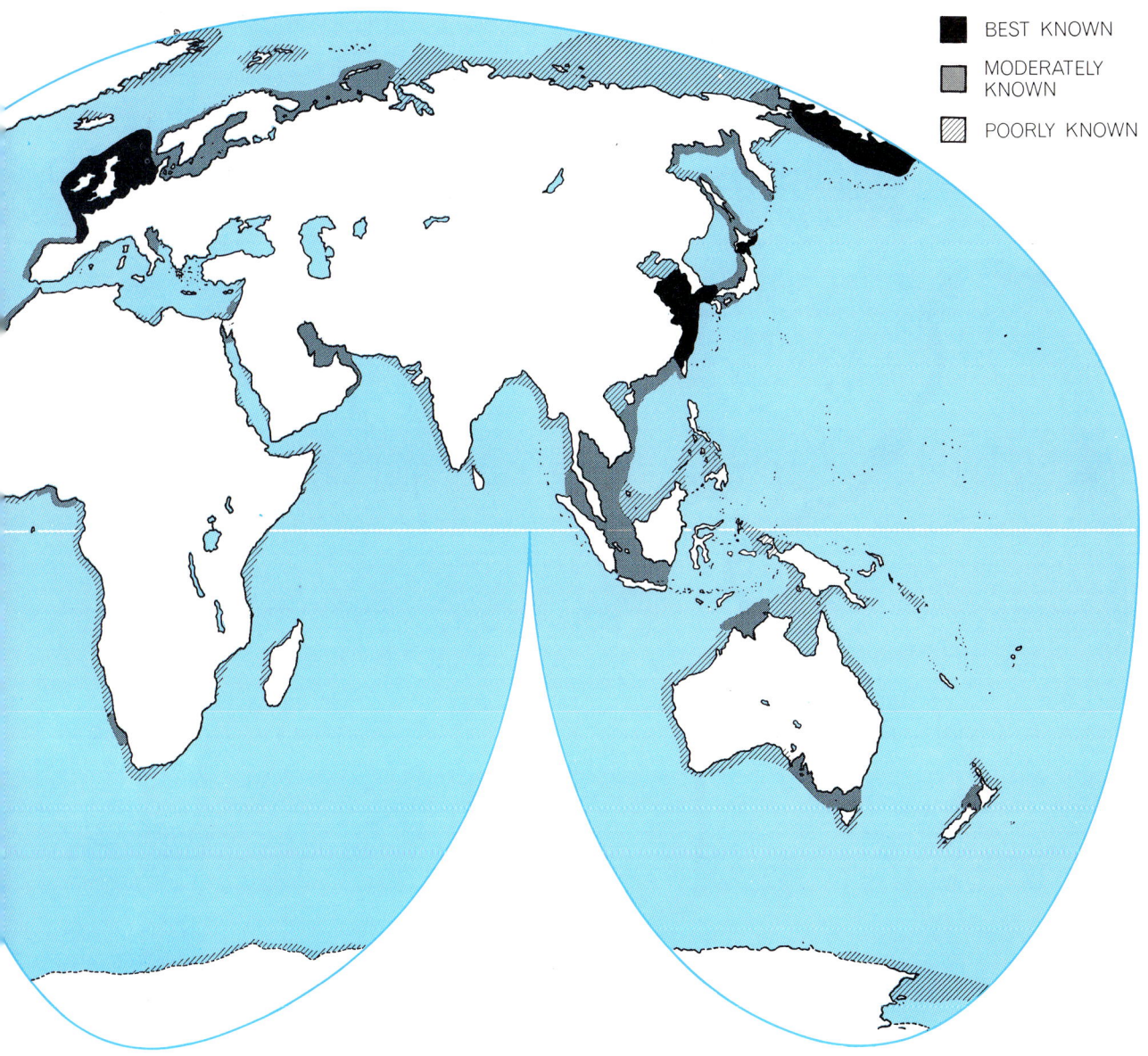

BEST KNOWN

MODERATELY KNOWN

POORLY KNOWN

the low-water line to the depth at which the ocean bottom slopes markedly downward. Conventionally the edge of the shelf is taken to lie at 100 fathoms, or 200 meters, but a more accurate average value for all continents is about 130 meters. Worldwide the shelf has an average width of 78 kilometers. The illustration indicates the present state of knowledge of the continental shelves of the world.

have risen in the last 500 million years or so, each dam in turn causing the ponding of sediments from the land. There was a similar dam off the entire length of the East Coast of the U.S. between 270 and 60 million years ago; in time the trench on its landward side was filled with sediments that subsequently spilled over the former dam to build a continental slope that is held in place only by the angle of rest of the sediments. That this angle is unstable is shown by numerous landslides and erosion features recorded in seismic profiles across the continental slope and rise.

The continental shelf off the western part of the Gulf Coast of the U.S. is held in place by a diapir dam: a dam formed by the upward movement of salt from a bed that is buried several kilometers deep and is about 150 million years old. Seismic profiles and dredgings show the presence of still another kind of dam in the eastern Gulf of Mexico and off the southeastern coast of the U.S. This is an algal reef that dates from 130 million years ago and was succeeded by a coral reef off Florida at some time before 25 million years ago; even today the Florida keys are bordered by a living coral reef. Similar tectonic and biogenic dams elsewhere in the world have trapped huge quantities of sediments in the geological past.

Shelf areas underlain by igneous and metamorphic rocks are found on top of the tectonic dams. Off Maine, however, glacial erosion has removed the sedimentary rocks that once covered such a dam [*see illustration on pages 8 and 9*]. Other shelves underlain by igneous and metamorphic rocks are known, but most of them appear to be at high latitudes where glacial erosion has been effective. Nevertheless, even at high altitudes probably most of the shelves are underlain by sedimentary rock. In a sense we can consider the shelves whose shape is chiefly due to glacial or wave erosion as youthful ones (or rejuvenated ones); the shelves that are mainly depositional, with a thick prism of sediments on top of igneous and metamorphic rocks, can be regarded as mature ones. The sediments have built the shelves upward during the concurrent sinking of the edges of the continents. Perhaps more important from the viewpoint of real estate, these sediments have increased the size of the continents, widening them as much as 800 kilometers in areas where rivers have brought much sediment from the land and where tectonic or other dams effectively prevent the escape of the sediment to the ocean basins. This dammed sediment, as well as the sediment held only by its angle of rest, has an estimated average thickness of about two kilometers, yielding a total

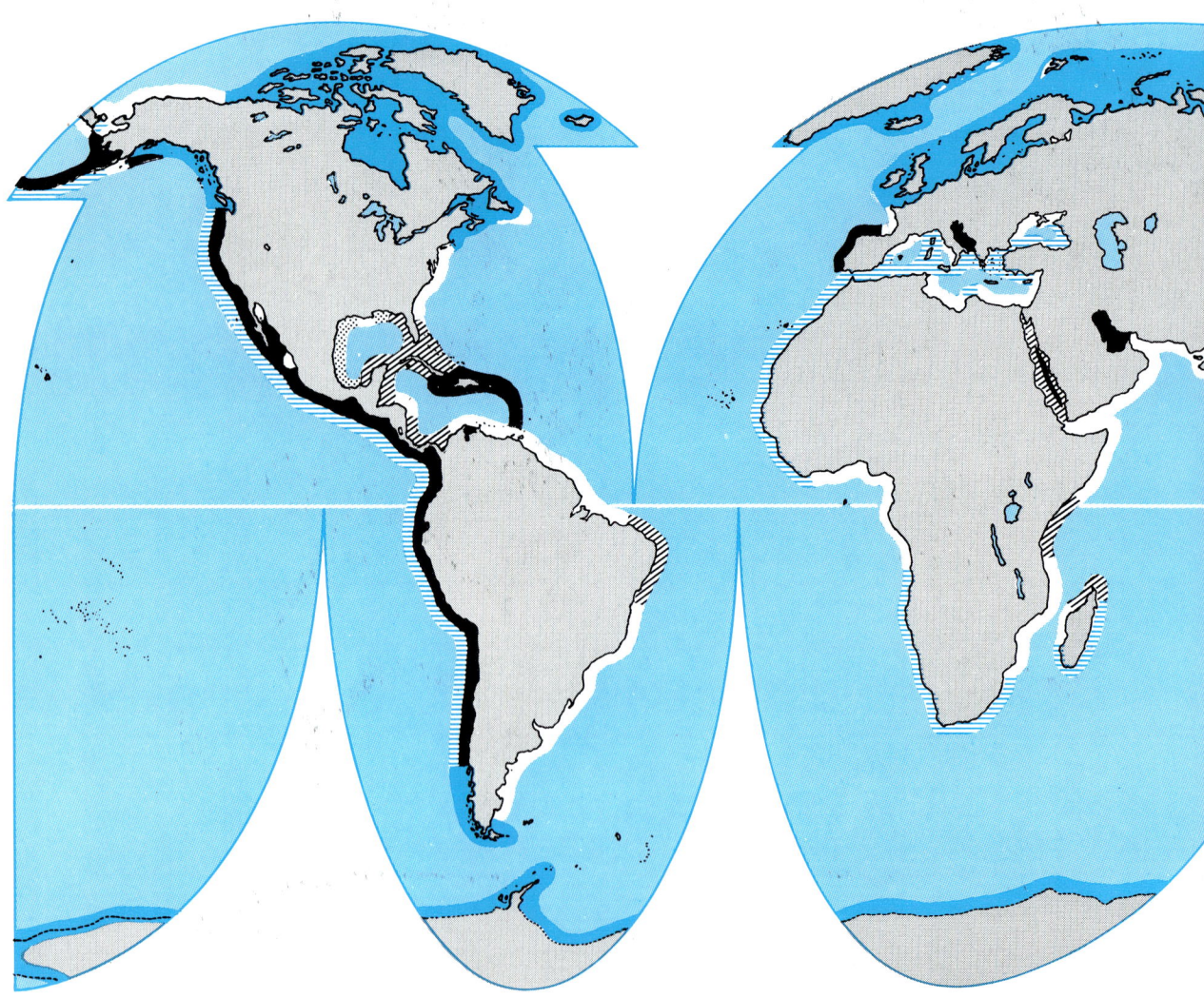

CHARACTER OF CONTINENTAL SHELF depends largely on how the shelf was formed. Six types of shelf, classified by origin, are indicated in this worldwide map. Many shelves are deposited behind three kinds of dams: tectonic dams, formed by geological uplift or upwelling of lava; reef dams, created by marine organisms, and diapir dams, which are pushed upward by salt domes.

volume of sedimentary strata under continental shelves of about 50 million cubic kilometers.

Perhaps the most dramatic period in the history of the continental shelves was the million-year passage of the Pleistocene epoch when the sea level changed in response to the waxing and waning of the continental glaciers. At their maximum the glaciers appear to have been so extensive as to have stored in the form of ice enough water to lower the surface of the ocean nearly 150 meters below the present level. Four major lowerings of the sea level were produced by the four main glaciations, with minor lowerings caused by secondary fluctuations of climate and ice volume. Limited investigations with special seismic equipment off the East Coast of the U.S. show four or five somewhat irregular acoustical reflecting surfaces near the top

of the shelf sediments. These reflecting surfaces probably can be explained by erosion and sand deposition at stages of low sea level. Cores from these beds probably would provide much interesting information about Pleistocene climates and Pleistocene chronology. For the present, however, our data for glacial effects on the continental shelf are restricted largely to surface sediments and topography.

About 50 years ago most textbooks of geology led the reader to believe that sediments became progressively finer in texture with distance from shore: gravel and sand at the shore, coarse sand grading to fine sand across the shelf, and finally silt and clay (the "mud line") at the shelf edge. Bottom-sediment charts compiled during World War II, however, showed that this simple pattern is very rare, and that the size of sediment grains is unrelated to the distance from

shore. The examination of actual samples showed that most of the shelves are floored with coarse sands that commonly are stained by iron and contain the empty shells of mollusks that live only close to shore in shallow depths. Broken shells or shell sand are particularly abundant at the outer edge of the shelf and on small submerged hills that are relatively inaccessible to detrital sediments. Some of these same areas contain glauconite and phosphorite, minerals that are precipitated from seawater, but so slowly that they are obscured or diluted beyond recognition where detrital minerals are present. The only areas that exhibit a consistent seaward decrease of grain size are those between the shore and depths of 10 or 20 meters—in short, whatever areas are shallow enough to be ruled by the waves. At greater depths the sediments are too deep to be reached by new supplies of sand. These sections

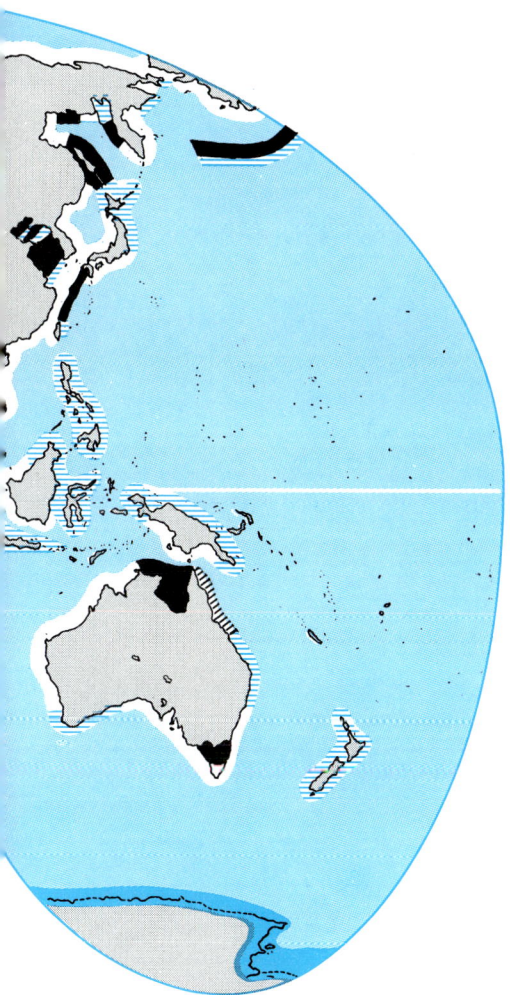

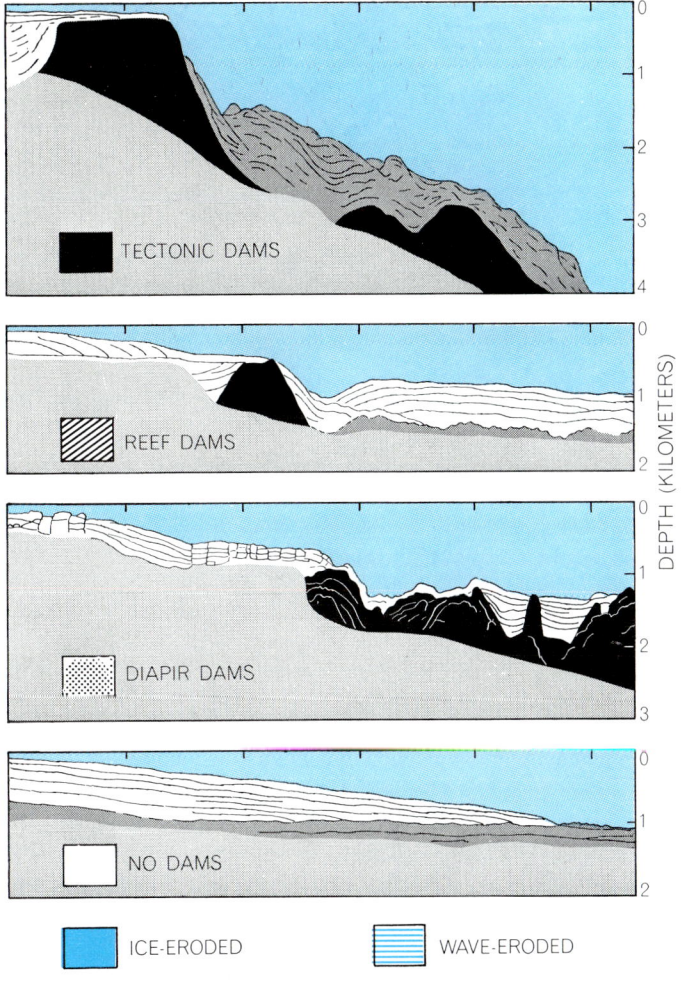

These three kinds of dams are shown in black in the typical shelf cross sections at the right; a simple damless shelf is also depicted. The vertical scale is exaggerated six times. Sediment deposited before formation of the dam is shaded gray; sediment deposited subsequently is unshaded. All four kinds of shelf structure may be eroded by waves (*hatched color on map*) or by ice (*solid color*).

of the shelves are also bypassed by contemporary silts and clays that remain in suspension en route to deeper or quieter waters.

The sediments on about 70 percent of the world's continental-shelf area have been laid down in the past 15,000 years, since the last glacial lowering of the sea level. The rest of the shelf is floored by silts at the mouths of large rivers, in quiet waters behind barriers and in shelf basins, by recent shell debris and by chemically deposited minerals. This means that when the sea level was low, the entire shelf was exposed and the rivers deposited sands on the then broader coastal plain and transported their silts and clays to the ocean. At that time ocean waves, with no shelf to reduce their height, were probably higher at the shore than they are today, with the result that shore sediments were probably coarser. The broad expanse of lowland favored the development of ponds and marshes, which were partly filled with debris from the forests and meadows that extended unbroken from the inland areas across what is now the sea floor. Freshwater peat now submerged in the ocean has been sampled at 10 sites off the eastern U.S. and at many other sites on the shore; similar peats have been found off Europe, Japan and elsewhere.

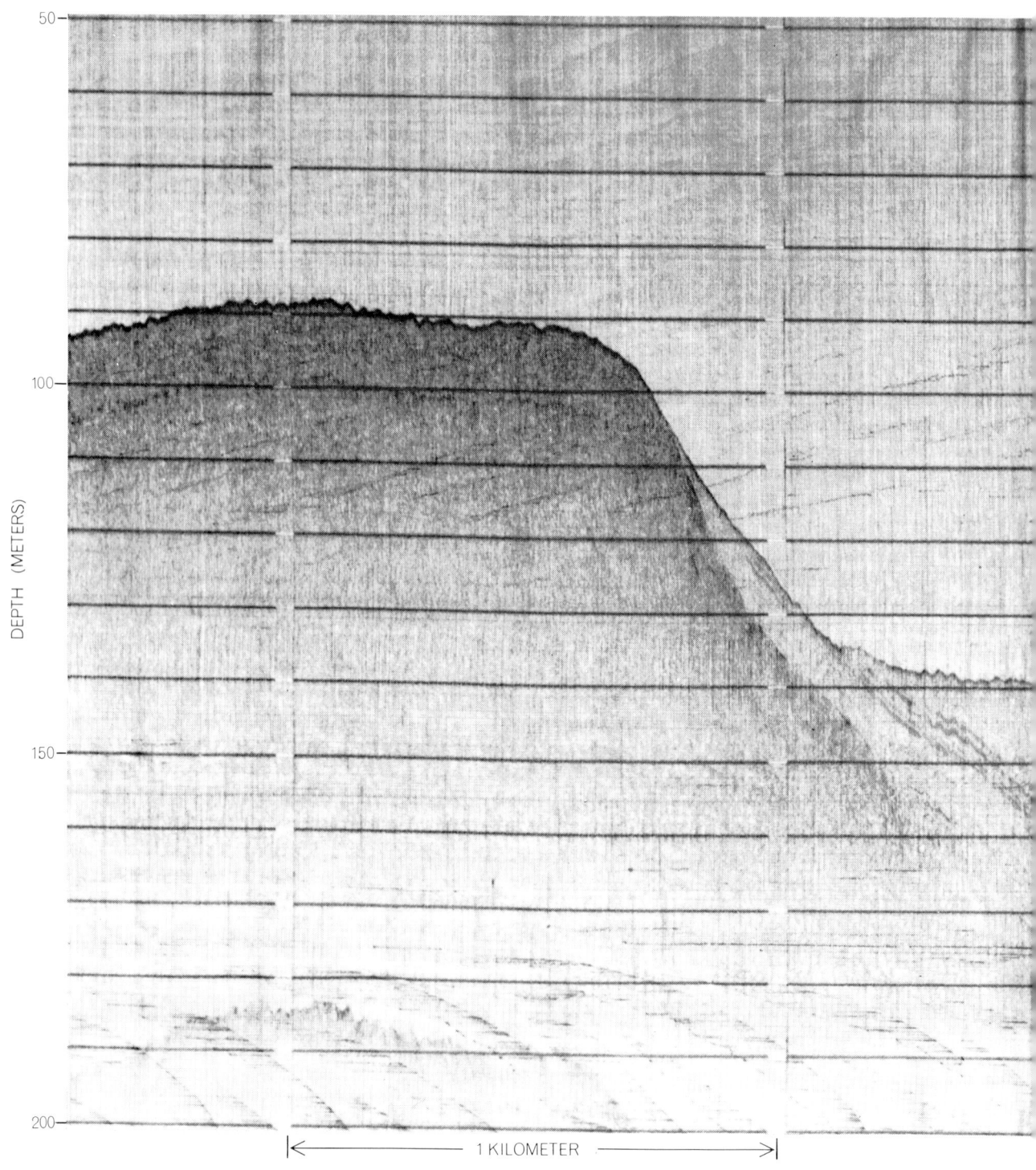

1 KILOMETER

ICE-ERODED SHELF about 100 kilometers off the coast of Maine, landward of Georges Bank, is shown in this seismic-reflection record. The deep trough was gouged out of the basement rock by ice some 15,000 years ago during the last glaciation. Subsequently the trough

Pollen analysis shows a succession from tundra to boreal spruce and pine some 12,000 years ago, followed by oak and other Temperate Zone deciduous trees about 8,500 years ago; the deciduous trees flourished until the site was submerged. Birds once flew among the trees in many areas where fish now swim.

The vegetation attracted many animals, but only their heavier bones are preserved or are readily detected by dredging. Nearly 50 teeth of mammoths and mastodons have been collected off the East Coast of the U.S., along with the bones of the musk ox, giant moose, horse, tapir and giant ground sloth. Similar finds have been reported off Europe and Japan.

Carbon-14 dates have been obtained for more than 50 samples of shallow-water material from the shelf off the East Coast of the U.S. The materials include salt-marsh peat, oölites (concentrically banded calcium carbonate pellets that typically form only in warm, shallow, agitated seawater) and the shells of oysters and other mollusks (which live in only a few meters of water but whose empty shells are found as deep as 130 meters). The dates and depths make it possible to draw a curve showing the changes of sea level in an-

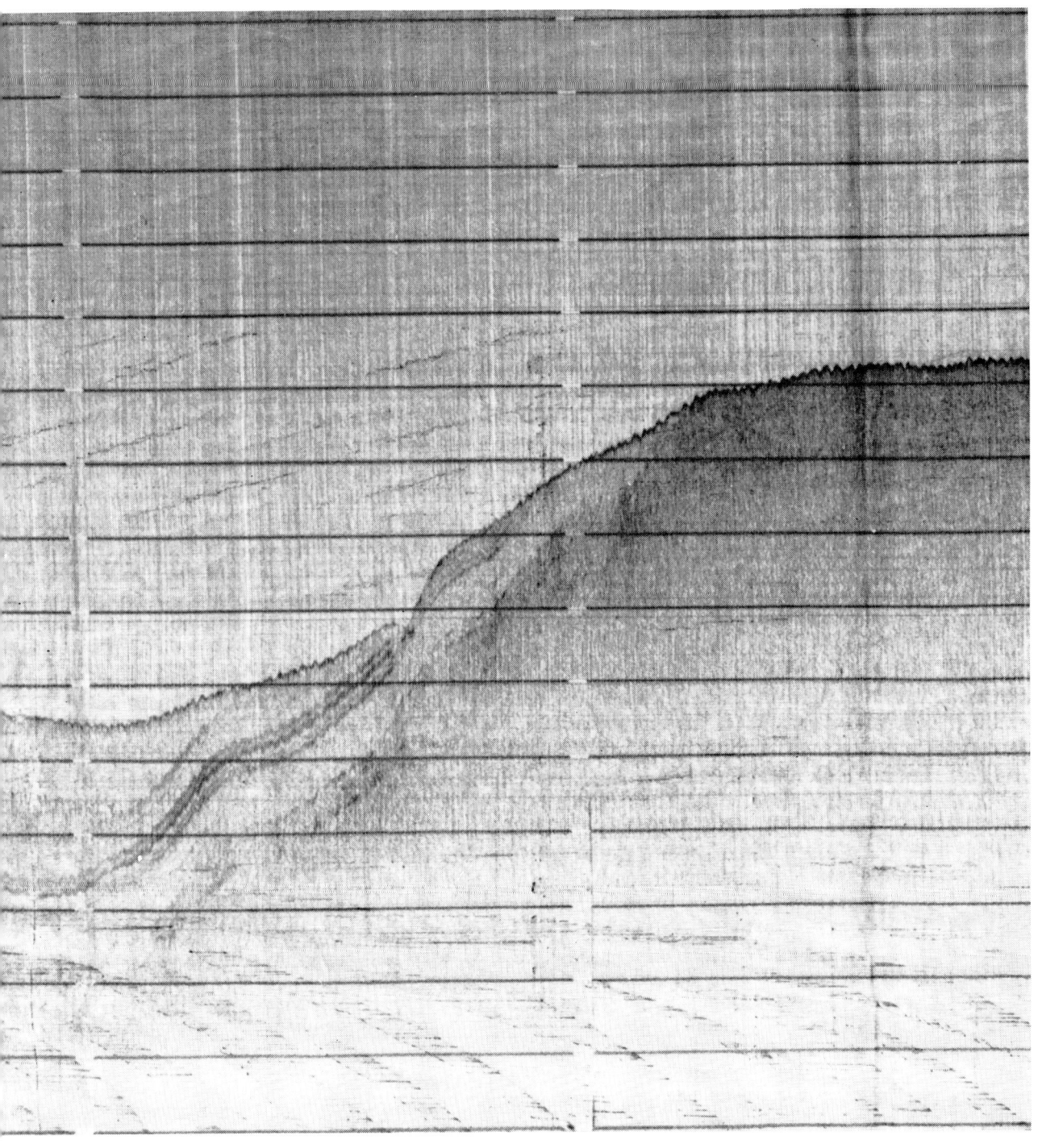

became filled with sediments about 30 meters thick. The recording was made this past July with high-frequency seismic equipment aboard the *Dolphin*, a vessel operated by the U.S. Geological Survey in cooperation with the Woods Hole Oceanographic Institution.

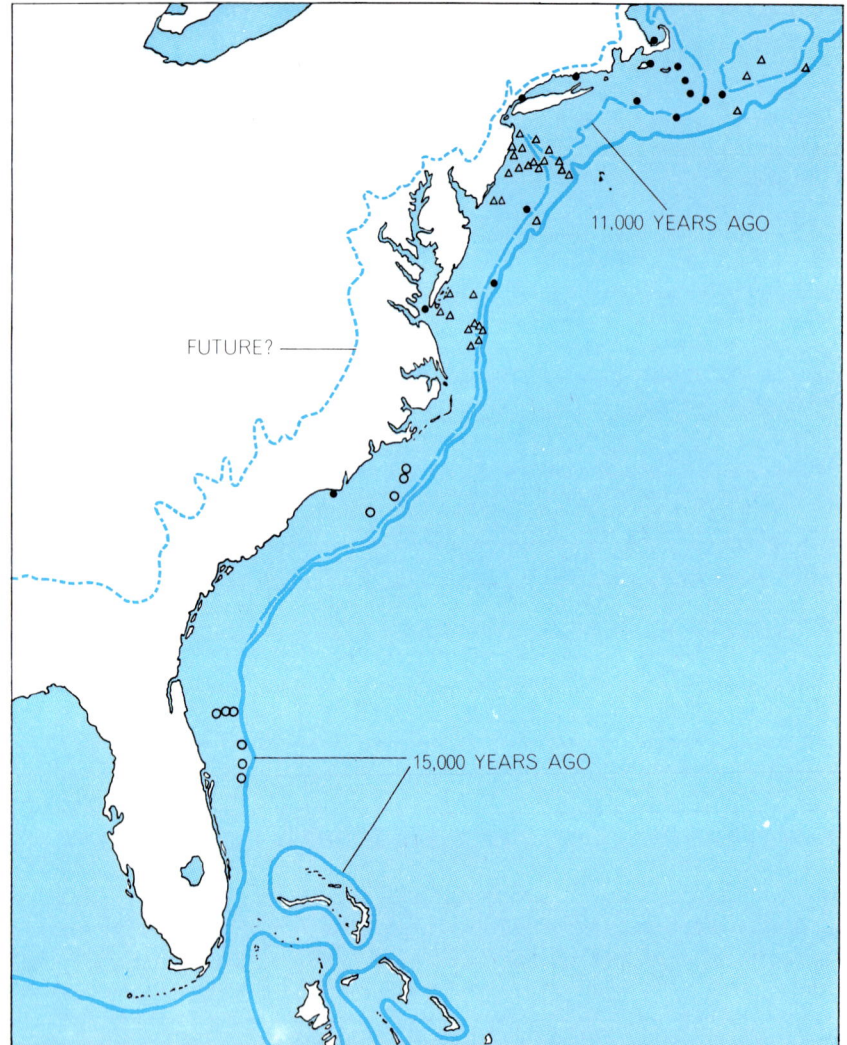

ATLANTIC COAST SHORELINE has varied greatly in the past and will undoubtedly continue to in the future. This illustration compares the shoreline of 15,000 and 11,000 years ago with the probable shoreline if all the ice at the poles were to melt. Confirmation that the continental shelf was once laid bare is found in discoveries of elephant teeth (*triangles*), freshwater peat (*dots*) and the shallow-water formations called oölites (*circles*).

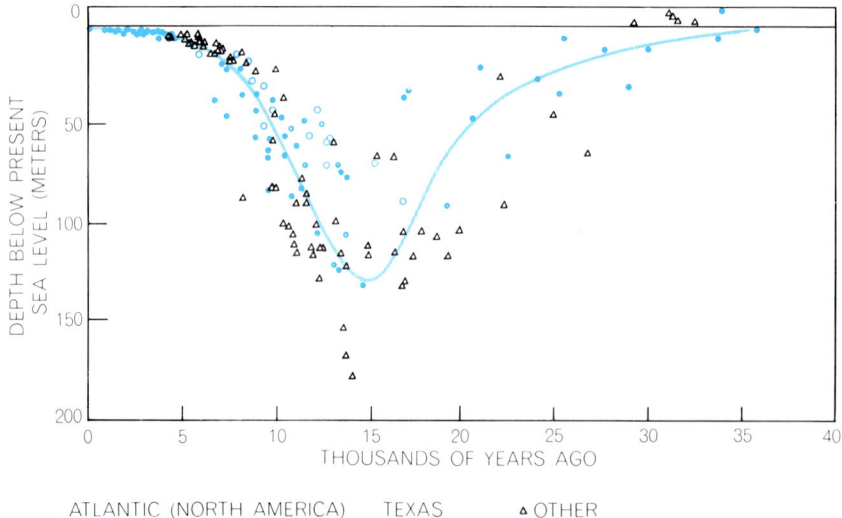

WORLDWIDE CHANGES IN SEA LEVEL can be inferred from the radiocarbon ages of shallow-water marine organisms and the depth at which they were recovered. Samples are from the Atlantic shelf of North America, the Texas shelf and other parts of the world. The depth inconsistency of the Texas samples implies that the shelf there has been uplifted.

cient times [*see bottom illustration at left*].

Apparently the sea was near its present level about 35,000 years ago and began to recede about 30,000 years ago. The level dropped by 130 meters or more 15,000 years ago; then it rose rather rapidly to within about five meters of the present level 5,000 years ago. The slow rise during the past 5,000 years has been documented by perhaps 100 carbon-14 dates for peat under existing salt marshes. Less complete sequences of dates for similar samples from elsewhere in the world show a sea-level curve resembling the one for the East Coast. Only the samples from the shelf in the western Gulf of Mexico provide a different curve, which suggests that this part of the gulf shelf was uplifted about 40 meters during the past 10,000 years.

Early men of the Clovis culture (characterized by fluted stone projectile points) appeared in North America some 12,000 years ago, when the sea level was still very low. What is more reasonable than to suppose such men ranged over the forested lowland that is now continental shelf? Game, fish and oysters were abundant. How were they to know or care that in a few thousand years the area was to be drowned by the advancing sea, any more than New Yorkers know or care that when the remaining glaciers melt, the ocean will rise to the 20th story of tall buildings? [*see top illustration at left*]. The search for traces of early man far out on the shelf began with the discovery of what may be the remains of an oyster dinner on a former beach off Chesapeake Bay, a site that is now 43 meters below sea level. This discovery was made from the Woods Hole Oceanographic Institution's research submarine *Alvin;* many similar discoveries will probably be made during the next decade.

Submerged barrier beaches are common on the continental shelf, but they are easily confused with the sand waves that are formed by strong currents. More spectacular and of certain origin are the submerged sea cliffs and terraces that mark the temporary stillstands of the sea level. Most of the shelves that have been studied have four to six such terraces, but the recognition of the terraces depends on their width and sharpness. On gently sloping shelves the terraces are almost imperceptible; on steep shelves they are narrow or absent; on shelves receiving a large supply of sediment they are buried. Variation in depth is to be expected in view of the large variation in depth of the most prominent terrace of all—the edge of the shelf. Passing through the terraces are channels cut by streams that flowed across the shelf when the sea level was low. Most of these channels have been filled by sediment; they can be recognized only by seismic profiling and by drilling on the shore at the mouths of stream valleys. Probably hundreds of channels cross the continental shelves of the U.S., but only a dozen are known. One channel, the one cut by the Hudson River off New York, is so large it is not yet filled with sediment.

At the seaward end of the channels, near the edge of the shelf, the channels are replaced by the heads of submarine canyons that continue down the continental slope to depths of several kilometers. The continuation of the submarine canyons to depths far below the maximum probable lowering of the sea level means that the canyons must have been formed by some process that operates under the ocean surface. Although the matter is still the subject of debate, most of the evidence favors the view that the canyons were excavated by turbidity

currents: currents that arise when sediment slips down a slope and becomes mixed with overlying water, thereby increasing its density so that it continues down the slope, often at high speed. Today the shelf off the East Coast of the U.S. is only slightly modified by submarine canyons; only the heads of the canyons indent the shelf edge. When the sea level was at its lowest, the canyons were probably important factors in sedimentation. The shelf off the West Coast of the U.S. is so narrow that the heads of many canyons reach almost to the shore. In those areas the canyons serve to trap and divert sand that is moved along in the shore zone under the influence of wind-driven waves and their associated currents. As a result the sand that is brought to the shore by streams and cliff erosion is only temporarily added to the beaches; eventually it moves seaward through the canyons in the form of slow sand glaciers or rapid turbidity currents.

The water above the continental shelves is complex in composition and movement because it is shallow and

close to the land. Large rivers contribute so much fresh water that they dilute the ocean, but they also increase the local concentration of calcium, phosphate, silica and nitrate—precisely those elements and compounds that elsewhere in the ocean have been reduced to low concentration by incorporation into marine plants and animals. Continental-shelf waters that are distant from river mouths are sometimes saltier than the open sea because their rate of evaporation is high. Local variations in salinity (and therefore density) control the direction of currents on the shelf. For example, the low salinity at the mouth of a river means a higher sea level near the shore than farther out on the shelf, leading to a flow toward the right (when one is facing the ocean in the Northern Hemisphere).

Just at the shore, however, the longshore currents are mainly controlled by the angle at which waves intersect the beach, which in turn is a function of the wind direction. As a result the cur-

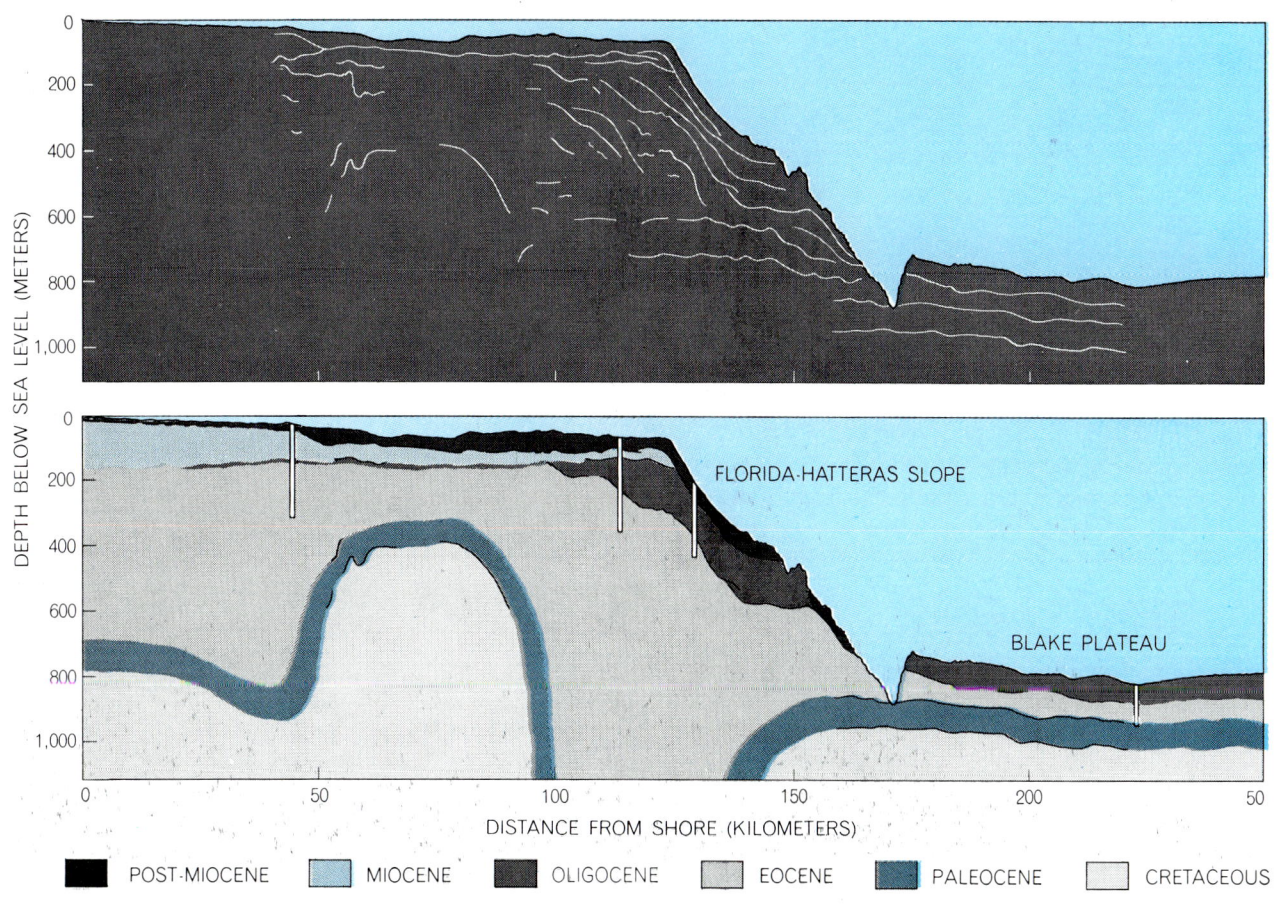

SHELF OFF JACKSONVILLE, FLA., has been studied by two geophysical methods: seismic reflection (top) and drilling (bottom). Seismic studies can show only the general nature of the stratigraphy. Cores obtained by the JOIDES project, a drilling study conducted by a consortium of institutions, made it possible to map the stratigraphy in considerable detail. The vertical scale is exaggerated 67 times. The approximate termination dates for the various geologic periods are as follows: Miocene, 10 million years ago; Oligocene, 25 million years ago; Eocene, 40 million years ago; Paleocene, 55 million years ago; Cretaceous, 65 million years ago.

rent in the wave zone may be northward, the current on the inner half of the shelf may be southward and the current on the outer shelf may be northward again (for example where an oceanic current such as the Gulf Stream runs along the edge of the shelf). Where rivers bring water to the ocean there must be a gen-

eral current component toward the ocean at the surface; this induces a return flow toward the land at the bottom [*see top illustration, page 713*]. Thus the sediment on the sea floor may be moved landward, often working its way into the mouths of estuaries. This means estuaries are truly ephemeral features, re-

ceiving sediments from both rivers and the open shelf.

Temperature zones on land are mainly a function of latitude, with secondary modifications resulting from winds whose direction may change seasonally or may be controlled by topography. Similarly, ocean water is cooled at high

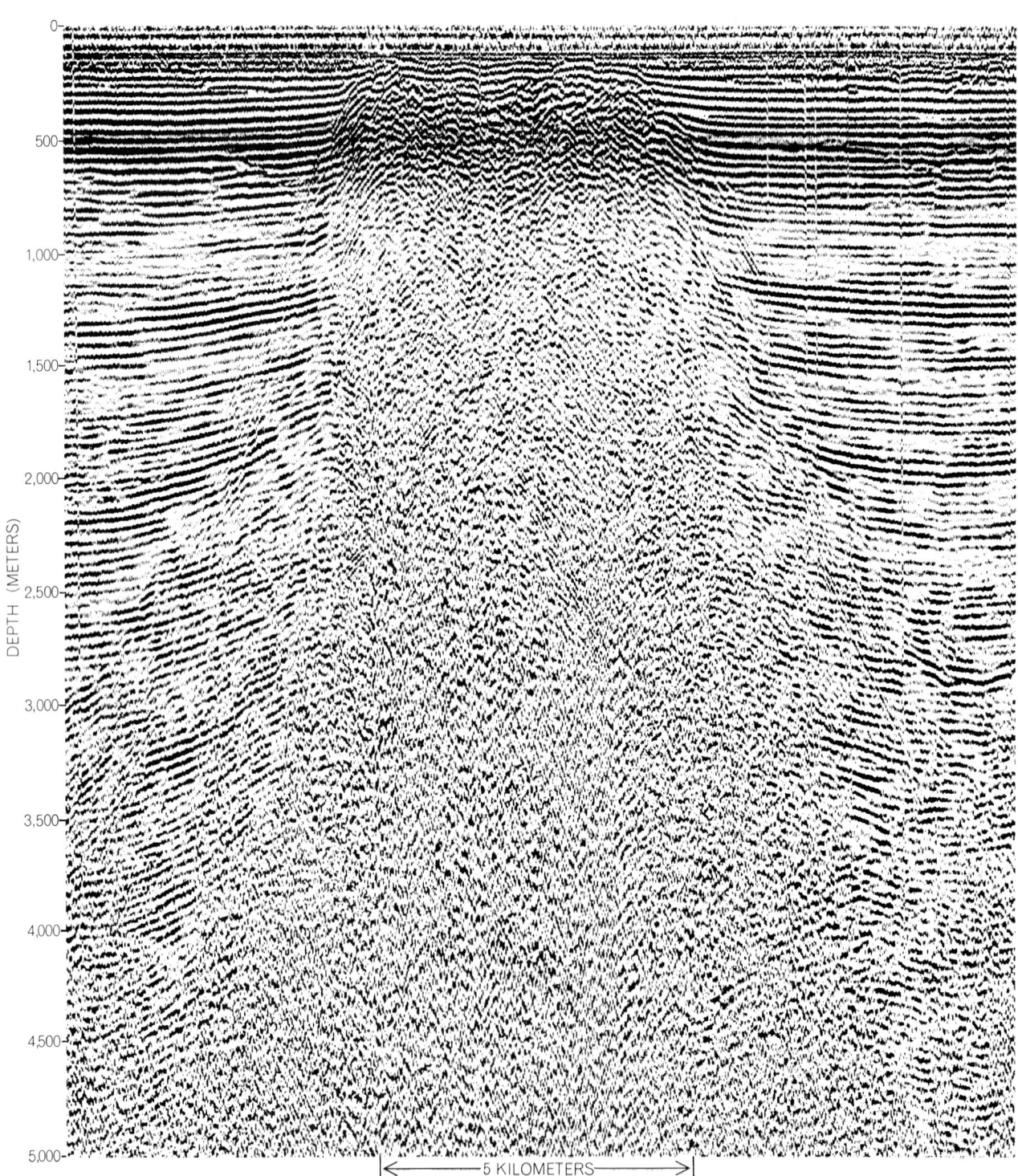

SALT DOME bulging upward into the continental shelf about 10 miles south of Galveston in the Gulf of Mexico is shown in this seismic record. The water is so shallow (between 10 and 20 meters) that the reflection from the surface of the shelf is virtually at the

top of the recording. Geologists can discern significant features in such a record down to a depth of about 3,000 meters. The record was made by the Teledyne Exploration Company. The salt dome was subsequently drilled and was found to contain hydrocarbons.

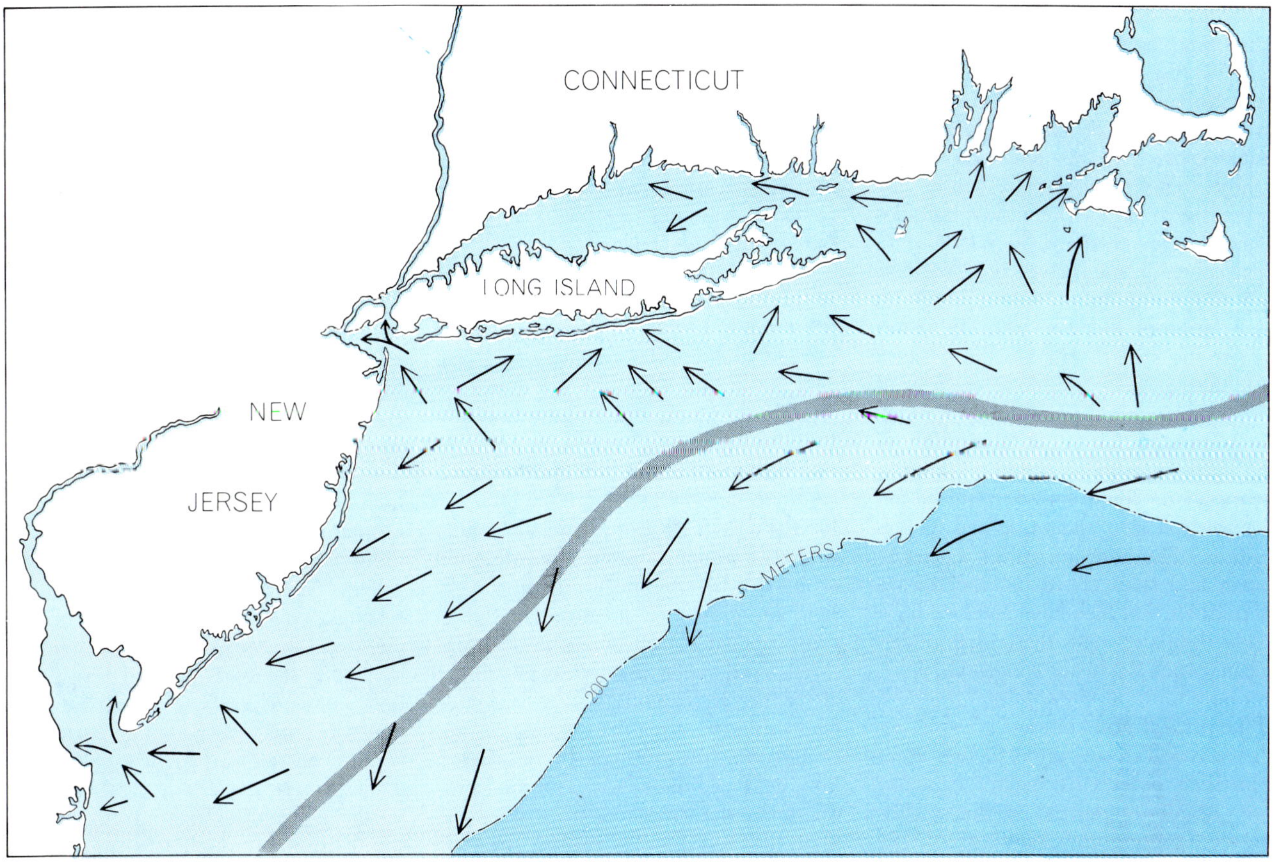

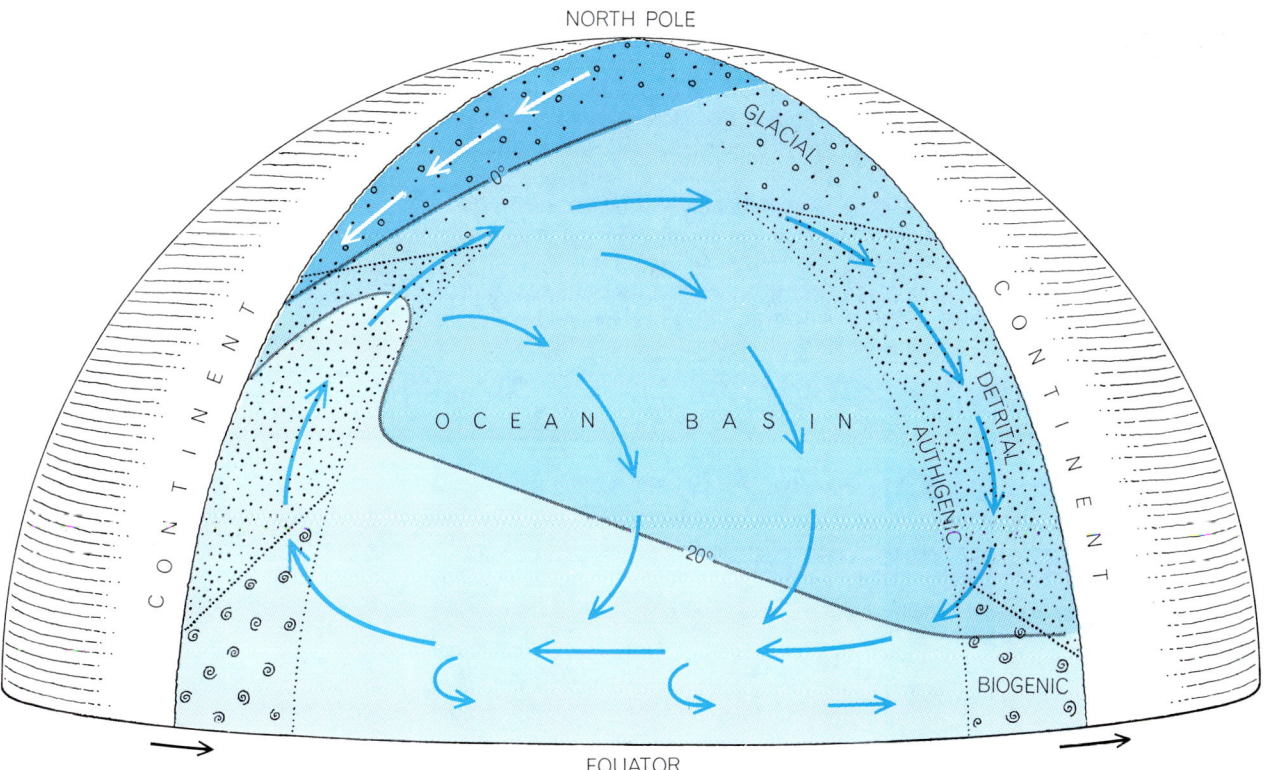

BOTTOM CURRENTS, indicated by arrows, can be traced with the help of simple plastic devices called bottom-drifters. The gray band marks the boundary between landward flow and seaward flow. The broken line represents the edge of the continental shelf.

CHARACTER OF SHELF SEDIMENTS around an ocean basin, shown here schematically, is heavily influenced by oceanic currents. The rotation of the earth produces a clockwise flow in the Northern Hemisphere, so that the western edge of the basin, up to a certain latitude, is warmer than the eastern edge. Thus biogenic sediments extend farther north on the west than on the east. The effect would be greater except for a counterflow of arctic water on the western side of the basin. Detrital sediments are the typical outwash of continents. Authigenic sediments are minerals that come out of solution under suitable conditions and fall to the ocean floor.

latitudes and heated at low ones. At the same time, however, the pattern of currents in the open ocean displaces the climatic zones in a clockwise direction in the Northern Hemisphere and counterclockwise in the Southern Hemisphere. The displacement causes the water above the shelf at middle latitudes to be warmer on the western side of an ocean than on the eastern side [see bottom illustration on preceding page]. At high latitudes the flow of arctic water makes the shelf colder on the western side than on the eastern one. As a result the correlation of animal species with shelf latitude shows a displacement on the opposite sides of oceans. Moreover, temperature zones are compressed on the western side and expanded on the eastern. The movements of currents, waves and tides above the shelves are so complex that they have received little study compared with those of the deep ocean. Much fieldwork is needed.

The present great interest in exploring the world's continental shelves flows from their potential economic exploitation. About 90 percent of the world's marine food resources, now extracted at the rate of $8 billion per year, comes from the shelves and adjacent bays [see "The Food Resources of the Ocean," by S. J. Holt; SCIENTIFIC AMERICAN Offprint 886]. Most of this is fish for human and animal consumption; the remainder is largely used for fertilizer.

Second in economic importance is petroleum and natural gas from the shelf; their present annual value is about $4 billion, representing nearly a fifth of the total world production of these substances [see "The Physical Resources of the Ocean," by Edward Wenk, Jr.; SCIENTIFIC AMERICAN Offprint 885].

Currently about $1 billion worth of oil and gas a year is extracted from the shelves off the U.S., and much of the rest was developed by American companies with interests abroad. It is safe to predict that the future production from the world's continental shelves will increase at a greater rate than production from wells drilled on land.

The third marine resource in terms of present annual production and future potential is lowly sand and gravel. At present about $200 million worth per year is mined for landfill and road construction in the U.S., for concrete aggregate in Britain and for both purposes elsewhere. As cities and megalopolises continue to grow and show a preference for the coastal regions, and as readily available stream deposits are exhausted or are overlain by houses, there is every prospect that the offshore production of sand and gravel will increase greatly.

We read much about the possibility of economic exploitation of valuable heavy minerals from the sea floor, namely ilmenite, rutile, zircon, tin, monazite, iron, gold and diamonds. The total production of these minerals from below the sea is now less than $50 million per year. Production may increase, particularly in the case of tin, but it is decreasing for iron. Prospects for gold are not very hopeful, and diamonds have never been mined profitably from the sea floor. The basic problem is that economic placer deposits of tin and gold are found only within a few kilometers of the original igneous sources, and few continental shelves contain metalliferous igneous sources. Similarly, ilmenite, rutile, zircon and monazite require the high-energy wave environment of beaches to form deposits that are concentrated enough and large enough to be mined at a profit.

When ancient beach deposits are submerged, even if they are not buried under worthless sediment or mixed with it, the cost of mining increases substantially. They will probably be mined in the future but not until they are economically competitive with shoreline deposits. This could come about either through a rise in prices, resulting from a diminution of known deposits on land, or when more efficient mining and separation methods are devised for the marine environment.

Phosphorite is present in large quantities on shelves off southern California, Peru, southeastern Africa, northeastern Africa and Florida. It can be mined off the U.S., but it has to compete with high-grade land deposits in Florida, Montana, Idaho and Wyoming (where there is about a 1,000-year supply at present rates of mining). Most investigators have concluded that the cost of mining at sea exceeds the cost of mining on land plus the costs of land transportation. Some deposits far from the U.S., however, may justify mining, particularly because some of them may be near places where there is a great need for fertilizer, such as India. Unfortunately the distribution of phosphorite in these areas is poorly known, and little or no effort is currently being expended on their investigation.

The would-be exploiter of the ocean will do well to remember the words of the old Newfoundland skipper, "We don't be takin' nothin' from the sea. We has to sneak up on what we wants and wiggle it away." Nevertheless, the continental shelves, when they are properly investigated, promise to greatly increase our knowledge of the earth's history and to become a steadily more important source of food and raw materials.

The Author

K. O. EMERY is at the Woods Hole Oceanographic Institution, where he is involved in a program sponsored jointly by the institution and the U.S. Geological Survey to study the geological history of the continental shelf and slope from Maine to Florida. Following his departure in 1941 from the University of Illinois, where he had received his bachelor's degree in 1937 and his Ph.D. in 1941, he spent two years with the Illinois State Geological Survey and two years with the Division of War Research of the University of California. From 1945 to 1962 he taught and did research at the University of Southern California, where his work centered on marine geology. During most of that time he also participated in field and laboratory studies by the U.S. Geological Survey of marine geology at Bikini Atoll and other atolls, at Guam and in the Persian Gulf. He went to Woods Hole in 1962; last year he served as dean of graduate studies there.

Bibliography

ANCIENT OYSTER SHELLS ON THE ATLANTIC CONTINENTAL SHELF. Arthur S. Merrill, K. O. Emery and Meyer Rubin in *Science*, Vol. 147, No. 3656, pages 398–400; January 22, 1965.

CHARACTERISTICS OF CONTINENTAL SHELVES AND SLOPES. K. O. Emery in *Bulletin of the American Association of Petroleum Geologists*, Vol. 49, No. 9, pages 1379–1384; September, 1965.

THE ATLANTIC CONTINENTAL MARGIN OF THE UNITED STATES DURING THE PAST 70 MILLION YEARS. K. O. Emery in *The Geological Association of Canada Special Paper No. 4, Geology of the Atlantic Region*, pages 53–70; November, 1967.

RELICT SEDIMENTS ON CONTINENTAL SHELVES OF WORLD. K. O. Emery in *The American Association of Petroleum Geologists Bulletin*, Vol. 52, No. 3, pages 445–464; March, 1968.

SCIENTIFIC
AMERICAN September 1969, Vol. 221, No. 3, pp. 126–142

OFFPRINT 883

THE DEEP-OCEAN FLOOR

by H. W. Menard

The discovery that it is growing outward from the mid-ocean
ridges has suggested that it is formed in huge plates that
act as units in the dynamic processes of the earth's crust

Oceanic geology is in the midst of a revolution. All the data gathered over the past 30 years—the soundings of the deep ocean, the samples and photographs of the bottom, the measurements of heat flow and magnetism—are being reinterpreted according to the concept of continental drift and two new concepts: sea-floor spreading and plate tectonics (the notion that the earth's crust consists of plates that are created at one edge and destroyed at the other). Discoveries are made and interpretations developed so often that the scientific literature cannot keep up with them; they are reported by preprint and wandering minstrel. At such a time any broad synthesis is likely to be short-lived, yet so many diverse observations can now be fitted into a coherent picture that it seems worthwhile to present it.

Before continental drift, sea-floor spreading and plate tectonics captured the imagination of geologists, most of them conceived the earth's crust as being a fairly stable layer enveloping the earth's fluid mantle and core. The only kind of motion normally perceived in this picture was isostasy: the tendency of crustal blocks to float on a plastic mantle. The horizontal displacement of any geologic feature by as much as 100 kilometers was considered startling. This view is no longer consistent with the ge-

PILLOW LAVA assumes its rounded shape because it cools rapidly in ocean water. This flow lies on the western slope of the mid-ocean ridge in the South Atlantic at a depth of 2,650 meters. Such flows erupt from the many volcanic vents and fissures that are created as the ocean floor spreads out from the mid-ocean ridges in the form of vast crustal plates. The photograph was made under the direction of Maurice Ewing of the Lamont-Doherty Geological Observatory.

ological evidence. Instead each new discovery seems to favor sea-floor spreading, continental drift and plate tectonics. These concepts are described elsewhere in this issue [see "The Origin of the Oceans," by Sir Edward Bullard; SCIENTIFIC AMERICAN Offprint 880]. Here I shall recapitulate them briefly to show how they are related to the actual features of the deep-ocean floor.

According to plate tectonics the earth's crust is divided into huge segments afloat on the mantle. When such a plate is in motion on the sphere of the earth, it describes a circle around a point termed the pole of rotation (not to be confused with the entire earth's pole of rotation). This motion has profound geological effects. When two plates move apart, a fissure called a spreading center opens between them. Through this fissure rises the hot, plastic material of the mantle, which solidifies and joins the trailing edge of each plate. Meanwhile the edge of the plate farthest from the spreading center—the leading edge—pushes against another plate. Where that happens, the leading edge may be deflected downward so that it sinks into a region of soft material called the asthenosphere, 100 kilometers or more below the surface. This process destroys the plate material at the same rate at which it is being created at its trailing edge. Many of the fissures where plate material is being created are in the middle of the ocean floor, which therefore spreads continuously from a median line. Where the plates float apart, the continents, which are embedded in them, also drift away from one another.

The most obvious consequence of this process on the ocean floor is the symmetrical seascape on each side of a spreading center. As two crustal plates

move apart (at a rate of one to 10 centimeters per year), the basaltic material that wells up through the spreading center between them splits down the middle. The upwelling in some way produces a ridge, flanked on each side by deep ocean basins and capped by long hills and mountains that run parallel to the crest. The flow of heat from the earth's interior is generally high along the crest because dikes of molten rock have been injected at the spreading center. A spreading center may also open under a continent. If it does, it produces a linear deep such as the Red Sea or the Gulf of California. If it continues to spread, or if the spreading center opens in an existing ocean basin, the same symmetrical seascape is ultimately formed.

This symmetry extends to less tangible features of the ocean floor such as the magnetic patterns in the basalt of the slopes on either side of the mid-ocean ridge. As the plastic material reaches the surface and hardens, it "freezes" into it the direction of the earth's magnetic field. The earth's magnetic field reverses from time to time, and as each band of new material moves outward across the ocean floor it retains a magnetic pattern shared by a corresponding band on the other side of the ridge. The result is a matching set of parallel bands on both sides. These patterns provide evidence of symmetrical flows and make it possible to date them, since they correspond to similar patterns on land that have been reliably dated by other means.

The steepness of the mid-ocean ridge is determined by a balance between the rate at which material moves outward from the spreading center and the rate at which it sinks as it ages after solidifying. The rate of sinking remains fairly constant throughout the ocean basin, and it seems to depend on the age of the

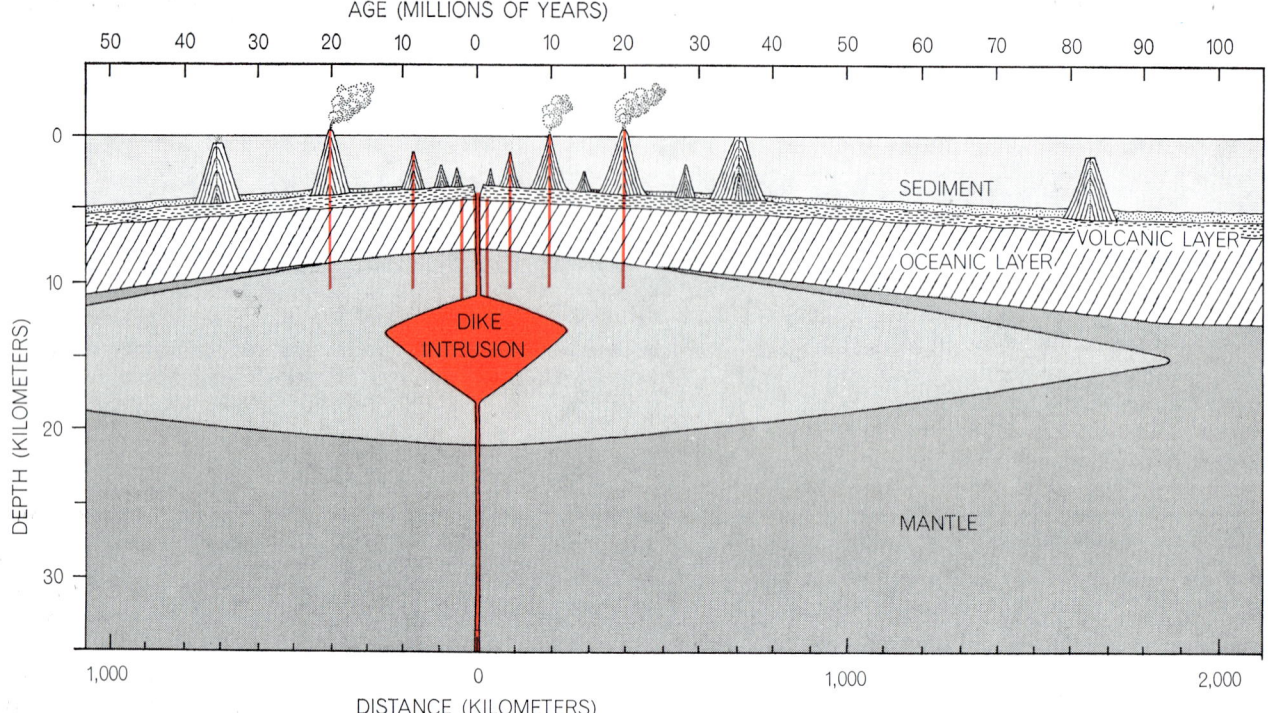

FAST SPREADING of the sea floor is revealed by gentle slopes. The sea floor is created at a spreading center that leaks molten rock from several dikes intruding from a pool in the low-density mantle (*light shading*). As the molten rock emerges it cools and adheres to the crust sliding away on each side of the fissure. If the crust moves at more than three centimeters per year, the slopes are gradual because spreading, which is horizontal, is rapid compared with the sinking of the crust. The balance between the two determines the steepness of the slope. Fast spreading also produces a thin volcanic layer because material moves so quickly from the fissure that it cannot accumulate. Islands built by eruptions are distant from the center because they grow on rapidly moving crust.

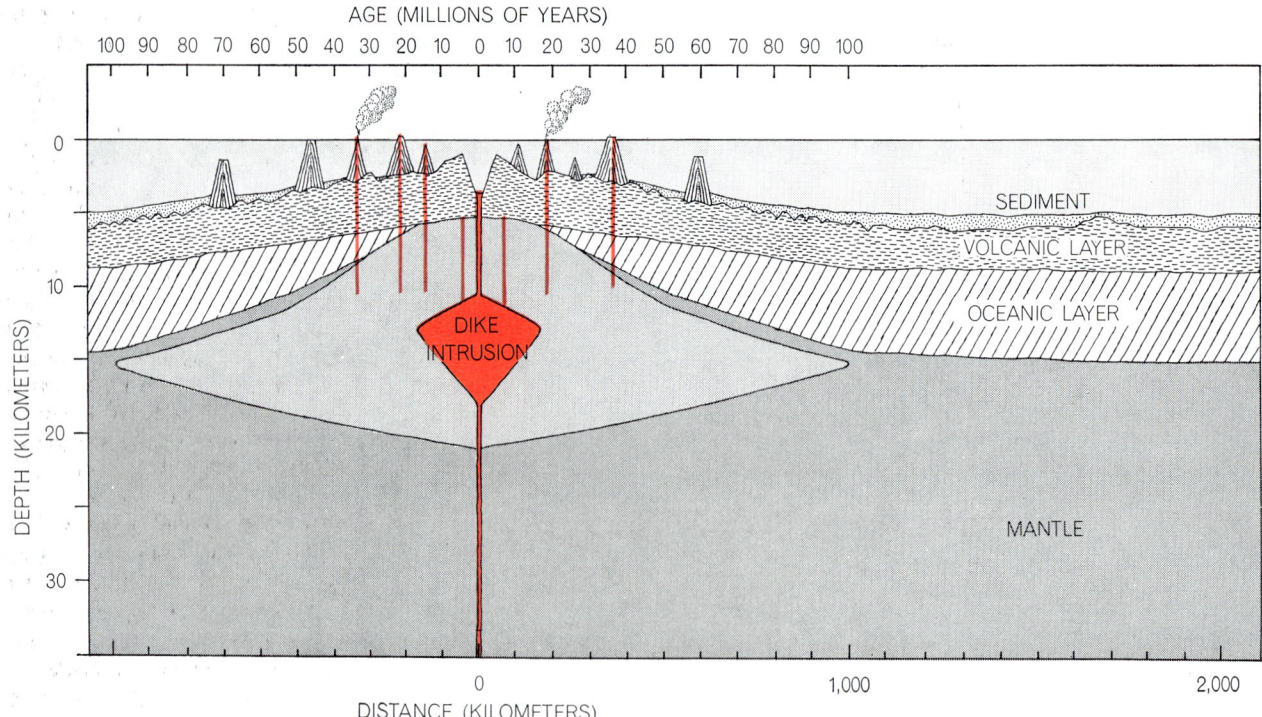

SLOW SPREADING produces steep slopes. Here the crust moves less than three centimeters per year; consequently sinking dominates the slope-forming process and produces steep escarpments. The volcanic layer is thicker at a slow center because material has time to accumulate. Mountains and volcanoes are high near the spreading center because the crust moves so slowly that the lava piles up. After 100 million years crust produced from a slow spreading center has strong similarity to crust from a fast center. Both kinds of crust have sunk to a depth of five kilometers. The oceanic layer is about five kilometers thick. Both fast- and slow-spreading crust are covered by the same kind of sediment. Slow spreading occurs mainly in the Atlantic, fast spreading in the Pacific.

crust. It can be calculated if the age of the oceanic crust (as indicated by the magnetic patterns) is divided into the depth at which a particular section lies. Such calculations show that the crust sinks about nine centimeters per 1,000 years for the first 10 million years after it forms, 3.3 centimeters per 1,000 years for the next 30 million years, and two centimeters per 1,000 years thereafter. Not all the crust sinks: on the southern Mid-Atlantic Ridge the sea floor has remained at the same level for as long as 20 million years.

The rate at which the sea floor spreads varies from one to 10 centimeters per year. Therefore fast spreading builds broad elevations and gentle slopes such as those of the East Pacific Rise. The steep, concave flanks of the Mid-Atlantic Ridge, on the other hand, were formed by slow spreading.

Whether the slopes are steep or gentle, the trailing edge of the plate at the spreading center is about three kilometers higher than the leading edge on the other side of the plate. The reason for this difference in elevation is not known. Heating causes some elevation and cooling some sinking, but the total relief appears much too great to be attributed to thermal expansion. Cooling might account for the relatively rapid sinking observed during the first 10 million years, but continued sinking remains a puzzle.

A decade ago scanty information suggested that the mid-ocean ridge in both the Atlantic and Pacific was continuous, with a few branches. More complete surveys have revealed that crustal plates have ragged edges. Instead of extending unbroken for thousands of kilometers, a mid-ocean ridge at the trailing edges of two crustal plates forms a zigzag line consisting of many short segments connected by fracture zones to other ridges, trenches, young mountain ranges or crustal sinks. The fracture zones connecting the ridge segments are associated with what are called "transform" faults. They provide important clues to the history of a plate. Because they form some of the edges of the plate, they delineate the circle around its pole of rotation, thereby indicating the direction in which it has been moving.

From what has been said so far it might appear that the spreading centers are fixed and stationary. The constantly repeated splitting of the new crust at the spreading center produces symmetrical continental margins, symmetrical magnetic patterns on the ocean floor, symmetrical ridge flanks and even

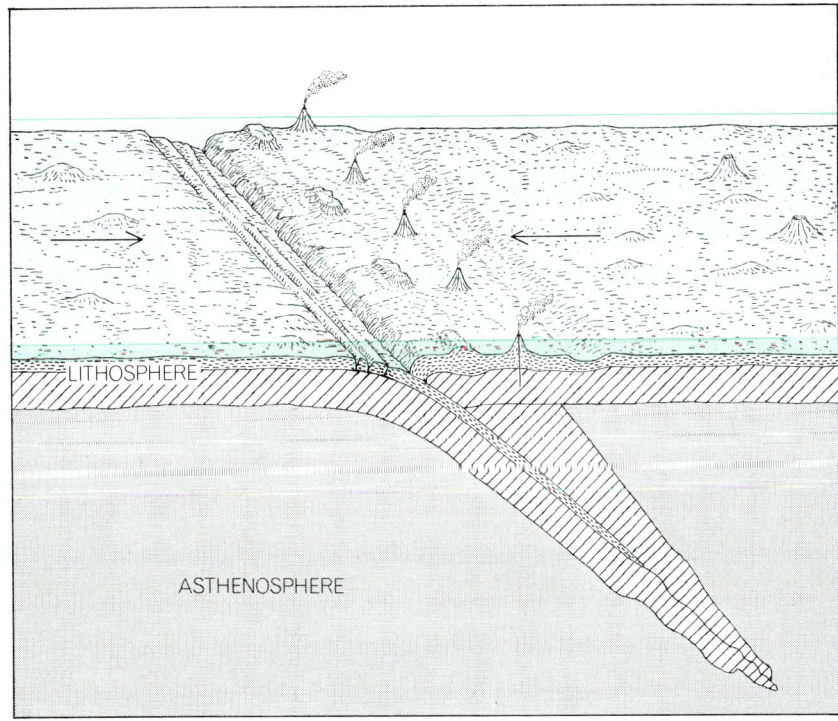

TRENCH IS CREATED where the leading edge of a plate that emerges from a fast spreading center collides with another plate. Because the combined speed of the two is more than six centimeters per year neither can absorb the impact by buckling. Instead one crustal plate (in lithosphere) plunges under the other to be destroyed in the asthenosphere, a hot, weak layer below. The impact produces volcanoes, islands and a deep, such as the Tonga Trench. Beside a trench are cracks that are produced by bending of the crust.

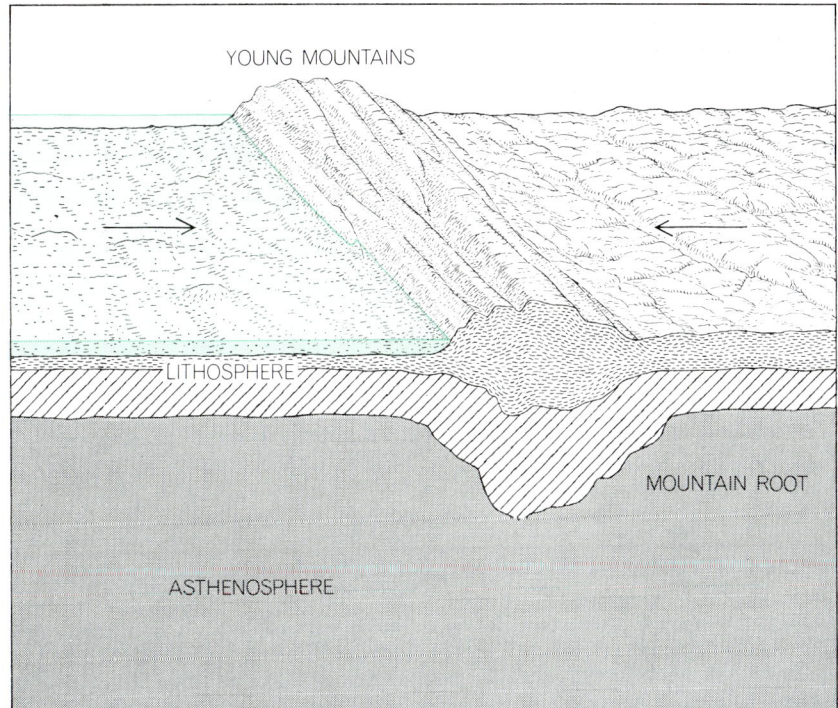

MOUNTAIN RANGE IS FORMED when the leading edges of two plates come together at less than six centimeters per year. Instead of colliding catastrophically, so that one plate slides under the other, both plates buckle, raising a young mountain range between them. The range consists of crustal material that folds upward under the compression exerted by the two plates (and also downward, forming the root of the mountain). Such ranges can be identified because they contain cherts and other material typical of the ocean bottom.

symmetrical mountain ranges. More often than not, however, it has been found that the spreading center itself moves. Oddly enough such movement gives rise to the same symmetrical geology. All that is required in order to maintain the symmetry is that the spreading center move at exactly half the rate at which the plates are separating. If it moved faster or slower, the symmetry of the magnetic patterns would be destroyed.

Imagine, for instance, that the plate to the east of a spreading center remains stationary as the plate to the west moves. Since the material welling up through the fissure splits down the middle, half of it adheres to the stationary plate and the other half adheres to the moving plate. The next flow of material to well up through the split thus appears half

the width of the spreading center away from the stationary plate. The flow after that appears a whole width of the spreading center away from the stationary plate, and so on. In effect the spreading center is migrating away from the stationary plate and following the moving one. If the speed of the spreading center exceeded half the speed of the migrating plate, however, a kind of geological Doppler effect would set in: the bands of the magnetic pattern would be condensed in the direction of the moving plate, and they would be stretched out in the direction of the stationary plate [see illustration below].

It might seem unlikely that the spreading center would maintain its even rate of speed and remain exactly between the two plates. W. Jason Morgan of Princeton University observes, how-

ever, that there is no impediment to such motion, provided only that the crust splits where it is weakest (which is where it split before, at the point where the hot dike was originally injected). As a result the spreading center is always exactly between two crustal plates whether it moves or not.

Moving spreading centers account for some of the major features of the ocean floor. The Chile Rise off the coast of South America and the East Pacific Rise are adjacent spreading centers. Since there is no crustal sink between them, and new plate is constantly being added on the inside edge of each rise, at least one of the centers must be moving, otherwise the basin between them might fold and thrust upward into a mountain range or downward into a trench. Similarly, the Carlsberg Ridge in the Indian Ocean and the Mid-Atlantic Ridge are not separated by a crustal sink and hence one of them must be moving.

A moving center may have created the ancient Darwin Rise on the western edge of the Pacific basin and also the modern East Pacific Rise. As in the case of the Carlsberg Ridge and the Mid-Atlantic Ridge, the existence of two vast spreading centers on opposite sides of the ocean with no intervening crustal sink has puzzled geologists. If such centers can move, however, it is possible that the spreading center in the western Pacific merely migrated all the way across the basin, leaving behind the ridges of the Darwin Rise. In this way one rise could simply have become the other. Many other examples exist, and Manik Talwani of the Lamont-Doherty Geological Observatory proposes that all spreading centers move.

As a plate forms at a spreading center it consists of two layers of material, an upper "volcanic" layer and a lower "oceanic" one. Lava and feeder dikes from the mantle form the volcanic layer; its rocks are oceanic tholeiite (or a metamorphosed equivalent), which is rich in aluminum and poor in potassium. The oceanic layer is also some form of mantle material, but its precise composition, density and condition are not known. Farther down the slope of the ridge the plate acquires a third layer consisting of sediment.

The sediment comes from the continents and sifts down on all parts of the basin, accumulating to a considerable depth. It is mixed with a residue of the hard parts of microorganisms that is called calcareous ooze. Below a certain depth (which varies among regions) this

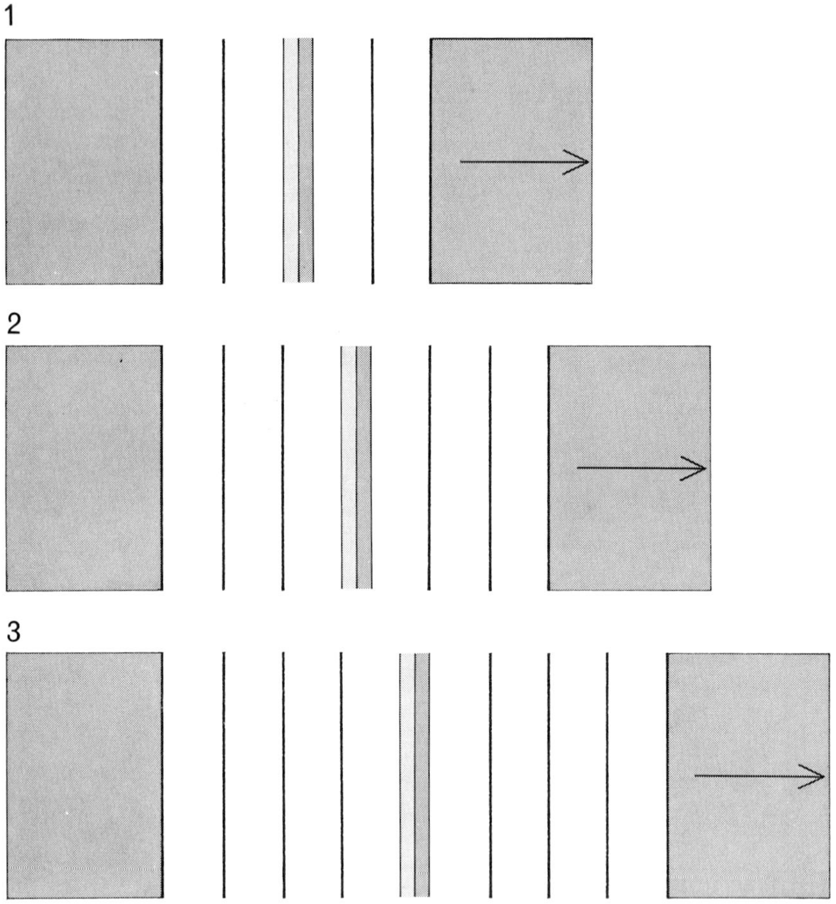

1

2

3

SPREADING CENTER MOVES, yet it can still leave a symmetrical pattern of magnetized rock. The molten material emerging from a spreading center becomes magnetized because as it cools it captures the prevailing direction of the earth's periodically changing magnetic field. In the instance illustrated here the right-hand plate moves out to the right while the left-hand plate remains stationary. In 1 hot material from a dike arrives at the surface, cools and splits down the middle. In 2 the next injection of material arrives in the crevice between the two halves of the preceding mass of rock. The new mass is therefore half the width of the preceding mass farther from the stationary plate than the preceding mass of material itself was. In 3 the new material has cooled and split in its turn and another mass has appeared that is a whole width farther from the left-hand plate. As long as the center moves at half the speed at which the right-hand plate moves away the magnetic bands remain symmetrical. If plate moved faster or more slowly, they would be jumbled.

material dissolves, and only red clay and other resistant components remain.

For reasons only partly known the sediment is not uniformly distributed. At the spreading center the newly created crust is of course bare of sediment, and within 100 kilometers of such a center the calcareous ooze is rarely thick enough to measure. The ooze accumulates at an average rate of 10 meters for each million years, during which time the plate moves horizontally from 10,000 to 100,000 meters and sinks 100 meters. Where the red clay appears, it accumulates at a rate of less than one meter per million years.

The puzzle deepens when one considers that sediment on oceanic crust older than 20 million years stops increasing in thickness after it sinks to the depth where the calcareous ooze dissolves. Indeed, in many places the age of the oldest sediment is about the same as the volcanic layer on which it lies. It would therefore seem that almost all the deep-ocean sediment accumulates in narrow zones on the flanks of the mid-ocean ridges. If this is correct, it has yet to be explained.

The volcanic layer forms mainly at the spreading center. Volcanoes and vents on the slopes of the mid-ocean ridge contribute a certain amount of oceanic tholeiite to it. It can be said in general that the thickness of the volcanic layer decreases as the spreading rate increases. If the crust spreads slowly, the material has time to accumulate. Fast spreading reduces this time and therefore the accumulation. The conclusion can be drawn that the rate at which the volcanic-layer material is discharged is nearly constant. These relations are based on only 10 observations, but they apply to spreading rates from 1.4 centimeters to 12 centimeters per year, and to thicknesses from .8 kilometer to 3.8 kilometers.

The total flow of volcanic material from all active spreading centers is about four cubic kilometers per year—four times the flow on the continents. Not the slightest sign of this volcanism on the ocean bottom can be detected at the surface of the ocean, with one possible exception: late in the 19th century a ship reported seeing smoke rising from the waters above the equatorial Mid-Atlantic Ridge. The British oceanographer Sir John Murray remarked that he hoped the smoke signified the emergence of an island, since the Royal Navy needed a coaling station at that point.

Like the volcanic layer, the oceanic

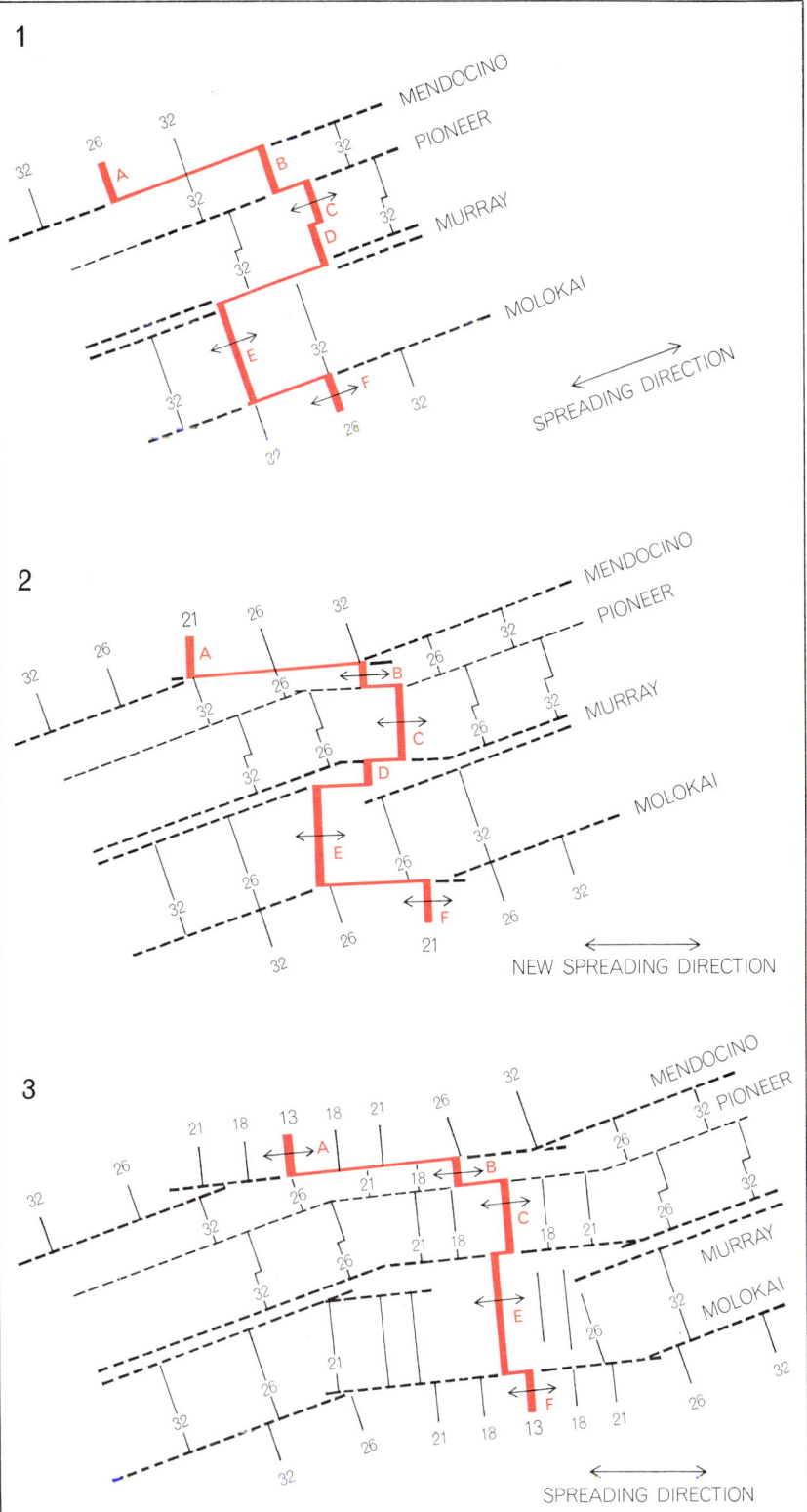

HOW A PLATE MOVES is revealed by the patterns of magnetic bands (time of formation is indicated by numbers) and by the relation between ridges and faults in the northeastern Pacific. At the time illustrated in *1* material from the Murray fault and other faults connected segments of the ridge (*indicated by letters*) offset from one another by plate motion. At the time shown in *2* the spreading direction changed. The readjustment of the plates has shortened the Pioneer-Mendocino ridge segment (*B*) while lengthening and reorienting the Pioneer, Mendocino and Molokai transform faults. In *3* the Mendocino fault remains the same length, but between most of the other ridges faults have been shortened or have almost disappeared as ridge segments tended to rejoin one another. Between Murray and Molokai faults ridge has jumped eastward, and one segment (*D*) has vanished.

layer forms at the spreading center. Acoustical measurements of the thickness of this layer at spreading centers, on the flanks of mid-ocean ridges and on the deep-ocean floor show, however, that at least part of the oceanic layer evolves slowly from the mantle rather than solidifying quickly and completely at the spreading center. At the spreading center the thickness of the layer depends on how fast the ocean floor moves. In regions such as the South Atlantic, where the floor spreads at a rate of two centimeters per year, no oceanic layer forms within a few hundred kilometers of the spreading center. Farther away from the spreading center the oceanic layer accumulates rapidly, reaching a normal thickness of four to five kilometers on the flank of the mid-ocean ridge. A spreading rate of three centimeters per year is associated with an oceanic layer roughly two kilometers deep at the center that thickens by one kilometer in 13 million years. A plate with a spreading rate of eight centimeters per year is three kilometers deep at the center and thickens by one kilometer in 20 million years. Thus the thinner the initial crust is, the faster the thickness increases as the crust spreads.

As a plate flows continuously from the spreading center, faulting, volcanic eruptions and lava flows along the length of the mid-ocean ridge build its mountains and escarpments. This process can be most easily observed in Iceland, a part of the Mid-Atlantic Ridge that has grown so rapidly it has emerged from the ocean. A central rift, 45 kilometers wide at its northern end, cuts the island parallel to the ridge. The sides of the rift consist of active, steplike faults. There are other step faults on the rift floor, which is otherwise dominated by a large number of longitudinal fissures. Some of these fissures are open and filled with dikes. Fluid lava wells up from the fissures and either buries the surrounding mountains, valleys and faults or forms long, low "shield" volcanoes. Two hundred such young volcanoes, which have been erupting about once every five years over the past 1,000 years, dot the floor of the rift. Thirty of them are currently active.

Just as a balance between spreading and sinking shapes the slopes of the mid-ocean ridges, so does a balance between lava discharge and spreading build undersea mountains, hills and valleys. High mountains normally form at slow centers where spreading proceeds at two to 3.5 centimeters per year. In contrast, a spreading center that opens at a rate of five to 12 centimeters per year produces long hills less than 500 meters high. This relationship is a natural consequence of the long-term constancy of the lava discharge. Over a short period of time, however, the lava discharge may fluctuate or pulsate, a picture suggested by the fact that the thick volcanic layer associated with slow spreading can consist of volcanic mountains (which represent copious flow) separated by valleys covered by a thinner volcanic layer.

The volcanic activity and faulting that first appear near the spreading centers decrease rapidly as the plate ages and material moves toward its center, but volcanic activity in some form is never entirely absent. Small conical volcanoes are found on crust only a few hundred thousand years old near spreading centers, and active circular volcanoes such as those of Hawaii exist even in the middle of a plate. It would appear that the great cracks that serve as conduits for dikes and lava flows are soon sealed as a plate ages and spreads. Volcanic activity is then concentrated in a few central vents, created at different times and places.

Many of these vents remain open for tens of millions of years, judging by the size and distribution of the different classes of marine volcanoes. First, the biggest volcanoes are increasingly big at greater distances from a spreading center, which means they must continue to erupt and grow as the crust ages and sinks, even when the age of the crust exceeds 10 million years. In most places, in fact, a volcano needs at least 10 million years in order to grow large enough to become an island. Volcanoes that discharge lava at a rate lower than 100 cubic kilometers per million years never become islands because the sea floor sinks too fast for them to reach the sea surface.

Other volcanoes drifting with a spreading ocean floor may remain active or become active on crust that is 100 million years old (as the volcanoes of the Canary Islands have). Normally, however, volcanoes become inactive by the time the crust is 20 to 30 million years old. This is demonstrated by the existence of guyots, drowned ancient island volcanoes that were submerged by the gradual sinking of the aging crust. Guyots are found almost entirely on crust that is more than 30 million years old, such as the floor of the western equatorial Pacific.

Traditionally it has been thought that marine volcanoes spew lava from a magma chamber located deep in the mantle. Some volcanoes have a top composed of alkali basalt, slopes with transitional basalt outcrops and a base of oceanic tholeiite, and it was therefore assumed that the lava in the magma chamber became differentiated into components that, rather like a pousse-café, separated into several layers of different

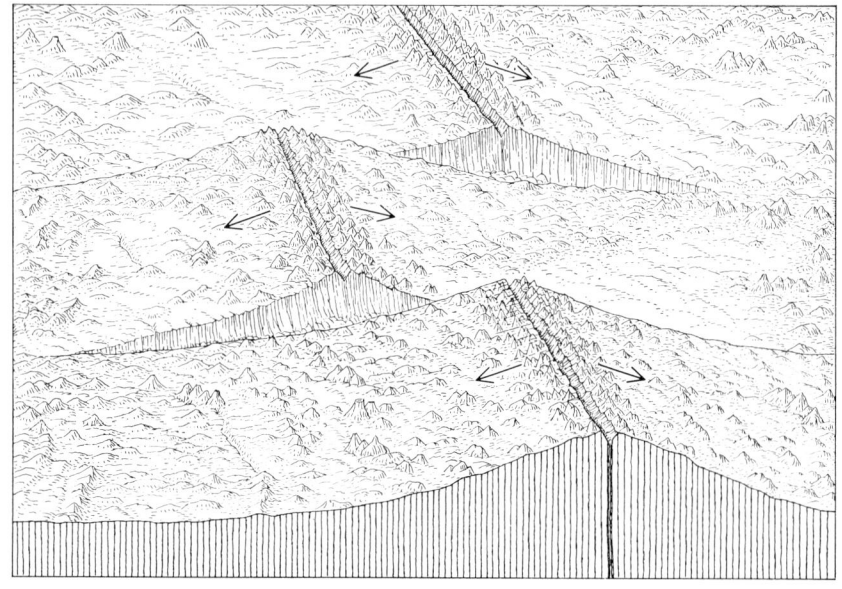

RIDGE-RIDGE TRANSFORM FAULT appears between two segments of ridge that are displaced from each other. Mountains are built, earthquakes shake the plate edges and volcanoes erupt in such an area because of the forces generated as plates, formed at the spreading centers under the ridges, slide past each other in opposite directions. On outer slopes of the mid-ocean ridges, however, this intense seismic activity appears to subside.

kinds of material, each of which followed the layer above it up the spout. The emergence of plate tectonics and continental drift as respectable concepts have now brought this view of volcanic action into question.

It remains perfectly possible for a volcano to drift for tens of millions of years over hundreds of kilometers while tapping a single magma chamber embedded deep in the mantle. The motion of the plates, however, suggests another hypothesis. According to this view, the volcano and its conduit drift along with the crust as the conduit continually taps different parts of a relatively stationary magma that is ready to yield various kinds of lava whenever a conduit appears. In actuality the composition of the lava usually changes only slightly after the first 10 to 20 kilometers of drifting. Although the older hypothesis is still reasonable, the newer one must also be considered because it explains the facts equally well.

In addition to their characteristic volcanoes and mountains, spreading centers are marked by median valleys, which in places such as the North Atlantic or the northwestern Indian Ocean are deeper than the surrounding region. These rifts are commonly found in centers opening at a rate of two to five centimeters per year. The deepest rifts, which may go as deep as 1,000 to 1,300 meters below the surrounding floor, are associated with spreading at three to four centimeters per year. Only one valley is known to be associated with spreading at five to 12 centimeters per year. Although rifts are not found in all spreading centers, they usually do appear in conjunction with a slow center. Both of these features are also associated with volcanic activity.

The mid-ocean ridges, as we have noted, seldom run unbroken for more than a few hundred kilometers. They are interrupted by fracture zones, and the segments are shifted out of line with respect to one another. These fracture zones run at right angles to the ridge and connect the segments. Where they lie between the segments they are termed ridge-ridge transform faults, which are the site of intense geological activity. As the two edges of the fault slide past each other they rub and produce earthquakes. The slope of a transform fault drops steeply from the crest of one ridge segment to a point halfway between it and the adjacent segment and then climbs to the top of the adjacent segment, reflecting the fact that the crust is elevated at the spreading center and subsides at some distance from it [see illustration on page 722].

Like spreading centers, fracture zones have their own complex geology. In these

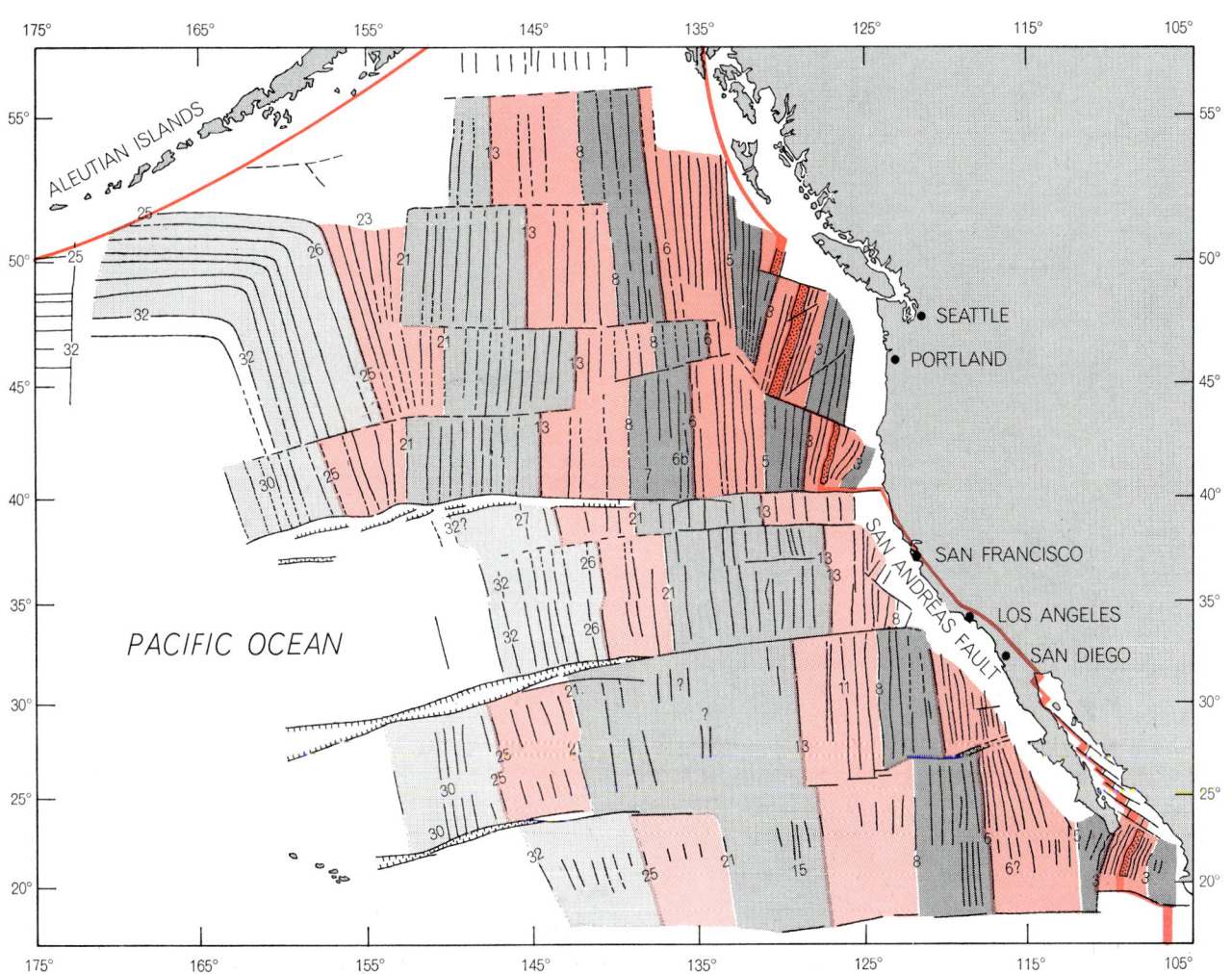

MAGNETIC PATTERNS reveal how the plate forming the floor of the northeastern Pacific has moved. Its active eastern edge now stretches from Alaska through California (where it forms the San Andreas fault) to the Gulf of California. In the gulf spreading centers break into short segments joined by active faults. Plate motion is opening the gulf and moving coastal California in the direction of the Aleutians. To the south lies the Great Magnetic Bight, formed by three plates that spread away from one another.

areas the ridges stand as much as several kilometers high, and the troughs are equally deep. It appears that the same volcanic forces that shape the main ridges produce the mountains and valleys of the faults. As fracture zones open they slowly leak lava from hot dikes. At the same time the crust sinks away from the fault line, and this balance produces high mountains.

Beyond the spreading centers the fracture zones become the inactive remains of earlier faulting. The different rates at which these outer flanks of the mid-ocean ridge sink do produce some vertical motion as the scarps of the fracture zone decay. This may account for the few earthquakes in these areas. I should emphasize that it is not known if horizontal motion is also absent from such dead fracture zones. It is not necessary, however, to postulate such motion in order to explain existing observations.

A fracture zone can become active again at any time, but if it does so, it becomes the side of a smaller new plate rather than part of the trailing edge of an old one. If the flank fracture zones are as quiescent as they appear to be, then the plates forming the earth's crust are large and long-lived. If these fracture zones were active, on the other hand, it could only be concluded that each one marked the flank of a small, elongated plate.

The direction the plate is moving can be deduced from the magnetic pattern that runs at right angles to the fractures in the fracture zone. When the direction of plate motion changes, the direction in which the fracture zone moves also changes. This change in direction can be most clearly seen in the northeastern Pacific, where our knowledge is most detailed. On this part of the ocean floor the changes of direction have taken place at the same time in many zones, indicating that the entire North Pacific plate has changed direction as a unit [see illustration on page 723].

On the bottom of the Pacific and the North Atlantic the magnetic patterns are sometimes garbled. Old transform faults may have vanished if short segments of spreading center have been united by reorientation. By the same token new transform faults may have formed if the change in plate motion has been too rapid to be accommodated by existing motions. Thus fracture zones may be discontinuous. They may start and stop abruptly, and the offset of the magnetic patterns may change from place to place along them without indicating any activity except at the former edges of plates.

Some patterns are even harder to interpret. Douglas J. Elvers and his colleagues in the U.S. Coast and Geodetic Survey discovered an abrupt boomerang-shaped bend in the magnetic pattern south of the Aleutians. The arms of this

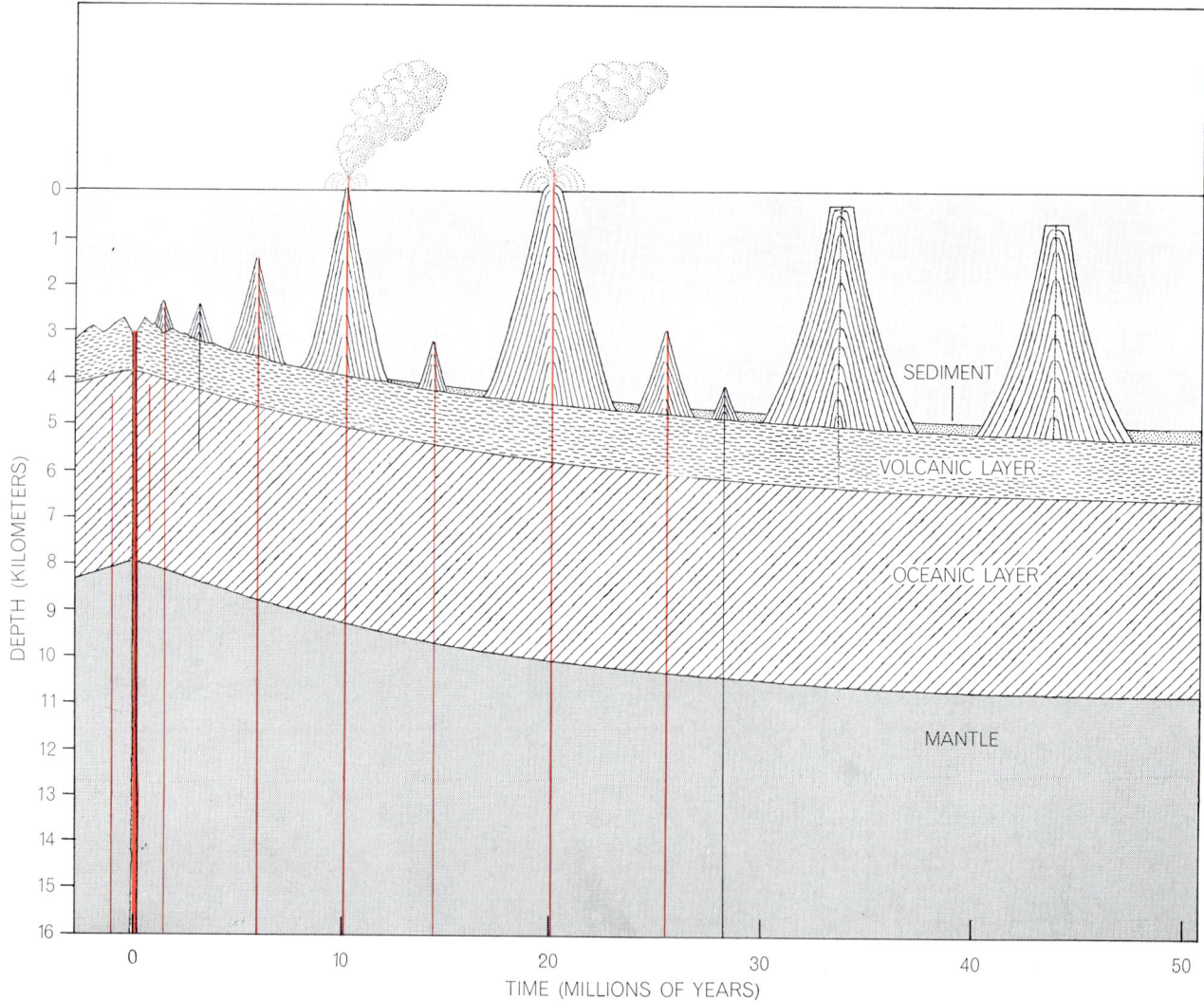

UNDERSEA VOLCANOES normally begin to rise near spreading centers. Then they ride along on the moving plate as they grow. If a volcano rises fast enough to surmount the original depth of the water and the sinking of the ocean floor, it emerges as an island such as St. Helena in the South Atlantic. To rise above the water an undersea volcano must grow to a height of about four kilometers in 10 million years. Island volcanoes sink after 20 to 30 million years and become the sediment-capped seamounts called guyots.

configuration, which Elvers calls the Great Magnetic Bight, are offset by fracture zones at right angles to the magnetic pattern. The Great Magnetic Bight seems to have required impossible forms of seafloor spreading and plate movement. However, Walter C. Pitman III and Dennis E. Hayes of the Lamont-Doherty Geological Observatory, among others, have been able to show that the configuration is fully understandable if it is assumed that the trailing edges of three plates met and formed a Y. Similar complex patterns have now been found in the Atlantic and the Pacific. Indeed, if the transform faults are perpendicular to the spreading centers and two spreading rates are known, both the orientation and the spreading rate of the third spreading center can be calculated even before it is mapped. One can also calculate the orientation and spreading rate for a third center that has already vanished in a trench.

As a crustal plate grows, its leading edge is destroyed at an equal rate. Sometimes this edge slides under the oncoming edge of another plate and returns to the asthenosphere. When this happens, a deep trench such as the Mariana Trench and the Tonga Trench in the Pacific is formed. In other areas the movement of the crust creates young mountain ranges. Xavier Le Pichon of the Lamont-Doherty Geological Observatory concludes that the occurrence of one event or the other is a function of the rate at which plates are moving together. If the rate is less than five to six centimeters per year, the crust can absorb the compression and buckles up into large mountain ranges such as the Himalayas. In these ranges folding and overthrusting deform and shorten the crust. If the rate is higher, the plate breaks free and sinks into the mantle, creating an oceanic trench in which the topography and surface structure indicate tension.

Several crustal sinks are no longer active. Their past, however, can be deduced from their geology. Large-scale folding and thrust-faulting can be taken as evidence of the former presence of a crustal sink, although such deformation can also arise in other ways. Certain types of rock may also indicate the formation of a trench. The arcs of islands that lie parallel to trenches, for instance, are characterized by volcanoes that produce andesitic lavas, which are quite different from the basaltic lavas of the ocean floor. The trenches themselves, and the deep-sea floor in general, are featured by deposits of graywackes and cherts. These rocks are commonly found exposed on land in the thick prisms of

sediment that lie in geosynclines: large depressed regions created by horizontal forces resembling those generated by a drifting crustal plate. Thus the presence of some or all of these types of rock may indicate the former existence of a crustal sink. This linking of marine geology at spreading centers with land geology at crustal sinks is becoming one of the most fruitful aspects of plate tectonics. Still, crustal sinks are by no means as informative about the history of the ocean floor as spreading centers are, because in such sinks much of the evidence of past events is destroyed. Even if the leading edge of a plate was once the side of a plate (or vice versa), there would be no way to tell them apart.

At the boundary between the land and the sea a puzzle presents itself. The sides of an oceanic trench move together at more than five centimeters per year, and it would seem that the sediment sliding into the bottom of the trench should be folded into pronounced ridges and valleys. Yet virtually undeformed sediments have been mapped in trenches by David William Scholl and his colleagues at the U.S. Naval Electronics Laboratory Center. Furthermore, the enormous quantity of deep-ocean sediment that has presumably been swept up to the margins of trenches cannot be detected on sub-bottom profiling records. There are many ingenious (but unpublished) explanations of the phenomenon in terms of plate tectonics. One of them may conceivably be correct. According to that hypothesis, the sediments are intricately folded in such a way that the slopes and walls of trenches cannot be detected by normal survey techniques, which look at the sediments from the ocean surface and along profiles perpendicular to the slopes. This kind of folding could be detected only by trawling a recording instrument across the trench much closer to the bottom or by crossing the slope at an acute angle.

The concepts of sea-floor spreading and plate tectonics allow a quantitative evaluation of the interaction of many important variables in marine geology. By combining empirical observation with theory it is possible not only to explain but also to predict the thickness and age of sediments in a given locality, the scale and orientation of topographic relief, the thickness of various crustal layers, the orientation and offsetting of magnetic patterns, the distribution and depth of drowned ancient islands, the occurrence of trenches and young mountain ranges, the characteristics of earthquakes, and many other previously unrelated and un-

predictable phenomena. This revolution in marine geology may take some years to run its course. Ideas are changing, and new puzzles present themselves even as the old ones are solved. The only certainty is that the subject will never be the same again.

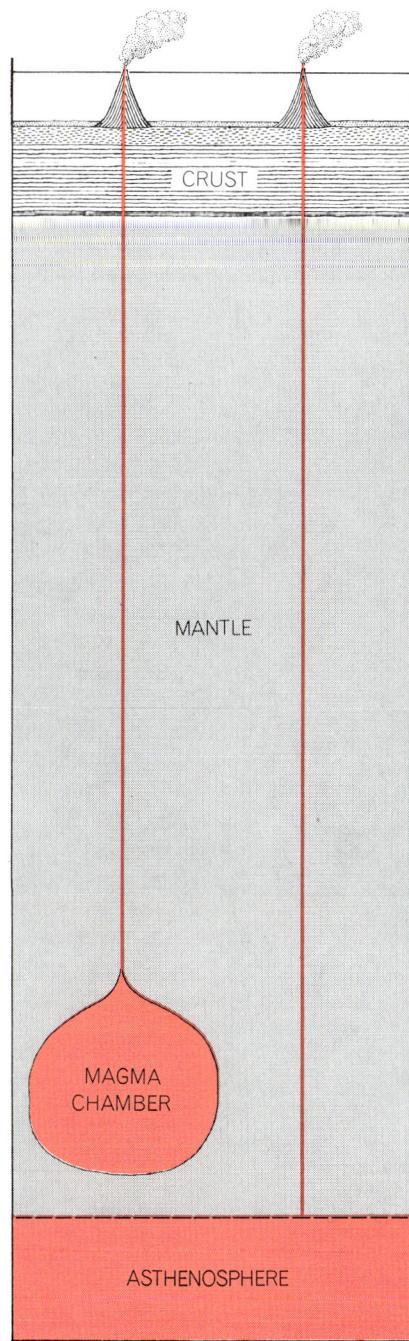

UNDERSEA ERUPTIONS can be explained in two ways, both consistent with observed facts. Since a volcano consists of different kinds of rock, it was originally thought that its conduit carried different forms of lava up from a magma chamber 50 kilometers or more down in the mantle. Now that crustal plates have been found to move, another theory must be considered. According to this idea, the conduit reaches through the mantle and taps several different kinds of magma at different places in the asthenosphere.

The Author

H. W. MENARD is professor of marine geology at the Institute of Marine Resources and the Scripps Institution of Oceanography of the University of California at San Diego. He received his bachelor's and master's degrees at the California Institute of Technology; in 1949 he obtained his Ph.D. from Harvard University. Since then he has been concerned with marine geology; his work has included participation in 17 oceanographic expeditions, mostly in the Pacific. He has also been a visiting professor at Cal Tech, a Guggenheim Fellow at the University of Cambridge and a staff member of the Office of Science and Technology in Washington. His article is the third he has written for SCIENTIFIC AMERICAN. He is also the author of a forthcoming book, *Anatomy of an Expedition,* describing modern scientific life at sea.

Bibliography

HISTORY OF OCEAN BASINS. H. H. Hess in *Petrologic Studies: A Volume in Honor of A. F. Buddington,* edited by A. E. J. Engel, Harold L. James and B. F. Leonard. The Geological Society of America, 1962.

A NEW CLASS OF FAULTS AND THEIR BEARING ON CONTINENTAL DRIFT. J. Tuzo Wilson in *Nature,* Vol. 207, No. 4995, pages 343–347; July 24, 1965.

SPREADING OF THE OCEAN FLOOR: NEW EVIDENCE. F. J. Vine in *Science,* Vol. 154, No. 3755, pages 1405–1415; December 16, 1966.

SEA FLOOR SPREADING, TOPOGRAPHY, AND THE SECOND LAYER. H. W. Menard in *Transactions American Geophysical Union,* Vol. 48, No. 1, page 217; March, 1967.

RISES, TRENCHES, GREAT FAULTS AND CRUSTAL BLOCKS. W. Jason Morgan in *Journal of Geophysical Research,* Vol. 73, No. 6, pages 1959–1982; March 15, 1968.

SCIENTIFIC
AMERICAN   September 1969, Vol. 221, No. 3, pp. 146–162   OFFPRINT 884

THE NATURE OF OCEANIC LIFE

by John D. Isaacs

The conditions of the marine environment have given rise
to a food web in which the dominant primary production
of organic matter is carried out by microscopic plants.

I plan to take the reader on a brief tour of marine life from the surface layers of the open sea, down through the intermediate layers to the deep-sea floor, and from there to the living communities on continental shelves and coral reefs. Like Dante, I shall be able to record only a scattered sampling of the races and inhabitants of each region and to point out only the general dominant factors that typify each domain; in particular I shall review some of the conditions, principles and interactions that appear to have molded the forms of life in the sea and to have established their range and compass.

The organisms of the sea are born, live, breathe, feed, excrete, move, grow, mate, reproduce and die within a single interconnected medium. Thus interactions among the marine organisms and interactions of the organisms with the chemical and physical processes of the sea range across the entire spectrum from simple, adamant constraints to complex effects of many subtle interactions.

Far more, of course, is known about the life of the sea than I shall be able even to suggest, and there are yet to be achieved great steps in our knowledge of the living entities of the sea. I shall mention some of these possibilities in my concluding remarks.

A general discussion of a living system should consider the ways in which plants elaborate basic organic material from inorganic substances and the successive and often highly intricate steps by which organisms then return this material to the inorganic reservoir. The discussion should also show the forms of life by which such processes are conducted. I shall briefly trace these processes through the regions I have indicated, returning later to a more detailed discussion of the living forms and their constraints.

Some organic material is carried to the sea by rivers, and some is manufactured in shallow water by attached plants. More than 90 percent of the basic organic material that fuels and builds the life in the sea, however, is synthesized within the lighted surface layers of open water by the many varieties of phytoplankton. These sunny pastures of plant cells are grazed by the herbivorous zooplankton (small planktonic animals) and by some small fishes. These in turn are prey to various carnivorous creatures, large and small, who have their predators also.

The debris from the activities in the surface layers settles into the dimly lighted and unlighted midlayers of the sea, the twilight mesopelagic zone and the midnight bathypelagic zone, to serve as one source of food for their strange inhabitants. This process depletes the surface layers of some food and particularly of the vital plant nutrients, or fertilizers, that become trapped below the surface layers, where they are unavailable to the plants. Food and nutrients are also actively carried downward from the surface by vertically migrating animals.

The depleted remnants of this constant "rain" of detritus continue to the sea floor and support those animals that live just above the bottom (epibenthic animals), on the bottom (benthic animals) and burrowed into the bottom. Here filter-feeding and burrowing (deposit-feeding) animals and bacteria rework the remaining refractory particles. The more active animals also find repast in mid-water creatures and in the occasional falls of carcasses and other larger debris. Except in unusual small areas there is an abundance of oxygen in the deep water, and the solid bottom presents advantages that allow the support of a denser population of larger creatures than can exist in deep mid-water.

In shallower water such as banks, atolls, continental shelves and shallow seas conditions associated with a solid bottom and other regional modifications of the general regime enable rich populations to develop. Such areas constitute about 7 percent of the total area of the ocean. In some of these regions added food results from the growth of larger fixed plants and from land drainage.

With the above bare recitation for general orientation, I shall now discuss these matters in more detail.

The cycle of life in the sea, like that on land, is fueled by the sun's visible light acting on green plants. Of every million photons of sunlight reaching the earth's surface, some 90 enter into the net production of basic food. Perhaps 50 of the 90 contribute to the growth of land plants and about 40 to the growth of the single-celled green plants of the sea, the phytoplankton [see illustration, pages 728–29]. It is this minute fraction of the suns radiant energy that supplies the living organisms of this planet not only with their food but also with a breathable atmosphere.

The terrestrial and marine plants and animals arose from the same sources, through similar evolutionary sequences and by the action of the same natural laws. Yet these two living systems differ greatly at the stage in which we now view them. Were we to imagine a terrestrial food web that had developed in a form limited to that of the open sea, we would envision the land populated predominantly by short-lived simple plant cells grazed by small insects, worms and snails, which in turn would support a sparse predaceous population of larger insects, birds, frogs and lizards. The population of still larger carnivores would be a small fraction of the populations of

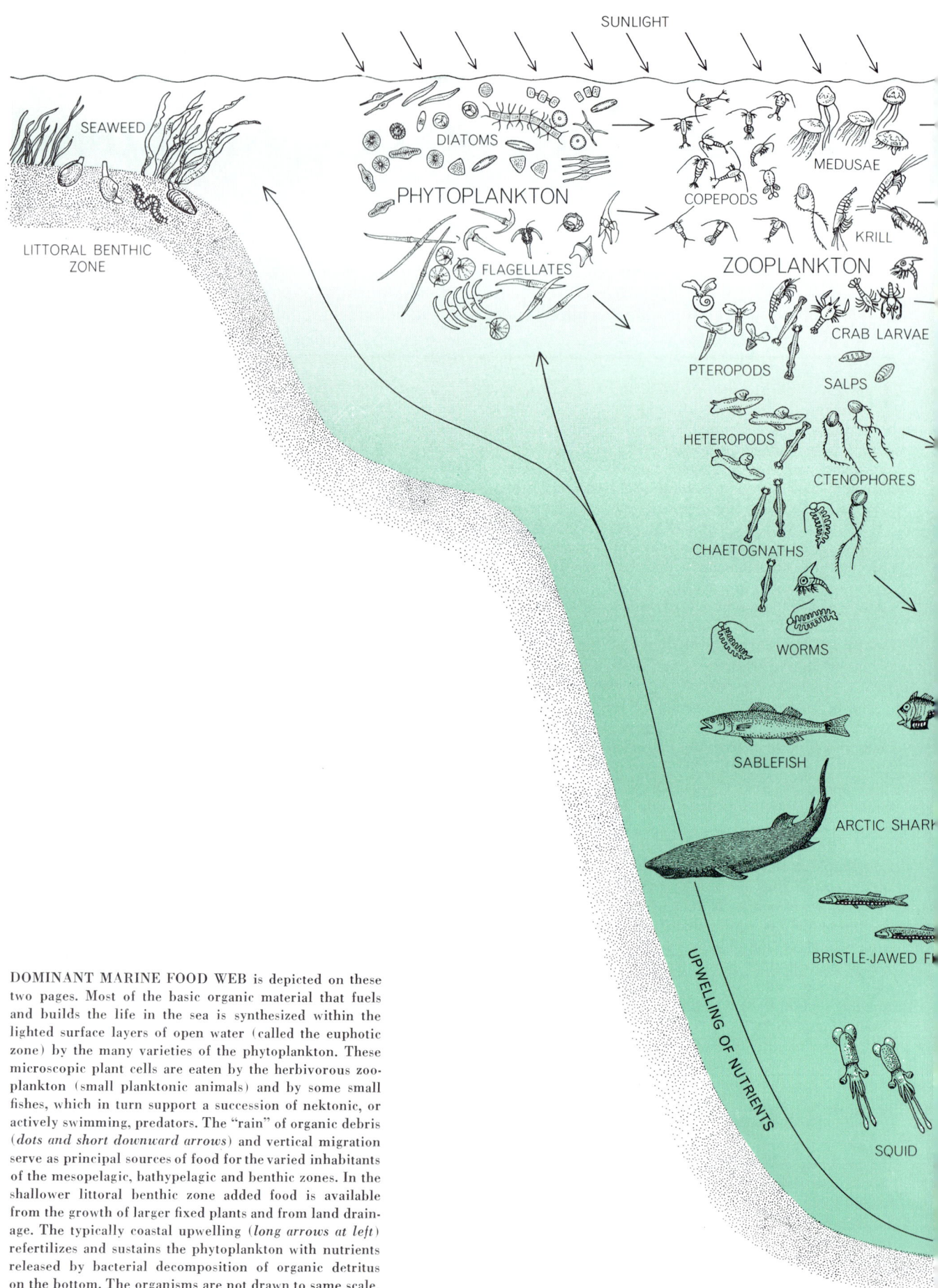

SUNLIGHT

SEAWEED

LITTORAL BENTHIC
ZONE

DIATOMS

PHYTOPLANKTON

FLAGELLATES

MEDUSAE

COPEPODS

KRILL

ZOOPLANKTON

CRAB LARVAE

PTEROPODS

SALPS

HETEROPODS

CTENOPHORES

CHAETOGNATHS

WORMS

SABLEFISH

ARCTIC SHARK

BRISTLE-JAWED F

UPWELLING OF NUTRIENTS

SQUID

DOMINANT MARINE FOOD WEB is depicted on these two pages. Most of the basic organic material that fuels and builds the life in the sea is synthesized within the lighted surface layers of open water (called the euphotic zone) by the many varieties of the phytoplankton. These microscopic plant cells are eaten by the herbivorous zooplankton (small planktonic animals) and by some small fishes, which in turn support a succession of nektonic, or actively swimming, predators. The "rain" of organic debris (*dots and short downward arrows*) and vertical migration serve as principal sources of food for the varied inhabitants of the mesopelagic, bathypelagic and benthic zones. In the shallower littoral benthic zone added food is available from the growth of larger fixed plants and from land drainage. The typically coastal upwelling (*long arrows at left*) refertilizes and sustains the phytoplankton with nutrients released by bacterial decomposition of organic detritus on the bottom. The organisms are not drawn to same scale.

FLYING FISH

HERRING-LIKE FISHES

BASKING SHARK

DOLPHINFISH

SEAL

PORPOISE

BALEEN WHALE

TUNA

SQUID

MACKEREL

BONITO

TOOTHED SHARK

SWORDFISH

NEKTON

SPERM WHALE

LANTERN FISH

HATCHETFISH

LARGE SQUID

OCTOPUS

SCARLET PRAWNS

VIPERFISH

ANGLERFISH

SWALLOWERS

GULPER

ANGLERFISH

CRINOIDS

GRENADIER

TRIPOD FISH

BRITTLE STARS

LAMP SHELLS

GLASS SPONGES

EUPHOTIC ZONE

MESOPELAGIC ZONE

BATHYPELAGIC ZONE

BENTHIC ZONE

large creatures that the existing land food web can nurture, because organisms in each of these steps pass on not more than 15 percent of the organic substance.

In some important respects this imaginary condition is not unlike that of the dominant food web of the sea, where almost all marine life is sustained by microscopic plants and near-microscopic herbivores and carnivores, which pass on only a greatly diminished supply of food to sustain the larger, more active and more complex creatures. In other respects the analogy is substantially inaccurate, because the primary marine food production is carried out by cells dispersed widely in a dense fluid medium.

This fact of an initial dispersal imposes a set of profound general conditions on all forms of life in the sea. For comparison, the concentration of plant food in a moderately rich grassland is of the order of a thousandth of the volume of the gross space it occupies and of the order of half of the mass of the air in which it is immersed. In moderately rich areas of the sea, on the other hand, food is hundreds of times more dilute in volume and hundreds of thousands of times more dilute in relative mass. To crop this meager broth a blind herbivore or a simple pore in a filtering structure would need to process a weight of water hundreds of thousands of times the weight of the cell it eventually captures. In even the densest concentrations the factor exceeds several thousands, and with each further step in the food web dilution increases. Thus from the beginnings of the marine food web we see many adaptations accommodating to this dilution: eyes in microscopic herbivorous animals, filters of exquisite design, mechanisms and behavior for discovering local concentrations, complex search gear and, on the bottom, attachments to elicit the aid of moving water in carrying out the task of filtration. All these adaptations stem from the conditions that limit plant life in the open sea to microscopic dimensions.

It is in the sunlit near-surface of the

open sea that the unique nature of the dominant system of marine life is irrevocably molded. The near-surface, or mixed, layer of the sea varies in thickness from tens of feet to hundreds depending on the nature of the general circulation, mixing by winds and heating [see "The Atmosphere and the Ocean," by R. W. Stewart; SCIENTIFIC AMERICAN Offprint 881]. Here the basic food production of the sea is accomplished by single-celled plants. One common group of small phytoplankton are the coccolithophores, with calcareous plates, a swimming ability and often an oil droplet for food storage and buoyancy. The larger microscopic phytoplankton are composed of many species belonging to several groups: naked algal cells, diatoms with complex shells of silica and actively swimming and rotating flagellates. Very small forms of many groups are also abundant and collectively are called nannoplankton.

The species composition of the phytoplankton is everywhere complex and varies from place to place, season to season and year to year. The various regions of the ocean are typified, however, by dominant major groups and particular species. Seasonal effects are often strong, with dense blooms of phytoplankton occurring when high levels of plant nutrients suddenly become usable or available, such as in high latitudes in spring or along coasts at the onset of upwelling. The concentration of phytoplankton varies on all dimensional scales, even down to small patches.

It is not immediately obvious why the dominant primary production of organic matter in the sea is carried out by microscopic single-celled plants instead of free-floating higher plants or other intermediate plant forms. The question arises: Why are there no pelagic "trees" in the ocean? One can easily compute the advantages such a tree would enjoy, with its canopy near the surface in the lighted levels and its trunk and roots extending down to the nutrient-rich waters under the mixed layer. The answer to this fundamental question probably has

PRODUCTIVITY of the land and the sea are compared in terms of the net amount of energy that is converted from sunlight to organic matter by the green cells of land and sea plants. Colored lines denote total energy reaching the earth's upper atmosphere (a), total energy reaching earth's surface (b), total energy usable for photosynthesis (c), total energy usable for photosynthesis at sea (d), total energy usable for photosynthesis on land (e), net energy used for photosynthesis on land (f), net energy used for photosynthesis at sea (g), net energy used by land herbivores (h) and net energy used by sea herbivores (i). Although more sunlight falls on the sea than on the land (by virtue of the sea's larger surface area), the total land area is estimated to outproduce the total sea area by 25 to 50 percent. This is primarily due to low nutrient concentrations in the euphotic zone and high metabolism in marine plants. The data are from Walter R. Schmitt of Scripps Institution of Oceanography.

several parts. The evolution of plants in the pelagic realm favored smallness rather than expansion because the mixed layer in which these plants live is quite homogeneous; hence small incremental extensions from a plant cell cannot aid it in bridging to richer sources in order to satisfy its several needs.

On land, light is immediately above the soil and nutrients are immediately below; thus any extension is of immediate benefit, and the development of single cells into higher erect plants is able to follow in a stepwise evolutionary sequence. At sea the same richer sources exist but are so far apart that only a very large ready-made plant could act as a bridge between them. Although such plants could develop in some other environment and then adapt to the pelagic conditions, this has not come about. It is difficult to see how such a plant would propagate anyway; certainly it could not propagate in the open sea, because the young plants there would be at a severe disadvantage. In the sea small-scale differential motions of water are rapidly damped out, and any free-floating plant must often depend on molecular diffusion in the water for the uptake of nutrients and excretion of wastes. Smallness and self-motion are then advantageous, and a gross structure of cells cannot exchange nutrients or wastes as well as the same cells can separately or in open aggregations.

In addition the large-scale circulation of the ocean continuously sweeps the pelagic plants out of the region to which they are best adapted. It is essential that some individuals be returned to renew the populations. More mechanisms for this essential return exist for single-celled plants than exist for large plants, or even for any conventional spores, seeds or juveniles. Any of these can be carried by oceanic gyres or diffused by large-scale motions of surface eddies and periodic counterflow, but single-celled plants can also ride submerged countercurrents while temporarily feeding on food particles or perhaps on dissolved organic material. Other mechanisms of distribution undoubtedly are also occasionally important. For example, living marine plant cells are carried by storm-borne spray, in bird feathers and by well-fed fish and birds in their undigested food.

No large plant has solved the many problems of development, dispersal and reproduction. There *are* no pelagic trees, and these several factors in concert therefore restrict the open sea in a profound way. They confine it to an initial food web composed of microscopic

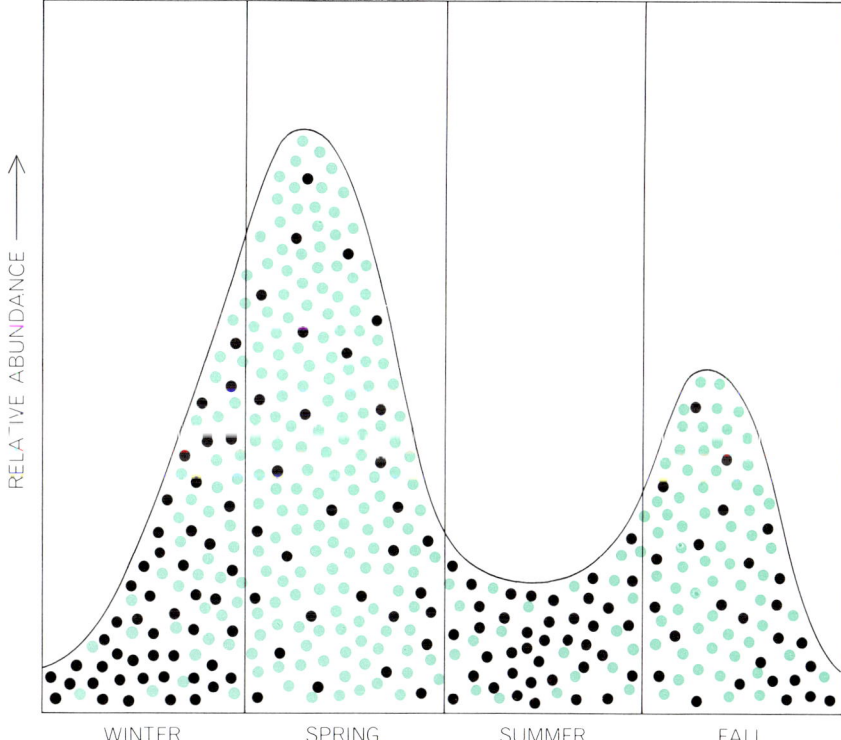

RELATIVE ABUNDANCE →

WINTER   SPRING   SUMMER   FALL

SPECIES COMPOSITION AND ABUNDANCE of the phytoplankton varies from season to season, particularly at high latitudes. During the winter the turbulence caused by storms replenishes the supply of nutrients in the surface layers. During this period flagellates (*black dots*) tend to dominate. In early spring the increase in the amount of sunlight reaching the surface stimulates plant growth, and diatoms (*colored dots*) are stimulated to grow. Later in spring grazing by zooplankton and a decrease in the supply of nutrients caused by calmer weather result in a general reduction in the phytoplankton population, which reaches a secondary minimum in midsummer, during which time flagellates again dominate. The increased mixing caused by early autumn storms causes a rise in the supply of nutrients and a corresponding minor surge in the population of diatoms. The decreasing sunlight of late fall and grazing by zooplankton again reduce the general level of the plant population.

forms, whereas larger plants live attached only to shallow bottoms (which comprise some 2 percent of the ocean area). Attached plants, unlike free-floating plants, are not subject to the aforementioned limitations. For attached plants all degrees of water motion enhance the exchange of nutrients and wastes. Moreover, their normal population does not drift, much of their reproduction is by budding, and their spores are adapted for rapid development and settlement. Larger plants too are sometimes found in nonreproducing terminal accumulations of drifting shore plants in a few special convergent deep-sea areas such as the Sargasso Sea.

Although species of phytoplankton will populate only regions with conditions to which they are adapted, factors other than temperature, nutrients and light levels undoubtedly are important in determining the species composition of phytoplankton populations. Little is understood of the mechanisms that give rise to an abundance of particular spe-

cies under certain conditions. Grazing herbivores may consume only a part of the size range of cells, allowing certain sizes and types to dominate temporarily. Little is understood of the mechanisms that give rise to an abundance of particular species under certain conditions. Chemical by-products of certain species probably exclude certain other species. Often details of individual cell behavior are probably also important in the introduction and success of a species in a particular area. In some cases we can glimpse what these mechanisms are.

For example, both the larger diatoms and the larger flagellates can move at appreciable velocities through the water. The diatoms commonly sink downward, whereas the flagellates actively swim upward toward light. These are probably patterns of behavior primarily for increasing exchange, but the interaction of such unidirectional motions with random turbulence or systematic convective motion is not simple, as it is with an inactive particle. Rather, we would expect diatoms to be statistically abundant in upward-moving water and to sink out of the near-surface layers when turbulence or upward convection is low.

Conversely, flagellates should be statistically more abundant in downwelling water and should concentrate near the surface in low turbulence and slow downward water motions. These effects seem to exist. Off some continental coasts in summer flagellates may eventually collect in high concentrations. As they begin to shade one another from the light, each individual struggles closer to the lighted surface, producing such a high density that large areas of the water are turned red or brown by their pigments. The concentration of flagellates in these "red tides" sometimes becomes too great for their own survival. Several species of flagellates also become highly toxic as they grow older. Thus they sometimes both produce and participate in a mass death of fish and invertebrates that has been known to give rise to such a high yield of hydrogen sulfide as to blacken the white houses of coastal cities.

Large diatom cells, on the other hand, spend a disproportionately greater time in upward-moving regions of the water and an unlimited time in any region where the upward motion about equals their own downward motion. (The support of unidirectionally moving objects by contrary environmental motion is observed in other phenomena, such as the production of rain and hail.) Diatom cells are thus statistically abundant in upwelling water, and the distribution of diatoms probably is often a reflection of the turbulent-convective regime of the water. Sinking and the dependence of the larger diatoms on upward convection and turbulence for support aids them in reaching upwelling regions, where nutrients are high; it helps to explain their dominance in such regions and such other features of their distribution as their high proportion in rich ocean regions and their frequent inverse occurrence with flagellates. Differences in adaptations to the physical and chemical conditions, and the release of chemical products, probably reinforce such relations.

In some areas, such as parts of the equatorial current system and shallow seas, where lateral and vertical circulation is rapid, the species composition of phytoplankton is perhaps more simply a result of the inherent ability of the species to grow, survive and reproduce under the local conditions of temperature, light, nutrients, competitors and herbivores. Elsewhere second-order effects of the detailed cell behavior often dominate. Those details of behavior that

give rise to concentrations on any dimensional scale are particularly important to all subsequent steps in the food chain.

All phytoplankton cells eventually settle from the surface layers. The depletion of nutrients and food from the surface layers takes place continuously through the loss of organic material, plant cells, molts, bodies of animals, fecal pellets and so forth, which release their content of chemical nutrients at various depths through the action of bacteria and other organisms. The periodic downward migration of zooplankton further contributes to this loss.

These nutrients are "trapped" below the level of light adequate to sustain photosynthesis, and therefore the water in

which plants must grow generally contains very low concentrations of such vital substances. It is this condition that is principally responsible for the comparatively low total net productivity of the sea compared with that of the land. The regions where trapping is broken down or does not exist—where there is upwelling of nutrient-rich water along coasts, in parts of the equatorial regions, in the wakes of islands and banks and in high latitudes, and where there is rapid recirculation of nutrients over shallow shelves and seas—locally bear the sea's richest fund of life.

The initial factors discussed so far have placed an inescapable stamp on the form of all life in the open sea, as irrev-

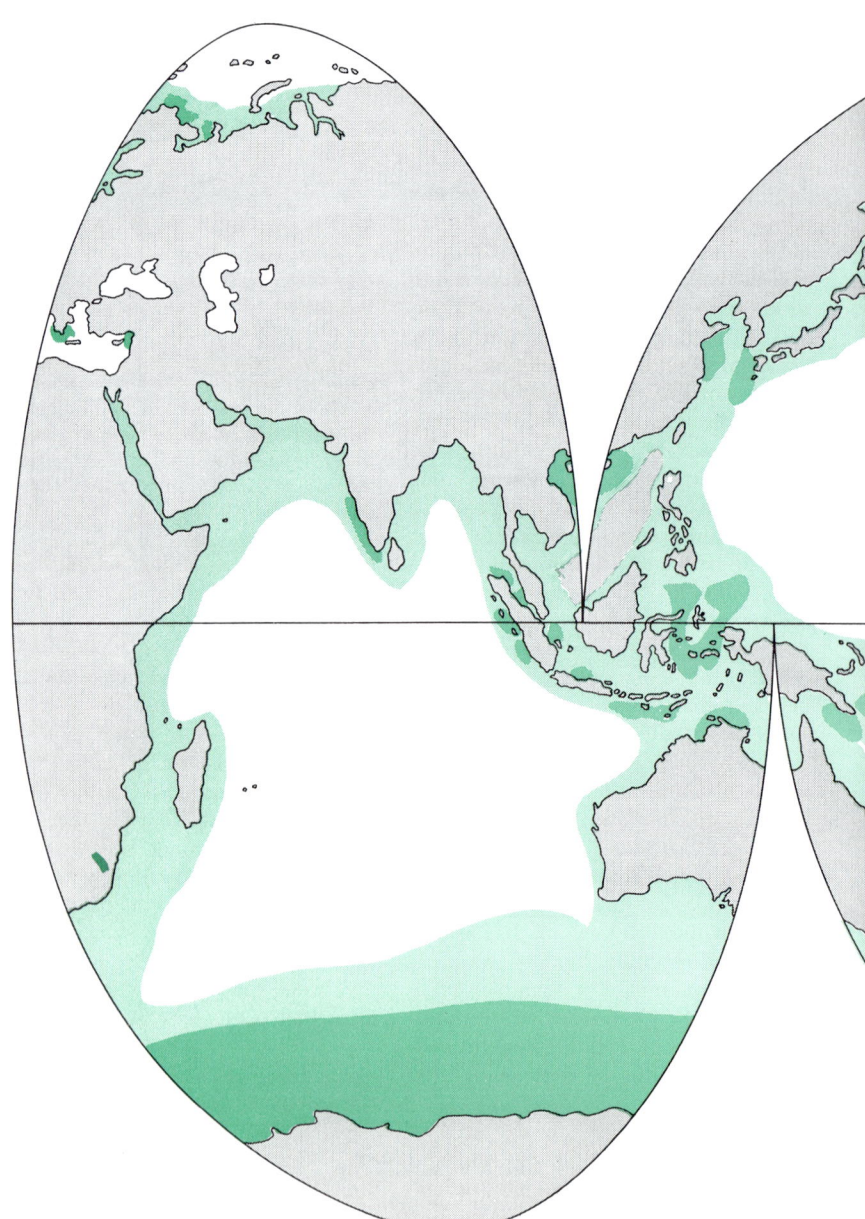

FAVORABLE CONDITIONS for the growth of phytoplankton occur wherever upwelling or mixing tends to bring subsurface nutrients up to the euphotic layer of the ocean. This map,

ocably no doubt as the properties and distribution of hydrogen have dictated the form of the universe. These factors have limited the dominant form of life in the sea to an initial microscopic sequence that is relatively unproductive, is stimulated by upwelling and mixing and is otherwise altered in species composition and distribution by physical, chemical and biological processes on all dimensional scales. The same factors also limit the populations of higher animals and have led to unexpectedly simple adaptations, such as the sinking of the larger diatoms as a tactic to solve the manifold problems of enhancing nutrient and waste exchange, finding nutrients, remaining in the surface waters and

repopulating.

The grazing of the phytoplankton is principally conducted by the herbivorous members of the zooplankton, a heterogeneous group of small animals that carry out several steps in the food web as herbivores, carnivores and detrital (debris-eating) feeders. Among the important members of the zooplankton are the arthropods, animals with external skeletons that belong to the same broad group as insects, crabs and shrimps. The planktonic arthropods include the abundant copepods, which are in a sense the marine equivalent of insects. Copepods are represented in the sea by some 10,-000 or more species that act not only

as herbivores, carnivores or detrital feeders but also as external or even internal parasites! Two or three thousand of these species live in the open sea. Other important arthropods are the shrimplike euphausiids, the strongest vertical migrators of the zooplankton. They compose the vast shoals of krill that occur in high latitudes and that constitute one of the principal foods of the baleen whales. The zooplankton also include the strange bristle-jawed chaetognaths, or arrowworms, carnivores of mysterious origin and affinities known only in the marine environment. Widely distributed and abundant, the chaetognaths are represented by a surprisingly small number of species, perhaps fewer than 50. Larvae

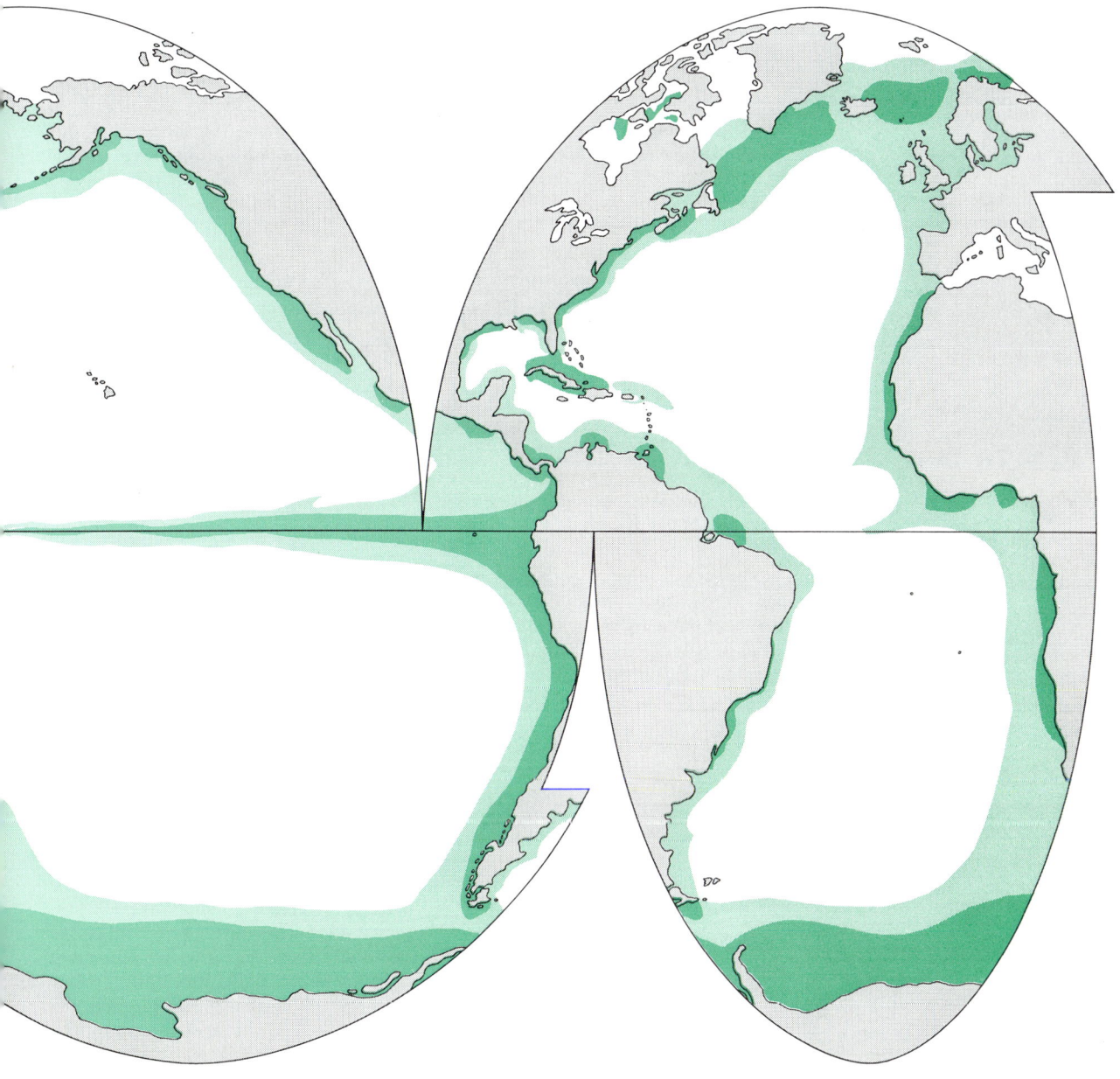

which is adapted from one compiled by the Norwegian oceanographer Harald U. Sverdrup, shows the global distribution of such

waters, in which the productivity of marine life would be expected to be very high (*dark color*) and moderately high (*light color*).

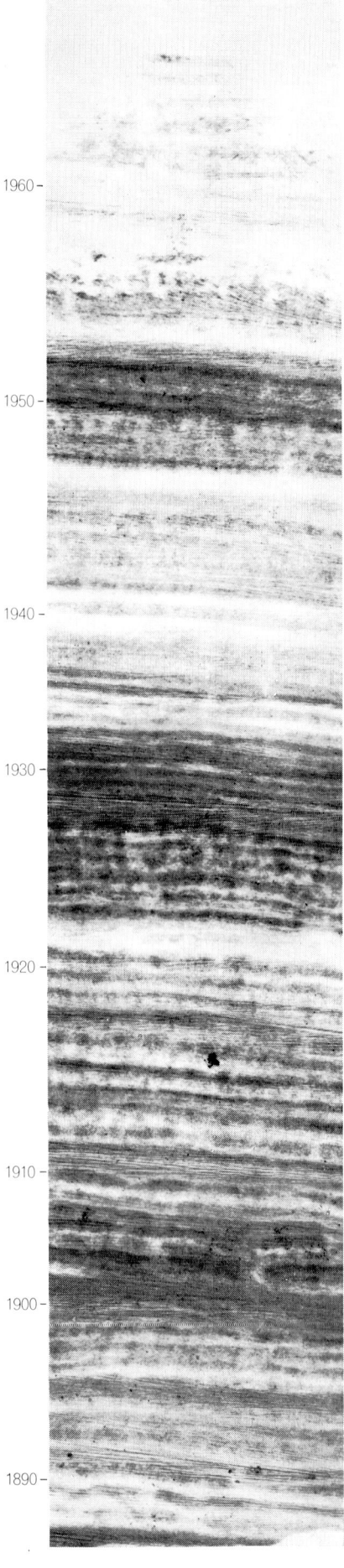

of many types, worms, medusae (jelly-fish), ctenophores (comb jellies), gastropods (snails), pteropods and heteropods (other pelagic mollusks), salps, unpigmented flagellates and many others are also important components of this milieu, each with its own remarkably complex and often bizarre life history, behavior and form.

The larger zooplankton are mainly carnivores, and those of herbivorous habit are restricted to feeding on the larger plant cells. Much of the food supply, however, exists in the form of very small particles such as the nannoplankton, and these appear to be available almost solely to microscopic creatures. The immense distances between plant cells, many thousands of times their diameter, place a great premium on the development of feeding mechanisms that avoid the simple filtering of water through fine pores. The power necessary to maintain a certain rate of flow through pores or nets increases inversely at an exponential rate with respect to the pore or mesh diameter, and the small planktonic herbivores, detrital feeders and carnivores show many adaptations to avoid this energy loss. Eyesight has developed in many minute animals to make possible selective capture. A variety of webs, bristles, rakes, combs, cilia and other structures are found, and they are often sticky. Stickiness allows the capture of food that is finer than the interspaces in the filtering structures, and it greatly reduces the expenditure of energy.

A few groups have developed extremely fine and apparently quite effective nets. One group that has accomplished this is the Larvacea. A larvacian produces and inhabits a complex external "house," much larger than its owner, that contains a system of very finely constructed nets through which the creature maintains a gentle flow [see illustration on page 736]. The Larvacea have apparently solved the problem of energy loss in filtering by having proportionately large nets; fine strong threads and a low rate of flow.

The composition of the zooplankton differs from place to place, day to night, season to season and year to year, yet most species are limited in distribution, and the members of the planktonic communities commonly show a rather stable representation of the modes of life.

The zooplankton are, of course, faced with the necessity of maintaining breeding assemblages and, like the phytoplankton, with the necessity of establishing a reinoculation of parent waters. In addition, their behavior must lead to a correspondence with their food and to the pattern of large-scale and small-scale spottiness already imposed on the marine realm by the phytoplankton. The swimming powers of the larger zooplankton are quite adequate for finding local small-scale patches of food. That this task is accomplished on a large scale is indirectly demonstrated by the observed correspondence between the quantities of zooplankton and the plant nutrients in the surface waters. How this large-scale task is accomplished is understood for some groups. For example, some zooplankton species have been shown to descend near the end of suitable conditions at the surface and to take temporary residence in a submerged countercurrent that returns them upstream.

There are many large and small puzzles in the distribution of zooplankton. As an example, dense concentrations of phytoplankton are often associated with low populations of zooplankton. These are probably rapidly growing blooms that zooplankton have not yet invaded and grazed on, but it is not completely clear that this is so. Chemical repulsion may be involved.

The concentration of larger zooplankton and small fish in the surface layers is much greater at night than during the day, because of a group of strongly swimming members that share their time between the surface and the mesopelagic region. This behavior is probably primarily a tactic to enjoy the best of two worlds: to crop the richer food developing in the surface layers and to minimize mortality from predation by remaining always in the dark, like timid rabbits emerging from the thicket to graze the nighttime fields, although still in the presence of foxes and ferrets. Many small zooplankton organisms also make a daily migration of some vertical extent.

RARE SEDIMENTARY RECORD of the recent annual oceanographic, meteorological and biological history of part of a major oceanic system is revealed in this radiograph of a section of an ocean-bottom core obtained by Andrew Soutar of the Scripps Institution of Oceanography in the Santa Barbara Basin off the California coast. In some near-shore basins such as this one the absence of oxygen causes refractory parts of the organic debris to be left undecomposed and the sediment to remain undisturbed in the annual layers called varves. The dark layers are the densest and represent winter sedimentation. The lighter and less dense layers are composed mostly of diatoms and represent spring and summer sedimentation.

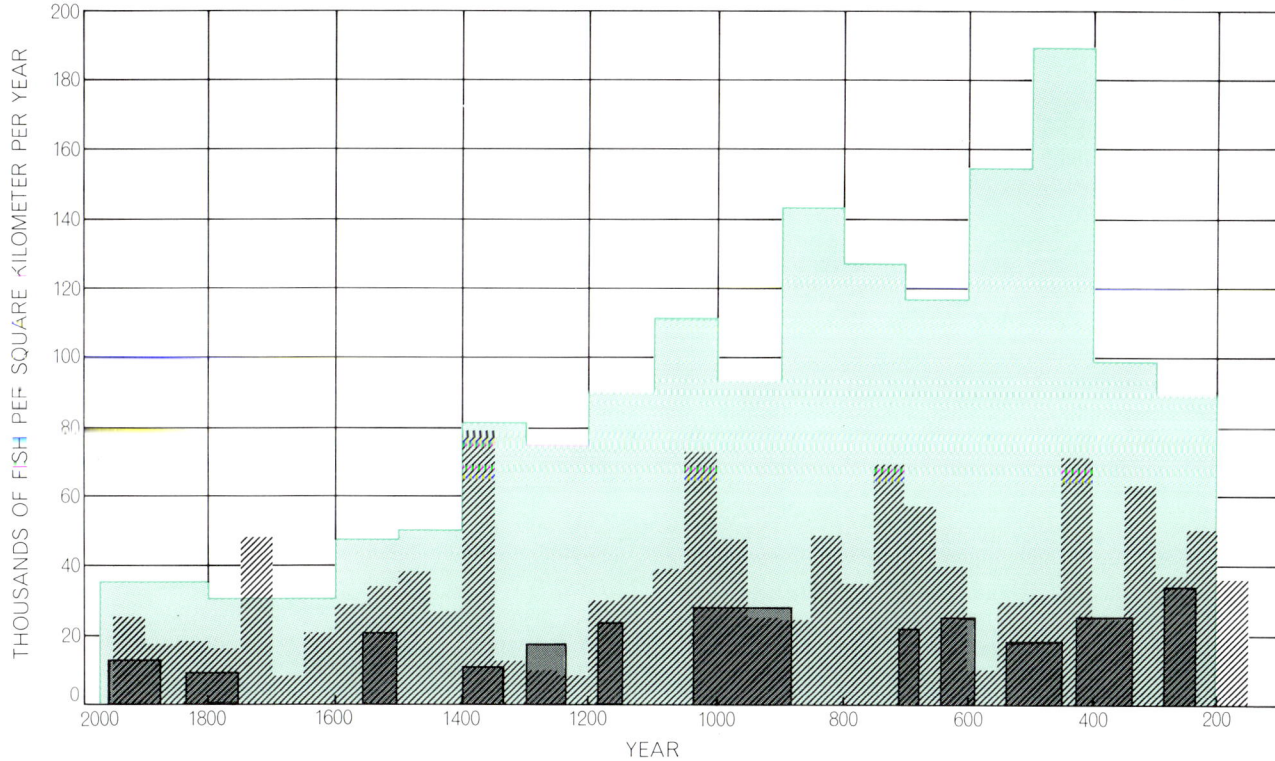

ESTIMATED FISH POPULATIONS in the Santa Barbara Basin over the past 1,800 years were obtained for three species by counting the average number of scales of each species in the varves of the core shown on the opposite page. Minimum population estimates for fish one year old and older are given for Pacific sardines (*gray*), northern sardines (*colored areas*) and Pacific hake (*hatched*).

In addition to its primary purpose daily vertical migration undoubtedly serves the migrating organisms in a number of other ways. It enables the creatures to adjust their mean temperature, so that by spending the days in cooler water the amount of food used during rest is reduced. Perhaps such processes as the rate of egg development are also controlled by these tactics. Many land animals employ hiding behavior for similar kinds of adjustment. Convincing arguments have also been presented to show that vertical migration serves to maintain a wide range of tolerance in the migrating species, so that they will be more successful under many more conditions than if they lived solely in the surface layers. This migration must also play an important part in the distribution of many species. Interaction of the daily migrants with the water motion produced by daily land-sea breeze alternation can hold the migrants offshore by a kind of "rectification" of the oscillating water motion. More generally, descent into the lower layers increases the influence of submerged countercurrents, thereby enhancing the opportunity to return upstream, to enter upwelling regions and hence to find high nutrient levels and associated high phytoplankton productivity.

Even minor details of behavior may strongly contribute to success. Migrants spend the day at a depth corresponding to relatively constant low light levels, where the movement of the water commonly is different from that at the surface. Most of the members rise somewhat even at the passage of a cloud shadow. Should they be carried under the shadow of an area rich in phytoplankton, they migrate to shallower depths, thereby often decreasing or even halting their drift with respect to this rich region to which they will ascend at night. Conversely, when the surface waters are clear and lean, they will migrate deeper and most often drift relatively faster.

We might simplistically view the distribution of zooplankton, and phytoplankton for that matter, as the consequence of a broad inoculation of the oceans with a spectrum of species, each with a certain adaptive range of tolerances and a certain variable range of feeding, reproducing and migrating behavior. At some places and at some times the behavior of a species, interacting even in detailed secondary ways with the variable conditions of the ocean and its other inhabitants, results in temporary, seasonal or persistent success.

There are a few exceptions to the microscopic dimensions of the herbivores in the pelagic food web. Among these the herrings and herring-like fishes are able to consume phytoplankton as a substantial component of their diet. Such an adaptation gives these fishes access to many times the food supply of the more carnivorous groups. It is therefore no surprise that the partly herbivorous fishes comprise the bulk of the world's fisheries [see "The Food Resources of the Ocean," by S. J. Holt; SCIENTIFIC AMERICAN Offprint 886].

The principal food supplies of the pelagic populations are passed on in incremental steps and rapidly depleted quantity to the larger carnivorous zooplankton, then to small fishes and squids, and ultimately to the wide range of larger carnivores of the pelagic realm. In this region without refuge, either powerful static defenses, such as the stinging cells of the medusae and men-o'-war, or increasing size, acuity, alertness, speed and strength are the requirements for survival at each step. Streamlining of form here reaches a high point of development, and in tropical waters it is conspicuous even in small fishes, since the lower viscosity of the warmer waters will enable a highly streamlined small prey to escape a poorly streamlined predator, an effect that exists only for fishes of twice the length in cold, viscous, arctic or deep waters.

The pelagic region contains some of the largest and most superbly designed creatures ever to inhabit this earth: the exquisitely constructed pelagic tunas; the multicolored dolphinfishes, capturers of flying fishes; the conversational porpoises; the shallow- and deep-feeding swordfishes and toothed whales, and the greatest carnivores of all, the baleen whales and some plankton-eating sharks, whose prey are entire schools of krill or small fishes. Seals and sea lions feed far into the pelagic realm. In concert with these great predators, large carnivorous sharks await injured prey. Marine birds, some adapted to almost continuous pelagic life, consume surprising quantities of ocean food, diving, plunging, skimming and gulping in pursuit. Creatures of this region have developed such faculties as advanced sonar, unexplained senses of orientation and homing, and extreme olfactory sensitivity.

These larger creatures of the sea commonly move in schools, shoals and herds. In addition to meeting the needs of mating such grouping is advantageous in both defensive and predatory strategy, much like the cargo-ship convoy and submarine "wolf pack" of World War II. Both defensive and predatory assemblages are often complex. Small fishes of several species commonly school together. Diverse predators also form loosely cooperative groups, and many species of marine birds depend almost wholly on prey driven to the surface by submerged predators.

At night, schools of prey and predators are almost always spectacularly illuminated by bioluminescence produced by the microscopic and larger plankton. The reason for the ubiquitous production of light by the microorganisms of the sea remains obscure, and suggested explanations are controversial. It has been suggested that light is a kind of inadvertent by-product of life in transparent organisms. It has also been hypothesized that the emission of light on disturbance is advantageous to the plankton in making the predators of the plankton conspicuous to *their* predators! Unquestionably it does act this way. Indeed, some fisheries base the detection of their prey on the bioluminescence that the fish excite. It is difficult, however, to defend the thesis that this effect was the direct factor in the original development of bioluminescence, since the effect was of no advantage to the individual microorganism that first developed it. Perhaps the luminescence of a microorganism also discourages attack by the light-avoiding zooplankton and is of initial survival benefit to the individual. As it then became general in the population, the effect of revealing plankton predators to their predators would also become important.

The fallout of organic material into the deep, dimly lighted mid-water supports a sparse population of fishes and invertebrates. Within the mesopelagic and bathypelagic zones are found some of the most curious and bizarre creatures of this earth. These range from the highly developed and powerfully predaceous intruders, toothed whales and swordfishes, at the climax of the food chain, to the remarkable squids, octopuses, eu-phausiids, lantern fishes, gulpers and anglerfishes that inhabit the bathypelagic region.

In the mesopelagic region, where some sunlight penetrates, fishes are often countershaded, that is, they are darker above and lighter below, as are surface fishes. Many of the creatures of this dimly lighted region participate in the daily migration, swimming to the upper layers at evening like bats emerging from their caves. At greater depths, over a half-mile or so, the common inhabitants are often darkly pigmented, weak-bodied and frequently adapted to unusual feeding techniques. Attraction of prey by luminescent lures or by mimicry of small prey, greatly extensible jaws and expansible abdomens are common. It is, however, a region of Lilliputian monsters, usually not more than six inches in length, with most larger fishes greatly reduced in musculature and weakly constructed.

There are some much larger, stronger and more active fishes and squids in this region, although they are not taken in trawls or seen from submersibles. Knowledge of their existence comes mainly from specimens found in the stomach of sperm whales and swordfish. They must be rare, however, since the slow, conservative creatures that are taken in trawls could hardly coexist with large numbers of active predators. Nevertheless, populations must be sufficiently large to attract the sperm whales and swordfish. There is evidence that the sperm whales possess highly developed long-range hunting sonar. They may locate their prey over relatively great distances, perhaps miles, from just such an extremely sparse population of active bathypelagic animals.

Although many near-surface organisms are luminescent, it is in the bathypelagic region that bioluminescence has reached a surprising level of development, with at least two-thirds of the species producing light. Were we truly marine-oriented, we would perhaps be more surprised by the almost complete absence of biological light in the land environment, with its few rare cases of fireflies, glowworms and luminous bacteria. Clearly bioluminescence can be valuable to higher organisms, and the creatures of the bathypelagic realm have developed light-producing organs and structures to a high degree. In many cases the organs have obvious functions. Some fishes, squids and euphausiids possess searchlights with reflector, lens and iris almost as complex as the eye. Others have complex patterns of small lights that may serve the functions of recogni-

LARVACIAN is representative of a group of small planktonic herbivores that has solved the problem of energy loss in filtering, apparently without utilizing "stickiness," by having proportionately large nets, strong fine threads and a low rate of water flow. The larvacian (*black*) produces and inhabits a complex external "house" (*color*), much larger than its owner, which contains a system of nets through which the organism maintains a gentle flow. In almost all other groups simple filters are employed only to exclude large particles.

tion, schooling control and even mimicry of a small group of luminous plankton. Strong flashes may confuse predators by "target alteration" effects, or by producing residual images in the predators' vision. Some squids and shrimps are more direct and discharge luminous clouds to cover their escape. The luminous organs are arranged on some fishes so that they can be used to countershade their silhouettes against faint light coming from the surface. Luminous baits are well developed. Lights may also be used for locating a mate, a problem of this vast, sparsely populated domain that has been solved by some anglerfishes by the development of tiny males that live parasitically attached to their relatively huge mates.

It has been shown that the vertebrate eye has been adapted to detect objects in the lowest light level on the earth's surface—a moonless, overcast night under a dense forest canopy—but not lower. Light levels in the bathypelagic region can be much lower. This is most probably the primary difference that accounts for the absence of bioluminescence in

higher land animals and the richness of its development in the ocean forms.

The densest populations of bathypelagic creatures lie below the most productive surface regions, except at high latitudes, where the dearth of winter food probably would exhaust the meager reserves of these creatures. All the bathypelagic populations are sparse, and in this region living creatures are less than one hundred-millionth of the water volume. Nevertheless, the zone is of immense dimensions and the total populations may be large. Some genera, such as the feeble, tiny bristle-jawed fishes, are probably the most numerous fishes in the world and constitute a gigantic total biomass. There are some 2,000 species of fishes and as many species of the larger invertebrates known to inhabit the bathypelagic zone, but only a few of these species appear to be widespread. The barriers to distribution in this widely interconnected mid-water region are not obvious.

The floor of the deep sea constitutes an environment quite unlike the mid-

water and surface environments. Here are sites for the attachment of the larger invertebrates that filter detritus from the water. Among these animals are representatives of some of the earliest multi-celled creatures to exist on the earth, glass sponges, sea lilies (crinoids)—once thought to have been long extinct—and lamp shells (brachiopods).

At one time it was also thought that the abyssal floor was sparsely inhabited and that the populations of the deep-ocean floor were supplied with food only by the slow, meager rain of terminal detrital food material that has passed through the surface and bathypelagic populations. Such refractory material requires further passage into filter feeders or through slow bacterial action in the sediment, followed by consumption by larger burrowing organisms, before it becomes available to active free-living animals. This remnant portion of the food web could support only a very small active population.

Recent exploration of the abyssal realm with a baited camera throws doubt on the view that this is the exclusive

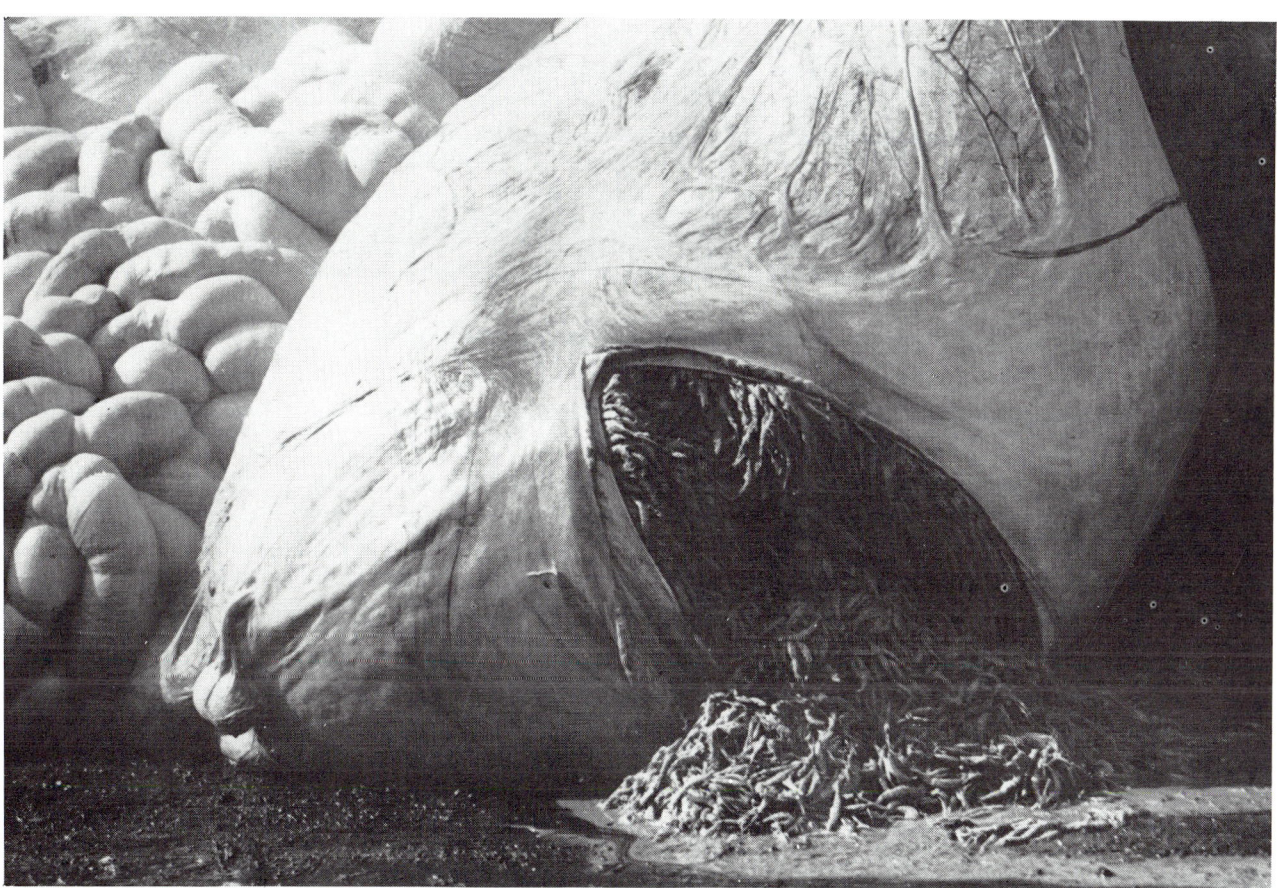

CHAMPION FILTER FEEDER of the world ocean in terms of volume is the blue whale, a mature specimen of which lies freshly butchered on the deck of a whaling vessel in this photograph. The whale's stomach has been cut open with a flensing knife to reveal its last meal: an immense quantity of euphausiids, or krill, each measuring about three inches in length. The baleen whales are not plankton-filterers in the ordinary sense but rather are great carnivores that seek out and engulf entire schools of small fish or invertebrates. The photograph was made by Robert Clarke of the National Institute of Oceanography in Wormley, England.

mechanism of food transfer to the deep bottom. Larger numbers of active fishes and other creatures are attracted to the bait almost immediately [see illustration at left]. It is probably true that several rather independent branches of the food web coexist in support of the deep-bottom creatures: one the familiar rain of fine detritus, and the other the rare, widely separated falls of large food particles that are in excess of the local feeding capacity of the broadly diffuse bathypelagic population. Such falls would include dead whales, large sharks or other large fishes and fragments of these, the multitude of remnants that are left when predators attack a school of surface fish and now, undoubtedly, garbage from ships and kills from underwater explosions. These sources result in an influx of high-grade food to the sea floor, and we would expect to find a population of active creatures adapted to its prompt discovery and utilization. The baited cameras have demonstrated that this is so.

Other sources of food materials are braided into these two extremes of the abyssal food web. There is the rather subtle downward diffusion of living and dead food that results initially from the daily vertical migration of small fish and zooplankton near the surface. This migration appears to impress a sympathetic daily migration on the mid-water populations down to great depths, far below the levels that light penetrates. Not only may such vertical migration bring feeble bathypelagic creatures near the bottom but also it accelerates in itself the flux of dead food material to the bottom of the deep sea.

There must also be some unassignable flux of food to the abyssal population resulting from the return of juveniles to their habitat. The larvae and young of many abyssal creatures develop at much shallower levels. To the extent that the biomass of juveniles returning to the deep regions exceeds the biomass of spawn released from it, this process, which might be called "Faginism," constitutes an input of food.

Benthic animals are much more abundant in the shallower waters off continents, particularly offshore from large rivers. Here there is often not only a richer near-surface production and a less hazardous journey of food to the sea floor but also a considerable input of food conveyed by rivers to the bottom. The deep slopes of river sediment wedges are typified by a comparatively rich population of burrowing and filtering animals that utilize this fine organic material. All the great rivers of the world save one, the Congo, have built sedimentary wedges along broad reaches of their coast, and in many instances these wedges extend into deep water. The shallow regions of such wedges are highly productive of active and often valuable marine organisms. At all depths the wedges bear larger populations than are common at similar depths elsewhere. Thus one wonders what inhabits the fan of the Congo. That great river, because of a strange invasion of a submarine canyon into its mouth, has built no wedge but rather is depositing a vast alluvial fan in the two-mile depths of the Angola Basin. This great deep region of the sea floor may harbor an unexplored population that is wholly unique.

In itself the pressure of the water at great depths appears to constitute no insurmountable barrier to water-breathing animal life. The depth limitations of many creatures are the associated conditions of low temperature, darkness, sparse food and so on. It should perhaps come as no surprise, therefore, that some of the fishes of high latitudes, which are of course adapted to cold dark waters, extend far into the deep cold waters in much more southern latitudes. Off the coast of Lower California, in water 1,200 to 6,000 feet deep, baited cameras have found an abundance of several species of fishes that are known at the near surface only far to the north. These include giant arctic sharks, sablefish and others. It appears that some of the fishes that have been called arctic species are actually fishes of the dark cold waters of the seas, which only "outcrop" in the Arctic, where cold water is at the surface.

I have discussed several of the benthic and epibenthic environments without pointing out some of the unique features the presence of a solid interface entails. The bottom is much more variable than the mid-water zone is. There are as a result more environmental niches for an organism to occupy, and hence we see organisms that are of a wider range of form and habit. Adaptations develop for hiding and ambuscade, for mimicry and controlled patterns. Nests and burrows can be built, lairs occupied and defended and booby traps set.

Aside from the wide range of form and function the benthic environment elicits from its inhabitants, there are more fundamental conditions that influence the nature and form of life there. For example, the dispersed food material settling from the upper layers becomes much concentrated against the sea floor. Indeed, it may become further concentrated by lateral currents moving it into depressions or the troughs of ripples.

In the mid-water environment most creatures must move by their own energies to seek food, using their own food stores for this motion. On the bottom, however, substantial water currents are present at all depths, and creatures can await the passage of their food. Although this saving only amounts to an added effectiveness for predators, it is of critical importance to those organisms that filter water for the fine food material it contains, and it is against the bottom interface that a major bypass to the microscopic steps of the dominant food web is achieved. Here large organisms can grow by consuming microscopic or even submicroscopic food particles. Clams, scallops, mussels, tube worms, barnacles and a host of other creatures that inhabit this zone have developed a wide range of extremely effective filtering mechanisms. In one step, aided by their attachment, the constant currents and the concentration of detritus against the interface, they perform the feat, most unusual in the sea, of growing large organisms directly from microscopic food.

Although the benthic environment enables the creatures of the sea to develop a major branch of the food web that is emancipated from successive microscopic steps, this makes little difference to the food economy of the sea. The sea is quite content with a large population of tiny organisms. From man's standpoint, however, the shallow benthic environment is an unusually effective producer of larger creatures for his

NEW EVIDENCE that an abundance of large active fishes inhabit the deep-sea floor was obtained recently by the author and his colleagues in the form of photographs such as the one on the opposite page. The photograph was made by a camera hovering over a five-gallon bait can at a depth of 1,400 meters off Lower California. The diagonal of the bait can measures a foot. The larger fish are mostly rat-tailed grenadiers and sablefish. The fact that large numbers of such fish are attracted almost immediately to the bait suggests that two rather independent branches of the marine food web coexist in support of the deep-bottom creatures by dead material: the rain of fine detritus, which supports a variety of attached filter-feeding and burrowing organisms, and rare, widely separated falls of large food fragments, which support active creatures adapted to the discovery and utilization of such food.

food, and he widely utilizes these resources.

Man may not have created an ideal environment for himself, but of all the environments of the sea it is difficult to conceive of one better for its inhabitants than the one marine creatures have created almost exclusively for themselves: the coral islands and coral reefs. In these exquisite, immense and well-nigh unbelievable structures form and adaptation reach a zenith.

An adequate description of the coral reef and coral atoll structure, environments and living communities is beyond the scope of this article. The general history and structure of atolls is well known, not only because of an inherent fascination with the magic and beauty of coral islands but also because of the wide admiration and publicity given to the prescient deductions on the origin of atolls by Charles Darwin, who foresaw much of what modern exploration has affirmed.

From their slowly sinking foundations of ancient volcanic mountains, the creatures of the coral shoals have erected the greatest organic structures that exist. Even the smallest atoll far surpasses any of man's greatest building feats, and a large atoll structure in actual mass approaches the total of all man's building that now exists.

These are living monuments to the success of an extremely intricate but balanced society of fish, invertebrates and plants, capitalizing on the basic advantages of benthic populations already discussed. Here, however, each of the reef structures acts almost like a single great isolated and complex benthic organism that has extended itself from the deep poor waters to the sunlit richer surface. The trapping of the advected food from the surface currents enriches the entire community. Attached plants further add to the economy, and there is considerable direct consumption of plant life by large invertebrates and fish. Some of the creatures and relationships that have developed in this environment are among the most highly adapted found on the earth. For example, a number of the important reef-building animals, the corals, the great tridacna clams and others not only feed but also harbor within their tissues dense populations of single-celled green plants. These plants photosynthesize food that is then directly available within the bodies of the animals; the plants in turn depend on the animal waste products within the body fluids, with which they are bathed, to derive their basic nutrients. Thus within the small environment of these plant-animal composites both the entire laborious nutrient cycle and the microscopic food web of the sea appear to be substantially bypassed.

There is much unknown and much to be discovered in the structure and ecology of coral atolls. Besides the task of unraveling the complex relationships of its inhabitants there are many questions such as: Why have many potential atolls never initiated effective growth and remained submerged almost a mile below the surface? Why have others lost the race with submergence in recent times and now become shallowly submerged, dying banks? Can the nature of the circulation of the ancient ocean be deduced from the distribution of successful and unsuccessful atolls? Is there circulation within the coral limestone structure that adds to the nutrient supply, and is this related to the curious development of coral knolls, or coral heads, within the lagoons? Finally, what is the potential of cultivation within these vast, shallow-water bodies of the deep open sea?

There is, of course, much to learn about all marine life: the basic processes of the food web, productivity, populations, distributions and the mechanisms of reinoculation, and the effects of intervention into these processes, such as pollution, artificial upwelling, transplantation, cultivation and fisheries. To learn of these processes and effects we must understand the nature not only of strong simple actions but also of weak complex interactions, since the forms of life or the success of a species may be determined by extremely small second- and third-order effects. In natural affairs, unlike human codes, *de minimis curat lex*—the law *is* concerned with trivia!

Little is understood of the manner in which speciation (that is, the evolution of new species) occurs in the broadly intercommunicating pelagic environment with so few obvious barriers. Important yet unexpected environmental niches may exist in which temporary isolation may enable a new pelagic species to evolve. For example, the top few millimeters of the open sea have recently been shown to constitute a demanding environment with unique inhabitants. Further knowledge of such microcosms may well yield insight into speciation.

As it has in the past, further exploration of the abyssal realm will undoubtedly reveal undescribed creatures including members of groups thought long extinct, as well as commercially valuable populations. As we learn more of the conditions that control the distribution of species of pelagic organisms, we shall become increasingly competent to read the pages of the earth's marine-biological, oceanographic and meteorological history that are recorded in the sediments by organic remains. We shall know more of primordial history, the early production of a breathable atmosphere and petroleum production. Some of these deposits of sediment cover even the period of man's recorded history with a fine time resolution. From such great records we should eventually be able to increase greatly our understanding of the range and interrelations of weather, ocean conditions and biology for sophisticated and enlightened guidance of a broad spectrum of man's activities extending from meteorology and hydrology to oceanography and fisheries.

Learning and guidance of a more specific nature can also be of great practical importance. The diving physiology of marine mammals throws much light on the same physiological processes in land animals in oxygen stress (during birth, for example). The higher flowering plants that inhabit the marine salt marshes are able to tolerate salt at high concentration, desalinating seawater with the sun's energy. Perhaps the tiny molecule of DNA that commands this process is the most precious of marine-life resources for man's uses. Bred into existing crop plants, it may bring salt-water agriculture to reality and nullify the creeping scourge of salinization of agricultural soils.

Routine upstream reinoculation of preferred species of phytoplankton and zooplankton might stabilize some pelagic marine populations at high effectiveness. Transplanted marine plants and animals may also animate the dead saline lakes of continental interiors, as they have the Salton Sea of California.

The possible benefits of broad marine-biological understanding are endless. Man's aesthetic, adventurous, recreational and practical proclivities can be richly served. Most important, undoubtedly, is the intellectual promise: to learn how to approach and understand a complex system of strongly interacting biological, physical and chemical entities that is vastly more than the sum of its parts, and thus how better to understand complex man and his interactions with his complex planet, and to explore with intelligence and open eyes a huge portion of this earth, which continuously teaches that when understanding and insight are sought for their own sake, the rewards are more substantial and enduring than when they are sought for more limited goals.

The Author

JOHN D. ISAACS is professor of oceanography and director of marine life research at the Scripps Institution of Oceanography. The titles reflect only a portion of his interests; his others include the mechanism of the heating of the earth's interior, the construction of harbors by atomic explosions and the disposal of radioactive wastes at sea. He was also a member of the group that suggested tethering a satellite to the earth by a cable, which would be attached near the Equator so that energy from the turning of the earth would drive loads up the cable to the satellite; the installation could be used to maintain an observatory or fuel spacecraft or even to support very tall structures on the earth's surface. Isaacs came to oceanography in a roundabout way. He went to college to study engineering but became bored and dropped out. Thereafter he worked as a logger, a fire lookout, a merchant seaman and a commercial fisherman. He returned to college in 1944 and obtained his bachelor's degree in engineering (from the University of California at Berkeley) at the age of 32. He has been at Scripps since 1948.

Bibliography

THE OPEN SEA: THE WORLD OF PLANKTON. Alister C. Hardy. Houghton Mifflin Company, 1957.

THE OPEN SEA: FISH AND FISHERIES. Alister C. Hardy. Houghton Mifflin Company, 1959.

OCEANS: AN ATLAS-HISTORY OF MAN'S EXPLORATION OF THE DEEP. Edited by G. E. R. Deacon. Paul Hamlyn, 1962.

BIOLOGY OF SUSPENSION FEEDING. C. B. Jorgensen. Pergamon Press, 1966.

SCIENTIFIC AMERICAN September 1969, Vol. 221, No. 3, pp. 166–176

OFFPRINT **885**

THE PHYSICAL RESOURCES OF THE OCEAN

by Edward Wenk, Jr.

They include not only the oil and minerals of the bottom
and the minerals dissolved in seawater but also seawater
itself and the shoreline carved by the action of the sea.

Men have caught fish in the ocean and extracted salt from its brine for thousands of years, but only within the past decade have they begun to appreciate the full potential of the resources of the sea. Three converging influences have been responsible for today's intensive exploration and development of these resources. First, scientific oceanography is generating new knowledge of what is in and under the sea. Second, new technologies make it feasible to reach and extract or harvest resources that were once inaccessible. Third, the growth of population and the industrialization of society are creating new demands for every kind of raw material.

The ocean's resources include the vast waters themselves, as a processing plant to convert solar energy into protein [see "The Food Resources of the Ocean," by S. J. Holt; SCIENTIFIC AMERICAN Offprint 886], a storehouse of dissolved minerals and fresh water, a receptacle for wastes, a source of tidal energy and a medium for new kinds of transportation. They also include the sea floor, sediments and rocks below the waters as sites of fossil fuel and mineral deposits; the seacoast as a unique resource that is vulnerable to rapid, irrevocable degradation by man.

Because the oceans are so wide and so deep, statistics on their gross resource potential are impressive. It is important to understand, however, that the immediate significance of these resources and

OFFSHORE OIL PLATFORM in the photograph on the opposite page is in Alaska's Cook Inlet, 60 miles southwest of Anchorage. Wells are drilled from derricks set over the massive legs, 14 feet in diameter. The plume of flame is burning natural gas, a waste product in this case. Oil and gas account for more than 90 percent of the value of minerals now being retrieved from the oceans.

their long-term relevance to society involve both exploration and development, and development depends on economic, social, legal and political considerations. One special feature of marine resources that may at first retard development may in the long run promote it: the fact that almost without exception sea-floor resources are in areas not subject to private ownership (although the resources will be largely privately developed). More than 85 percent of the ocean bottom lies beyond the present boundaries of national jurisdictions, and in the areas that are subject to national control the resources are considered common property. This circumstance may uniquely invoke a balancing of public and private interests, disciplined resource management and enhanced international cooperation.

The 350 million cubic miles of ocean water constitute the earth's largest continuous ore body. Dissolved solids amount to 35,000 parts per million, so that each cubic mile (4.7 billion tons of water) contains about 165 million tons of solids. Although most chemical elements have been detected (and probably all are present) in seawater, only common salt (sodium chloride), magnesium and bromine are now being extracted in significant amounts. The production of salt (which can be traced back to Neolithic times and resulted in the first U.S. patent) is currently valued at $175 million per year worldwide. Magnesium, the third most abundant element in the oceans, is by far the most valuable mineral extracted from seawater in this country, with annual production worth about $70 million. Although the ocean contains bromine in concentrations of only 65 parts per million, it is the source of 70 percent of the world's production of this element, which is used principal-

ly in antiknock compounds for gasoline. The economic recovery of other chemicals from seawater is questionable because of extraction costs. In a cubic mile of seawater the value of 17 critical metals (including cobalt, copper, gold, silver, uranium and zinc) is less than $1 million at current prices; a plant to handle a cubic mile of water per year would have to process 2.1 million gallons per minute every minute of the year, and operating it would cost significantly more than the value of all its products.

One of the potential resources of seawater that has been most difficult to extract economically is water itself—fresh water. As requirements for water for domestic use, agriculture and industry rise sharply, however, desalting the sea becomes increasingly attractive. More than 680 desalination plants with a capacity of more than 25,000 gallons of fresh water per day are now in operation or under construction around the globe, and the growth rate is projected at 25 percent per year over the next decade. The cost of desalting has been decreased by new technology to less than 85 cents per 1,000 gallons, but this is still generally prohibitive in the U.S., where the cost of 1,000 gallons of fresh water is about 35 cents. In water-deficient areas or where the local water supply is unfit for consumption, however, desalted water is competitive. This accounts for the presence of more than 50 plants in Kuwait, 22 on Ascension Island and the 2.6-million-gallon facility at Key West, the first U.S. city to obtain its water supply directly from the ocean. Considerably lower costs will be attained within the next decade where large-scale desalting operations are combined with nuclear-fueled power plants to take advantage of their output of waste heat.

Once upon a time man could safely utilize the waters of the sea as a recep-

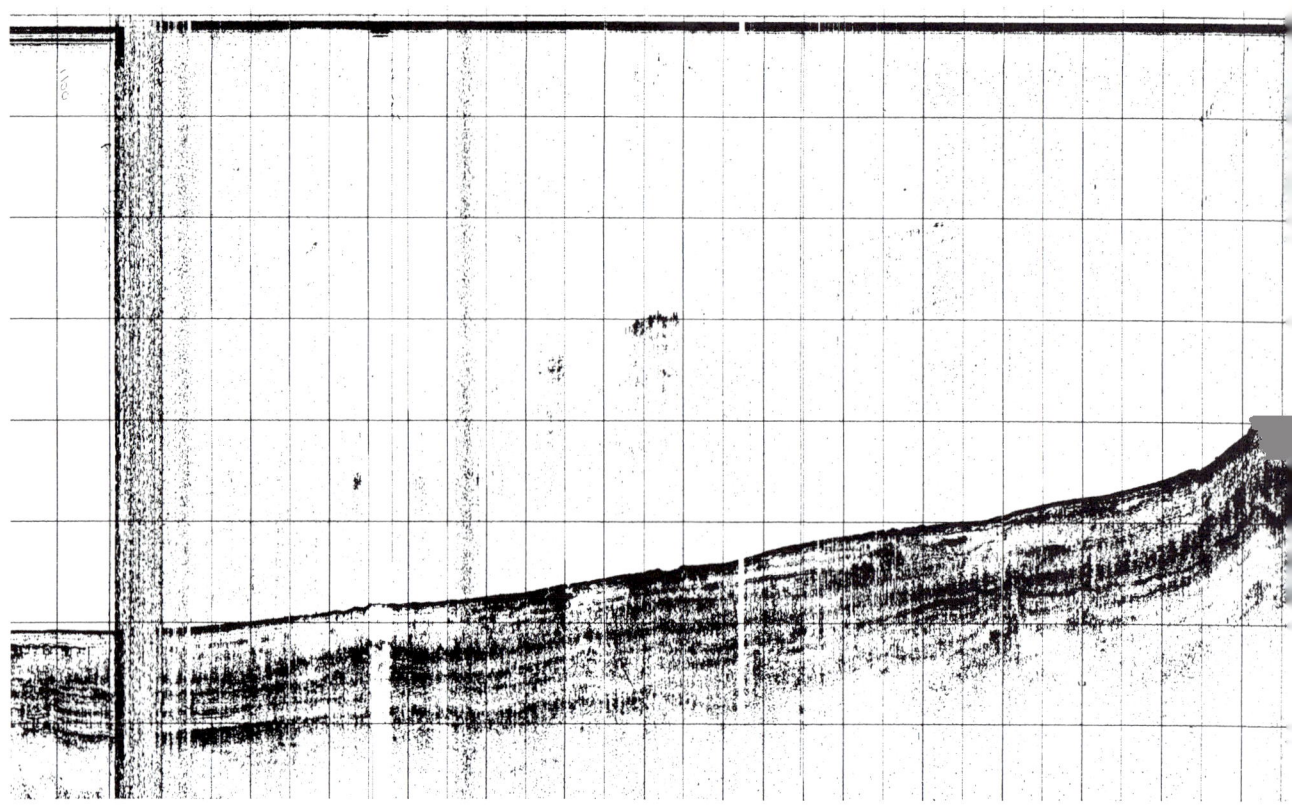

CONTINENTAL RISE, which may be rich in resources, is evident in this seismic profile made off Liberia by the Global Ocean Floor Analysis and Research Center of the U.S. Naval Oceanographic Office. The hard, straight line across the top of the record is a water reflection. The abyssal plain (*left*) is about 15,000 feet below sea level. From this plain a thick apron of land-derived layered sediments comprising the continental rise slants gently up to the toe of the continental slope. The continental slope, which is here

tacle for sewage and other effluents from municipalities and industries, confident that the wastes would rapidly be diluted, dispersed and degraded. With the growth of population and the concentration of coastal industry that is no longer possible. The sheer bulk of the material disposed of and the presence of new types of nondegradable waste products are a special threat to coastal waters— the same waters that, as we shall see, are subject to increasing demands from a wide range of competing activities. In addition pollutants are now beginning to concentrate at an alarming rate far from shore in the open ocean. Since tetraethyl lead was introduced into gasoline 45 years ago, lead concentrations in Pacific Ocean waters have jumped tenfold. Toxic DDT residues have been detected in the Bay of Bengal, having drifted with

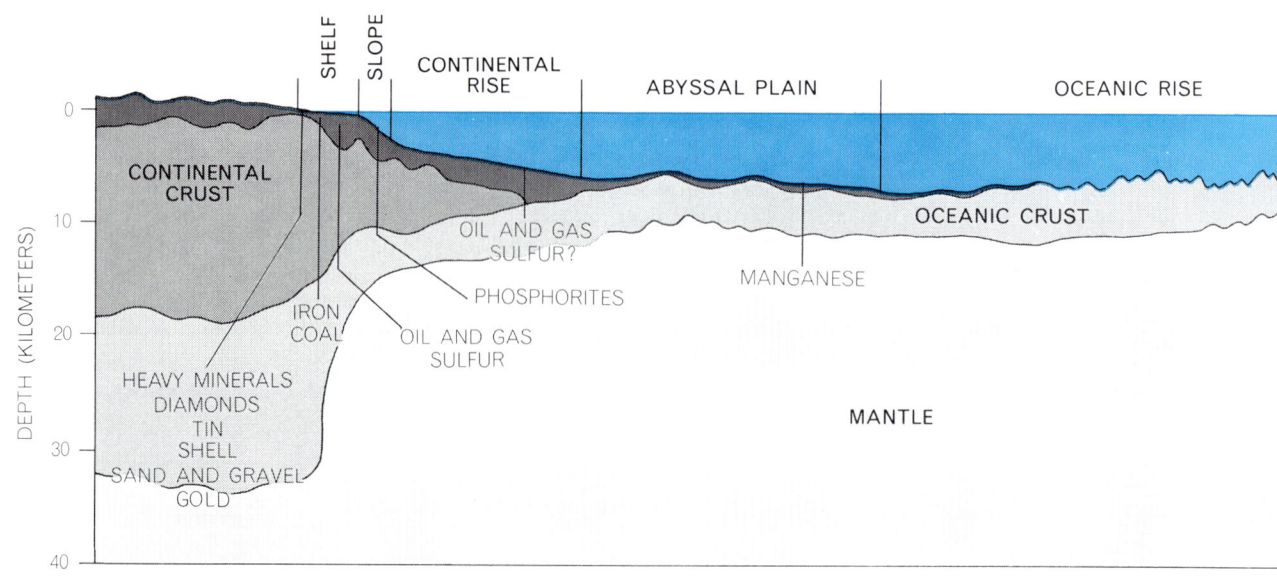

OCEAN-FLOOR RESOURCES that are known or believed to exist in the various physiographic provinces are indicated on a schematic cross section of a generalized ocean basin extending from a continent out to a mid-ocean ridge. Some of these resources are now

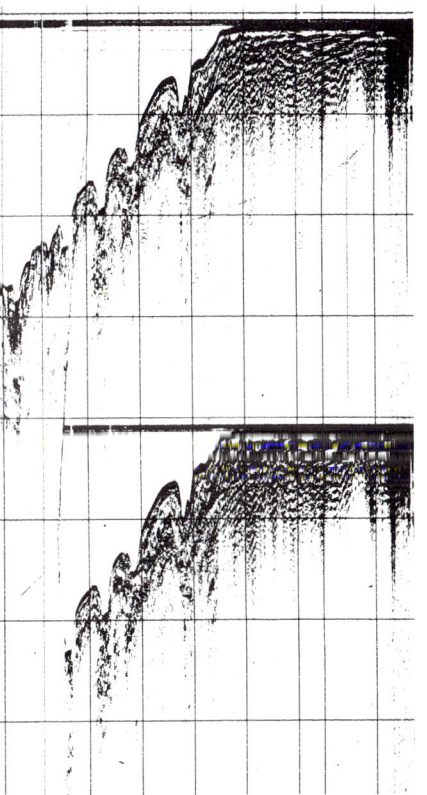

marked by large sedimentary ridges, ascends more steeply to the shallow continental shelf. The "multiple" (*right*) is in effect an echo of the structures shown above it.

the wind from as far away as Africa. And man-made radioactivity from nuclear fallout can be isolated in any 50-gallon water sample taken anywhere in the ocean.

The mineral resources of the seabed, unlike those of the essentially uniform overlying waters, occur primarily in scattered, highly localized deposits and structures on top of and within the sediments and rocks of the ocean floor. They include (1) fluids and soluble minerals, such as oil, gas, sulfur and potash, that can be extracted through boreholes; (2) consolidated subsurface deposits, such as coal, iron ore and other metals found in veins, which are so far mined only from tunnels originating on land, and (3) unconsolidated surface deposits that can be dredged, such as heavy metals in ancient beaches and stream beds, oyster shell, sand and gravel, diamonds, and "authigenic" minerals, nodules of manganese and phosphorite that have been formed by slow precipitation from seawater. Economic exploitation has so far been confined to the continental shelves in waters less than 350 feet deep and within 70 miles of the coastline.

Oil and gas represent more than 90 percent by value of all minerals obtained from the oceans and have the greatest potential for the near future. Offshore sources are responsible for 17 percent of the oil and 6 percent of the natural gas produced by non-Communist countries. Projections indicate that by 1980 a third of the oil production—four times the present output of 6.5 million barrels a day—will come from the ocean; the increase in gas production is expected to be comparable. Subsea oil and gas are now produced or are about to be produced by 28 countries; another 50 are engaged in exploratory surveys. Since 1946 more than 10,000 wells have been drilled off U.S. coasts and more than $13 billion has been invested in petroleum exploration and development. The promise of large oil reserves has stimulated industry to invest more than $1.7 billion since mid-1967 to obtain Federal leases off Louisiana, Texas and California that guarantee only the right to search for and develop unproved reserves. To date more than 6.5 million acres of the outer continental shelf off the U.S. have been leased, which is half of the acreage offered, resulting in lease income to the Federal Government of $3.4 billion. With more than 90 percent of the most favorable inland areas explored and less than 10 percent of the U.S. shelf areas surveyed in detail, the prospects are encouraging for additional large oil finds off U.S. coasts.

Sulfur, one of the world's prime industrial chemicals, is found in the cap rock of salt domes buried within continental and sea-floor sediments. The sulfur is recovered rather inexpensively by melting it with superheated water piped down from the surface and then forcing it up with compressed air. Only 5 percent of the explored salt domes contain commercial quantities of sulfur, and offshore production has been limited to two mines off Louisiana that supply two million tons, worth $37 million, a year. Now a critical shortage of sulfur and the recent discovery by the deep-drilling ship *Glomar Challenger* of sulfur-bearing domes in the deepest part of the Gulf of Mexico have stimulated an intensive search for offshore sulfur deposits.

Undersea subfloor mining can be traced back to 1620, when coal was ex-

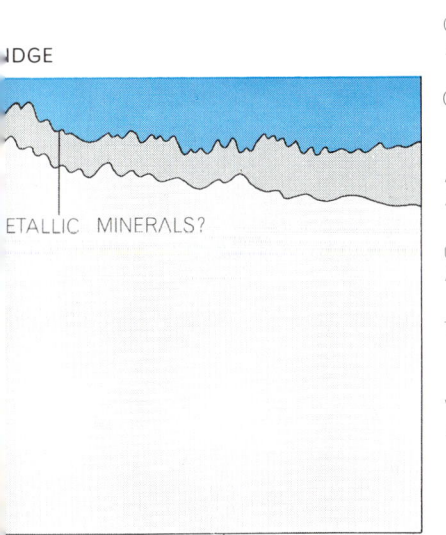

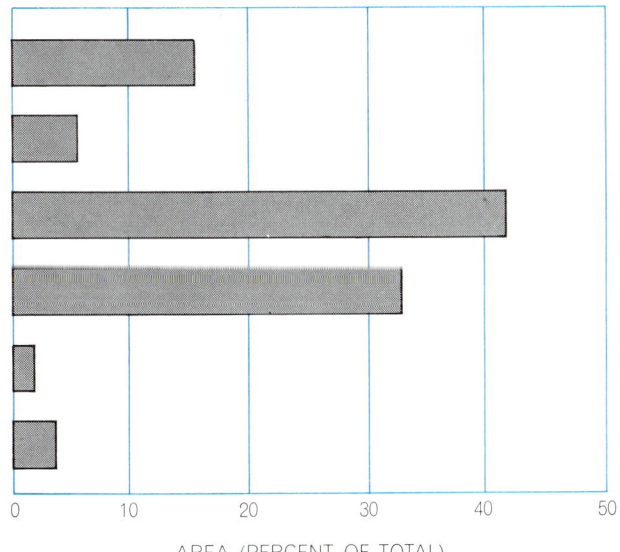

being exploited but others may not be economic for years. Sedimentary layers (*black*) are the most likely site of recoverable raw materials. The chart (*right*) shows what percent of the ocean floor's 140 million square miles of area is occupied by each province.

tracted in Scotland through shafts that were driven seaward from an offshore island. To date 100 subsea mines with shaft entries on land have recovered coal, iron ore, nickel-copper ores, tin and limestone off a number of countries in all parts of the world. Coal extracted from as deep as 8,000 feet below sea level accounts for almost 30 percent of Japan's total production and more than 10 percent of Britain's. With present technology subsea mining can be conducted economically as far as 15 miles offshore, given mineral deposits that are worth $10 to $15 per ton and occur in reserves of more than $100 million. The economically feasible distance should increase to 30 miles by 1980 with the development of new methods for rapid underground excavation. Eventually shafts may be driven directly from the seabed

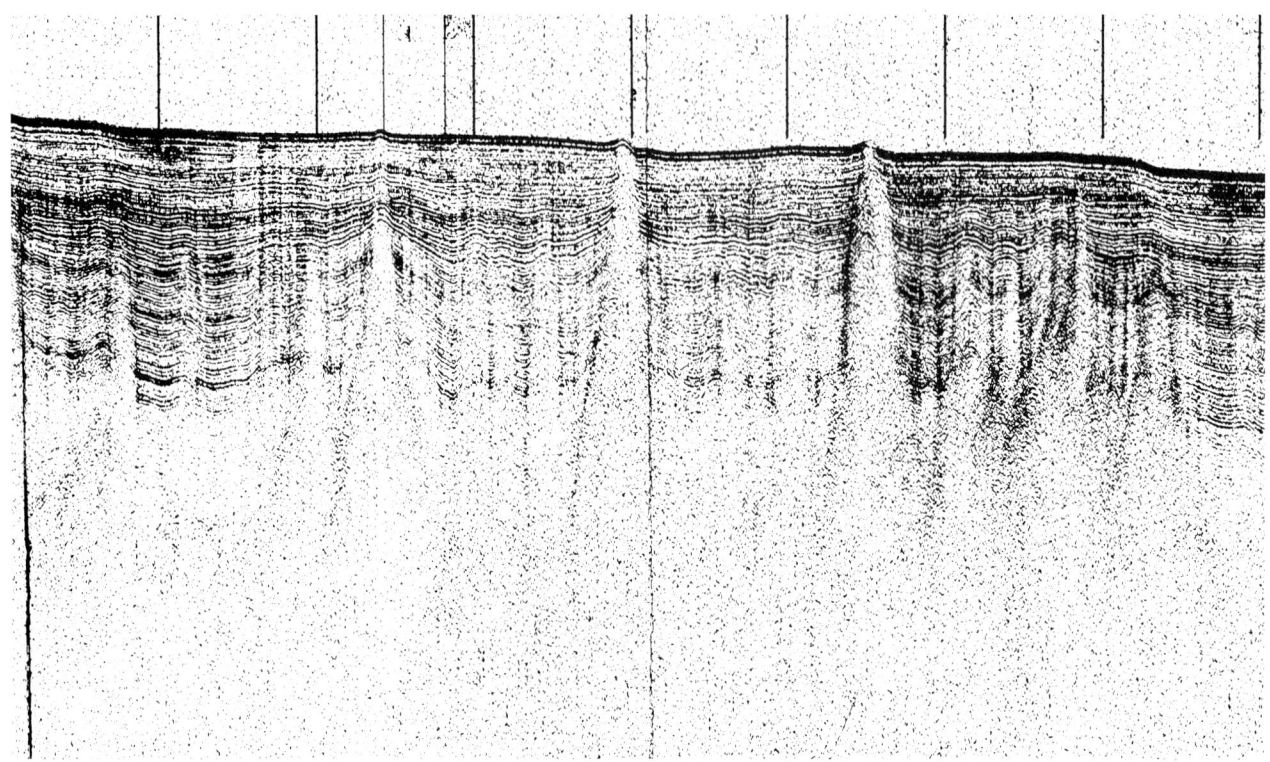

POSSIBILITY OF OIL in the deep-sea floor was revealed by this record from some 250 miles northwest of the Cape Verde Islands. The record, like the one on the preceding pages, was made by the research ship *Kane* of the Naval Oceanographic Office. The tall narrow structures appear to be salt domes, along the flanks of which surface-seeking oil is often trapped in tilted sedimentary layers.

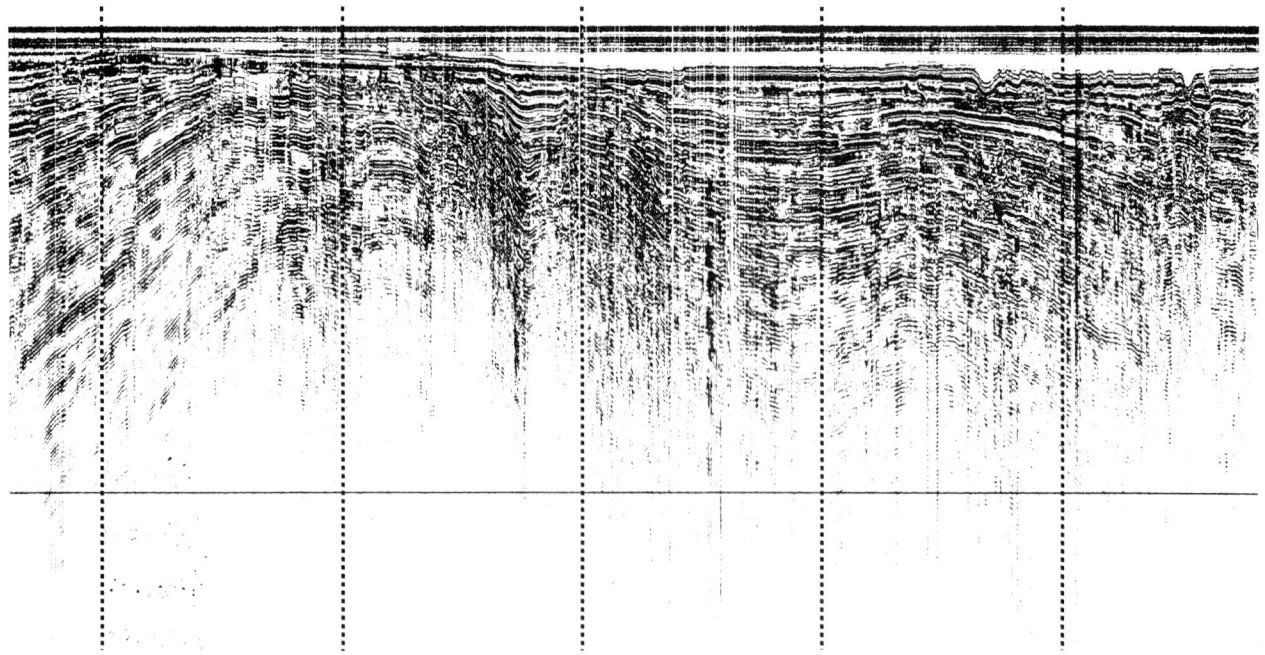

FOLDED SEDIMENTARY LAYERS are shown in this *Kane* record made on the continental shelf north of Trinidad. The band at top is from water reflections. The record shows anticlines (arches) and synclines (troughs); oil is often trapped in crests of anticlines.

if ore deposits are located in ocean-floor rock far from land.

Seventy percent of the world's continental shelves consist of ancient unconsolidated sediments from which are dredged such commodities as sand, gravel, oyster shell, tin, heavy-mineral sands and diamonds. Dredging is an attractive mining technique because of low capital investment, quick returns and high profits and the operational mobility offered by floating dredges. So far it has been limited to nearshore waters less than 235 feet deep and protected from severe weather effects. As knowledge of resources in deeper water increases, industry will undoubtedly upgrade its dredging technology.

Of the many potentially valuable surface deposits, sand and gravel are the most important in dollar terms, and only these and oyster shells are now mined off the U.S. coast. Some 20 million tons of oyster shells are extracted from U.S. continental shelves annually as a source of lime; sand and gravel run about 50 million cubic yards. As coastal metropolitan areas spread out, they cover dry-land deposits of the very construction materials required to sustain their expansion; in these circumstances sea-floor sources such as one recently found off New Jersey, which is thought to contain a billion tons of gravel, become commercially valuable.

In the deeper waters of the continental shelves, on the upper parts of the slopes and on submarine banks and ridges widespread deposits of marine phosphorite nodules are found at depths between 100 and 1,000 feet. The best-known large deposits are off southern California, where total reserves are estimated at 1.5 billion tons, and off northwestern Mexico, Peru and Chile, the southeastern U.S. and the Union of South Africa. The only major attempt at mining was made in 1961, when a company leased an area off California, but that lease was returned unexploited to the Federal Government four years later. With large land sources generally available to meet the demand for phosphates for fertilizer and other products, offshore exploitation of this resource is not likely to occur soon, except possibly in phosphate-poor countries.

The only known minerals on the floor of the deep ocean that appear to be of potential economic importance are the well-publicized manganese nodules, formed by the precipitation from seawater of manganese oxides and other mineral salts, usually on a small nucleus such as a bit of stone or a shark's tooth.

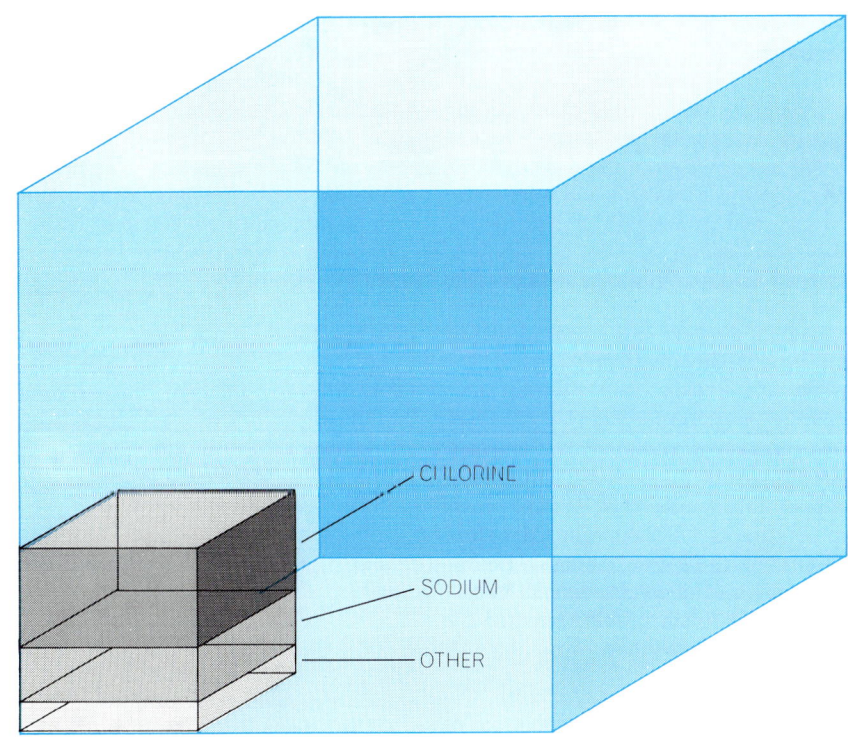

SEAWATER contains an average of 35,000 parts per million of dissolved solids. In a cubic mile of seawater, weighing 4.7 billion tons, there are therefore about 165 million tons of dissolved matter, mostly chlorine and sodium (*gray cube*). The volume of the ocean is about 350 million cubic miles, giving a theoretical mineral reserve of about 60 quadrillion tons.

ELEMENT	TONS PER CUBIC MILE	ELEMENT	TONS PER CUBIC MILE
CHLORINE	89,500,000	NICKEL	9
SODIUM	49,500,000	VANADIUM	9
MAGNESIUM	6,400,000	MANGANESE	9
SULFUR	4,200,000	TITANIUM	5
CALCIUM	1,900,000	ANTIMONY	2
POTASSIUM	1,800,000	COBALT	2
BROMINE	306,000	CESIUM	2
CARBON	132,000	CERIUM	2
STRONTIUM	38,000	YTTRIUM	1
BORON	23,000	SILVER	1
SILICON	14,000	LANTHANUM	1
FLUORINE	6,100	KRYPTON	1
ARGON	2,800	NEON	.5
NITROGEN	2,400	CADMIUM	.5
LITHIUM	800	TUNGSTEN	.5
RUBIDIUM	570	XENON	.5
PHOSPHORUS	330	GERMANIUM	.3
IODINE	280	CHROMIUM	.2
BARIUM	140	THORIUM	.2
INDIUM	94	SCANDIUM	.2
ZINC	47	LEAD	.1
IRON	47	MERCURY	.1
ALUMINUM	47	GALLIUM	.1
MOLYBDENUM	47	BISMUTH	.1
SELENIUM	19	NIOBIUM	.05
TIN	14	THALLIUM	.05
COPPER	14	HELIUM	.03
ARSENIC	14	GOLD	.02
URANIUM	14		

CONCENTRATION of 57 elements in seawater is given in this table. Only sodium chloride (common salt), magnesium and bromine are now being extracted in significant amounts.

They are widely distributed, with concentrations of 31,000 tons per square mile on the floor of the Pacific Ocean. Although commonly found at depths greater than 12,500 feet, nodules exist in 1,000 feet of water on the Blake Plateau off the southeastern U.S. and were located last year at a depth of 200 feet in the Great Lakes.

The nodules average about 24 percent manganese, 14 percent iron, 1 percent nickel, .5 percent copper and somewhat less than .5 percent cobalt. Since ore now being mined from land deposits in a number of countries averages 35 to 55 percent manganese, it may be the minor constituents of the nodules, particularly copper, cobalt and nickel, that first prove to be attractive economically. Many experts think the key to profitable exploitation is the solution of a difficult metallurgical separation problem created by the unique combination of minerals in the nodules.

Few discoveries have created more excitement among earth scientists than the location, by different expeditions in 1964, 1965 and 1966, of three undersea pools of hot, high-density brines in the middle of the Red Sea. The brines contain minerals in concentrations as high as 300,000 parts per million—nearly 10 times as much solid matter as is commonly dissolved in ocean water—and overlie sediments rich in such heavy metals as zinc, copper, lead, silver and gold. Similar deposits may be characteristic of other enclosed basins associated, as is the Red Sea, with rift valleys.

As this decade ends resource exploration is advancing on many fronts. Chromite has been found by Russian oceanographers in sea-floor rifts in the Indian Ocean, and zirconium, titanium and other heavy minerals have been detected in sediments from extensive areas off the Texas coast. Methane deposits sufficient to supply Italy's needs for at least six years have been confirmed in the Adriatic Sea. New oil fields of economic value have been discovered off Mexico, Trinidad, Brazil, Dahomey and Australia. Surveys of the Yellow Sea and the East China Sea indicate that the continental shelf between Taiwan and Japan may contain one of the richest oil reserves in the world. It is now becoming clear that the continental rises, which lie at depths ranging from about 5,000 to 18,000 feet and contain a far larger total volume of sediments than the shelves, may hold significant petroleum reserves. Within the past year the *Glomar Challenger* has drilled into oil-bearing sediments lying under 11,700 feet of water in the

Gulf of Mexico, and seismic surveys have revealed what appear to be typical oil-bearing structures under the deep ocean-basin floor [*see upper illustration on page 746*].

As on land, resource development of a frontier requires a mixture of public and private entrepreneurship. Historically basic exploration has been sponsored by government; this broad-ranging exploration reveals opportunities that are followed up by detailed privately funded surveys. This pattern is likely to persist, and as the International Decade of Ocean Exploration gets under way a wide variety of new opportunities

for marine resource development will surely come to light.

Limitations on the exploitation of the oceans stem partly from lack of knowledge about the distribution of resources and the state of the art of undersea technology. The major limits, however, are set by venture economics, the motivating factor for the profit sector. That factor is influenced by the availability of competing land deposits, by extraction technology and the legal situation and, most critically, by market demand. On the basis of projections of world population and gross national products to the year 2000, which indicate respective in-

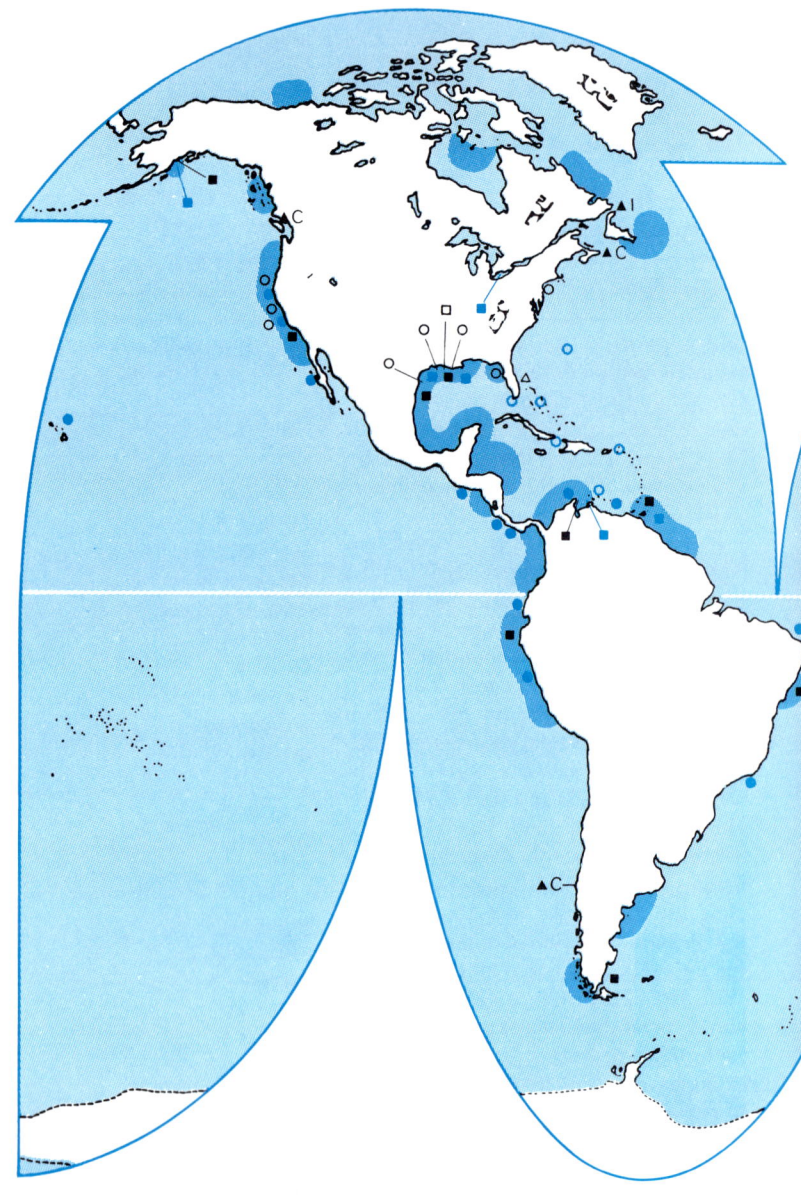

CURRENT PRODUCTION of major ocean resources (except sand, gravel and shell) is mapped with areas of oil and gas exploration. Data come from U.S. Geological Survey, *Oil &*

creases of almost 100 and 500 percent over 1965, a sharp rise in total resource demand can be anticipated, and with it a greater role for the sea.

Other major impediments to the rapid development of ocean resources arise from social and legal constraints. Damage to beaches and wildlife from oil leaks, as in the Santa Barbara Channel, and uncertainty about the effect of dredging on marine organisms have brought public awareness that offshore development may have detrimental consequences. The public, the owner of the resources, is demanding greater safeguards, questioning the wisdom of re-source development in areas where it may threaten the environment. In deeper waters seabed development comes up against the potent issue of ownership. There are major questions about the boundaries of national jurisdictions and about the jurisdiction over the seabed beyond such boundaries [see "The Ocean and Man," by Warren S. Wooster; SCIENTIFIC AMERICAN Offprint 888].

The coastal margin—the ribbon of land and water where people and oceans meet and are profoundly influenced by each other—has only recently come to be recognized and treated as a valuable and perishable resource. It is actually a complex of unique physical resources: estuaries and lagoons, marshes, beaches and cliffs, bays and harbors, islands and spits and peninsulas.

In the year 2000 half of the estimated 312-million population of the U.S. will live on 5 percent of the land area in three coastal urban belts: the megalopolises of the Atlantic, the Pacific and the Great Lakes. Along with the people will come an intensification of competing demands for the limited resources of the narrow, fragile coastal zone. To make matters worse, the coastal resource is shrinking under the pressure of natural forces (hur-

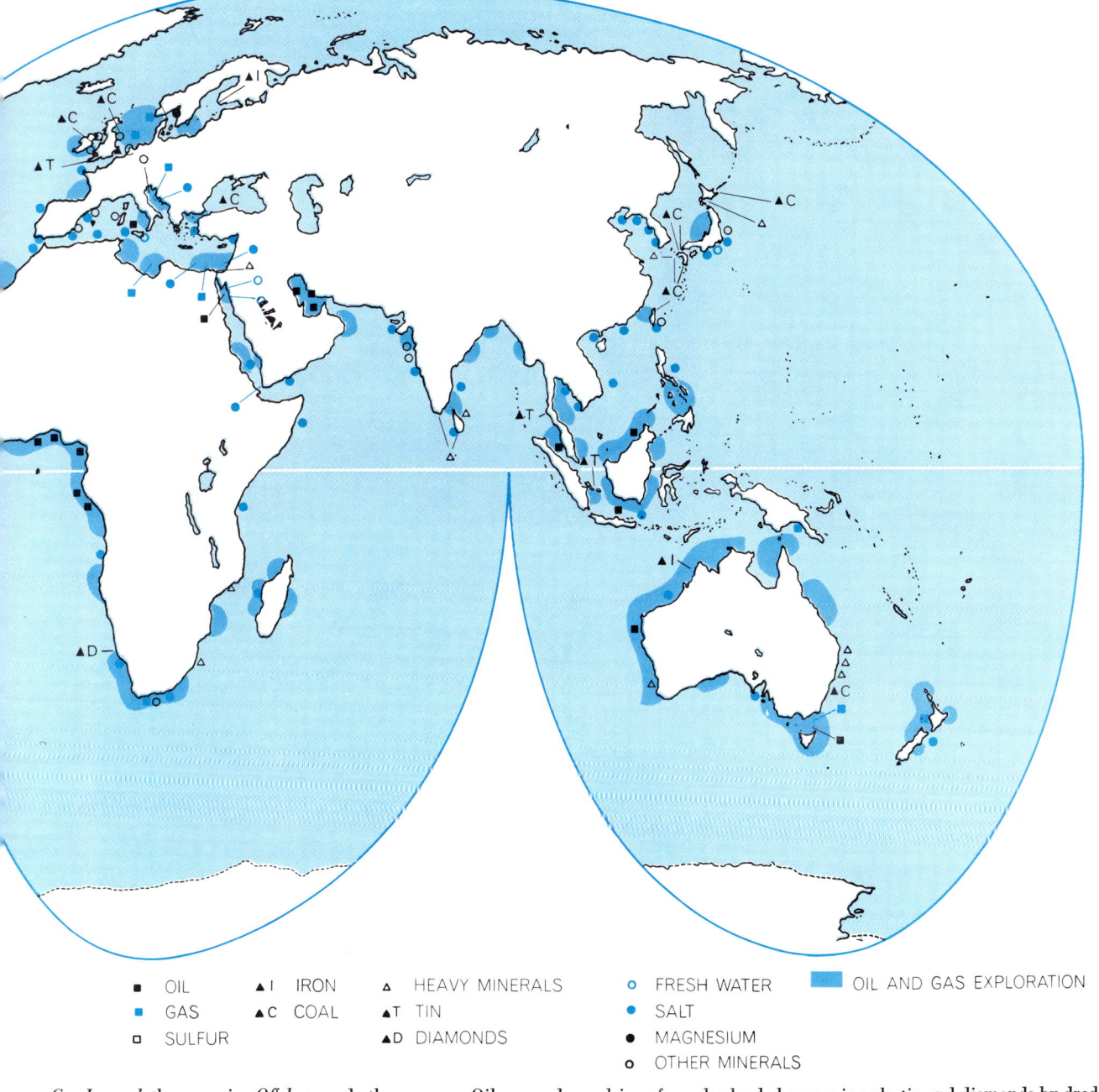

■	OIL	▲I	IRON	△	HEAVY MINERALS	○	FRESH WATER		OIL AND GAS EXPLORATION
■	GAS	▲C	COAL	▲T	TIN	●	SALT		
▢	SULFUR			▲D	DIAMONDS	●	MAGNESIUM		
						○	OTHER MINERALS		

Gas Journal, the magazine *Offshore* and other sources. Oil, gas and sulfur are produced by drilling; coal and iron ore from mines driven from dry land; heavy minerals, tin and diamonds by dredging; fresh water, salt, magnesium and other minerals from seawater.

MANGANESE NODULES, formed by precipitation from seawater, are generally found on the deep-sea floor. These nodules were photographed on the Blake Plateau off the southeastern U.S., less than 3,000 feet deep, by a prospecting ship operated by Deepsea Ventures, Inc. They average two inches in diameter, about a quarter-pound in weight. The manganese content is between 15 and 30 percent, the nickel and copper content about 1 percent each.

ricanes have caused $5 billion in damage to the U.S. economy in the past 15 years) and human exploitation and neglect.

More than a tenth of the 10.7 million square miles of shellfish-producing waters bordering the U.S. is now unusable because of pollution. Dredging, drainage projects and even chemical mosquito-control programs are having devastating effects on fish and other aquatic life. The amount of industrial waste reaching the oceans will increase sevenfold within the decade. Whereas 14 nuclear-powered generating plants are operating in the U.S. today, more than 100 are scheduled by 1975, with nine planned for Long Island Sound alone. Thermal pollution from the discharge of hot water is therefore a potential threat to coastal waters as well as inland lakes and rivers.

In the competition for the zone's resources among different uses—industrial and housing development, ports, shipbuilding, recreation, commercial fisheries and waste disposal—natural wetlands and estuarine open spaces are losing out. Of the tidal wetlands along the Atlantic coast from Maine to Delaware, 45,000 acres were lost between 1955 and 1964. An inventory shows that 34 percent of that area was dried up by being used as a dumping ground for dredging operations; 27 percent was filled for housing developments; 15 percent went to recreational developments (parks, beaches and marinas) and 10 percent to bridges, roads, parking lots and airports; 7 percent was turned into industrial sites and 6 percent into garbage and trash dumps. (In Maryland 176 acres of submerged

land in Chesapeake Bay were sold recently for $100 an acre and, after being filled with dredged bay-bottom muck, were subdivided into lots selling for between $4,000 and $8,000 each.)

With the demand for marine recreation growing with the coastal population, pressure is increasing on the one-third of the coastal zone that has recreational potential. Only about 6.5 percent of this is now in public ownership, yet in order to meet the projected demand it is considered essential that about 15 percent be accessible to the public. The mere fact that coastal land with recreational potential exists, moreover, is far from meaning that it will ever be put to recreational use. Swimming, boating and skin diving are often incompatible with competing alternative uses, many of which appear to have equally valid claims. In the face of conflicts between public and private, and long-term and short-term, benefits, how and by whom will the ultimate decisions be made on the proper utilization of coastal land?

Management of the coastal zone is unwieldy because the environment is almost hopelessly fragmented by political subdivisions: 24 states, more than 240 counties, some 600 coastal cities, townships, towns and villages and numerous regional authorities and special districts with their own regulatory powers. Superimposed on the many public jurisdictions there is another tapestry of private ownership. Because the states hold coastal resources in trust out to the three-mile limit, the Federal Government has a restricted role in resolving disputes, but it may be able to exert leadership in defining the issues.

Thoreau once admonished: "What is the use of a house if you haven't got a tolerable planet to put it on." Unless rational alternatives among competing uses are evaluated, the trend will continue to be toward single-purpose uses, motivated by short-term advantages to individuals, industry or local governments. Such exploitation may actually dissipate resources. Private beach development restricts public access; dredging and filling downgrade commercial fishing; offshore drilling rigs limit freedom of navigation. Each single-purpose use may seem justifiable on its own, but the overall effect of piecemeal development can be chaos.

In this technological age man can do many more of the things he wants to do. The oceans place before him a vast store of little-developed material resources; the tools of science and technology are at his disposal. This combination of a new frontier, new knowledge and new technical capability may be unique in the human experience. We are accumulating the basic information with which to define the ecological base from which we operate, to understand the natural forces at work and to predict the consequences of each insult to the environment. With this new comprehension it will soon be possible to develop the engineering with which to harvest mineral wealth, maintain water quality, inhibit beach erosion, create modern ports and harbors—and to establish the criteria for making necessary choices among courses of action and the law and institutions to effectuate them. In time we may even be able to correct mistakes that were made long ago in ignorance or that occur in the future because of man's stupidity, neglect or greed.

TIDAL WETLANDS, an important coastal resource, are disappearing rapidly. The top photograph shows Boca Ciega Bay, near St. Petersburg on the west coast of Florida, as it was in 1949. The bottom photograph shows the same area filled and developed, in 1969.

The Author

EDWARD WENK, JR. is executive secretary of the National Council on Marine Resources and Engineering Development, a cabinet-level body consisting of the Vice President, who is chairman, and the heads of the eight Federal agencies with programs in the marine sciences. He describes himself as "a research engineer with experience in marine affairs, laboratory management and public administration." Wenk was graduated from Johns Hopkins University in 1940 with a degree in civil engineering. He studied architecture at Harvard University's Graduate School of Design, received a master's degree in applied mechanics from Harvard in 1947 and obtained his doctorate in civil engineering from Johns Hopkins in 1950. From 1942 to 1956 he was responsible for the U.S. Navy's program of structural research on ships. From 1956 to 1959 he was chairman of the department of engineering mechanics at the Southwest Research Institute, designing while there the deep-diving research submersible *Aluminaut*. In 1959 Wenk joined the Legislative Reference Service of the Library of Congress as the first adviser to Congress on science and technology. He was appointed to the White House staff in 1961 as assistant to the President's science adviser and served as executive secretary of the Federal Council for Science and Technology. From 1964 until he was named to his present position in 1966 he was with the Library of Congress as head of the Science Policy Research Division of the Legislative Reference Service.

Bibliography

THE MINERAL RESOURCES OF THE SEA. John L. Mero. Elsevier Publishing Company, 1965.

MINERAL RESOURCES OF THE WORLD OCEAN: PROCEEDINGS OF A SYMPOSIUM HELD AT THE NAVAL WAR COLLEGE, NEWPORT, RHODE ISLAND, JULY 11–12, 1968. Edited by Elisabeth Keiffer. Graduate School of Oceanography, University of Rhode Island, Occasional Publication No. 4, 1968.

USES OF THE SEAS. Edited by Edmund A. Gullion. Prentice-Hall, Inc., 1968.

ENCOURAGING DEVELOPMENT OF NONLIVING RESOURCES. *Marine Science Affairs—a Year of Broadened Participation: The Third Report of the President to the Congress on Marine Resources and Engineering Development.* U.S. Government Printing Office, January, 1969.

ENHANCING BENEFITS FROM THE COASTAL ZONE. *Marine Science Affairs—a Year of Broadened Participation: The Third Report of the President to the Congress on Marine Resources and Engineering Development.* U.S. Government Printing Office, January, 1969.

PETROLEUM RESOURCES UNDER THE OCEAN FLOOR. National Petroleum Council, 1969.

SCIENTIFIC
AMERICAN September 1969, Vol. 221, No. 3, pp. 178–194

OFFPRINT 886

THE FOOD RESOURCES
OF THE OCEAN

by S. J. Holt

The present harvest of the oceans is roughly 55 million tons a year,
half of which is consumed directly and half converted into fish meal
A well-managed world fishery could yield more than 200 million tons.

I suppose we shall never know what was man's first use of the ocean. It may have been as a medium of transport or as a source of food. It is certain, however, that from early times up to the present the most important human uses of the ocean have been these same two: shipping and fishing. Today, when so much is being said and written about our new interests in the ocean, it is particularly important to retain our perspective. The annual income to the world's fishermen from marine catches is now roughly $8 billion. The world ocean-freight bill is nearly twice that. In contrast, the wellhead value of oil and gas from the seabed is barely half the value of the fish catch, and all the other ocean mineral production adds little more than another $250 million.

Of course, the present pattern is likely to change, although how rapidly or dramatically we do not know. What is certain is that we shall use the ocean more intensively and in a greater variety of ways. Our greatest need is to use it wisely. This necessarily means that we use it in a regulated way, so that each ocean resource, according to its nature, is efficiently exploited but also conserved. Such regulation must be in large measure of an international kind, particularly insofar as living resources are concerned. This will be so whatever may be the eventual legal regime of the high seas and the underlying bed. The obvious fact about most of the ocean's living resources is their mobility. For the most part they are lively animals, caring nothing about the lines we draw on charts.

The general goal of ecological research, to which marine biology makes an important contribution, is to achieve an understanding of and to turn to our advantage all the biological processes that give our planet its special character. Marine biology is focused on the prob-

lems of biological production, which are closely related to the problems of production in the economic sense. Our most compelling interest is narrower. It lies in ocean life as a renewable resource: primarily of protein-rich foods and food supplements for ourselves and our domestic animals, and secondarily of materials and drugs. I hope to show how in this field science, industry and government need each other now and will do so even more in the future. First, however, let me establish some facts about present fishing industries, the state of the art governing them and the state of the relevant science.

The present ocean harvest is about 55 million metric tons per year. More than 90 percent of this harvest is finfish; the rest consists of whales, crustaceans and mollusks and some other invertebrates. Although significant catches are reported by virtually all coastal countries, three-quarters of the total harvest is taken by only 14 countries, each of which produces more than a million tons annually and some much more. In the century from 1850 to 1950 the world catch increased tenfold—an average rate of about 25 percent per decade. In the next decade it nearly doubled, and this rapid growth is continuing [see illustration on page 757]. It is now a commonplace that fish is one of the few major foodstuffs showing an increase in global production that continues to exceed the growth rate of the human population.

This increase has been accompanied by a changing pattern of use. Although some products of high unit value as luxury foods, such as shellfish, have maintained or even enhanced their relative economic importance, the trend has been for less of the catch to be used directly as human food and for more to be reduced to meal for animal feed. Just be-

fore World War II less than 10 percent of the world catch was turned into meal; by 1967 half of it was so used. Over the same period the proportion of the catch preserved by drying or smoking declined from 28 to 13 percent and the proportion sold fresh from 53 to 31 percent. The relative consumption of canned fish has hardly changed but that of frozen fish has grown from practically nothing to 12 percent.

While we are comparing the prewar or immediate postwar situation with the present, we might take a look at the composition of the catch by groups of species. In 1948 the clupeoid fishes (herrings, pilchards, anchovies and so on), which live mainly in the upper levels of the ocean, already dominated the scene (33 percent of the total by weight) and provided most of the material for fish meal. Today they bulk even larger (45 percent) in spite of the decline of several great stocks of them (in the North Sea and off California, for example). The next most important group, the gadoid fishes (cod, haddock, hake and so on), which live mainly on or near the bottom, comprised a quarter of the total in 1948. Although the catch of these fishes has continued to increase absolutely, the proportion is now reduced to 15 percent. The flounders and other flatfishes, the rosefish and other sea perches and the mullets and jacks have collectively stayed at about 15 percent; the tunas and mackerels, at 7 percent. Nearly a fifth of the total catch continues to be recorded in statistics as "Unsorted and other"—a vast number of species and groups, each contributing a small amount to a considerable whole.

The rise of shrimp and fish meal production together account for another major trend in the pattern of fisheries development. A fifth of the 1957 catch was sold in foreign markets; by 1967, two-fifths were entering international trade and export values totaled $2.5 billion. Furthermore, during this same period the participation of the less developed countries in the export trade grew from a sixth to well over 25 percent. Most of these shipments were destined for markets in the richer countries, particularly shrimp for North America and fish meal for North America, Europe and Japan. More recently several of the less developed countries have also become importers of fish meal, for example Mexico and Venezuela, South Korea and the Republic of China.

The U.S. catch has stayed for many years in the region of two million tons, a low figure considering the size of the country, the length of the coastline and the ready accessibility of large resources on the Atlantic, Gulf and Pacific seaboards. The high level of consumption in the U.S. (about 70 pounds per capita) has been achieved through a steady growth in imports of fish and fish meal: from 25 percent of the total in 1950 to more than 70 percent in 1967. In North America 6 percent of the world's human population uses 12 percent of the world's catch, yet fishermen other than Americans take nearly twice the amount of fish that Americans take from the waters most readily accessible to the U.S.

There has not been a marked change in the broad geography of fishing [see illustration on these two pages]. The Pacific Ocean provides the biggest share (53 percent) but the Atlantic (40 percent, to which we may add 2 percent for the Mediterranean) is yielding considerably more per unit area. The Indian Ocean is still the source of less than 5 percent of the catch, and since it is not a biologically poor ocean it is an obvious target for future development. Within the major ocean areas, however, there have been significant changes. In the Pacific particular areas such as the waters off Peru and Chile and the Gulf of Thailand have rapidly acquired importance. The central and southern parts of the Atlantic, both east and west, are of growing interest to more nations. Al-

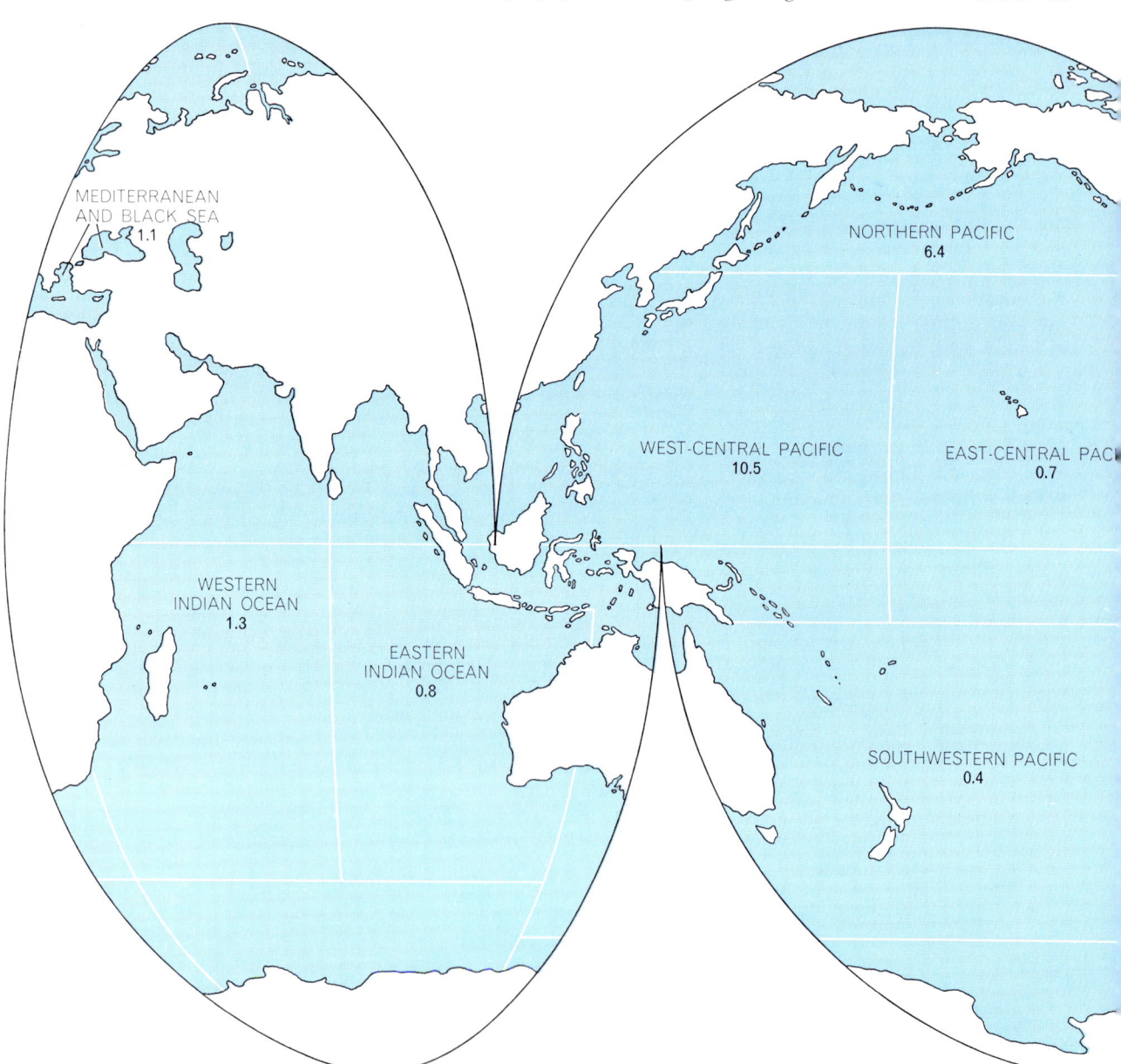

MAJOR MARINE FISHERY AREAS are 14 in number: two in the Indian Ocean (*left*), five in the Pacific Ocean (*center*) and six in the Atlantic (*right*). Due to the phenomenal expansion of the Peru fishery, the total Pacific yield is now a third larger than the Atlantic total. The bulk of Atlantic and Pacific catches, however, is still taken well north of the Equator. The Indian Ocean, with a

though, with certain exceptions, the traditional fisheries in the colder waters of the Northern Hemisphere still dominate the statistics, the emergence of some of the less developed countries as modern fishing nations and the introduction of long-range fleets mean that tropical and subtropical waters are beginning to contribute significantly to world production.

Finally, in this brief review of the trends of the past decade or so we must mention the changing importance of countries as fishing powers. Peru has become the leading country in terms of sheer magnitude of catch (although not of value or diversity) through the development of the world's greatest one-species fishery: 10 million tons of anchovies

per year, almost all of which is reduced to meal [see illustration on page 759]. The U.S.S.R. has also emerged as a fishing power of global dimension, fishing for a large variety of products throughout the oceans of the world, particularly with large factory ships and freezer-trawlers.

At this point it is time to inquire about the future expectations of the ocean as a major source of protein. In spite of the growth I have described, fisheries still contribute only a tenth of the animal protein in our diet, although this proportion varies considerably from one part of the world to another. Before such an inquiry can be pursued, however, it is necessary to say something

about the problem of overfishing.

A stock of fish is, generally speaking, at its most abundant when it is not being exploited; in that virgin state it will include a relatively high proportion of the larger and older individuals of the species. Every year a number of young recruits enter the stock, and all the fish—but particularly the younger ones—put on weight. This overall growth is balanced by the natural death of fish of all ages from disease, predation and perhaps senility. When fishing begins, the large stock yields large catches to each fishing vessel, but because the pioneering vessels are few, the total catch is small.

Increased fishing tends to reduce the level of abundance of the stock progressively. At these reduced levels the losses accountable to "natural" death will be less than the gains accountable to recruitment and individual growth. If, then, the catch is less than the difference between natural gains and losses, the stock will tend to increase again; if the catch is more, the stock will decrease. When the stock neither decreases nor increases, we achieve a sustained yield. This sustained yield is small when the stock is large and also when the stock is small; it is at its greatest when the stock is at an intermediate level—somewhere between two-thirds and one-third of the virgin abundance. In this intermediate stage the average size of the individuals will be smaller and the age will be younger than in the unfished condition, and individual growth will be highest in relation to the natural mortality.

The largest catch that on the average can be taken year after year without causing a shift in abundance, either up or down, is called the maximum sustainable yield. It can best be obtained by leaving the younger fish alone and fishing the older ones heavily, but we can also get near to it by fishing moderately, taking fish of all sizes and ages. This phenomenon—catches that first increase and then decrease as the intensity of fishing increases—does not depend on any correlation between the number of parent fish and the number of recruits they produce for the following generation. In fact, many kinds of fish lay so many eggs, and the factors governing survival of the eggs to the recruit stage are so many and so complex, that it is not easy to observe any dependence of the number of recruits on the number of their parents over a wide range of stock levels.

Only when fishing is intense, and the stock is accordingly reduced to a small

NORTH-WESTERN ATLANTIC
4.0

NORTHEASTERN ATLANTIC
10.2

WEST-CENTRAL ATLANTIC
1.3

EAST-CENTRAL ATLANTIC
1.6

SOUTHEASTERN PACIFIC
11.2

SOUTHWESTERN ATLANTIC
1.3

SOUTHEASTERN ATLANTIC
2.5

total catch of little more than two million metric tons, live weight, is the world's major underexploited region. The number below each area name shows the millions of metric tons landed during 1967, as reported by the UN Food and Agriculture Organization.

fraction of its virgin size, do we see a decline in the number of recruits coming in each year. Even then there is often a wide annual fluctuation in this number. Indeed, such fluctuation, which causes the stock as a whole to vary greatly in abundance from year to year, is one of the most significant characteristics of living marine resources. Fluctuation in number, together with the considerable variation in "availability" (the change in the geographic location of the fish with respect to the normal fishing area), largely account for the notorious riskiness of fishing as an industry.

For some species the characteristics of growth, natural mortality and recruitment are such that the maximum sustainable yield is sharply defined. The catch will decline quite steeply with a change in the amount of fishing (measured in terms of the number of vessels, the tonnage of the fleet, the days spent at sea or other appropriate index) to either below or above an optimum. In other species the maximum is not so sharply defined; as fishing intensifies above an optimum level the sustained catch will not significantly decline, but it will not rise much either.

Such differences in the dynamics of different types of fish stock contribute

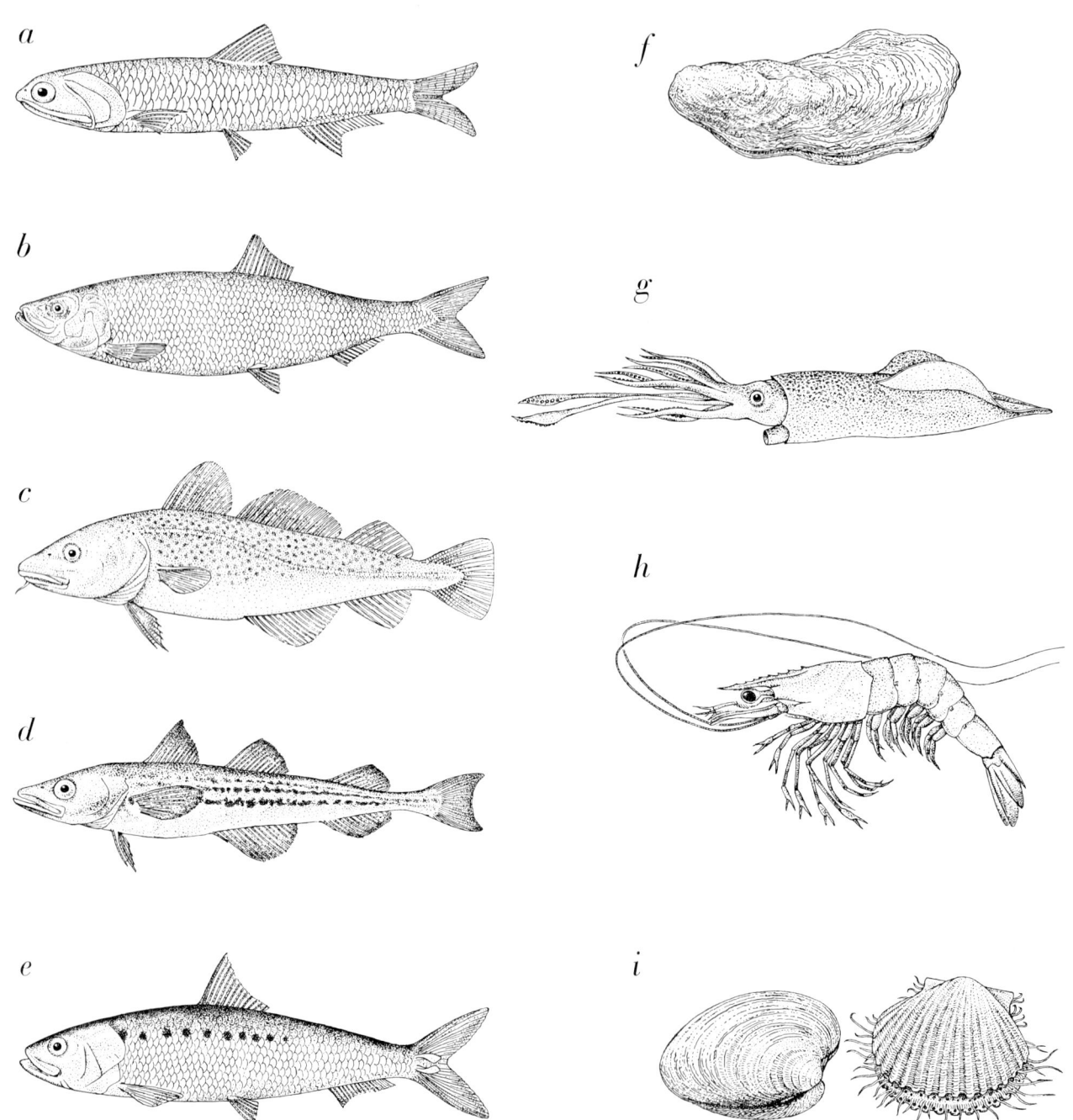

LARGEST CATCHES of individual fish species include the five fishes shown here (left). They are, according to the most recent detailed FAO fishery statistics (1967), the Peruvian anchoveta (a), with a catch of more than 10.5 million metric tons; the Atlantic herring (b), with a catch of more than 3.8 million tons; the Atlantic cod (c), with a catch of 3.1 million tons; the Alaska walleye pollack (d), with a catch of 1.7 million metric tons, and the South African pilchard (e), with a catch of 1.1 million tons. No single invertebrate species (right) is harvested in similar quantities. Taken as a group, however, various oyster species (f) totaled .83 million tons in 1967; squids (g), .75 million tons; shrimps and prawns (h), .69 million tons; clams and cockles (i), .48 million tons.

to the differences in the historical development of various fisheries. If it is unregulated, however, each fishery tends to expand beyond its optimum point unless something such as inadequate demand hinders its expansion. The reason is painfully simple. It will usually still be profitable for an individual fisherman or ship to continue fishing after the *total* catch from the stock is no longer increasing or is declining, and even though his own rate of catch may also be declining. By the same token, it may continue to be profitable for the individual fisherman to use a small-meshed net and thereby catch young as well as older fish, but in doing so he will reduce both his own possible catch and that of others in future years. Naturally if the total catch is declining, or not increasing much, as the amount of fishing continues to increase, the net economic yield from the fishery—that is, the difference between the total costs of fishing and the value of the entire catch—will be well past its maximum. The well-known case of the decline of the Antarctic baleen whales provides a dramatic example of overfishing and, one would hope, a strong incentive for the more rational conduct of ocean fisheries in the future.

There is, then, a limit to the amount that can be taken year after year from each natural stock of fish. The extent to which we can expect to increase our fish catches in the future will depend on three considerations. First, how many as yet unfished stocks await exploitation, and how big are they in terms of potential sustainable yield? Second, how many of the stocks on which the existing fisheries are based are already reaching or have passed their limit of yield? Third, how successful will we be in managing our fisheries to ensure maximum sustainable yields from the stocks?

The first major conference to examine the state of marine fish stocks on a global basis was the United Nations Scientific Conference on the Conservation and Utilization of Resources, held in 1949 at Lake Success, N.Y. The small group of fishery scientists gathered there concluded that the only overfished stocks at that time were those of a few high-priced species in the North Atlantic and North Pacific, particularly plaice, halibut and salmon. They produced a map showing 30 other known major stocks they believed to be underfished. The situation was reexamined in 1968. Fishing on half of those 30 stocks is now close to or beyond that required for maximum yield. The fully fished or overfished stocks include some tunas in most ocean areas, the herring, the cod and ocean perch

in the North Atlantic and the anchovy in the southeastern Pacific. The point is that the history of development of a fishery from small beginnings to the stage of full utilization or overutilization can, in the modern world, be compressed into a very few years. This happened with the anchovy off Peru, as a result of a massive local fishery growth, and it has happened to some demersal, or bottom-dwelling, fishes elsewhere through the large-scale redeployment of long-distance trawlers from one ocean area to another.

It is clear that the classical process of fleets moving from an overfished area to another area, usually more distant and less fished, cannot continue indefinitely.

It is true that since the Lake Success meeting several other large resources have been discovered, mostly in the Indian Ocean and the eastern Pacific, and additional stocks have been utilized in fishing areas with a long history of intensive fishing, such as the North Sea. In another 20 years, however, very few substantial stocks of fish of the kinds and sizes of commercial interest and accessible to the fishing methods we know now will remain underexploited.

The Food and Agriculture Organization of the UN is now in the later stages of preparing what is known as its Indicative World Plan (IWP) for agri-

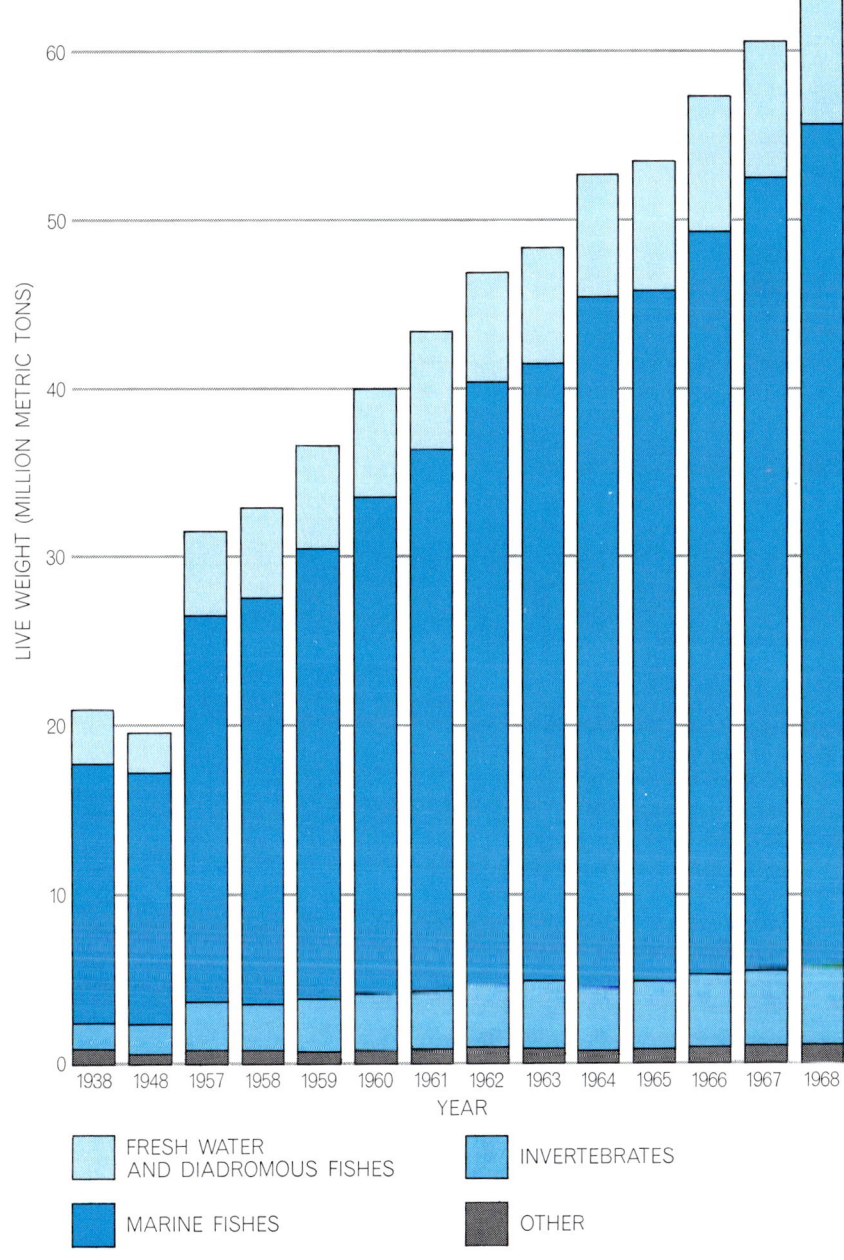

WORLD FISH CATCH has more than tripled in the three decades since 1938; the FAO estimate of the 1968 total is 64 million metric tons. The largest part consists of marine fishes. Humans directly consume only half of the catch; the rest becomes livestock feed.

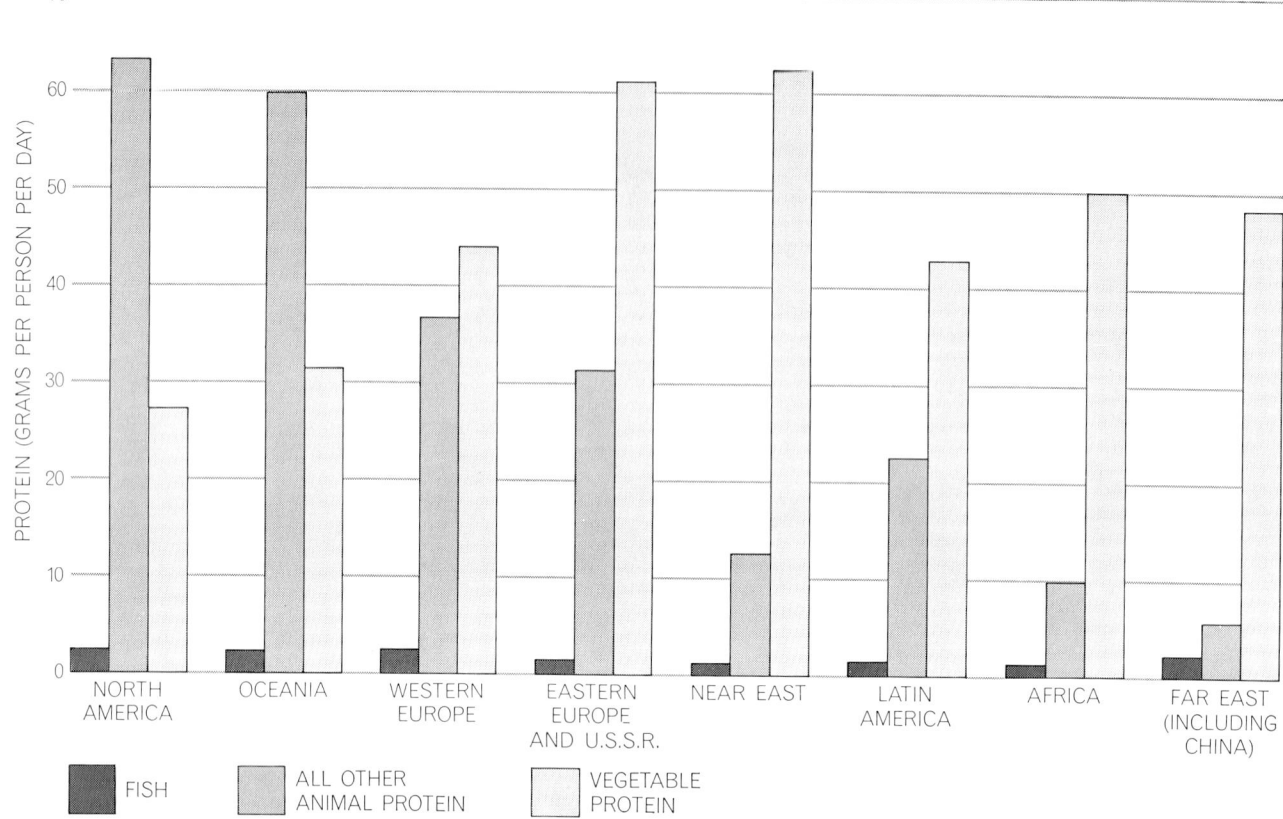

RELATIVELY MINOR ROLE played by fish in the world's total consumption of protein is apparent when the grams of fish eaten per person per day in various parts of the world (*left column in each group*) is compared with the consumption of other animal protein (*middle column*) and vegetable protein (*right column*). The supply is nonetheless growing more rapidly than world population.

cultural development. Under this plan an attempt is being made to forecast the production of foodstuffs in the years 1975 and 1985. For fisheries this involves appraising resource potential, envisioning technological changes and their consequences, and predicting demand. The latter predictions are not yet available, but the resource appraisals are well advanced. With the cooperation of a large number of scientists and organizations estimates are being prepared in great detail on an area basis. They deal with the potential of known stocks, both those fished actively at present and those exploited little or not at all. Some of these estimates are reliable; others are naturally little more than reasonable guesses. One fact is abundantly clear: We still have very scrappy knowledge, in quantitative terms, of the living resources of the ocean. We can, however, check orders of magnitude by comparing the results of different methods of appraisal. Thus where there is good information on the growth and mortality rates of fishes and measures of their numbers in absolute terms, quite good projections can be made. Most types of fish can now in fact virtually be counted individually by the use of specially cali-

brated echo sounders for area surveys, although this technique is not yet widely applied. The size of fish populations can also be deduced from catch statistics, from measurements of age based on growth rings in fish scales or bands in fish ear stones, and from tagging experiments. Counts and maps of the distribution of fish eggs in the plankton can in some cases give us a fair idea of fish abundance in relative terms. We can try to predict the future catch in an area little fished at present by comparing the present catch with the catch in another area that has similar oceanographic characteristics and basic biological productivity and that is already yielding near its maximum. Finally, we have estimates of the food supply available to the fish in a particular area, or of the primary production there, and from what we know about metabolic and ecological efficiency we can try to deduce fish production.

So far as the data permit these methods are being applied to major groups of fishes area by area. Although individual area and group predictions will not all be reliable, the global totals and subtotals may be. The best figure seems to be that the potential catch is about three times the present one; it might be as lit-

tle as twice or as much as four times. A similar range has been given in estimates of the potential yield from waters adjacent to the U.S.: 20 million tons compared with the present catch of rather less than six million tons. This is more than enough to meet the U.S. demand, which is expected to reach 10 million tons by 1975 and 12 million by 1985.

Judging from the rate of fishery development in the recent past, it would be entirely reasonable to suppose that the maximum sustainable world catch of between 100 and 200 million tons could be reached by the second IWP target date, 1985, or at least by the end of the century. The real question is whether or not this will be economically worth the effort. Here any forecast is, in my view, on soft ground. First, to double the catch we have to more than double the amount of fishing, because the stocks decline in abundance as they are exploited. Moreover, as we approach the global maximum more of the stocks that are lightly fished at present will be brought down to intermediate levels. Second, fishing will become even more competitive and costly if the nations fail to agree, and agree soon, on regulations to cure overfishing situations. Third, it is quite uncertain what will happen in the long run to the costs of production and the price of protein of marine origin in relation to other protein sources, particularly from mineral or vegetable bases.

In putting forward these arguments I am not trying to damp enthusiasm for the sea as a major source of food for coming generations; quite the contrary. I do insist, however, that it would be dangerous for those of us who are interested in such development to assume that past growth will be maintained along familiar lines. We need to rationalize present types of fishing while preparing ourselves actively for a "great leap forward." Fishing as we now know it will need to be made even more efficient; we shall need to consider the direct use of the smaller organisms in the ocean that mostly constitute the diet of the fish we now catch; we shall need to try harder to improve on nature by breeding, rearing and husbanding useful marine animals and cultivating their pasture. To achieve this will require a much larger scale and range of scientific research, wedded to engineering progress; expansion by perhaps an order of magnitude in investment and in the employment of highly skilled labor, and a modified legal regime for the ocean and its bed not only to protect the investments but also to ensure orderly development and provide for the safety of men and their installations.

To many people the improvement of present fishing activities will mean increasing the efficiency of fishing gear and ships. There is surely much that could be done to this end. We are only just beginning to understand how trawls, traps, lines and seines really work. For example, every few years someone tries a new design or rigging for a deep-sea trawl, often based on sound engineering and hydrodynamic studies. Rarely do these "improved" rigs catch more than the old ones; sometimes they catch much less. The error has been in thinking that the trawl is simply a bag, collecting more or less passive fish, or at least predictably active ones. This is not so at all. We really have to deal with a complex, dynamic relation between the lively animals and their environment, which includes in addition to the physical and biological environment the fishing gear itself. We can expect success in understanding and exploiting this relation now that we can telemeter the fishing gear, study its hydrodynamics at full scale as well as with models in towing tanks, monitor it (and the fish) by means of underwater television, acoustic equipment and divers, and

observe and experiment with fish behavior both in the sea and in large tanks. We also probably have something to learn from studying, before they become extinct, some kinds of traditional "primitive" fishing gear still used in Asia, South America and elsewhere—mainly traps that take advantage of subtleties of fish behavior observed over many centuries.

Successful fishing depends not so much on the size of fish stocks as on their concentration in space and time. All fishermen use knowledge of such concentrations; they catch fish where they have gathered to feed or to reproduce, or where they are on the move in streams or schools. Future fishing methods will surely involve a more active role for the fishermen in causing the fish to congregate. In many parts of the world lights or sound are already used to attract fish. We can expect more sophistication in the employment of these and other stimuli, alone and in combination.

Fishing operations as a whole also depend on locating areas of concentration and on the efficient prediction, or at least the prompt observation, of changes in these areas. The large stocks of pelagic, or open-sea, fishes are produced mainly in areas of "divergencies," where water is rising from deeper levels toward the surface and hence where surface waters are flowing outward. Many such areas are the "upwellings" off the western coasts of continental masses, for example off western and southwestern Africa, western India and western South America. Here seasonal winds, currents and continental configurations combine to cause a periodic enrichment of the surface waters.

Divergencies are also associated with certain current systems in the open sea. The classical notion is that biological production is high in such areas because nutrient salts, needed for plant growth and in limited supply, are thereby renewed in the surface layers of the water. On the other hand, there is a view that the blooming of the phytoplankton is associated more with the fact that the water coming to the surface is cooler than it is associated with its richness in nutrients. A cool-water regime is characterized by seasonal peaks of primary production; the phytoplankton blooms are followed, after a time lag, by an abundance of herbivorous zooplankton that

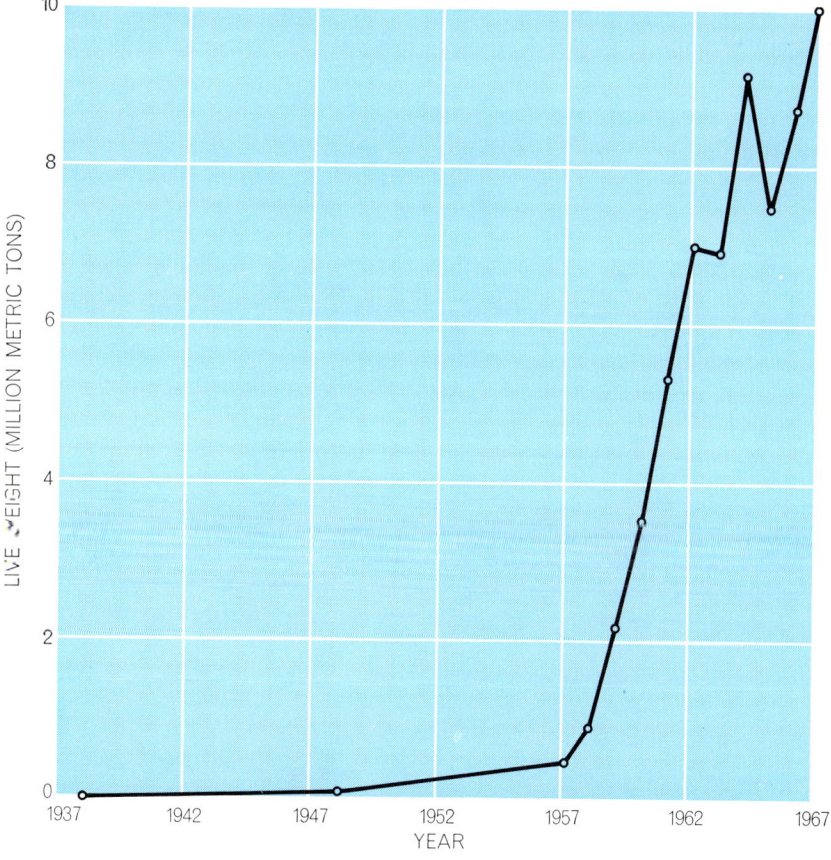

EXPLOSIVE GROWTH of the Peruvian anchoveta fishery is seen in rising number of fish taken between 1938 and 1967. Until 1958 the catch remained below half a million tons. By 1967, with more than 10.5 million tons taken, the fishery sorely needed management.

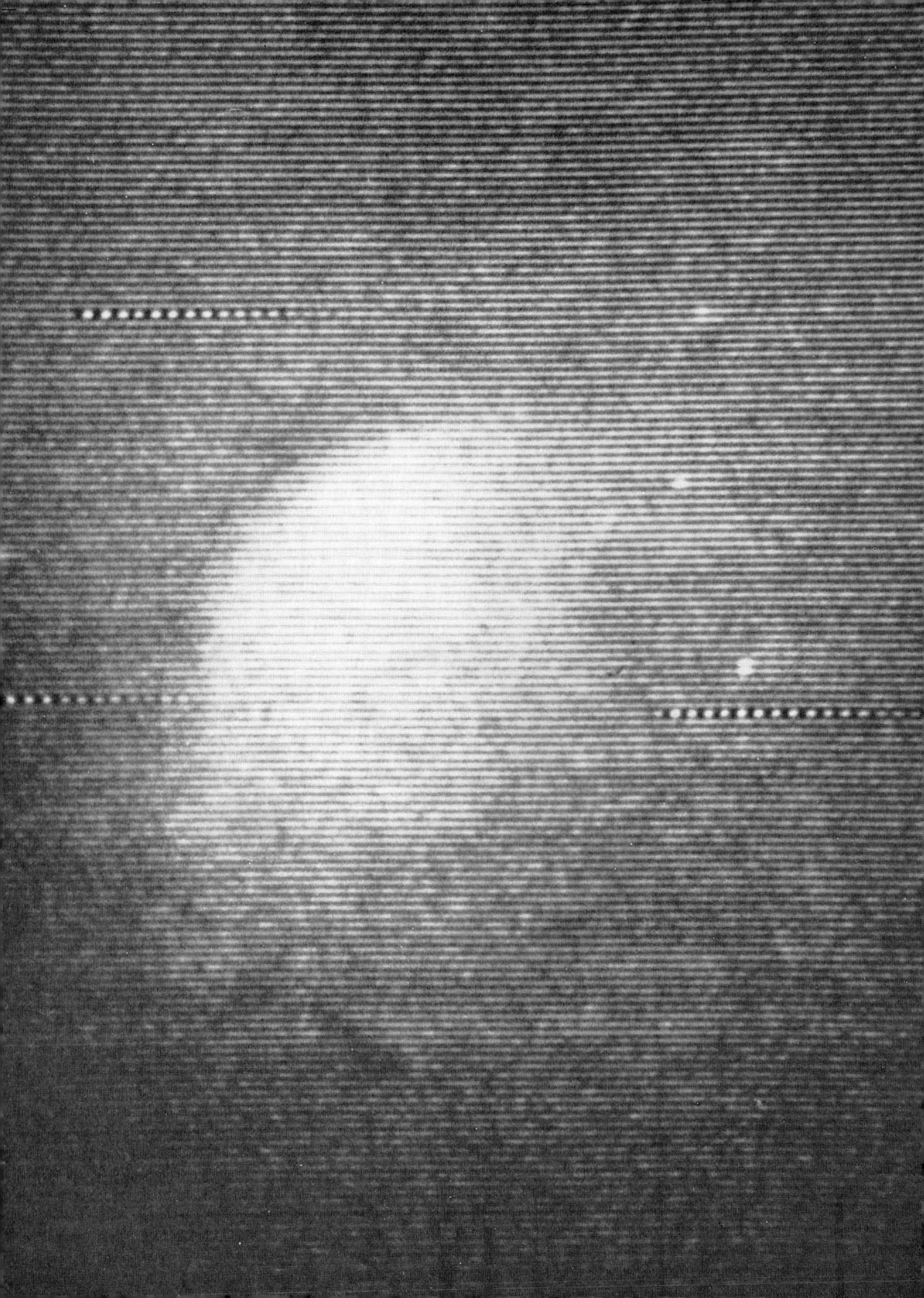

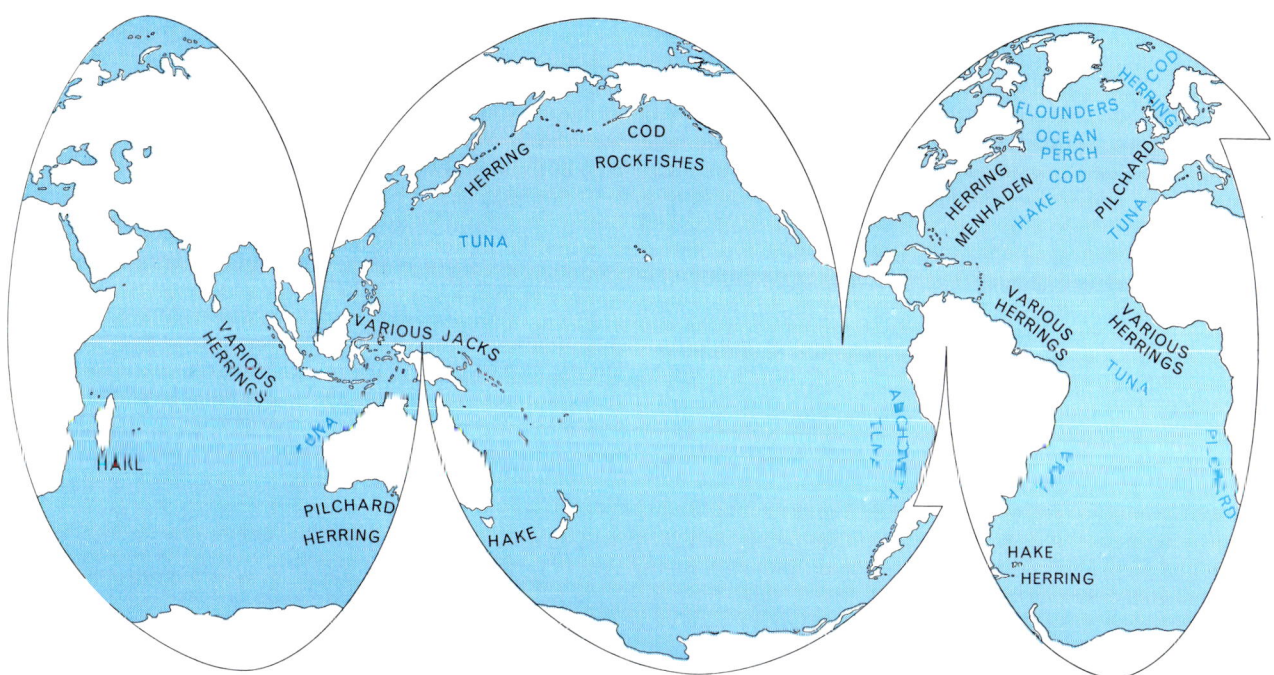

EXPLOITATION OF FISHERIES during the past 20 years is evident from this map, which locates 30 major fish stocks that were thought to be underfished in 1949. Today 14 of the stocks (*color*) are probably fully exploited or in danger of being overfished.

provides concentrations of food for large schools of fish. Fish, like fishermen, thrive best not so much where their prey are abundant as where they are most aggregated. In any event, the times and places of aggregation vary from year to year. The size of the herbivore crop also varies according to the success of synchronization with the primary production cycle.

There would be great practical advantage to our being able to predict these variations. Since the weather regime plays such a large part in creating the physical conditions for high biological production, the World Weather Watch, under the auspices of the World Meteorological Organization, should contribute much to fishery operations through both long-range forecasting and better short-term forecasting. Of course our interest is not merely in atmospheric forecasts, nor in the state of the sea surface, but in the deeper interaction of atmosphere and ocean. Thus, from the point of view of fisheries, an equal and complementary partner in the World Weather Watch will be the Integrated Global Ocean Station System (IGOSS) now being developed by the Intergovernmental

Oceanographic Commission. The IGOSS will give us the physical data, from networks of satellite-interrogated automatic buoys and other advanced ocean data acquisition systems (collectively called ODAS), by which the ocean circulation can be observed in "real time" and the parameters relevant to fisheries forecast. A last and much more difficult link will be the observation and prediction of the basic biological processes.

So far we have been considering mainly the stocks of pelagic fishes in the upper layers of the open ocean and the shallower waters over the continental shelves. There are also large aggregations of pelagic animals that live farther down and are associated particularly with the "deep scattering layer," the sound-reflecting stratum observed in all oceans. The more widespread use of submersible research vessels will tell us more about the layer's biological nature, but the exploitation of deep pelagic resources awaits the development of suitable fishing apparatus for this purpose.

Important advances have been made in recent years in the design of pelagic trawls and in means of guiding them in

three dimensions and "locking" them onto fish concentrations. We shall perhaps have such gear not only for fishing much more deeply than at present but also for automatically homing on deep-dwelling concentrations of fishes, squids and so on, using acoustic links for the purpose. The Indian Ocean might become the part of the world where such methods are first deployed on a large scale; certainly there is evidence of a great but scarcely utilized pelagic resource in that ocean, and around its edge are human populations sorely in need of protein. The Gulf of Guinea is another place where oceanographic knowledge and new fishing methods should make accessible more of the large sardine stock that is now effectively exploited only during the short season of upwelling off Ghana and nearby countries, when the schools come near the surface and can be taken by purse seines.

The bottom-living fishes and the shellfishes (both mollusks and crustaceans) are already more fully utilized than the smaller pelagic fishes. On the whole they are the species to which man attaches a particularly high value, but they cannot have as high a global abundance as the pelagic fishes. The reason is that they are living at the end of a longer food chain. All the rest of ocean life depends on an annual primary production of 150 billion tons of phytoplankton in the 2 to 3 percent of the water mass into which light penetrates and photosynthesis can occur. Below this "photic" zone dead

SCHOOL OF FISH is spotted from the air at night by detecting the bioluminescent glow caused by the school's movement through the water. As the survey aircraft flew over the Gulf of Mexico at an altitude of 3,500 feet, the faint illumination in the water was amplified some 55,000 times by an image intensifier before appearing on the television screen seen in the photograph on the opposite page. The fish are Atlantic thread herring. Detection of fish from the air is one of several means of increasing fishery efficiency being tested at the Pascagoula, Miss., research base of the U.S. Bureau of Commercial Fisheries.

and dying organisms sink as a continual rain of organic matter and are eaten or decompose. Out in the deep ocean little, if any, of this organic matter reaches the bottom, but nearer land a substantial quantity does; it nourishes an entire community of marine life, the benthos, which itself provides food for animals such as cod, ocean perch, flounder and shrimp that dwell or visit there.

Thus virtually everywhere on the bed of the continental shelf there is a thriving demersal resource, but it does not end there. Where the shelf is narrow but primary production above is high, as in the upwelling areas, or where the zone of high primary production stretches well away from the coast, we may find considerable demersal resources on the continental slopes beyond the shelf, far deeper than the 200 meters that is the average limiting depth of the shelf itself. Present bottom-trawling methods will work down to 1,000 meters or more, and it seems that, at least on some slopes, useful resources of shrimps and bottom-dwelling fishes will be found even down to 1,500 meters. We still know very little about the nature and abundance of these resources, and current techniques of acoustic surveying are not of much use in evaluating them. The total area of the continental slope from, say, 200 to 1,500 meters is roughly the same as that of the entire continental shelf, so that when we have extended our preliminary surveys there we might need to revise our IWP ceiling upward somewhat.

Another problem is posed for us by the way that, as fishing is intensified throughout the world, it becomes at the same time less selective. This may not apply to a particular type of fishing operation, which may be highly selective with regard to the species captured. Partly as a result of the developments in processing and trade, and partly because of the decline of some species, however, we are using more and more of the species that abound. This holds particularly for species in warmer waters, and also for some species previously neglected in cool waters, such as the sand eel in the North Sea. This means that it is no longer so reasonable to calculate the potential of each important species stock separately, as we used to do. Instead we need new theoretical models for that part of the marine ecosystem which consists of animals in the wide range of sizes we now utilize: from an inch or so up to several feet. As we move toward fuller utilization of all these animals we shall need to take proper account of the interactions among them. This will mean devising quantitative methods for evaluating the competition among them for a common food supply and also examining the dynamic relations between the predators and the prey among them.

These changes in the degree and quality of exploitation will add one more dimension to the problems we already face in creating an effective international system of management of fishing activities, particularly on the high seas. This system consists at present of a large number—more than 20—of regional or specialized intergovernmental organizations established under bilateral or multilateral treaties, or under the constitution of the FAO. The purpose of each is to conduct and coordinate research leading to resource assessments, or to promulgate regulations for the better conduct of the fisheries, or both. The organizations are supplemented by the 1958 Geneva Convention on Fishing and Conservation of the Living Resources of the High Seas. The oldest of them, the International Council for the Exploration of the Sea, based in Copenhagen and concerned particularly with fishery research in the northeastern Atlantic and the Arctic, has had more than half a century of activity. The youngest is the International Commission for the Conservation of Atlantic Tunas; the convention that establishes it comes into force this year.

For the past two decades many have hoped that such treaty bodies would ensure a smooth and reasonably rapid approach to an international regime for ocean fisheries. Indeed, a few of the organizations have fair successes to their credit. The fact is, however, that the fisheries have been changing faster than the international machinery to deal with them. National fishery research budgets and organizational arrangements for guiding research, collecting proper statistics and so on have been largely inadequate to the task of assessing resources. Nations have given, and continue to give, ludicrously low-level support to the bodies of which they are members, and the bodies themselves do not have the powers they need properly to manage the fisheries and conserve the resources. Add to this the trend to high mobility and range of today's fishing fleets, the problems of species interaction and the growing number of nations at various stages of economic development participating in international fisheries, and the regional bodies are indeed in trouble! There is some awareness of this, yet the FAO, having for years been unable to give adequate financial support to the fishery bodies it set up years ago in the Indo-Pacific area, the Mediterranean and the southwestern Atlantic, has been pushed, mainly through the enthusiasm of its new intergovernmental Committee on Fisheries, to establish still other bodies (in the Indian Ocean and in the east-central and southeastern Atlantic) that will be no better supported than the ex-

RUSSIAN FACTORY SHIP *Polar Star* lies hove to in the Barents Sea in June, 1968, as two vessels from its fleet of trawlers unload their catch for processing. The worldwide activities of the Russian fishing fleet have made the U.S.S.R. the third-largest fishing nation.

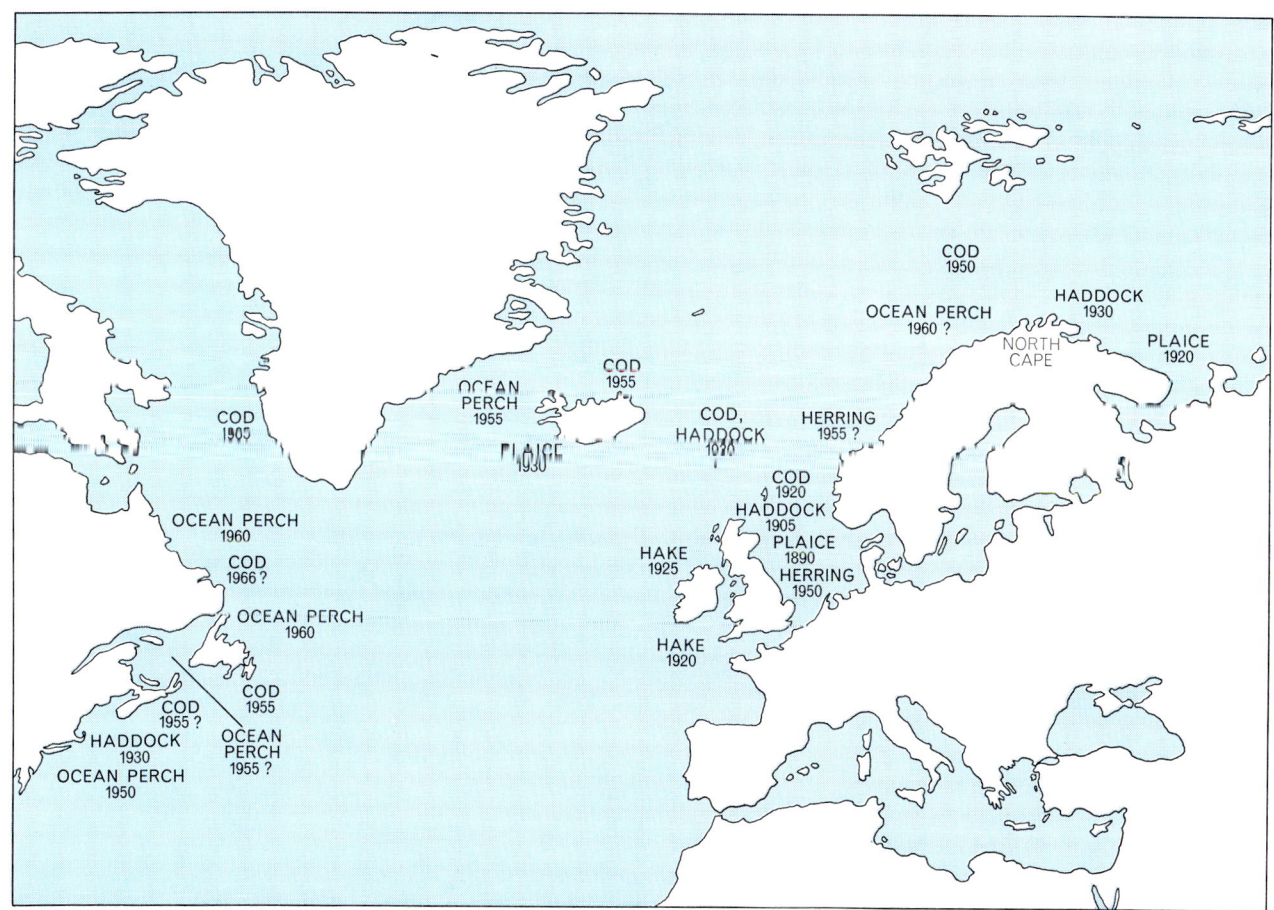

OVERFISHING in the North Atlantic and adjacent waters began some 80 years ago in the North Sea, when further increases in fishing the plaice stock no longer produced an increase in the catch of that fish. By 1950 the same was true of North Sea cod, haddock and herring, of cod, haddock and plaice off the North Cape and in the Barents Sea, of plaice, haddock and cod south and east of Iceland and of the ocean perch and haddock in the Gulf of Maine. In the period between 1956 and 1966 the same became true of ocean perch off Newfoundland and off Labrador and of cod west of Greenland. It may also be true of North Cape ocean perch and Labrador cod.

isting ones. A grand plan to double the finance and staff of the FAO's Department of Fisheries (including the secretariats and working budgets of the associated regional fishery bodies) over the six-year period 1966–1971, which member nations endorsed in principle in 1965, will be barely half-fulfilled in that time, and the various nations concerned are meanwhile being equally parsimonious in financing the other international fishery bodies.

Several of these bodies are now facing a crucial, and essentially political, problem: How are sustainable yields to be shared equitably among participating nations? It is now quite evident that there is really no escape from the paramount need, if high yields are to be sustained; this is to limit the fishing effort deployed in the intensive fisheries. This could be achieved by setting total quotas for each species in each type of fishery, but this only leads to an unseemly scramble by each nation for as large a share as possible of the quota. This can only be avoided by agreement on national al-

locations of the quotas. On what basis can such agreement be reached? On the historical trends of national participation? If so, over what period: the past two years, the past five, the past 20? On the need for protein, on the size or wealth of the population or on the proximity of coasts to fishing grounds? Might we try to devise a system for maximizing economic efficiency in securing an optimum net economic yield? How can this be measured in an international fishery? Would some form of license auction be equitable, or inevitably loaded in favor of wealthy nations? The total number or tonnage of fishing vessels might be fixed, as the United Kingdom suggested in 1946 should be done in the North Sea, but what flags should the ships fly and in what proportion? Might we even consider "internationalizing" the resources, granting fishing concessions and using at least a part of the economic yield from the concessions to finance marine research, develop fish-farming, police the seas and aid the participation of less developed nations?

Some of my scientific colleagues are optimistic about the outcome of current negotiations on these questions, and indeed when the countries participating are a handful of nations at a similar stage of economic and technical development, as was the case for Antarctic whaling, agreement can sometimes be reached by hard bargaining. What happens, however, when the participating countries are numerous, widely varying in their interests and ranging from the most powerful nations on earth to states most newly emerged to independence? I must confess that many of us were optimistic when 20 years ago we began proposing quite reasonable net-mesh regulations to conserve the young of certain fish stocks. Then we saw these simple—I suppose oversimple—ideas bog down in consideration of precisely how to measure a mesh of a particular kind of twine, and how to take account of the innumerable special situations that countries pleaded for, so that fishery research sometimes seemed to be becoming perverted from its earlier clarity and broad perspective.

Apprehension and doubt about the ultimate value of the concept of regulation through regional commissions of the present type have, I think, contributed to the interest in recent years in alternative regimes: either the "appropriation" of high-seas resources to some form of international "ownership" instead of today's condition of no ownership or, at the other extreme, the appropriation of increasingly wide ocean areas to national ownership by coastal states. As is well known, a similar dialectic is in progress in connection with the seabed and its mineral resources. Either solution would have both advantages and disadvantages, depending on one's viewpoint, on the time scale considered and on political philosophy. I do not propose to discuss these matters here, although personally I am increasingly firm in the conclusion that mankind has much more to gain in the long run from the "international" solution, with both seabed and fishery resources being considered as our common heritage. We now at least have a fair idea of what is economically at stake.

Here are some examples. The wasted effort in capture of cod alone in the northeastern Atlantic and salmon alone in the northern Pacific could, if rationally deployed elsewhere, increase the total world catch by 5 percent. The present catch of cod, valued at $350 million per year, could be taken with only half the effort currently expended, and the annual saving in fishing effort would amount to $150 million or more. The cost of harvesting salmon off the West Coast of North America could be re-duced by three-quarters if management policy permitted use of the most efficient fishing gear; the introduction of such a policy would increase net economic returns by $750,000 annually.

The annual benefit that would accrue from the introduction and enforcement of mesh regulations in the demersal fishery—mainly the hake fishery—in the east-central Atlantic off West Africa is of the order of $1 million. Failure to regulate the Antarctic whaling industry effectively in earlier years, when stocks of blue whales and fin whales were near their optimum size, is now costing us tens of millions of dollars annually in loss of this valuable but only slowly renewable resource. Even under stringent regulation this loss will continue for the decades these stocks will need to recover. Yellowfin tuna in the eastern tropical Pacific are almost fully exploited. There is an annual catch quota, but it is not allocated to nations or ships, with the classic inevitable results: an increase in the catching capacity of fleets, their use in shorter and shorter "open" seasons and an annual waste of perhaps 30 percent of the net value of this important fishery.

Such regulations as exist are extremely difficult to enforce (or to be seen to be enforced, which is almost as important). The tighter the squeeze on the natural resources, the greater the suspicion of fishermen that "the others" are not abiding by the regulations, and the greater the incentive to flout the regulations oneself. There has been occasional provision in treaties, or in *ad hoc* ar-rangements, to place neutral inspectors or internationally accredited observers aboard fishing vessels and mother ships (as in Antarctic whaling, where arrangements were completed but never implemented!). Such arrangements are exceptional. In point of fact the effective supervision of a fishing fleet is an enormously difficult undertaking. Even to know where the vessels are going, let alone what they are catching, is quite a problem. Perhaps one day artificial satellites will monitor sealed transmitters compulsorily carried on each vessel. But how to ensure compliance with minimum landing-size regulations when increasing quantities of the catch are being processed at sea? With factory ships roaming the entire ocean, even the statistics reporting catches by species and area can become more rather than less difficult to obtain.

Some of these considerations and pessimism about their early solution have, I think, played their part in stimulating other approaches to harvesting the sea. One of these is the theory of "working back down the food chain." For every ton of fish we catch, the theory goes, we might instead catch say 10 tons of the organisms on which those fish feed. Thus by harvesting the smaller organisms we could move away from the fish ceiling of 100 million or 200 million tons and closer to the 150 billion tons of primary production. The snag is the question of concentration. The billion tons or so of "fish food" is neither in a form of direct interest to man nor is it so concentrated in space as the animals it nourishes. In fact, the 10-to-one ratio of fish food to fish represents a use of energy—perhaps a rather efficient use—by which biomass is concentrated; if the fish did not expend this energy in feeding, man might have to expend a similar amount of energy—in fuel, for example—in order to collect the dispersed fish food. I am sure the technological problems of our using fish food will be solved, but only careful analysis will reveal whether or not it is better to turn fish food, by way of fish meal, into chickens or rainbow trout than to harvest the marine fish instead.

There are a few situations, however, where the concentration, abundance and homogeneity of fish food are sufficient to be of interest in the near future. The best-known of these is the euphausiid "krill" in Antarctic waters: small shrimp-like crustaceans that form the main food of the baleen whales. Russian investigators and some others are seriously charting krill distribution and production, relating them to the oceanographic features of the Southern Ocean, experiment-

JAPANESE MARICULTURE includes the raising of several kinds of marine algae. This array of posts and netting in the Inland Sea supports a crop of an edible seaweed, *Porphyra.*

ing with special gear for catching the krill (something between a mid-water trawl and a magnified plankton net) and developing methods for turning them into meal and acceptable pastes. The krill alone could produce a weight of yield, although surely not a value, at least as great as the present world fish catch, but we might have to forgo the whales. Similarly, the deep scattering layers in other oceans might provide very large quantities of smaller marine animals in harvestable concentration.

An approach opposite to working down the food chain is to look to the improvement of the natural fish resources, and particularly to the cultivation of highly valued species. Schemes for transplanting young fish to good high-seas feeding areas, or for increasing recruitment by rearing young fish to viable size, are hampered by the problem of protecting what would need to be quite large investments. What farmer would bother to breed domestic animals if he were not assured by the law of the land that others would not come and take them as soon as they were nicely fattened? Thus mariculture in the open sea awaits a regime of law there, and effective management as well as more research.

Meanwhile attention is increasingly given to the possibilities of raising more fish and shellfish in coastal waters, where the effort would at least have the protection of national law. Old traditions of shellfish culture are being reexamined, and one can be confident that scientific bases for further growth will be found. All such activities depend ultimately on what I call "productivity traps": the utilization of natural or artificially modified features of the marine environment to trap biological production originating in a wider area, and by such a biological route that more of the production is embodied in organisms of direct interest to man. In this way we open the immense possibilities of using mangrove swamps and productive estuarine areas, building artificial reefs, breeding even more efficient homing species such as the salmon, enhancing natural production with nutrients or warm water from coastal power stations, controlling predators and competitors, shortening food chains and so on. Progress in such endeavors will require a better predictive ecology than we now have, and also many pilot experiments with corresponding risks of failure as well as chances of success.

The greatest threat to mariculture is perhaps the growing pollution of the sea. This is becoming a real problem for fisheries generally, particularly coastal ones,

AUSTRALIAN MARICULTURE includes the production of some 60 million oysters per year in the brackish estuaries of New South Wales. The long racks in the photograph have been exposed by low tide; they support thousands of sticks covered with maturing oysters.

and mariculture would thrive best in just those regions that are most threatened by pollution, namely the ones near large coastal populations and technological centers. We should not expect, I think, that the ocean would not be used at all as a receptacle for waste—it is in some ways so good for such a purpose: its large volume, its deep holes, the hydrolyzing, corrosive and biologically degrading properties of seawater and the microbes in it. We should expect, however, that this use will not be an indiscriminate one, that this use of the ocean would be internationally registered, controlled and monitored, and that there would be strict regulation of any dumping of noxious substances (obsolete weapons of chemical and biological warfare, for example), including the injection of such substances by pipelines extending from the coast. There are signs that nations are becoming ready to accept such responsibilities, and to act in concert to overcome the problems. Let us hope that progress in this respect will be faster than it has been in arranging for the management of some fisheries, or in a few decades there may be few coastal fisheries left worth managing.

I have stressed the need for scientific research to ensure the future use of the sea as a source of food. This need seems to me self-evident, but it is undervalued by many persons and organizations concerned with economic development. It is relatively easy to secure a mil-

lion dollars of international development funds for the worthy purpose of assisting a country to participate in an international fishery or to set up a training school for its fishermen and explore the country's continental shelf for fish or shrimps. It is more difficult to justify a similar or lesser expenditure on the scientific assessment of the new fishery's resources and the investigation of its ocean environment. It is much more difficult to secure even quite limited support for international measures that might ensure the continued profitability of the new fishery for all participants.

Looking back a decade instead of forward, we recall that Lionel A. Walford of the U.S. Fish and Wildlife Service wrote, in a study he made for the Conservation Foundation: "The sea is a mysterious wilderness, full of secrets. It is inhabited only by wild animals and, with the exception of a few special situations, is uncultivated. Most of what we know about it we have had to learn indirectly with mechanical contrivances to probe, feel, sample, fish." There are presumably fewer wild animals now than there were then—at least fewer useful ones—but there seems to be a good chance that by the turn of the century the sea will be less a wilderness and more cultivated. Much remains for us and our children to do to make sure that by then it is not a contaminated wilderness or a battlefield for ever sharper clashes between nations and between the different users of its resources.

The Author

S. J. HOLT is with the Food and Agriculture Organization of the United Nations, serving temporarily as marine science and fishery coordinator with the United Nations Educational, Scientific and Cultural Organization on leave from his position as director of the Division of Fishery Resources and Exploitation in the Department of Fisheries of the FAO. He was born in London and was graduated from the University of Reading, which in 1966 awarded him a D.Sc. on the basis of his published work. From 1946 to 1953 he was with the British Fisheries Laboratory at Lowestoft, serving also from 1950 to 1953 with the Nature Conservancy in Edinburgh. He has been with the FAO since 1954. Holt writes: "My research has been on animal population dynamics, mainly of fish, with a particular interest in the application of such knowledge to the rational management of international fisheries. I am also concerned with problems of international organization and programming in marine science; with the question of ensuring the participation of small and developing countries in marine research and in its benefits, and with the problems of creating a scientifically competent element in the international civil service."

Bibliography

LIVING RESOURCES OF THE SEA: OPPORTUNITIES FOR RESEARCH AND EXPANSION. Lionel A. Walford. The Ronald Press Company, 1958.

FISHERIES BIOLOGY: A STUDY IN POPULATION DYNAMICS. D. H. Cushing. University of Wisconsin Press, 1968.

MARINE SCIENCE AND TECHNOLOGY: SURVEY AND PROPOSALS. REPORT OF THE SECRETARY-GENERAL. United Nations Economic and Social Council, E/4487, 1968.

THE STATE OF WORLD FISHERIES: WORLD FOOD PROBLEMS, No. 7. Food and Agriculture Organization of the United Nations, 1968.

WORK OF FAO AND RELATED ORGANIZATIONS CONCERNING MARINE SCIENCE AND ITS APPLICATIONS. FAO Fisheries Technical Paper No. 74. Food and Agriculture Organization of the United Nations, September, 1968.

SCIENTIFIC
AMERICAN   September 1969, Vol. 221, No. 3, pp. 198–217   OFFPRINT **887**

TECHNOLOGY AND THE OCEAN

by Willard Bascom

The materials, machines and techniques that can be employed
in the ocean have advanced greatly during the past decade.
Major developments include superships and deep-sea drilling.

Without technology, meaning knowledge fortified by machinery and tools, men would be ineffective against the sea. During the past decade the technology that can be brought to bear in the oceans has improved enormously and in many ways. The improvements have not only increased knowledge of the oceans but also speeded the flow of commerce while decreasing its cost, brought new mineral provinces within reach and made food from the sea more readily available.

With today's technology it is possible, given a sufficient investment of time and money, to design and build marine hardware that can do almost anything. The problem is to decide whether it is sensible to make a given investment. Industry decides on the basis of whether a proposed technological step will solve a specific problem and improve the firm's competitive position. Government has more latitude: it does not need to show a prompt return on investment, and it can better afford the high risk of developing expensive and exotic devices for which there may be no immediate or clearly defined need. The gains in ocean technology have resulted from the largely independent efforts of both industry and government.

This article will deal broadly with the progress in ocean technology over the past decade, concentrating on developments that seem to be the most important at present. I shall begin by making my own selection of the five most important advances. The main criterion in this selection is that each advance represents an order-of-magnitude improvement: in one way or another it is a tenfold step forward since the beginning of the decade. I have also given weight to the social and economic significance of these developments and to the degree of

engineering imagination and perseverance that each one required.

The first development is the supership. Not long ago a "supertanker" carried 35,000 deadweight tons. Now a fleet of ships with nearly 10 times that capacity is coming into being. For these vessels the Panama Canal and the Suez Canal are obsolete. By the same token the ships are making large new demands on the technology that provides the terminal facilities.

Second is the deep-diving submarine. Man can now go to the deep-ocean bottom in an "underwater balloon" submersible such as the *Trieste*, which reached a depth of 36,000 feet in the Mariana Trench 200 miles southwest of Guam. Somewhat more conveniently he can go to a depth of about 6,000 feet in any of several small submarines. This rapidly developing technology still has a long way to go, but it has certainly improved by an order of magnitude in the past decade. Several techniques have been employed to solve the problem of how to make a submarine hull that is strong enough to resist great pressure and still light enough to return to the surface.

The third development is the ability to drill in deep water. This category includes both the drilling that is done in very deep water for scientific purposes and the use of full-scale drilling equipment on a floating platform to obtain oil from the continental shelf. The first deep-ocean drilling, which was carried out eight years ago by the National Academy of Sciences in water 12,000 feet deep near Guadalupe Island off the west coast of Mexico, improved on four previous records by an order of magnitude: the ship held its position at sea for a month without anchors, drilled in wa-

ter 20 times deeper than that at earlier marine drilling sites, penetrated 600 feet of the deep-sea floor and lifted weights of 40 tons from the bottom. These records have since been improved on even more. In fact, virtually all floating drilling equipment, including semisubmersible platforms and self-propelled vessels, has been designed and built in the past decade.

Fourth is the ability to navigate precisely. A ship in mid-ocean has rarely known its position within a mile; indeed, five miles is probably closer to the truth, notwithstanding assertions to the contrary. Now a ship 1,000 miles from land can fix its position within .1 mile. If the vessel is within 500 miles of land, the position can be ascertained within .01 mile. The position of a ship within 10 miles of land can be fixed to an accuracy of 10 feet. The techniques for these determinations include orbiting satellites, inertial guidance systems and a number of electronic devices that compare phases of radio waves.

Finally I would cite the ability to examine the ocean bottom in detail from the surface by means of television and side-looking sonar. These techniques, together with their recording devices and the capacity for precise navigation, have made it possible to inspect the sea floor much as land areas have been examined by aerial photography. New television tubes that amplify light by a factor of 30,000 make it possible to eliminate artificial lighting, thereby eliminating also the backscatter of light by small particles in the water.

The supership and the improvement in drilling are mainly industrial developments. The evolution of navigation technology has resulted largely from government efforts. Both industry and government have figured prominently

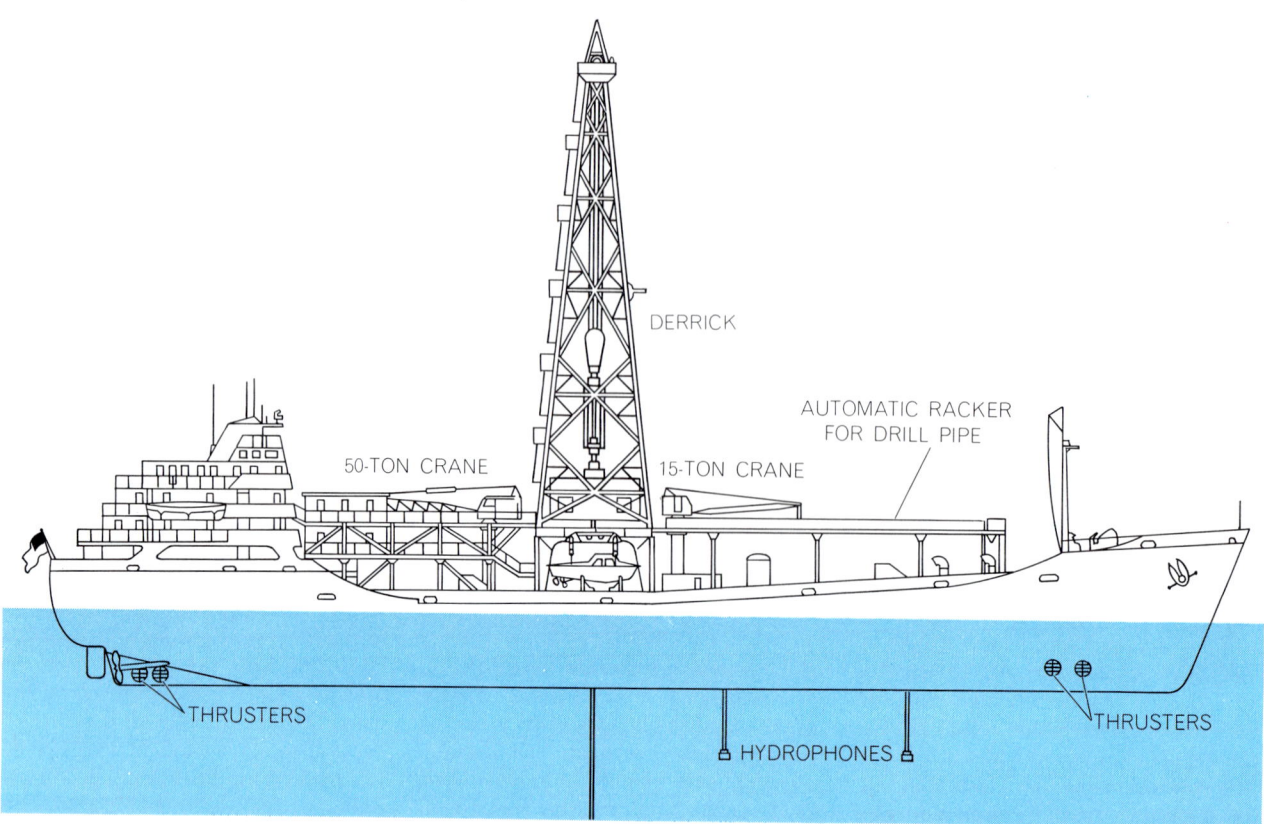

GLOMAR CHALLENGER has a 142-foot derrick as her most conspicuous feature. Her automatic pipe racker can hold 23,000 feet of drill pipe. Positioning equipment includes two tunnel thrusters at the bow and two near the stern to provide for sidewise maneuvers. When the ship is on station (above), four hydrophones are extended under the hull to receive signals from a sonar beacon on the ocean floor. The signals are fed into a computer that controls the thrusters to maintain the ship's position over the drill hole. At the sea bottom (below), as much as four miles under the ship, the drill penetrates as much as 2,500 feet of sediment and basement rock.

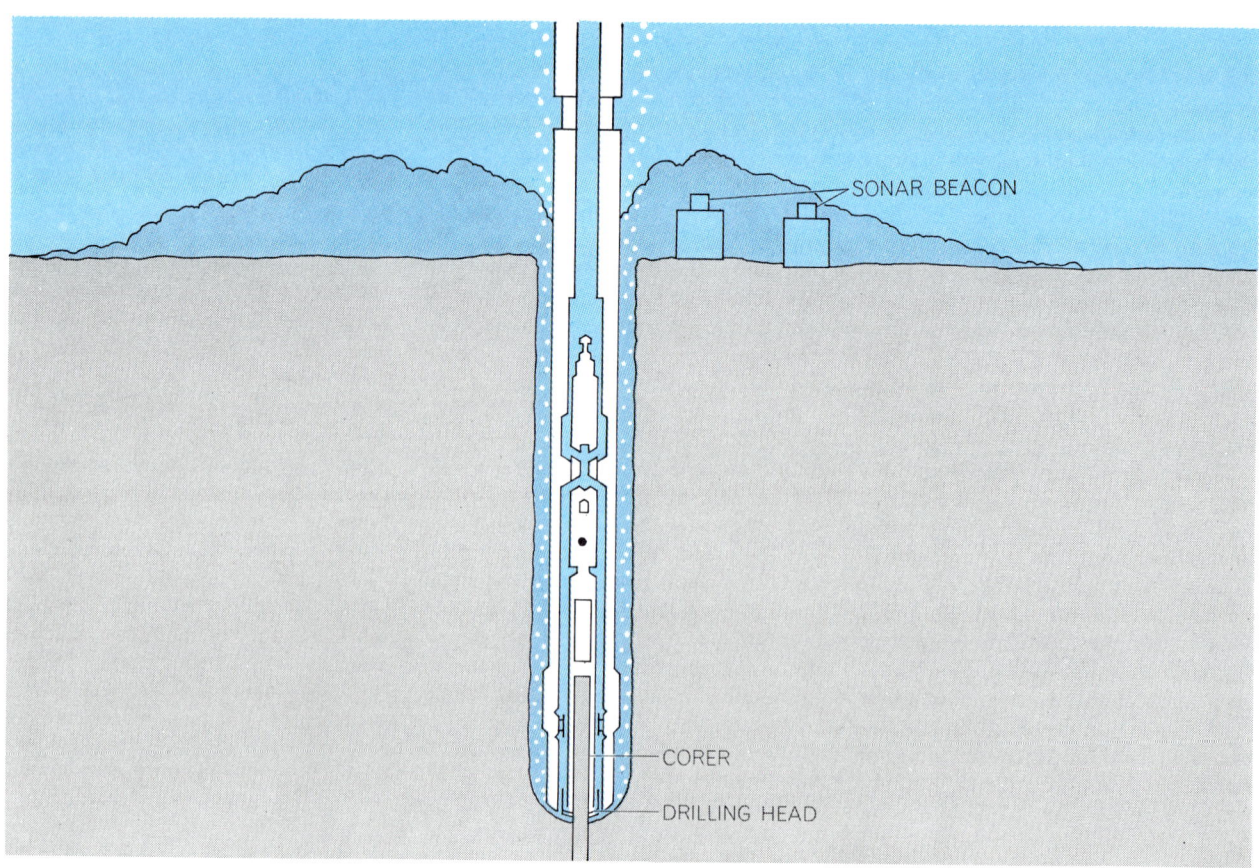

in the development of deep-diving submarines and techniques for examining the bottom with television and sonar.

In considering the application of these and other techniques one might classify them according to who uses them. For example, scientific investigators use research ships and submarines, instruments, buoys, samplers and computers. Industry constantly seeks better methods for mining, fishing, salvage and the production of oil. Waterborne commerce needs better ships, better cargo-handling methods and better port facilities. Exploration becomes more efficient as improved navigational systems, vehicles, geophysical tools and communications equipment become available. Adventure and recreation offer new toys such as air-cushion vehicles and scuba equipment.

The entire area of military technology, which is the most sophisticated of all, must be outside the scope of this article. The best of modern seaborne military technology is done in secrecy, with budgets far in excess of those spent for any of the other areas. Thus we shall not go into such matters as the duel between the submarine builder, who endlessly tries to make submarines go deeper, faster and quieter, and the antisubmarine expert, who tries to detect, identify and destroy the steadily improving submarines.

In any case, the classification of marine technology according to users is somewhat impractical because there is so much overlap. For example, certain kinds of diving and television equipment might be used by all the groups. Therefore I shall discuss marine technology in terms of materials, vehicles, instruments and systems.

What characteristics should a marine material have? It should be light, strong, easy to form and connect, rigid or flexible as desired and inexpensive. The difficulty ocean design engineers have in finding a material that meets most of the requirements for a given task has led them to speak whimsically of an ideal material called "nonobtainium." The problem is that characteristics such as lightness and strength are relative. Nonetheless, engineers and manufacturers recall that not many years ago fiber glass, Dacron and titanium were not obtainable, and so they are optimistic about the development of materials that come ever closer to the qualities of nonobtainium.

In the past decade the steel available for marine purposes has improved substantially under the spur of demands for

submarines that can withstand the pressure of great depth, drill pipe (unsupported by the hole wall that pipe in a land well has) that must survive high bending stresses, and great lengths of oceanographic cable that must not twist. For example, a new kind of maraging steel with a high nickel content is tougher, more resistant to notching and less subject to corrosion fatigue than the steel formerly available. The minimum yield strength of conveniently available steel shell plate has risen from 80,000 pounds per square inch to 130,000 pounds per square inch and more for the shells of deep-diving submarines. Steel in wire form now attains a strength of 350,000 pounds per square inch. Steel is becoming more uniform and reliable as the processes of mixing and rolling are subjected to better quality control. Indeed, some metallurgists believe nearly any metal requirement can be met by properly alloyed steel.

Also available for marine purposes are new, high-strength aluminum alloys, such as 5456 (a designation indicating the mix of metals in the alloy), that have a strength of more than 30,000 pounds per square inch after welding and are resistant to corrosion. They can also be cut with ordinary power saws instead of torches and welded by a technique that is easily taught. With this material small boats, ships up to 2,000 tons and superstructures for much larger ships can be built, as can a number of other structures where lightness and flexibility are important.

Titanium is becoming more readily available. When special properties of lightness, strength (as high as 120,000 pounds per square inch) and good resistance to corrosion are required, its relatively high cost becomes acceptable.

Glass, fiber glass and plastics are the glamorous materials of oceanography. They are virtually free of the problems of corrosion and electrolysis that have afflicted most materials in a marine environment, and they are easily formed

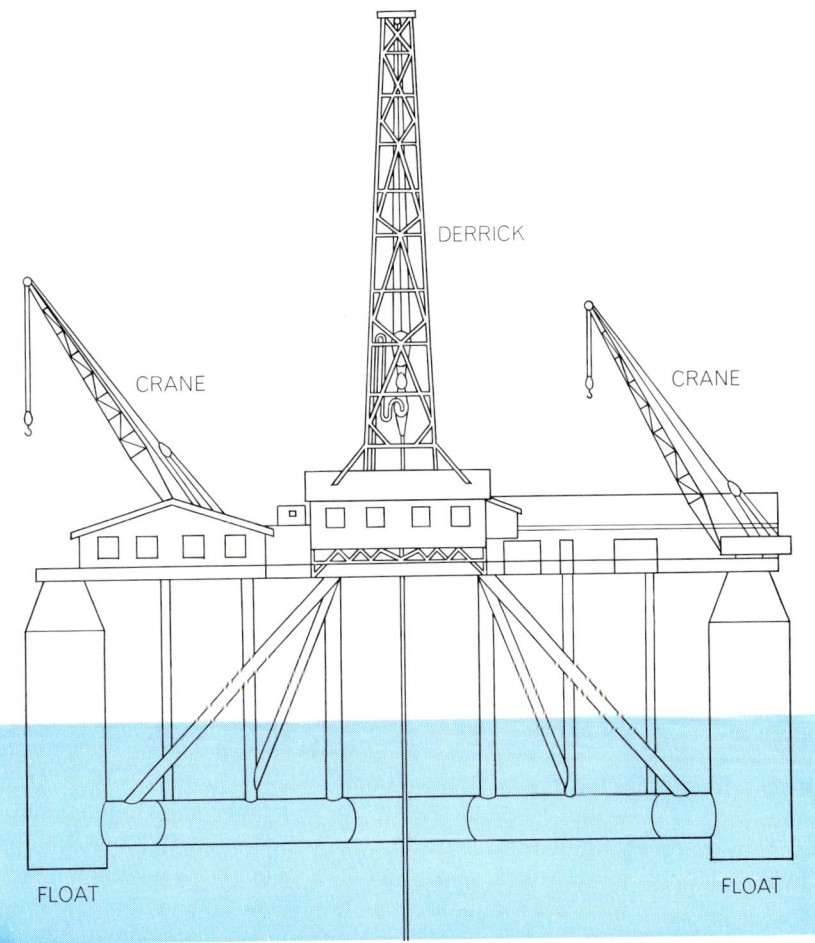

SEMISUBMERSIBLE PLATFORM, *Blue Water 3*, is now drilling for oil off Trinidad. When the platform has been towed to its position, water is drawn into the four corner cylinders to make the structure submerge enough so that wave motions have little effect on it. The platform, which is 220 feet by 198 feet, was designed for work in the open ocean.

into complex shapes. Constant research is improving the strength and versatility of these materials.

Glass is less fragile than most people think and has excellent properties in compression. It is finding increasing favor among the designers of small submarines, who want a glass-bubble pressure hull that is also a superwindow. Glass microspheres, which do the same thing as a submarine hull but on a microscopic scale, are packaged in blocks of epoxy and used to furnish incompressible flotation at depths of as much as 20,000 feet. The best such material to withstand the pressure at that depth so far weighs about 40 pounds per cubic foot; in seawater at 64 pounds per cubic foot the material therefore has a net buoyancy of 24 pounds per cubic foot.

The many remarkable characteristics

SUPERTANKER *UNIVERSE IRELAND* is seen at her loading berth in the Persian Gulf. The vessel, which is 1,135 feet in length, carries 312,000 deadweight tons of oil from the Persian Gulf to Ireland, going around Africa at an average speed of 15 knots.

of fiber glass are widely known. Its outstanding virtue as a marine material is that it enables precisely shaped hulls with complex lines to be reproduced easily. The result has been a revolution in the construction of small craft over the past decade. Fiber-glass hulls, which are light and strong without a rib structure, are a major contribution to marine technology and a boon to the small-boat owners, whose maintenance problems are reduced accordingly.

Among the plastics polyvinyl chloride has found use in marine pipelines subject to severe internal corrosion. It is light, inexpensive and easily joined. Nylon and polypropylene for cordage and fishing nets and Dacron for sails are appreciated by all sailors because the materials are light and elastic and do not rot. Ship-bottom paints, designed to reduce fouling by marine organisms, have been greatly improved. Inorganic zinc underpaints promise to decrease substantially the pitting of hulls and decks, which should increase the life of ships and the time between dry-dockings.

A remarkable collection of marine vehicles and equipment has made its appearance in the past 10 years. Ships now exist that can go up, down and sideways and can flip. They skim, fly and dive. Some of them are amphibious and some go through ice, over ice or under it.

This versatility is important; a ship cannot be efficient unless it has been designed to do exactly what the user wants. Widely varying requirements mean very different sea vehicles. A distinct place on the spectrum is occupied by the superships I mentioned earlier. They are bigger than anyone dreamed of only a few years ago; in fact, they are almost the largest man-made structures. The largest vessel now in service is the *Universe Ireland*, a tanker with a capacity of 312,000 deadweight tons. The ship is 1,135 feet long and delivers 37,400 shaft horsepower. It plies between the Persian Gulf and Ireland, going around Africa at 15 knots and pushing a 12-foot breaking wave ahead of it. Even the enormous vessels of the *Universe Ireland* class will soon be surpassed in size by ships being built in Japan and West Germany. They will be so large they will not be able to dry-dock in any yard except where they were built, and they will not be able to enter any ordinary harbor because they will draw up to 80 feet of water.

A variation in the tanker field is the conversion of the comparatively small (114,000 deadweight-ton capacity) *Man-*

ALUMINAUT, a mobile submersible capable of carrying two crew and four passengers to depths of 15,000 feet and of probing the bottom or moving heavy objects with manipulator arms attached to the hull, is photographed during a dive. The craft is 51 feet in length.

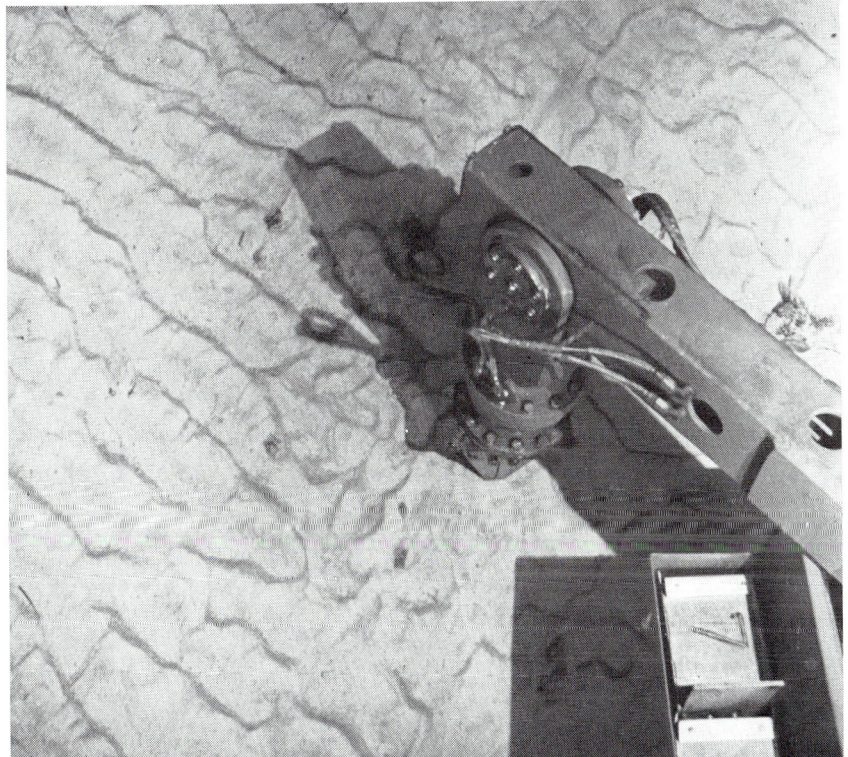

MANIPULATOR ARM of *Aluminaut* explores bottom off Bimini at a depth of nearly 1,800 feet. Numbered sample boxes are nearby. Thin layer of sand is rippled by a current moving it over a rock base; the dark areas are debris that are caught in filamentous organisms attached to rock. Photograph was made by A. Conrad Neumann of the University of Miami.

hattan to a supericebreaker. The purpose is to move the petroleum from the large new oil fields on the northern slope of Alaska to more moderate climates. A fleet of such ships may be able to keep open a northwest passage from the U.S. East Coast to Alaska. From ships of the *Manhattan* class it is only a small step conceptually to a ship five times larger that could cross the Arctic Ocean at will,

treating the ice, which averages eight feet in thickness, as an annoying scum.

The ships that go up include both the ground-effect machine, which can rise a few feet above the surface of the sea on a cushion of air, and the hydrofoil, which has a hull that flies above the surface at high speed with the support of small, precisely shaped underwater foils. The newest versions of these "flying

boats" represent substantial technical achievements, and yet neither vehicle seems likely to become a very important factor in marine affairs because each has basic problems, such as the danger to the hydrofoil of hitting heavy flotsam and the inability of the ground-effect machine to carry large loads or to operate in high waves. The ground-effect machine does have a potential, not much exploit-

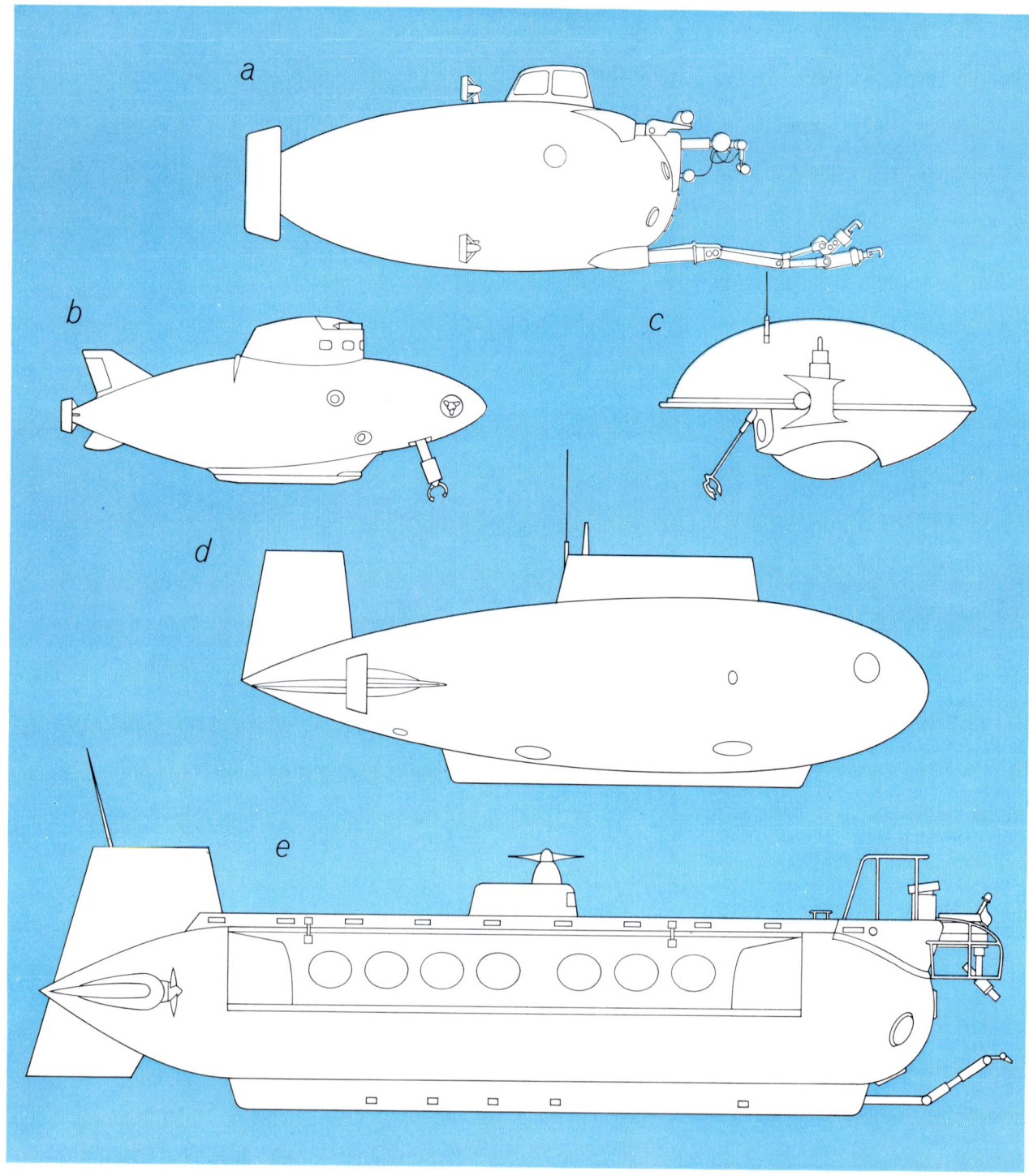

MANNED RESEARCH SUBMARINES are designed for deep diving. They include (*a*) *Beaver IV,* which can dive to 2,000 feet; (*b*) *Star III,* to 2,000 feet; (*c*) *Deepstar IV,* to 4,000 feet; (*d*) *Deep* *Quest,* to 8,000 feet, and (*e*) *Aluminaut,* which is designed to go to 15,000 feet with a staff of six. Vessels are drawn to scale. *Aluminaut* is made primarily of aluminum; the others are steel craft.

ed as yet, stemming from its ability to run up a beach and cross mud flats, ice and smooth land surfaces.

The ships that go down are of course submarines. Nuclear power in military submarines dates back further than the decade under discussion, but large advances in nuclear propulsion have been made during the decade. The circumnavigation of the earth without surfacing

and the trip under ice to the North Pole were both made possible by nuclear power and highly developed life-support systems for keeping the crews alive and well on the long missions.

Quite a number of small, deep-diving submarines are in existence. Of them the *Aluminaut*, designed to go to 15,000 feet while carrying six people, has accomplished the most. (Because of the prob-

lem of obtaining life insurance for its crew its deepest dive has been about 6,000 feet.) Among the many other small submarines are the ones of the *Alvin* class, which can dive to 6,000 feet; *Deep Quest*, to 8,000 feet; *Deepstar IV*, to 4,000 feet; *Beaver IV* and *Star III*, to 2,000 feet, and *Deep Diver*, to 1,000 feet. There is therefore a considerable choice of vehicles, instruments and sup-

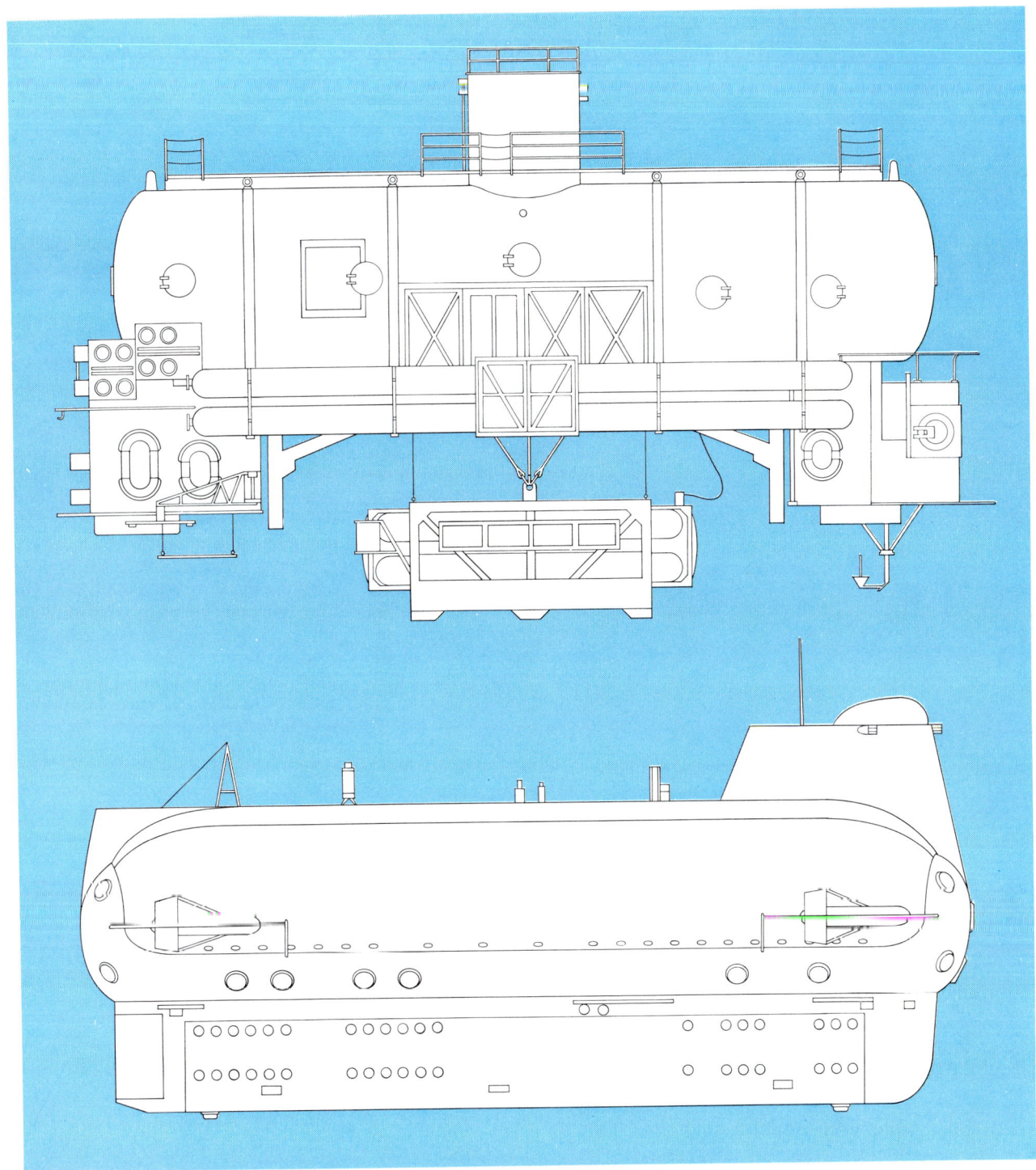

UNDERWATER LABORATORIES include the Navy's *Sealab III* (*top*) and the Grumman-Piccard submersible *Ben Franklin* (*bottom*). *Sealab* is designed to operate on the sea bottom, where it will provide living quarters for divers who will venture forth periodically in heated diving suits to explore the bottom. The first mission of *Ben Franklin* was a submerged drift up the Gulf Stream.

porting facilities. The problem is that there are few customers with the inclination to employ these vehicles at $1,000 or more per hour of diving.

The ships that move sideways are those with trainable propellers or vertical-axis propellers or tunnel thrusters. (A tunnel thruster enables a pilot to move the ship's bow sideways.) Vessels so equipped have found use in self-docking situations and in such waterways as the St. Lawrence Seaway. Dynamic positioning, which means holding position without anchors, is possible with ships that have precise local-navigation systems and central control of several maneuvering propellers.

The ship that flips is *FLIP*, operated by the Scripps Institution of Oceanography. It has two positions of stability. While it is under tow it lies on the surface and looks like a barge made from a big piece of pipe. On station it ballasts itself so as to float on end, much like a big, habitable buoy. In this position *FLIP*, because of its size, is detuned from the motion of the sea surface: it does not move vertically under the influence of ordinary waves and swell. As a result it is an excellent platform for making underwater sound measurements.

Among instruments and tools the now venerable sonar, the sound-ranging device, still figures prominently. It has been improved substantially. Frequencies have risen steadily, making it possible to narrow the beam width to searchlight dimensions, with the result that the distance to (or depth of) discrete areas of the ocean bottom can be measured more accurately. Sonars employing the Doppler effect, which is the change of pitch of a sound resulting from relative motion between the source and the observer, make possible the direct measurement of a ship's speed over the bottom—a measurement that is essential to the high-quality navigation required for such purposes as determining gravity at sea by means of a shipboard gravity meter. Frequency-scanning sonars are now available that better match the signal with the reflector.

Hydrophone arrays, sometimes a mile long, make it possible to use the low-frequency sound created by a series of gas explosions to examine the rocks under the sea bottom in great detail. The result is a continuous picture of a vertical geologic section. Such continuous-reflection seismic profiles have revealed folds and faults in sub-bottom rocks to depths of as much as 15,000 feet and have found many new undersea oil deposits.

Satellites are valuable ocean instruments. They are the essential elements in the system that makes it possible to determine a ship's position accurately wherever it may be. Other satellites transmit photographs of weather patterns, cloud cover and the state of the sea. By combining the picture with other weather data, meteorologists can

produce accurate charts with reliable and up-to-date information. The information is useful in routing ships and forecasting waves.

Buoys moored in the deep ocean hold instruments for measuring, recording and transmitting sea and weather conditions. A number of buoys are already in use, producing an abundance of hitherto unavailable information at minimal expense. It seems likely that hundreds of additional buoys will be put to work in the next few years.

Shipboard computers are becoming an accepted convenience. Such a computer plots the ship's position continuously and matches it with accumulating data of other kinds so that investigators aboard the ship have an information system describing the pulse of the sea below them.

Occasionally a single device can revolutionize an industry. Such a device is the Puretic power block, which handles fishing nets. It is, like many good inventions, basically simple: it is a wide-mouthed, rubber-lined pulley driven by a small hydraulic motor. During the past decade it has been adopted by many fishing fleets and now accounts for some 40 percent of the world's catch. With the block it is possible to handle much larger nets with fewer men. One result is that the tuna industry has shifted almost entirely from line fishing to net fishing.

Another trend in fishing has been to put fish-processing equipment on boats. The equipment includes automatic filleting machines, quick-freeze boxes and even packaging machines so that a finished frozen product can be delivered at dockside. Scallops, for example, can now be shucked and eviscerated on shipboard, so that the scalloper can remain at sea for a week at a time and return with a cargo of ready-to-eat scallops.

Barge-mounted cranes capable of lift-

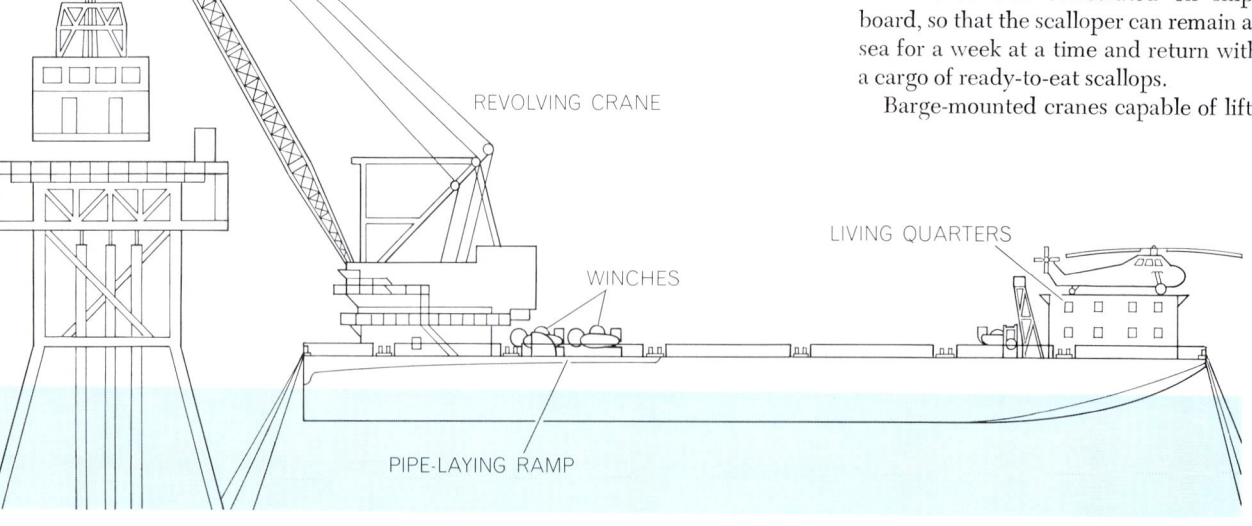

CONSTRUCTION BARGE, the *William Denny*, is about to be put in ocean operation by Raymond International Inc. It is 350 feet long and 25 feet deep and has a 100-foot beam. Its revolving crane, which has a 250-foot boom, can lift 500 tons at a 70-foot radius and 100 tons at a 215-foot radius. The craft can lift 750 tons over the stern. It can build structures, drive piles and lay pipelines.

ing 600-ton loads as much as 200 feet above the water are now available along many coasts. The result is a change in construction techniques. For example, a bridge can be built in large sections, which are then hoisted into place by the crane.

Shipyards are using elevators called Syncrolifts to lift and launch ships weighing as much as 6,000 tons. The machines are replacing dry docks and marine railways. The Syncrolift is simply a big platform that can be lowered below keel depth. A ship is then floated in, and a dozen or more synchronized winches hoist it up to the level of a transfer railway, which moves it to a position in the yard. In this way the yard can work on several ships simultaneously.

The first undersea dredge has just made its appearance. The machine moves along the bottom on crawler tracks in depths of as much as 200 feet. Hence it is not affected by the wave action that makes life on floating dredges hard. The machine was designed to replace eroding beaches with sand from offshore: the dredge has a 700-horsepower pump that moves the sand slurry a mile to shore.

It is fashionable now to speak of the "systems approach," which is a way of expressing the obvious idea that all the elements in the solution of a problem should fit together and be headed toward the same goal. All the ships and instru-

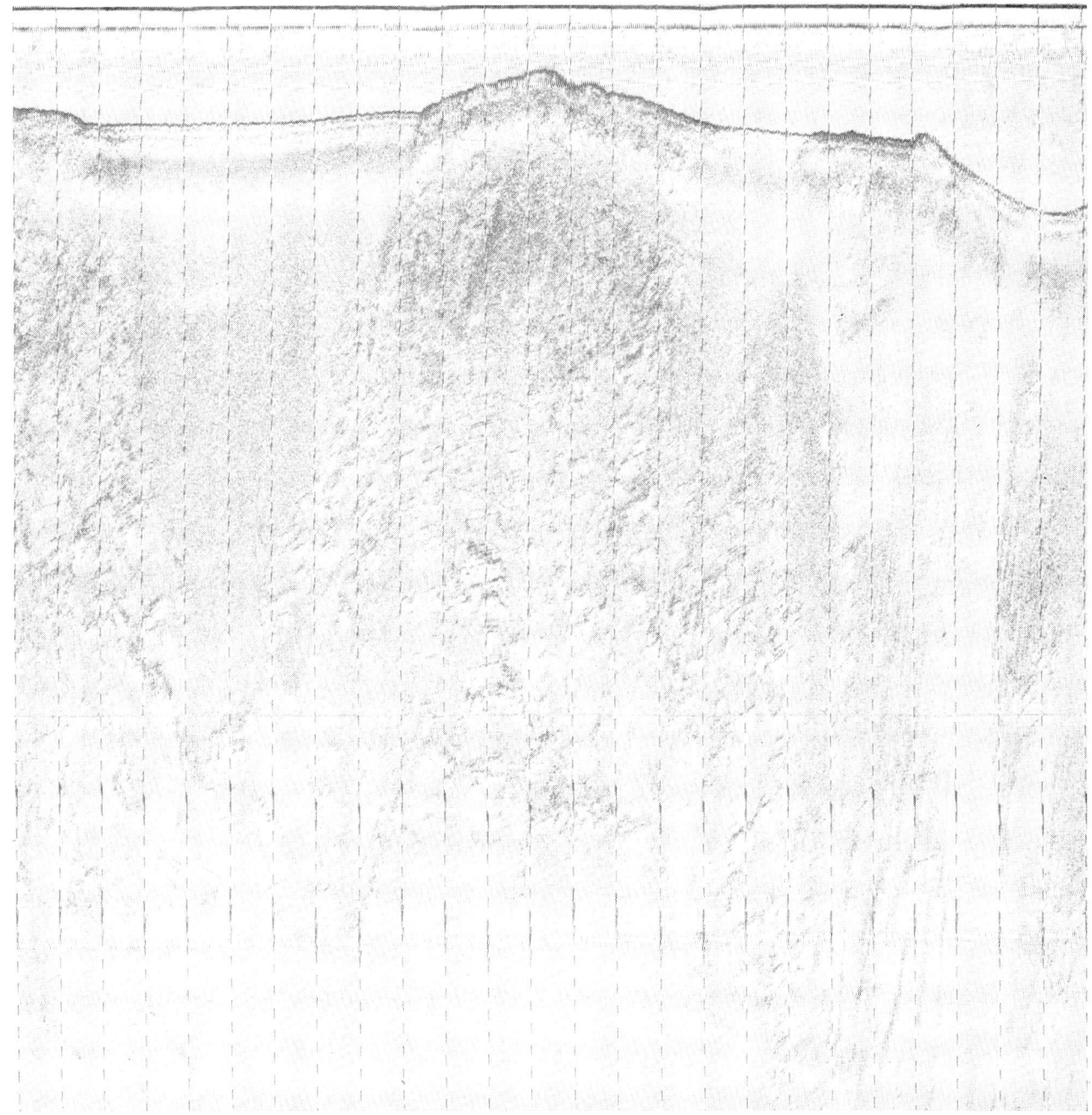

SIDE-LOOKING SONAR produced this view of the ocean bottom on the continental shelf northeast of Boston. From right to left the record covers a distance of about two kilometers along the ship's track. Broken lines show one-minute intervals. Irregular line near top is a profile view of the sea bottom as it appeared at the instrument's horizon. Irregular portions of the photograph are bedrock; darker flat areas are sand waves and gravel; lighter flat areas are smooth sand. Bottom was 60 to 140 meters below the ship. Record was made by John E. Sanders of Barnard College and K. O. Emery and Elazar Uchupi of Woods Hole Oceanographic Institution.

ments I have described are employed as parts of systems. There are, however, several integrated combinations of technology that are best described under the heading of systems. They include containerized cargo-handling, desalination of seawater, deep-ocean drilling and deep diving.

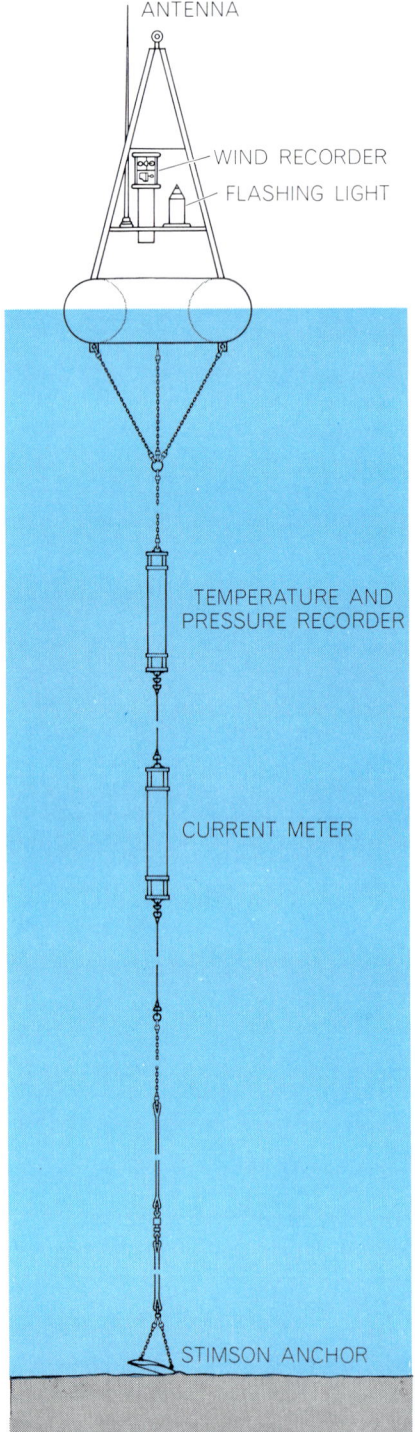

ANTENNA

WIND RECORDER

FLASHING LIGHT

TEMPERATURE AND PRESSURE RECORDER

CURRENT METER

STIMSON ANCHOR

BIG RESEARCH BUOY employed by the Woods Hole Oceanographic Institution gathers and transmits data from the surface to the sea floor. At the surface it records wind speed and direction; below the surface it measures temperature, pressure and current.

Containerization has become a magic word on the waterfront. The basic idea is that a shipper can move his goods from his inland manufacturing point to an inland customer overseas in a private container. A container moves by train or truck to a marshaling yard on the waterfront. There it is picked up by a straddle truck and moved within reach of a gantry crane, which sets it in slots on a container ship. The contents are safe from pilferage, weather and damage. A harbor facility, dealing with containers of standardized size, can semiautomatically unload and reload a large cargo ship in less than 24 hours. Labor cost is lowered; the ship spends more time at sea and less alongside a dock, and the freight moves faster and more cheaply than on a breakbulk cargo ship [see "Cargo-handling," by Roger H. Gilman; SCIENTIFIC AMERICAN, October, 1968].

Methods for desalting water have been improved substantially. The worldwide use of desalted water from the sea is now almost 100 million gallons per day. Most of the water is obtained by various distillation processes; the average cost is estimated to be about 75 cents per 1,000 gallons. Other means of desalination, such as vacuum freezing and reverse osmosis, are being developed. Major nuclear plants that would produce both fresh water and electricity are under study. Most of the desalted water now obtained or in prospect is for household and industrial purposes. The day of cheap irrigation water in large quantities is still far away.

Offshore drilling from floating platforms is less than a decade old and has evolved rapidly. The self-propelled drilling ship and the semisubmersible platform, both of which drill while anchored, represent the two ends of the spectrum. The ship emphasizes speed of movement to the drilling site; the platform provides more steadiness and room for working. A semisubmersible platform is towed to its drilling site, where it takes on enough water ballast to submerge its lower portions. In that position it floats on large cylindrical columns. The arrangement is such that the platform is little affected by waves or other motions of the sea.

The unanchored deep-ocean drilling system, which consists of a drilling rig in a ship hull, has so far been used only for scientific work. It has improved substantially the ability of geologists and other investigators to explore the strata under the deep ocean. The technology dates from 1961 and includes dynamic positioning, the control of stress in a long and unsupported drill pipe, placement of conductor pipe (leading the drill through the soft sea bottom to bedrock) in deep water and the use of a seawater turbo-drill to drill hard rock in more than 12,000 feet of water.

Later developments have led to the system employed on the *Glomar Challenger*, which is operated by the Scripps Institution of Oceanography in a National Science Foundation program involving the coring of deep-ocean sediments in water depths of up to 20,000 feet [see *illustration on page 768*]. The developments include acoustical position-sensing equipment and automatic control of the propulsion units, so that dynamic positioning is much more reliable. The ship has successfully drilled several dozen holes in water depths to 17,000 feet, penetrating as much as 2,500 feet of the sea bottom. The cores thus obtained have yielded much valuable information. Moreover, the discovery of hydrocarbons on Sigsbee Knolls deep in the Gulf of Mexico has done much to modify geological thinking about the possibility of oil in deep water.

Offshore oil production is moving steadily into deeper and rougher water and more remote areas. If it is to be profitable, several producing wells must be established in each cluster and the capital cost should not exceed the present cost of producing oil in 200 feet of water. Probably it is possible to build stationary platforms that would resemble existing ones for depths of up to 600 feet. The cost would be high, however, particularly for a system that involved completing the well atop the platform (installing the pipes and valves and related equipment needed to put the well into production after drilling has reached oil). The current trend is toward the use of floating drilling platforms such as the semisubmersible ones, with completion of the well being made on the sea floor. A system of this kind would include remotely controlled valves and flow lines to central collecting points. In depths of 1,000 feet or more submarine work chambers analogous to pressure-resistant elevators will lower workmen to the bottom; while they remain inside at normal surface pressure they will be able to remove and replace heavy components, make flow-line and electrical connections and inspect the machinery.

Deep diving is receiving increasing attention in ocean technology. Men go into increasingly deeper water by means of systems that grow ever more complicated, involving a variety of chambers, hoists, gases and instruments.

There are two competing methods: the bounce dive and the saturation dive. In a bounce dive the diver goes from atmospheric pressure to the required depth in a chamber, breathing gaseous mixtures that change in accordance with depth and physiological requirements.

He works for a few minutes (perhaps 10) and then returns to the surface in a fully pressurized chamber for slow decompression on the deck of the mother ship. The saturation-dive system makes possible multiple dives. In it the diver's body is saturated with inert gases while

he lives in a pressure chamber on shipboard. When it is time to dive, he moves into a similarly pressurized capsule that is lowered to the bottom. Since his body is already prepared for the pressure, he can immediately go to work. He can work much longer than the bounce diver

HOISTING DEVICE called the Syncrolift allows the Canadian submarine *Ojibwa* to be lifted out of the water (*above*) and pulled ashore on rails (*below*) for repair in a dockyard at Halifax, Nova Scotia. The platform is lowered under the keel of a vessel, the ship is floated in and the platform is raised by the array of winches visible in the photograph. The device has a capacity of 6,000 tons.

before he returns to the capsule and thence to the chamber on deck. Since he lives on the surface but at the pressure of the bottom, the procedure can be repeated for many days. Then the diver takes a slow decompression to atmospheric pressure. Divers using each of these systems have reached 1,000 feet.

Another scheme has divers living on the bottom in shallow water at ambient pressures in undersea chambers. Examples are the experiments that have been carried out in such chamber systems as the U.S. Navy's *Sealab II*, the University of Miami's *Tektite* and the French *Conshelf*.

Doubtless oceanographers can point to elements of ocean technology I have overlooked. The developments have been too rapid and profuse for one man to be familiar with them all. Sometimes one feels that the first question to be asked on hearing about a new oceanographic device is: "Is it obsolete yet?" The answer should be: "We're working on that problem and it soon will be."

DIVING GEOLOGISTS of the Shell Oil Company probe the ocean floor at a depth of about 20 feet on the Bahama Banks. The pipe and hose slanting across the center are parts of an air-lift apparatus that they are using to obtain information helpful in the search for oil.

The Author

WILLARD BASCOM is president and chairman of Ocean Science and Engineering, Inc. He began his career as a mining engineer but switched to oceanography in 1945 as a research engineer for the University of California—first at Berkeley and later at the Scripps Institution of Oceanography. In 1954 he joined the staff of the National Academy of Sciences, where he became widely known for his work in organizing and directing the first phase of the Mohole Project to drill in deep water through the earth's crust. Bascom founded Ocean Science and Engineering, Inc., in 1962. The company and its subsidiaries own several scientific-exploration ships and are involved in fishing, ocean engineering work, ship repair, dredging and the design, development and construction of oil-production equipment. Bascom is the author of many scientific and technical papers and the author or editor of three nontechnical books: *A Hole in the Bottom of the Sea, Waves and Beaches* and *Great Sea Poetry.*

Bibliography

THE SUBMERSIBLE AS A SCIENTIFIC INSTRUMENT. A. Conrad Neumann in *Oceanology International,* Vol. 3, No. 5, pages 39–43; July–August, 1968.

MARINE SCIENCE AFFAIRS—A YEAR OF BROADENED PARTICIPATION: THE THIRD REPORT OF THE PRESIDENT TO THE CONGRESS ON MARINE RESOURCES AND ENGINEERING DEVELOPMENT. U.S. Government Printing Office, January, 1969.

SCIENTIFIC
AMERICAN September 1969, Vol. 221, No. 3, pp. 218–234

OFFPRINT **888**

THE OCEAN AND MAN

by Warren S. Wooster

The increase of human activity in the sea presents deep problems
of relations among nations. Who owns the resources of the sea,
and how is the worldwide exploration of the sea to be organized?

The quickening of human activity in the oceans that has been chronicled in the preceding articles brings the problem of international cooperation in ocean study and management increasingly to the fore. Most of the world is ocean, and most of the ocean lies beyond the limits of national jurisdiction and is subject to no specific ownership. How, then, are the steadily more complex techniques for studying the ocean to be organized, financed and applied? How are the resources of the sea to be apportioned? Who is to regulate human activity in the sea so that it brings maximum benefits to mankind and at the same time does not result in harmful effects such as pollution?

The problem has a number of facets. One is the vastness of the area in question. The ocean waters cover four-fifths of the Southern Hemisphere and more than three-fifths of the Northern Hemisphere. The world ocean contains about 1,350 million cubic kilometers of water at an average depth of almost four kilometers.

Another consideration arises from the unity of the ocean. The waters and their contents mingle freely. The processes operating in the waters are of large scale and are driven by forces of planetary dimension. Life in the ocean is affected by these processes, so that the type, number and distribution of organisms may be controlled by events occurring in distant places. Because of this immense unity, investigation of the world ocean is inherently an international affair, requiring cooperation ranging from the simplest exchange of information to the most complex integration of research programs.

Man lives on continental islands embedded in the vast surrounding sea, and many of his activities involve it or are affected by it. To him it is a means of transport and a font of resources. It affects him also by being a reservoir of heat, so that weather and climate across the continents are determined to a large extent by events at the ocean surface.

Conversely, man now has the ability to affect the ocean, sometimes deliberately and sometimes inadvertently and often far from the site of the intervention. A film of oil coats the animals and beaches of southern California, pesticides applied on the African continent appear in the Bay of Bengal and a healthy, succulent oyster becomes a rarity. More subtle forms of intervention are being practiced. For example, completion of the Aswan Dam has almost halted the annual release of water from the Nile into the eastern Mediterranean. The nutrient elements the river water supplied every year gave rise to a proliferation of fish that supported a commercial fishery. Another effect of the flow was an annual decrease in the surface salinity of the Mediterranean for several hundred miles north of the river. Now these waters are no longer enriched, and the increased salinity and density of the surface water may affect mixing processes along much of the eastern shore of the Mediterranean in a way that is difficult to predict.

Another form of intervention is the linking of separate bodies of water by man-made straits. What oceanographic consequences might a sea-level canal across the Isthmus of Panama have if current proposals to build one are carried out? Evidence provided by the link between the Red Sea and the eastern Mediterranean through the Suez Canal is suggestive. During the first decades after the canal was opened the high salinity of the Bitter Lakes barred the exchange of animals between the two seas. As these lakes have gradually freshened by being connected with the eastern Mediterranean through the canal the barrier has been losing effectiveness, and Red Sea organisms are slowly invading the Mediterranean.

Most of the ocean research now being done is conducted by investigators in a small number of affluent countries. The major participants are Canada, France, Germany, Japan, the United Kingdom, the U.S. and the U.S.S.R. Many more countries use the ocean, particularly for fishing, and the technical capability of the developing nations for exploiting ocean resources is increasing. (Peru, for example, has in recent years operated the world's largest fishery.) The ability of such nations to conduct research has not advanced so rapidly, which is regrettable because of the importance to a developing country of participating in ocean research off its coasts. The coastal state has the most direct and immediate need to know the natural resources available to it. In the case of living resources it needs to know the relation between the ocean environment and the abundance, distribution and availability of the organisms. When a nation participates in ocean research, it can develop a corps of investigators who, through their own work and their interaction with workers from other nations, can interpret data on marine resources and the ocean environment and so contribute to the formulation by the country of a rational policy for the management of ocean resources.

It is plain enough that nations ought to cooperate in investigating the ocean. What kinds of international action are most desirable, and how can they be most effectively arranged?

The most basic form of scientific cooperation is the exchange of data and other information. Until about 20 years ago the community of oceanographers was so small that most of the members knew one another and could exchange information on a personal basis. Now the number of oceanographers is much larger, so that personal evaluation of reported work is often impossible. Moreover, the flow of data has increased enormously. For example, determination of the vertical distribution of temperature and salinity in the upper kilometer of the ocean is a common requirement. Formerly a measurement might be made with 20 sampling devices; the basic observation would require about an hour, and about two more hours would be needed to determine and plot the 60 corrected values (temperature, salinity and pressure at each of 20 depths) the apparatus had obtained. The data rate was 20 values per hour. Now observations are often made with *in situ* devices capable of giving values of temperature, salinity and pressure at one-meter intervals. The total observation may require 45 minutes; by then the oceanographer has available, already plotted on a strip chart, some 3,000 values. The data rate has become 4,000 values per hour— higher than the previous rate by a factor of 200.

Related to the increase of oceanographic activity is the problem of making data comparable. Many investigations require the pooling of data from a number of sources. The synthesis is particularly difficult when different methods and standards have been used. In recent years a considerable effort has been made, by experiments in intercomparison and by the development of methods of reference, to improve comparability.

Increased oceanographic activity has led to the broadening and improvement of the international system of data exchange. Such a system had existed on a regional basis for many years in north-

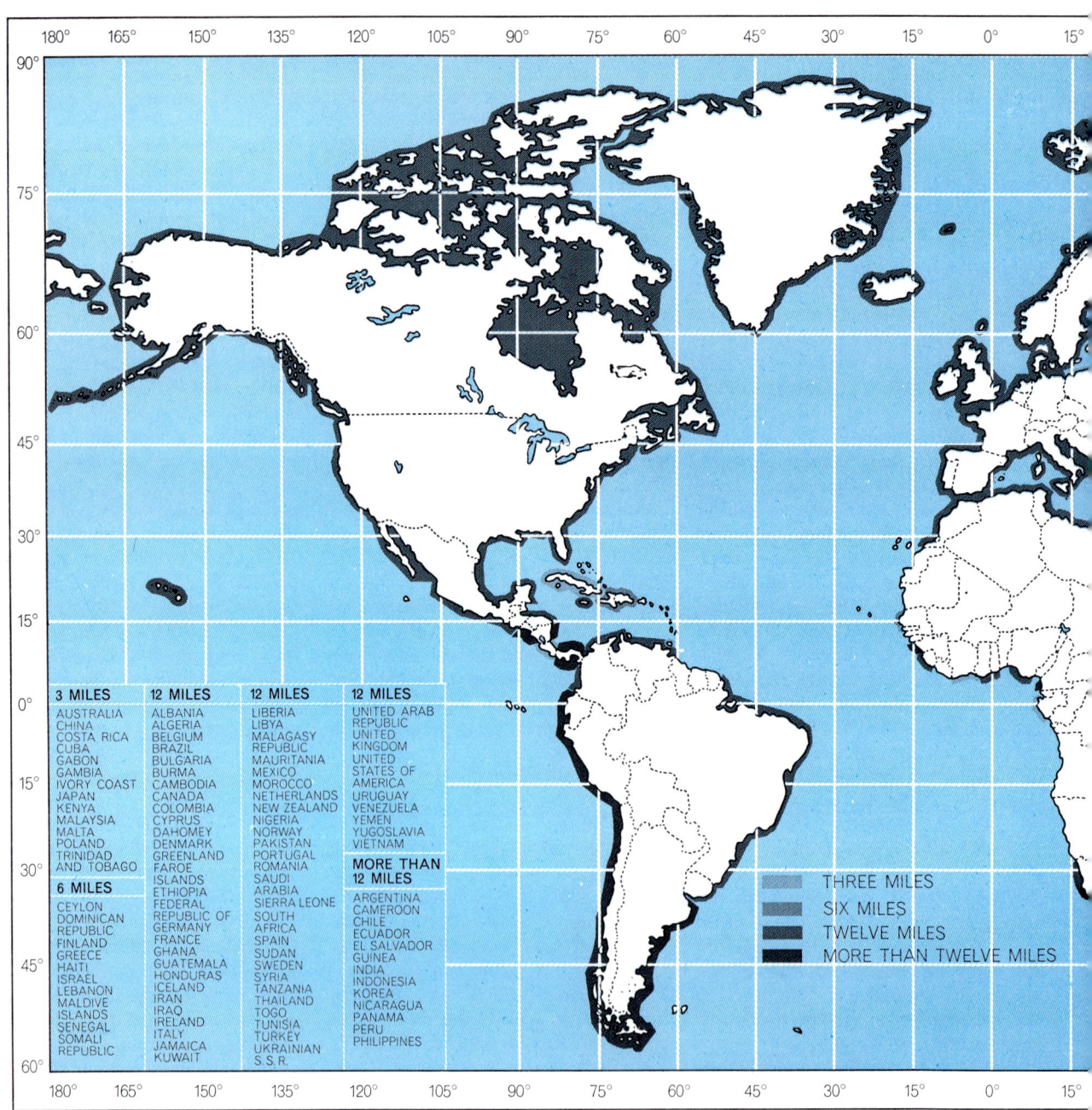

3 MILES	12 MILES	12 MILES	12 MILES
AUSTRALIA	ALBANIA	LIBERIA	UNITED ARAB
CHINA	ALGERIA	LIBYA	REPUBLIC
COSTA RICA	BELGIUM	MALAGASY	UNITED
CUBA	BRAZIL	REPUBLIC	KINGDOM
GABON	BULGARIA	MAURITANIA	UNITED
GAMBIA	BURMA	MEXICO	STATES OF
IVORY COAST	CAMBODIA	MOROCCO	AMERICA
JAPAN	CANADA	NETHERLANDS	URUGUAY
KENYA	COLOMBIA	NEW ZEALAND	VENEZUELA
MALAYSIA	CYPRUS	NIGERIA	YEMEN
MALTA	DAHOMEY	NORWAY	YUGOSLAVIA
POLAND	DENMARK	PAKISTAN	VIETNAM
TRINIDAD	GREENLAND	PORTUGAL	
AND TOBAGO	FAROE	ROMANIA	**MORE THAN**
	ISLANDS	SAUDI	**12 MILES**
6 MILES	ETHIOPIA	ARABIA	ARGENTINA
CEYLON	FEDERAL	SIERRA LEONE	CAMEROON
DOMINICAN	REPUBLIC OF	SOUTH	CHILE
REPUBLIC	GERMANY	AFRICA	ECUADOR
FINLAND	FRANCE	SPAIN	EL SALVADOR
GREECE	GHANA	SUDAN	GUINEA
HAITI	GUATEMALA	SWEDEN	INDIA
ISRAEL	HONDURAS	SYRIA	INDONESIA
LEBANON	ICELAND	TANZANIA	KOREA
MALDIVE	IRAN	THAILAND	NICARAGUA
ISLANDS	IRAQ	TOGO	PANAMA
SENEGAL	IRELAND	TUNISIA	PERU
SOMALI	ITALY	TURKEY	PHILIPPINES
REPUBLIC	JAMAICA	UKRAINIAN	
	KUWAIT	S.S.R.	

THREE MILES
SIX MILES
TWELVE MILES
MORE THAN TWELVE MILES

FISHING JURISDICTION claimed by various nations ranges from three miles to 200 miles. The claims shown are based on information published by the American Assembly last year. Additional rights have been claimed by Cambodia, Ceylon, Ghana, Korea, Nic-

ern Europe: the Hydrographic Service of the International Council for the Exploration of the Sea had collected and published data from its members since 1902. By the 1950's, however, most data were coming from regions other than the North Atlantic and a more general system of exchange was required.

The International Geophysical Year, conducted during 1957 and 1958 by the International Union of Geodesy and Geophysics, provided an opportunity to establish a series of world data centers in a number of fields of geophysics. Two oceanographic data centers, established in Washington and Moscow, are still in

operation. By now their catalogues list many thousands of useful observations, although the time has come, if the centers are to keep abreast of requirements, when the facilities should be transformed from archives to modern computer-based operations.

The international exchanges during the International Geophysical Year were made on a nongovernmental basis. Since 1961 the desirability of governmental commitments to exchange data has been recognized. Governments now make such commitments through the Intergovernmental Oceanographic Commission, an adjunct of the United Nations Educa-

tional, Scientific and Cultural Organization.

With the postwar expansion of oceanographic facilities it became possible to look beyond local waters and consider extending investigations over much larger portions of the ocean. The first manifestation of the trend was a series of single-ship circumnavigations, including those of the Swedish *Albatross*, the Danish *Galathea* and the British *Challenger*. Similar expeditions continue to be mounted, although one would think they would be difficult to justify in the absence of some newly recognized phenomenon or new observational capability that warranted a global exploration.

An important recent trend has been toward more comprehensive investigations made cooperatively by ships from a number of laboratories. The ships may make reasonably simultaneous observations over large areas of the ocean. A case in point is the EASTROPAC Expedition, which derived its name from its area of operation, the eastern tropical Pacific. During one two-month period of this expedition eight ships from the U.S., Mexico, Peru and Chile took readings at some 1,700 stations, sampling an area about a fourth as large as the entire Atlantic Ocean. Such investigations have ranged from the loosely coordinated International Indian Ocean Expedition to the more synchronized International Cooperative Investigations of the Tropical Atlantic.

The requirement for such multiship investigations arises in the study of processes of very large dimensions. When the rates of these processes are comparable with the time it takes to observe them, an adequate description demands simultaneous observations over the entire region affected by the process. It is unlikely that a single laboratory or even a single nation commands the necessary observational resources. Even when simultaneous observations are not required, because the rates of processes are slow (as in deep water or on the ocean floor), the need to obtain the desired spatial resolution calls for a large number of observations. A case in point would be an attempt to describe the present form of the sea floor. In such undertakings a collective effort makes possible the completion of the description in a reasonable period of time.

In the mounting of such efforts it is important to take into account the enormous range of oceanic phenomena—in time from seconds to millenniums and in space from centimeters to thousands of kilometers. One year may differ dramati-

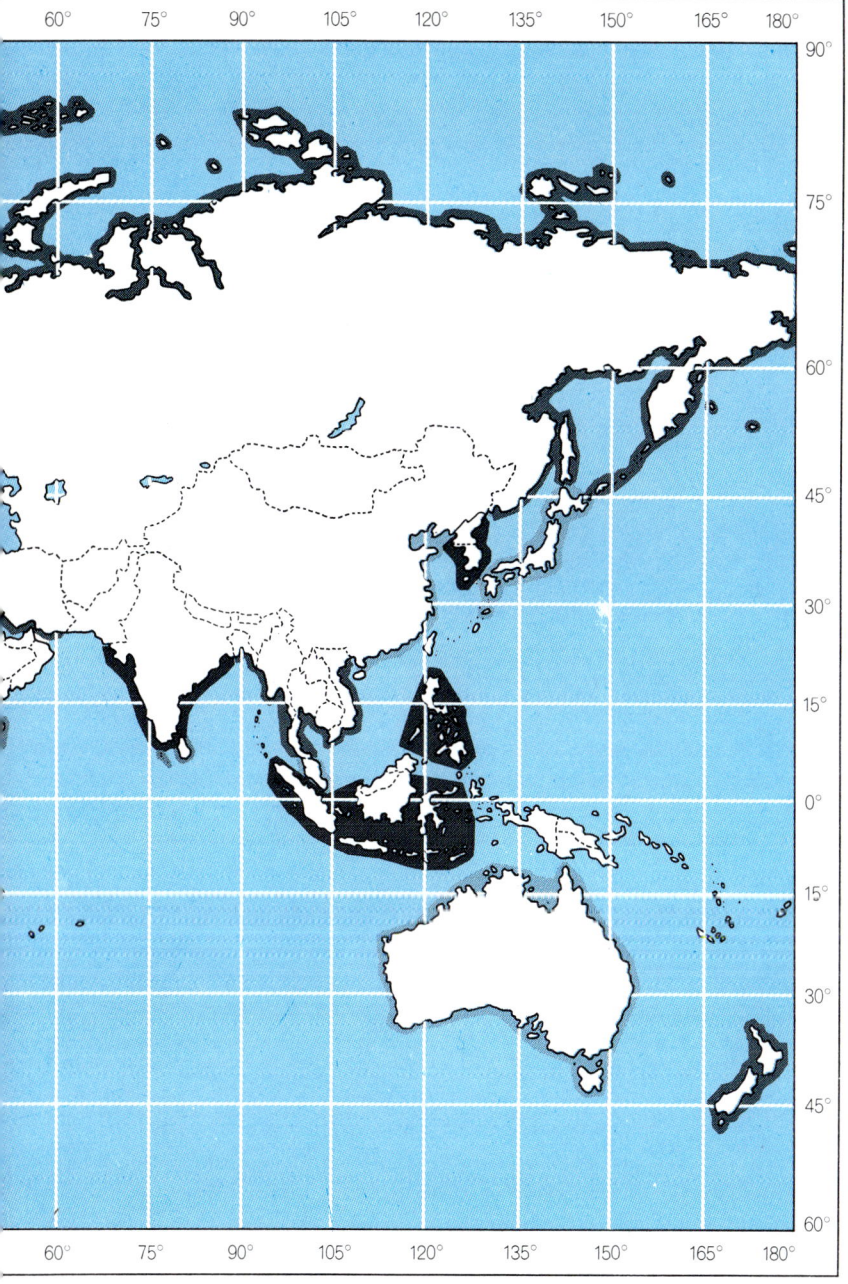

aragua, Pakistan and Tunisia for the continental shelf and superjacent waters. The claims indicate the difficulties involved in obtaining international agreement on use of the ocean.

cally from another. For example, the International Geophysical Year, which included much oceanographic work, took place during a period of unusual solar activity (indeed, the timing of the program was chosen for that reason), and it is not surprising that the period turned out to be aberrant in the ocean and the atmosphere. Much of the Pacific Ocean in particular was highly anomalous in behavior, with strong warming in the eastern North Pacific and a well-developed El Niño (the appearance of a warm surface layer over the cold Peru Current, usually with far-reaching effects on ma-

rine life and the weather of coastal regions) off the coast of South America. Thus if one were to draw conclusions about the oceanography of the eastern Pacific solely from the observations made during 1957 and 1958, the conclusions could be highly misleading.

The mounting of multiship expeditions requires a much higher level of cooperation than the exchange of information. As recently as 10 years ago most cooperative studies of the ocean were organized directly by the investigators concerned; governments had the passive role of providing funds. It became

evident, however, that the financial requirements were growing and that international cooperation might be more effective if formal commitments were made by the participating governments. Such commitments are now made through the Intergovernmental Oceanographic Commission, an organization created for that purpose in 1960. This commission has already acquired considerable experience in a number of large-scale cooperative investigations. It seems likely to play a central role in organizing and coordinating the International Decade of Ocean Exploration that

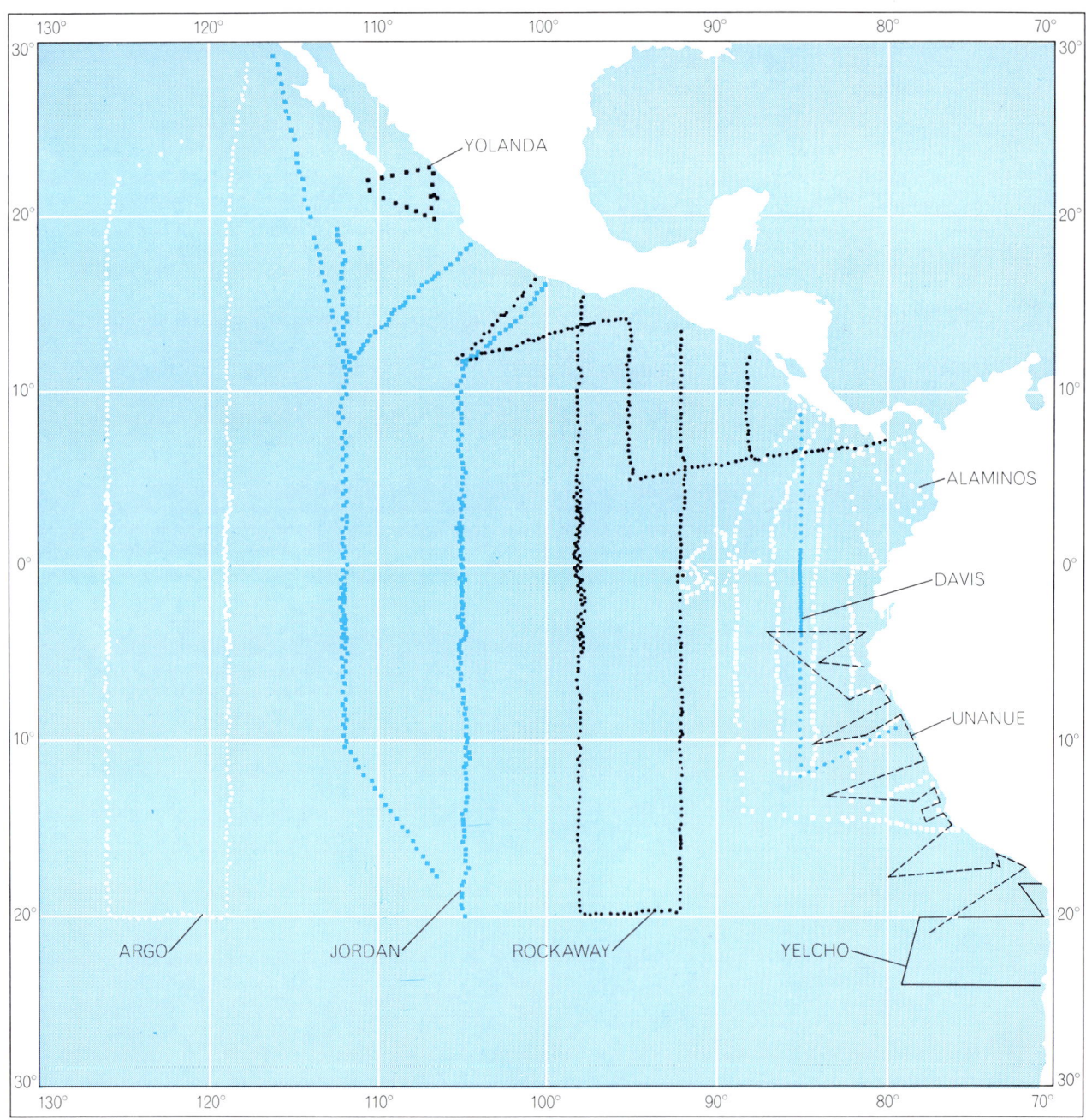

EASTROPAC EXPEDITION, which took its name from its area of operations in the eastern tropical Pacific, was an international surveying and sampling venture that involved eight ships from four nations. One of seven two-month cruises made during the expedition is mapped; dots show where each named ship took samples. Expedition was carried out in 1967 by U.S., Mexico, Chile and Peru.

was proposed by President Johnson in 1968 to be conducted during the 1970's.

Another field that depends on international cooperation is the provision of global services. Many of the services required by oceanographers are extensions of those provided for a number of years to other kinds of mariners. An example is the bathymetric chart, or depth chart, prepared for coastal waters by the various national hydrographic offices coordinated by the International Hydrographic Organization in Monaco. Traditionally mariners have been interested only in the topography of coastal regions and other places where they might run aground. Now the interests of ocean scientists, fishermen, oil companies and governments require detailed knowledge of the entire continental margin. Soon charts of the deep-sea floor will be needed for the same broad spectrum of users.

The need for a precise, reliable and broadly available global system of navigation is becoming urgent. Satellite systems and other techniques now give promise of precise positioning anywhere in the ocean at any time. This capability is as necessary for the development of sea-floor mineral resources as it is for research.

The needs of weather forecasting have given rise to a network of ships and island stations that make observations at the sea surface and transmit them to central stations for use in weather analyses. Even though the coverage is inadequate over most of the ocean, it is possible to get a crude picture of conditions on a given day, particularly in the Northern Hemisphere. Gaps in the observational coverage will be filled during the next decade by satellites and patterns of fixed and drifting buoys, measuring near-surface conditions in both the ocean and the atmosphere and transmitting their data via satellite to central stations. The instrumentation of the ocean in this fashion is being considered for the World Weather Watch of the World Meteorological Organization (a UN agency) and for its oceanographic counterpart, the Integrated Global Ocean Station System of the Intergovernmental Oceanographic Commission. Both operations at sea and the investigation of time-dependent oceanic phenomena will benefit from the improved descriptions and forecasts made possible by these monitoring systems.

So far I have confined my discussion largely to the scientific aspects of international cooperation in the ocean. The nonscientific problems—be they po-

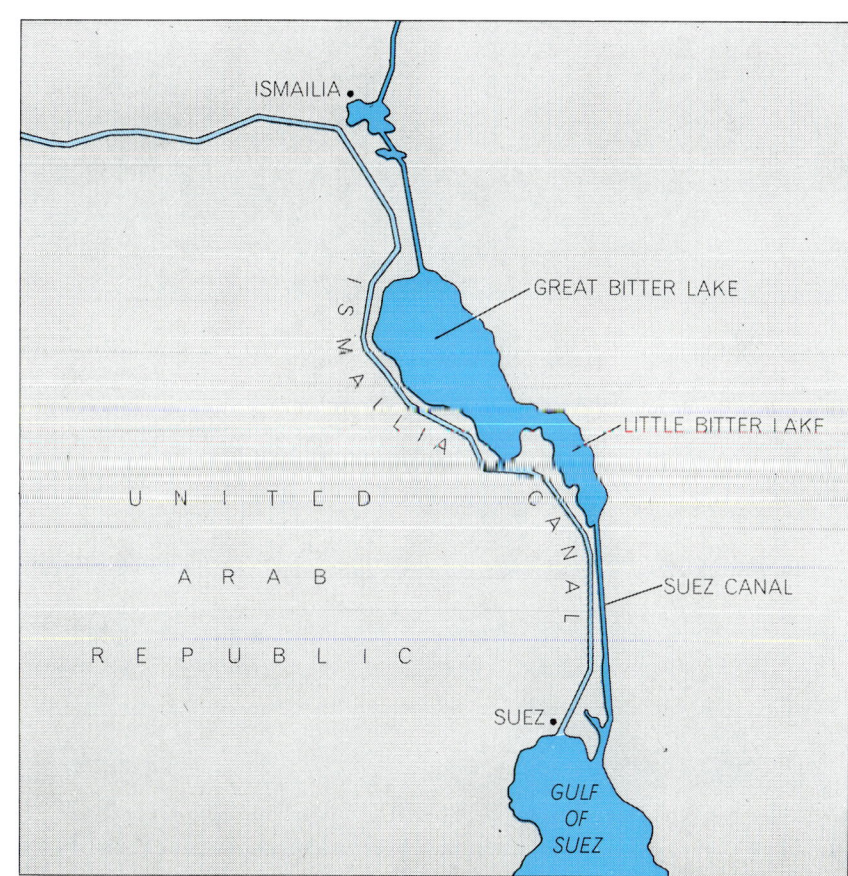

SUEZ REGION has been affected by a decline of salinity of the Bitter Lakes in the 100 years since the opening of the Suez Canal. The high salinity of the lakes originally prevented the exchange of animals between the Red Sea and the eastern Mediterranean through the canal, but as the lakes have freshened, the barrier has gradually lost its effectiveness.

litical, legal, organizational, social or economic—are far more complex. They deserve a more penetrating analysis than is possible here, but a few examples will illustrate their nature.

An important legal and political problem with a direct impact on ocean science is that of the freedom of research.

Oceanic features and processes are little affected by boundary lines drawn on charts. Investigations must be made on the basis of phenomena rather than political boundaries.

Up to the present there have been no significant restrictions on research on the high seas. On the continental margins,

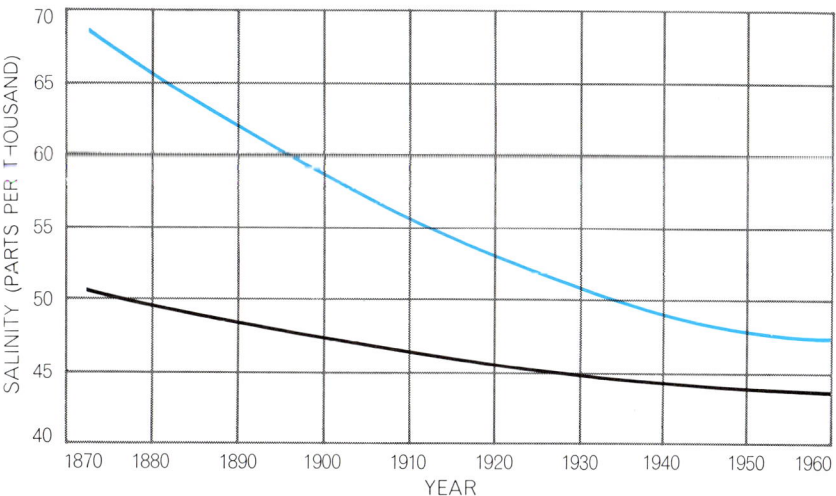

DECLINE OF SALINITY in Great Bitter Lake resulting from infusion of fresher waters by way of the Suez Canal is charted for the lake bottom (color) and for the surface (black).

al state can participate fully in the research.

Agreement on principles and workable rules for the rational use of the ocean and its resources will require international cooperation of the highest degree. Considerable interest in this problem has already been shown by the UN. In 1967 Malta proposed a treaty that would vest ownership of the mineral resources of the sea beyond national jurisdiction in an international regime and would assign the profits from the exploitation of resources to the development of poor nations. The General Assembly responded by establishing an *ad hoc* committee to study the peaceful uses of the ocean floor. A year later the Assembly established a permanent committee of 42 nations to continue the work. The Assembly also initiated consideration of the possible harmful effects that might arise from exploitation of seabed resources and welcomed the concept of an International Decade of Ocean Exploration.

It must be recognized that the potential resources of the deep-sea bed cannot be exploited economically today and that for the foreseeable future only a few countries will have the technical capability even for prospecting and pilot exploitation. There remains, however, a strong feeling among the disadvantaged nations that the resources of the deep-sea bed should ultimately benefit those with the greatest needs. These feelings account for the many efforts that have been made in recent years to establish an international regime over this vast territory, which has been estimated to represent five-sevenths of the earth's area. For such a regime to be effective it must stimulate rather than inhibit the development of the resources, provide for their rational management and conservation and facilitate an equitable distribution of the benefits received from them.

Attempts to manage international fisheries have already shown some of the difficulties of obtaining international agreement on the oceans. More than 20 international fishery conventions are in effect. Even the most successful of them faces problems in obtaining and evaluating the scientific information required

GIANT BUOY, part of a network of buoys involved in the North Pacific Experiment sponsored by the Navy's Office of Naval Research, is towed to its station 1,000 miles north of Hawaii by the Coast Guard cutter *Yocona*. The buoy gathers and transmits oceanographic and meteorological data on surface winds, barometric pressure, waves, surface current, and temperature and salinity, pressure and temperature at various depths. The radial array at top of buoy and the cone of wires descending therefrom form the buoy's antenna system.

on the other hand, the 1958 Geneva Convention on the Continental Shelf states that "the consent of the coastal State shall be obtained in respect of any research concerning the continental shelf and undertaken there." The convention defines the shelf as "adjacent to the coast but outside the area of the territorial sea" and extending "to a depth of 200 meters or, beyond that limit, to where the depth of the superjacent waters admits of the exploitation of the natural resources of the said area." The open-ended nature of this definition leaves in doubt how widely the restriction of obtaining consent will be applied.

A further difficulty arises from the convention's statement that "the coastal State exercises over the continental shelf sovereign rights for the purpose of ex-

ploring it and exploiting its resources." Presumably exploration in this context means the detailed pre-exploitation survey that might better be called prospecting. True exploration, however, is a fundamental element of oceanic research, and scientific exploration must not be excluded from the freedom of research.

As coastal states contemplate the resource potential of their shelves and nearshore waters they become more reluctant to permit foreign scientists to work there. Unless the research is done, however, neither the resources nor the associated environmental factors are likely to be revealed. Hence it would appear generally advantageous for permission to be granted under the conditions that all findings are made freely available and that scientists of the coast-

SPREAD OF OIL over the Santa Barbara Channel from the offshore well that began leaking last January is shown in the aerial photograph on the opposite page. The photograph was made during peak flow in February, when the amount of oil escaping was estimated at several thousand barrels a day.

for establishing effective management procedures and in receiving the full co-operation of all affected nations in implementing the procedures. Significant improvement in the use of the ocean and its resources must be accompanied by a considerable improvement in the mechanisms set down for regulating these activities.

There is already a bewildering variety of international organizations concerned with ocean science or the utilization of ocean resources. I have mentioned the numerous intergovernmental fishery organizations. Within the UN system are the UN itself, its Development Program, its Committee on the Peaceful Uses of the Sea-Bed (dealing with social, economic and political problems and the development of marine resources), UNESCO and its Intergovernmental Oceanographic Commission (scientific investigation of the ocean), the Food and Agriculture Organization (development of fishery research and resources) and the World Meteorological Organization (meteorological aspects of ocean research and services). Intergovernmental organizations outside the UN system include the International Hydrographic Organization (coordination of hydrographic services) and regional groups such as the International Council for the Exploration of the Sea (principally the North Atlantic and adjacent seas) and the International Commission for the Scientific Exploration of the Mediterranean Sea.

Nongovernmental scientific organizations form another important category. They include the Scientific Committee on Oceanic Research, the International Association for the Physical Sciences of the Ocean, the International Association of Biological Oceanography and the Commission of Marine Geology, which are all components of the International Council of Scientific Unions. Among their functions are scientific discussion and the providing of advice to governmental bodies.

A field such as oceanography may require a multiplicity of international organizations to meet its complex and diverse requirements. It is certain, however, that progress toward mastery and intelligent use of the ocean domain would benefit from a strengthening, simplification and consolidation of the organizations that now exist.

The Author

WARREN S. WOOSTER is professor of oceanography at the Scripps Institution of Oceanography, president of the Scientific Committee on Oceanic Research (a unit of the International Council of Scientific Unions) and chairman of the National Academy of Sciences–National Academy of Engineering committee to study the proposal for an International Decade of Ocean Exploration. Wooster began his career as a chemist, obtaining his bachelor's degree at Brown University in 1943 and his master's degree at the California Institute of Technology in 1947. "After having been out in the world as a naval officer," he writes, "I found the laboratory rather restrictive. I learned that my freshman chemistry professor was at Scripps and decided that oceanography would be an interesting career." Wooster received his Ph.D. in oceanography from Scripps in 1953. He has been affiliated with the institution since 1947 except for a year in Lima as director of investigations for the Council of Hydrobiological Investigations of Peru, which he describes as his "introduction to the world of biogeopolitics," and two and a half years in Paris as director of the Office of Oceanography of the United Nations Educational, Scientific and Cultural Organization and secretary of the Intergovernmental Oceanographic Commission. "I have been principally concerned," he writes, "with trying to understand the relation between the physical environment and the biosphere. This involves studying the physical processes as they occur in the ocean and are influenced by atmospheric events, and their ultimate effect on the primary production of organic matter."

Bibliography

INTERGOVERNMENTAL OCEANOGRAPHIC COMMISSION (FIVE YEARS OF WORK). Intergovernmental Oceanographic Commission Technical Series No. 2. UNESCO, 1966.

MARINE SCIENCE AND TECHNOLOGY: SURVEY AND PROPOSALS. REPORT OF THE SECRETARY-GENERAL. United Nations Economic and Social Council, E/4487, 1968.

THE LAW OF THE SEA: INTERNATIONAL RULES AND ORGANIZATION FOR THE SEA. PROCEEDINGS OF THE THIRD ANNUAL CONFERENCE, 1968. Law of the Sea Institute, University of Rhode Island, 1969.

AN OCEANIC QUEST: THE INTERNATIONAL DECADE OF OCEAN EXPLORATION. National Academy of Sciences Publication No. 1709, 1969.

SCIENTIFIC
AMERICAN October 1969, Vol. 221, No. 4, pp. 54–72

OFFPRINT 889

THE EXPLORATION OF THE MOON

by Wilmot Hess, Robert Kovach, Paul W. Gast and Gene Simmons

The successful mission of Apollo 11 opens an epoch of manned lunar
exploration. What questions should this exploration seek to answer,
and what areas of the moon should be visited to best confront them?

The success of the *Apollo 11* mission in putting men on the moon and bringing them back safely with samples of lunar material marks the beginning of what promises to be a period of fruitful exploration of the moon by men and machines. The objective will be to answer a large number of questions about the origin and evolution of the moon, its geology, its chemical and physical structure and what light its history can shed on the history of other bodies in the solar system. Our purpose in this article is to discuss the major questions the coming manned expeditions to the moon will be taking up and to describe the techniques likely to be employed on such missions.

In an astronomical sense the moon is usually considered to be a satellite of the earth. From the viewpoint of planetary processes, however, the moon can be regarded as the smallest of the "terrestrial" planets (the others being the earth, Mars, Venus and Mercury). Because its distance from the sun is about equal to that of the earth, the moon is subject to external influences similar to those affecting the earth. The moon's smaller size, however, implies a history quite different from the earth's.

Most planet-wide processes result from internal sources of energy and the means of its dissipation. The amount of internal energy and the means of dissipation are dependent on the size of the planet. On the earth the dissipation of energy has been accompanied by the transport of volcanic fluids from the interior of the earth to the surface, by the long-term development of a light crust and a dense core and by large-scale movements of the earth's crust, mantle and core. These processes, together with erosion and the chemical interaction of materials on the surface with the atmo-sphere and the hydrosphere, continually destroy the earth's surface features. For example, even the largest volcanoes are leveled by erosion within a few million years after volcanic activity ceases. It is extremely unlikely that any of the earth's original surface features still exist unchanged. No examples are known, and almost certainly none will be found.

It was once thought that many of the surface features of the moon date back to the moon's formation. The detailed views of the lunar surface provided by the photographs from the Lunar Orbiter space vehicles have somewhat diminished this possibility, because they indicate that erosion and other processes of change do take place on the moon. Preliminary analysis of the samples returned by *Apollo 11* nonetheless indicates that the material is very old, perhaps three billion years old. The highlands may be even older. The possibility that some of the material lying on the lunar surface is chemically unchanged since the formation of the planet remains high.

The Major Questions

Fundamental scientific questions about the moon are often stated in terms of terrestrial characteristics, which are of course more familiar. In inquiring about the gross chemical and physical structure of the moon, for instance, one wonders if the moon is chemically and mineralogically differentiated as the earth is. If processes such as volcanism have occurred on the moon, what has been their history over long periods of time? Did the moon ever have an atmosphere? Have protobiological materials ever existed or evolved on the moon?

Answers to such questions call for the recovery and analysis of samples of lunar materials from a variety of regions on the moon. Most of the analysis will have to be done on the earth. Determining the age of a sample of lunar material or making a chemical and mineralogical analysis of it requires instruments that cannot be deployed on the lunar surface within the next few years, particularly with little or no prior knowledge of the character of the materials to be analyzed.

The study of returned lunar materials will in fact provide one of the most intriguing challenges ever faced by natural scientists. How much of the moon's history and how many of the lunar surface processes can be understood from a few isolated samples of lunar material, aided by the fairly detailed knowledge of the surface morphology obtained from photographs? The possibilities are considerable, because the lunar surface is not subject to many of the chemical processes that occur on the earth's surface, such as the changes accompanying erosion and sedimentation. Furthermore, the distribution of material over the surface of the moon by the impact of meteorites suggests that a substantial amount of material in any given place may have come from great distances without significant changes in chemical composition.

Efforts to trace the evolution of the moon, to understand its gross internal structure and to explain the characteristics of its major morphological features will require knowledge of the kind and amount of internal energy released by moonquakes, heat flow at the surface and volcanism. The occurrence of moonquakes would reveal something of the distribution of stress with depth. The seismic waves arising from moonquakes would provide a powerful tool for deducing the distribution of basic physical properties with depth. Measurement of heat flow at the surface, combined with estimates of the distribution of radioactive elements in the lunar rocks, would make possible a determination of whether or not internal energy is in fact the cause of volcanism on the moon. Data for attacking these problems will be needed from a number of widely distributed points on the moon.

The Problem of the Mascons

The space vehicles employed in the Lunar Orbiter missions not only made excellent photographs of the lunar surface but also yielded a startling discovery having to do with the gravitational field of the moon. If the moon were a symmetrical spheroid, internally as well as externally, a satellite would move around it in a well-defined elliptical orbit at a smoothly varying speed. In actuality the moon, like the earth, is not quite a symmetrical spheroid, which introduces perturbations in satellite orbits. Over and above these perturbations, however, there are others introduced by lateral variations in the moon's density. As the Lunar Orbiter vehicles were tracked in their orbits it was noted that they gained speed whenever they passed over one of the moon's ringed maria, or dark circular "seas." Analysis of these motions by Paul M. Muller and William L. Sjogren of the Jet Propulsion Laboratory led to the finding that over the major circular maria (Imbrium, Serenitatis, Crisium, Humorum and Nectaris) there is a substantial excess of gravity [see illustration on page 789].

What is the cause of these gravitational variations? The large positive anomalies associated with the maria im-

AREAS DISCUSSED in this article are identified. Possible sites for future manned exploration include the fresh craters Censorinus and Mösting C, the major craters Copernicus and Tycho, the Marius Hills region and the Apennine Mountains. Broken line shows a possible route for a traverse by an unmanned vehicle guided from earth. *Apollo 12* explorers are scheduled to land near *Surveyor 3*.

ply concentrations of mass, now abbreviated as "mascons." An example of the concentration involved is provided by the estimate that the gravitational anomaly over Mare Imbrium is equivalent to one produced by a sphere of nickel-iron 70 kilometers in diameter centered at a depth of 50 kilometers.

The discovery of lunar mascons has given rise to much speculation and debate about their origin. It has also revived interest in exploring the lunar maria, which many investigators had dismissed as unlikely to be as rewarding sci-entifically as other areas of the moon. Do the mascons represent remnants of giant iron asteroids that struck the moon and subsequently were buried and fragmented, or were they formed by some other mechanism?

Most students of the moon favor the latter possibility. The debate centers on what the mechanism might have been. Several mechanisms have been proposed: the filling of a low-density, fragmented lunar crust with lava; a flow of lava into an impact crater; the upwelling of denser material from the lunar depths into giant impact basins; even the deposition of sediment in the maria by flowing water that later dried up. The last hypothesis carries the intriguing implication that water not only existed on the moon at one time but also played an important role in lunar history. In any case, the analysis of samples from the moon takes on added significance as a result of the mascon phenomenon. An exciting result from the preliminary measurements of *Apollo 11* samples is that their density of 3.2 to 3.4 grams per cubic centimeter, which would be high

MASS CONCENTRATIONS revealed by perturbations in the orbits of spacecraft orbiting the moon are represented by contours showing gravitational acceleration in tens of milligals. One milligal is equal to an acceleration of .001 centimeter per second per second. The mascons, as they have come to be known, are large under such circular maria as Imbrium, Serenitatis, Crisium and Nectaris.

for terrestrial rocks, may be related to the existence of the mascons.

Sites for Exploration

The present plan of the National Aeronautics and Space Administration is to make nine more manned explorations of the moon over the next three or four years. The sites for the first few will probably be determined on the basis of constrainst similar to those that were in effect for the *Apollo 11* mission, namely that a landing place must be on the side of the moon facing the earth, so that constant radio communication can be maintained between the earth and the landing party, that the site be in a region free of obstacles, and that it be accessible from a free-return orbit meaning an orbit that will enable the astronauts to return to the earth with a minimum of power if the main engine in the command module should fail. These constraints restrict the next few landings to mare sites near the lunar equator.

Later it should be possible to venture farther afield and to land at or near other sites of particular scientific interest. A number of places are under discussion as possibilities for these landings. Instead of describing them all, we shall focus on four candidate sites and a long-traverse

CRATER COPERNICUS was photographed by the low-resolution camera of the spacecraft *Lunar Orbiter V.* The prominent mountains in the floor of the crater are of particular interest for exploration. The boxed area appears in high resolution on opposite page.

area that we believe offer significant clues for deciphering lunar history. The five areas are identified in the illustration on page 788. No particular significance should be attached to the order in which we discuss the sites.

The first site is the small, extremely fresh crater Censorinus. A landing here could be expected to achieve three objectives: to establish the age of what is clearly a very young feature on the lunar surface, to investigate and characterize an unquestioned impact feature and to obtain samples of material from a region in the highlands. An alternative site, which would offer similar possibilities, is the crater Mösting C.

The second site represents the much more ambitious goal of exploring one of the major craters, Such a crater is Copernicus, which is about 70 kilometers in diameter and has prominent central peaks within it. The ejecta from this relatively young crater cover more than a tenth of the front face of the moon. The relief within the crater is more than 15,000 feet, making it comparable to the most mountainous areas of the earth. An alternative site, with quite similar characteristics is the crater Tycho. These large craters are of interest not only because they represent major events in the history of the moon but also be-

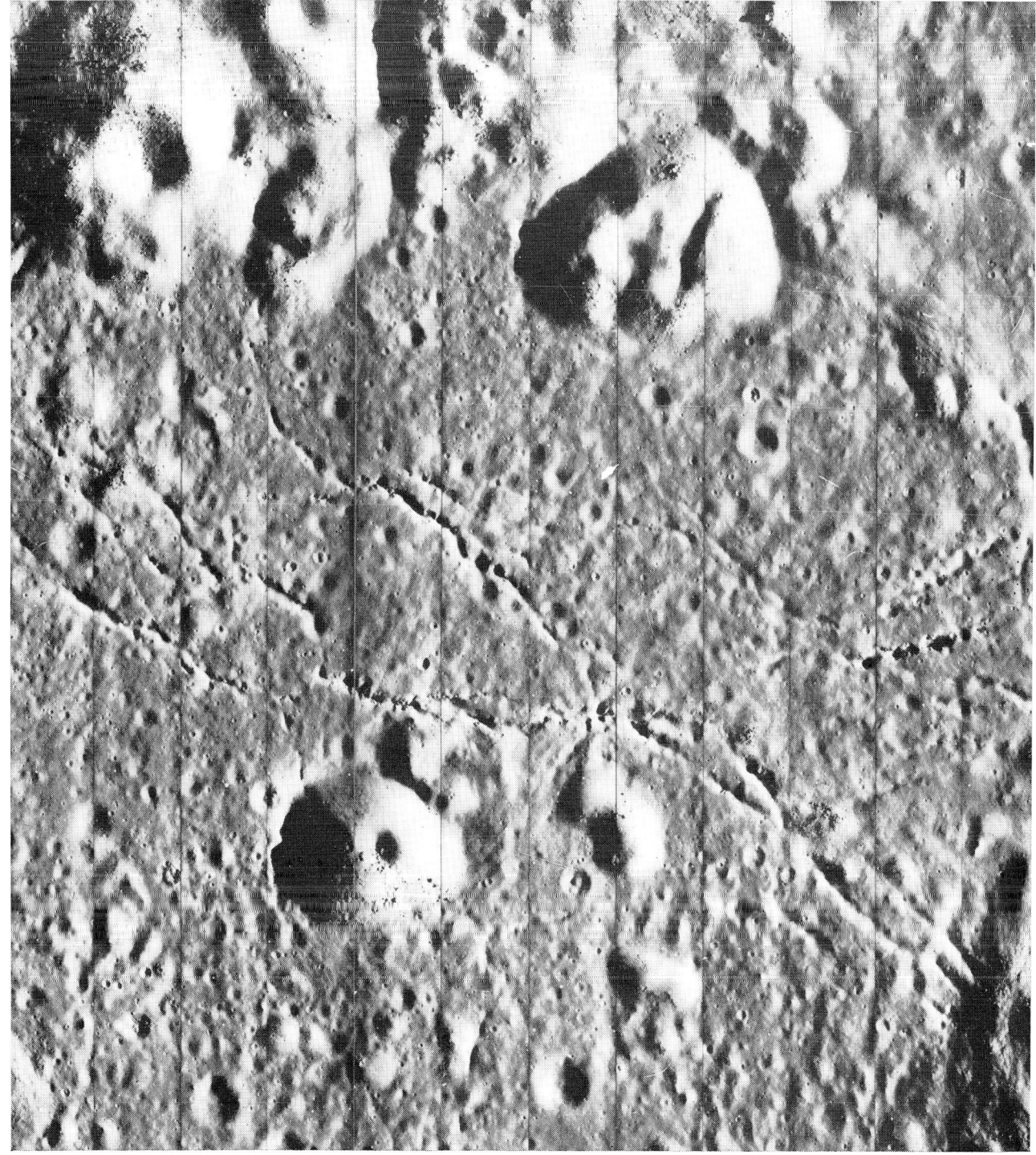

HIGH-RESOLUTION *ORBITER* VIEW of the area of the floor of Copernicus that is outlined in the photograph on the opposite page shows hills and apparent fractures. Large blocks of material are visible in several places, particularly on the slopes of the hills.

cause, by analogy with much smaller terrestrial craters, they should expose material from a range of depths up to 10 kilometers, and perhaps even more. It has been suggested that the central peaks in these craters may consist of material now at the surface that has come from depths of 10 to 15 kilometers or more. Thus even though the material in a crater may be jumbled, broken and deformed by shock processes, it should provide a diverse sample of the outer few kilometers of the moon and a basis for interpreting its history.

Third, we point to the extremely interesting Marius Hills region. It is one of several areas where constructional features such as domes and built-up cones are more numerous than craters of a comparable size. The region is also associated with one of the longest lunar

MARIUS HILLS REGION takes its name from the crater Marius, which is at upper right in this oblique photograph by *Lunar Or-* *biter II*. The region is of interest because of its many domelike structures, which are more conspicuous here than craters. The

ridge systems, which crosses a large expanse of Oceanus Procellarum on the western half of the moon. The tectonic setting of the region is similar to that of terrestrial volcanic fields such as Iceland and the Azores. The setting and structure of the Marius Hills region suggest that it is an area of volcanic activity where igneous material has been added to the surface through vents.

The origin and age of the seemingly volcanic features in the Marius Hills region are of considerable importance in understanding the evolution of the lunar surface. Terrestrial volcanic features are built up in very short times compared with the entire history of the earth. Even an extensive region such as the volcanic chain constituting the Hawaiian Islands represents a period of less than 70 million years. The absolute age and the

domes are believed to be of volcanic origin. If they are, material on surface near them may have issued from great depths and so could reveal information on moon's internal composition and temperature. Region is near equator at far left of moon as seen from earth.

CRATER CENSORINUS, the fresh crater near the center of this *Apollo 10* photograph, is of notable interest for exploration. It is one of the freshest craters on the moon's near side and appears as a bright spot in infrared photographs made during eclipses of the moon.

CLOSER VIEW of Censorinus was obtained by the high-resolution camera of *Lunar Orbiter V.* Large blocks that are as much as 100 meters in diameter appear on the crater's rim.

length of time involved in building up the Marius Hills domes will be of great interest in the characterization of lunar volcanism.

The Marius Hills region is far too extensive to be covered in a single manned expedition to the moon. Fortunately a number of characteristic features of smaller scale can be visited in several areas that are no more than 70 kilometers in diameter. A mission to such an area would be able to sample and study a number of small domes 50 to 100 meters in elevation with convex slopes; steep-sided domes with rough, intricate surfaces; steeply convex or bulbous domes that are smooth and generally symmetrical; steep-sided cones with linear depressions at the summit; narrow, steep-sided ridges, and a variety of impact features.

The fourth candidate site is the region of the Apennine Mountains, which roughly form the southeastern boundary of Mare Imbrium and also the northwestern leg of a triangular highland area bounded by Mare Imbrium, the southwestern boundary of Mare Serenitatis and the northern part of Sinus Aestuum. The Apennines are among the most impressive of the lunar mountain ranges. The Apennine front rises 4,800 meters above the adjacent mare level to the west.

What can be learned about the moon by visiting this area? The Apennine front is a major physical feature of the moon, exposing an extensive vertical section several thousand meters thick for sampling and examination. Here is an opportunity to assess what may be a long period of lunar history. Are the rocks uniform or physically and chemically heterogeneous? How old are they? Are they stratified? Answers to such questions could have a profound effect on our understanding of lunar history.

Two landing sites have been proposed near the Apennine front that are within five kilometers of important lunar features. One such feature is the rille, or canyon-like configuration, known as

FRESH CRATER in the center of the photograph on the opposite page was photographed from the *Apollo 10* spacecraft during lunar orbit; the craft was about 70 miles above Mare Spumans, which is close to the lunar equator and almost at the right edge of the moon as it is seen from the earth. The unnamed crater is about one kilometer in diameter. It is identifiable as a recent crater by the fact that the light-colored material ejected from it covers a number of adjacent surface features. Plans for future Apollo missions include exploration of such a crater.

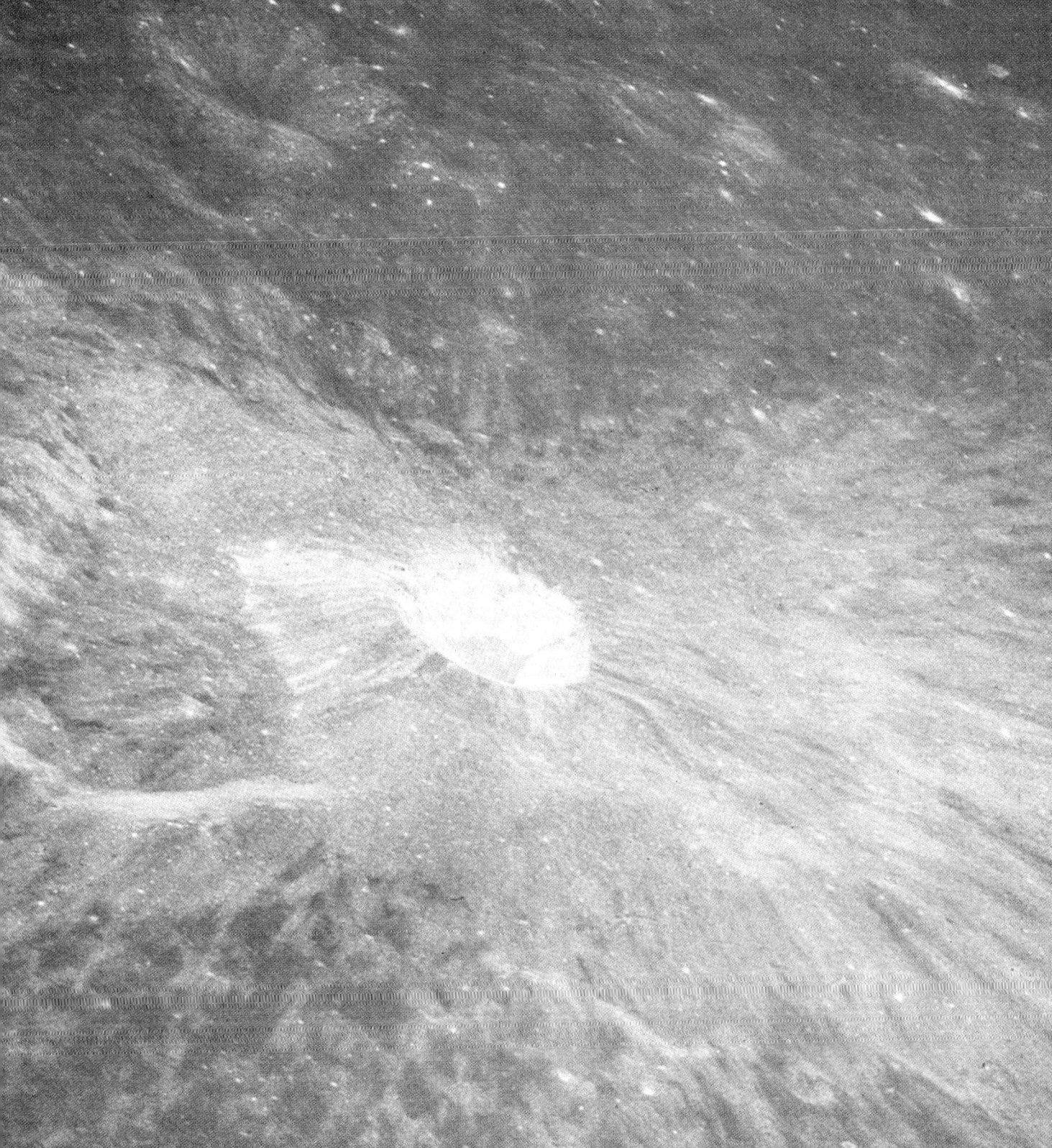

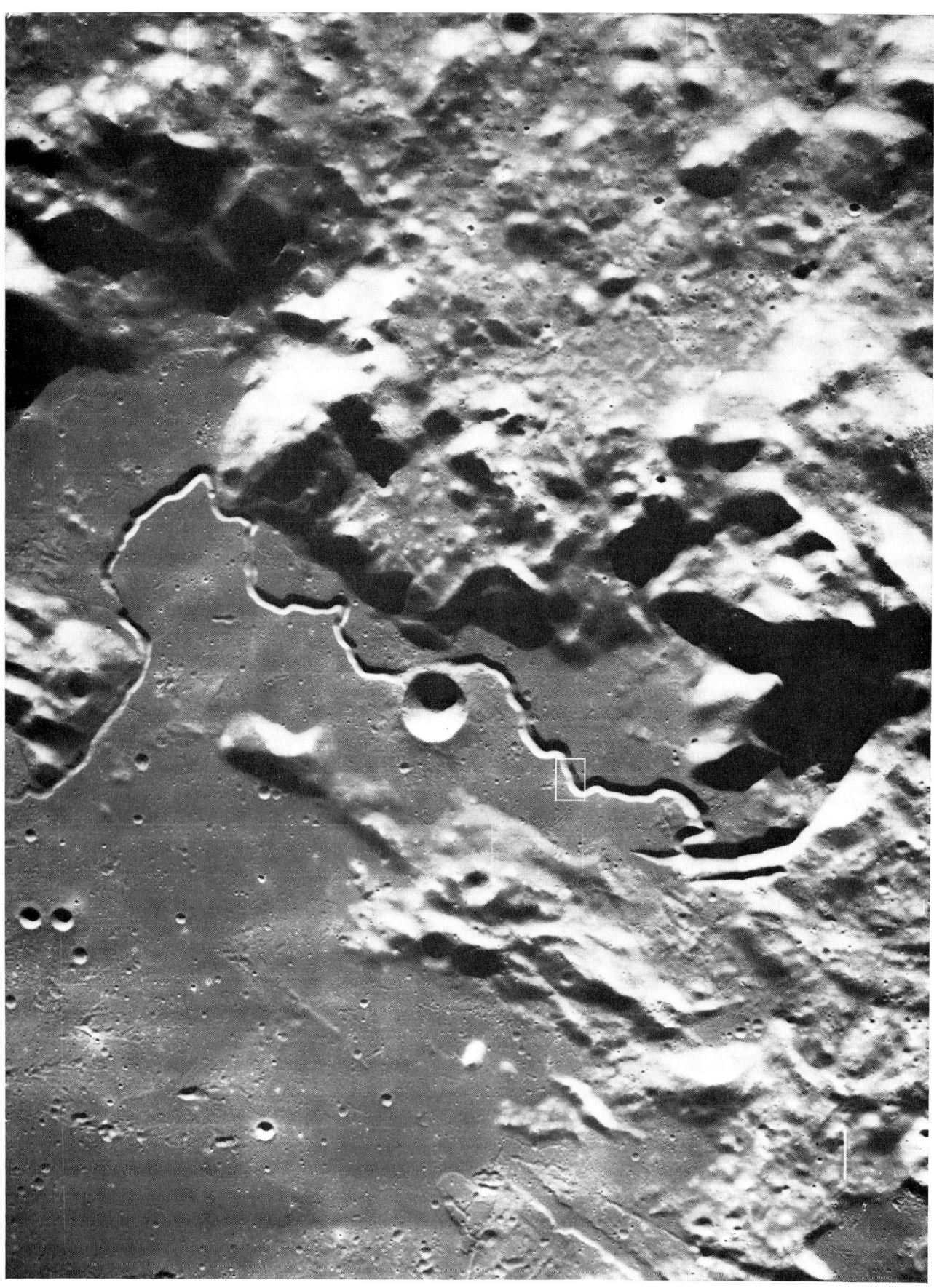

HADLEY'S RILLE is a conspicuous feature of the Apennine Mountain region. The rille, also known as Rima Hadley, is the riverlike structure visible across the center of the photograph. Whether it is a collapsed lava tube or a channel resulting from the surface flow of a liquid might be determined by a manned exploration. Outlined area is shown at higher resolution on the next page.

Rima Hadley [*see illustration on preceding page*]. Is it a surface-flow channel or a collapsed lava tube? If it was formed by water, as has been speculated, where did the water come from and what prevented its immediate evaporation?

The Significance of Rima Hadley

Close examination of the Lunar Orbiter photographs of this rille reveals that fresh exposures of rock are visible along its walls and that blocks have fallen down the walls to the floor of the rille [*see illustration next page*]. Rima Hadley cuts into the floor of a mare and thereby yields a depth and perhaps a cross section of the history of a major lunar feature. Hence it might provide answers to such questions as whether the maria are bedded deposits of lava or ash flows, sedimentary deposits that contain a sequential history of formation or simply an agglomeration of cold particulate matter accreted from space.

The location of the proposed landing site at the boundary between a highland and a mare provides the opportunity for another promising investigation. Deployment of a multiple-axis seismometer and recording of seismic waves from different directions should reveal something about any deep structural differences between the maria and the highlands. Thus one might answer the question of whether or not the maria and highlands are analogous to the oceans and continents on the earth, which show major structural differences.

The Long Traverse

After the early fixed-station landings at a wide variety of lunar sites some form of long-range, mobile surface exploration will be necessary to overcome the limitations of men on foot. The answer lies in vehicular traverses, which would make it possible to study cross-country variations on the moon and so form the bridge between the intensive observations that can be made in the vicinity of a landing site and the extensive averaging observations that can be made from orbit. The technique that is attracting particular interest at present is called the dual-mode lunar surface roving vehicle system. The term dual mode refers to the fact that the vehicle can be used by the astronauts while they are working on the lunar surface and can be operated remotely from the earth after they depart. The present plan entails two separate lunar landings 500 kilometers apart at sites chosen to maximize the amount of information returned from the unmanned, automatic

traverse.

Such an operation would proceed as follows. Near the end of surface activities in the first landing the men would start their unmanned vehicle on an automatic traverse. The vehicle, guided from the earth, would move across the moon toward a distant point that is within the second landing area. There the men participating in the second landing would meet the vehicle several months after it had started its journey. During its trav-

erse the vehicle would collect samples of rock, transmit television pictures and conduct geophysical experiments yielding data that would be transmitted to the earth by telemetry. After the rocks had been retrieved by the astronauts at the second site the vehicle could be used by them in the exploration of that site. If the vehicle was still in satisfactory condition, they could start it on another long traverse.

A typical traverse might go from Rima

HIGH-RESOLUTION VIEW of the portion of Rima Hadley outlined in the photograph on the opposite page was obtained by *Lunar Orbiter V.* Evidence of a ledge of bedrock appears on the right-hand wall at top center. Large rocks are abundant on the slopes of the rille. The average distance from rim to rim is 1.5 kilometers; depth is between 300 and 500 meters.

MARE AND MOUNTAINS are contrasted in this photograph made near the western edge of the Sea of Tranquility from *Apollo 11*. The view is westward. It emphasizes how flat the maria are in comparison with the terrain around them. The wrinkled ridges, which are common features of maria, may be places where volcanic rocks have come to the lunar surface.

Hadley into Mare Imbrium and thence into Mare Serenitatis. Along the way it would provide continuous profiles of the variations in gravity, magnetic and electric fields and depths of the surface layer. This particular traverse crosses one of the largest of the mascon areas and would cover enough ground to explore the phenomenon adequately with geophysical techniques.

The continuous monitoring of gravity along the traverse would provide information on the regional isostatic balance on the moon, that is, whether the higher topographic features are compensated by a deficiency of mass below them or whether they represent loads on the surface. An answer to this question would tell a great deal about the mechanism of formation of such features. Gravity information will also yield clues about the maximum depth of variations in density. If the moon has a crust analogous to the earth's, how does it vary between the lunar highlands near Rima Hadley and the center of the Imbrium basin?

The value of gravity measurements is increased if they can be combined with seismic information. Seismic measurements could be expected to resolve details of any layering in the lunar substrata. Along the traverse we have been describing a properly executed seismic experiment would quickly reveal the presence of giant iron asteroids buried in the mascons.

Improved Capabilities

Implicit in the accomplishment of the missions we have discussed is a considerable improvement of the already substantial capabilities shown in the *Apollo 11* mission. We shall enumerate a number of improvements that we think can be expected by 1972, although we should point out that the estimates are somewhat uncertain. The present capability is for a mission lasting 10.8 days, with a total of 22 hours spent on the moon; by 1972 a 16-day mission with 78 hours on the moon should be possible. The payload of scientific instruments delivered to the surface should increase from 300 to 600 pounds and the amount of material of scientific interest returned to the earth from 150 to 300 pounds. Landings, which are now limited to the equatorial zone, may be possible on most of the front face of the moon, and it may also be possible to land within .5 kilometer of a target area instead of within 10 kilometers as now. For men outside the lunar module the walking radius on the moon should increase from 100 meters to four kilometers, and the total distance covered during a single extravehicular activity from 500 meters to several kilometers.

The capacity of the life-support pack worn by the astronauts as they move about the lunar surface may be increased by as much as 50 percent from the present 4,800 B.T.U.'s. It may also be that the command module will be able, while it is in orbit around the moon, to launch a subsatellite that could make additional measurements.

These new capabilities seem within reach when considered individually. It will not be possible to have them all, because a few are mutually exclusive. For example, if it were decided to land a 600-pound load of scientific instruments on the moon, it probably would not be possible to equip the astronauts to traverse as much as 10 kilometers of the lunar surface.

It also seems probable that a constant-volume suit will be available for lunar astronauts soon. With the present variable-volume suit the astronaut has to do a considerable amount of work against the suit. He also cannot bend his waist or his ankles. The constant-volume suit will require about 30 percent less work for equivalent tasks because almost no work will have to be done against the suit. It will also be flexible at the waist and the ankles. With this suit the astronaut should have considerably more mobility on the lunar surface.

Even so, several of the sites we have discussed cannot be explored adequately unless the astronauts have more mobility than walking provides. Indeed, the radius of mobility ought to be about 30 kilometers from the landing site. Obviously a vehicle will be needed. Two approaches are possible: the vehicle could crawl along the lunar surface at a few miles per hour or it could fly over the surface at low altitude. The ground vehicle has the advantage of enabling the occupants to stop and look at interesting objects, whereas a rocket-powered flying platform makes it possible to move rapidly from one point of major interest to another. A flying vehicle could also move vertically, as will be desirable at certain lunar sites. Although both flying and crawling vehicles have distinct advantages, both are expensive, and it may well be that only one capability will be developed.

Inasmuch as the most that can be expected of a landing party is the exploration of 10 to 100 square kilometers in the vicinity of the landing site, the nine additional landings now planned will cover only about one part in 10,000 of the front face of the moon. In order to obtain a more comprehensive picture of the surface, including the far side, and to look for classes of features missed at the land-

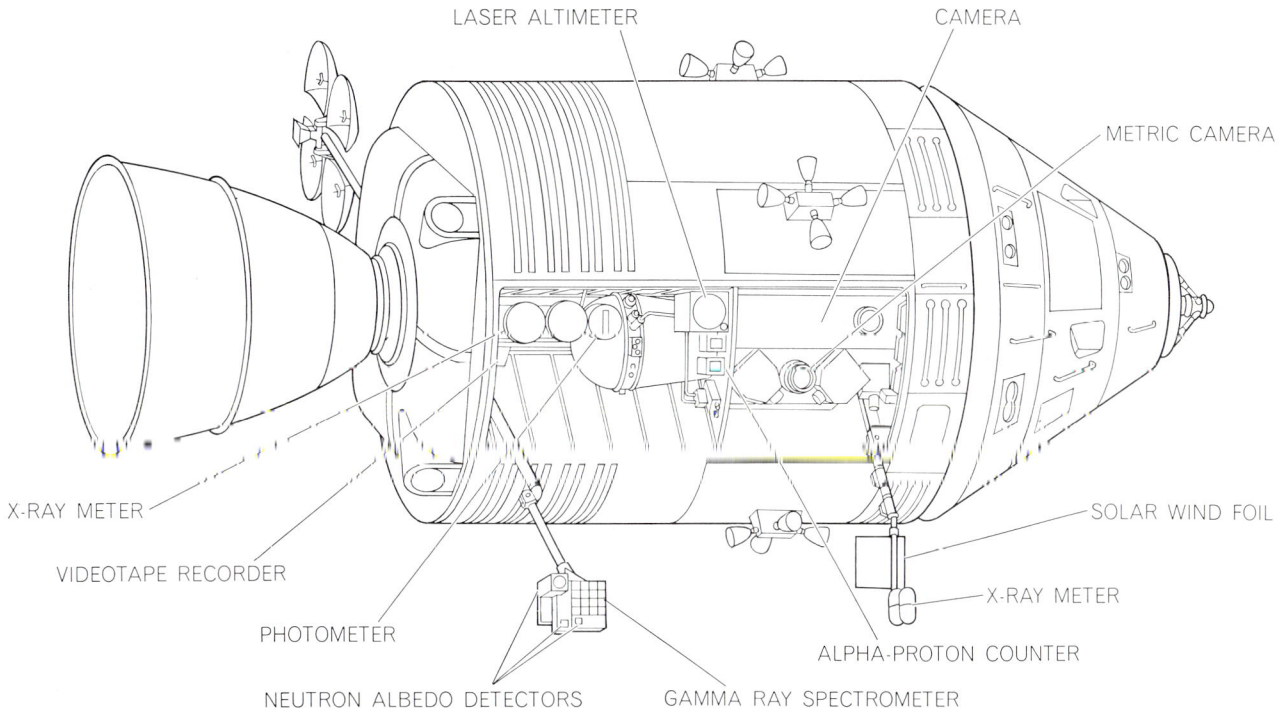

LASER ALTIMETER   CAMERA   METRIC CAMERA   X-RAY METER   VIDEOTAPE RECORDER   PHOTOMETER   NEUTRON ALBEDO DETECTORS   GAMMA RAY SPECTROMETER   ALPHA-PROTON COUNTER   X-RAY METER   SOLAR WIND FOIL

REMOTE SENSING by instruments carried in the service module is planned as a supplement to manned exploration. The nature of the experiments it would perform to measure the lunar surface and assess the structure of the moon is indicated by the instruments.

ing sites, NASA plans to use instruments mounted in the service module for remote sensing from orbit. The sensors will be put into service starting at about the sixth landing.

Orbital Sensing Instruments

Eight types of instrument are under consideration for the remote-sensing activity. Spectrometers measuring gamma rays, X rays and alpha particles emerging from the lunar surface would be able to detect several elements. The gamma ray instrument could ascertain the amounts of iron, potassium, thorium and uranium in the top foot of the lunar surface. The X-ray instrument would receive radiation excited by the sun in a very thin surface layer and give information on the concentration of major elements such as silicon, magnesium and aluminum. The alpha instrument would reveal if there were any extensive leakage of radon gas from the lunar interior, such as often accompanies volcanic or hot-spring activity on the earth. An infrared radiometer would measure infrared emission from the surface and thus would be able to find hot spots and volcanic activity. A gas mass spectrometer would measure the number and type of atoms around the service module, thereby determining the density and composition of the van-

ishingly small amounts of gas at lunar orbital altitudes. An electromagnetic sounder would bounce pulses of radio waves (10 kilohertz to 100 megahertz) off the moon and measure how much came back, thereby finding out about subsurface layering and determining whether there is chemical differentiation or even possibly a layer of ice. A metric camera would photograph most of the moon with good geometric control in order to determine how out-of-round the moon is and whether the centers of maria are lower than the edges. A laser altimeter would bounce a light beam off the lunar surface to measure altitude accurately; such measurements, taken together with orbital data and information from the metric camera, would help to determine the moon's shape.

This kind of broad coverage would mesh well with the detailed coverage astronauts on the surface would make of small areas. Each landing would provide a standard for the orbital experiments by measuring in detail what the instruments should see from orbit. The orbiting instruments then would yield a far broader coverage of surface characteristics than could be obtained from manned landings alone.

Attainment of the goals we have described will still leave several exciting frontiers for lunar exploration. They in-

clude visits to Mare Orientale, the polar region and the far side of the moon. Such visits will require the development of a new technology.

Long-Term Goals

Mare Orientale, the huge "bull's-eye" feature discovered in Lunar Orbiter photographs, is on the far western edge of the moon as viewed from the earth. It is a splendidly preserved, concentrically layered feature probably formed by the impact of a giant meteorite. The feature offers an unparalleled challenge for exploration, but it also presents large operational difficulties for a landing. The Cordillera Mountains, which ring the Orientale basin to form a circular outer scarp some 960 kilometers in diameter, are among the most massive on the moon, rising some 18,000 feet above the adjacent terrain. Perhaps this site, of all the possible ones on the moon, offers the best opportunity for studying the evolution and history of the moon.

A polar landing is a particularly fascinating prospect. Areas near the poles are in permanent shade, so that one might hope to find frozen ammonia, carbon dioxide, water and similar volatiles that otherwise would have escaped from the moon long ago.

The Authors

WILMOT HESS, ROBERT KO-
VACH, PAUL W. GAST and GENE
SIMMONS have been involved in the
planning of Apollo missions and (except
for Kovach) in the analysis of lunar samples from *Apollo 11.* Hess is director of
the research laboratory of the Environmental Science Services Administration.
He was graduated from Columbia College in 1946, took his master's degree at
Oberlin College in 1949 and obtained his
Ph.D. at the University of California in
1954. Kovach, who is associate professor
of geophysics at Stanford University, was
graduated from the Colorado School of
Mines, obtained his master's degree at
Columbia University and received his
Ph.D. at the California Institute of Technology. Gast is professor of geology at
Columbia University, where he obtained
his master's degree in 1954 and his Ph.D.
in 1957. He was graduated from Whea-
ton College in 1952. Simmons, who is
professor of geophysics at the Massachusetts Institute of Technology, was graduated from Texas A. & M. University
and received his master's degree at
Southern Methodist University and his
Ph.D. at Harvard University.

Bibliography

PHYSICS AND ASTRONOMY OF THE MOON.
Edited by Zdeněk Kopal. Academic
Press, 1962.

THE NATURE OF THE LUNAR SURFACE:
PROCEEDINGS OF THE 1965 IAU-NASA
SYMPOSIUM. Edited by Wilmot N.
Hess, Donald H. Menzel and John A.
O'Keefe. Johns Hopkins University
Press, 1966.

LUNAR ORBITER PHOTOGRAPHS: SOME
FUNDAMENTAL OBSERVATIONS. N. J.
Trask and L. C. Rowan. *Science*, Vol.
158, No. 3808, pages 1529–1535;
December 22, 1967.

PRELIMINARY EXAMINATION OF LUNAR
SAMPLES FROM APOLLO 11. *Science*,
Vol. 165, No. 3899, pages 1211–1227;
September 19, 1969.

SCIENTIFIC AMERICAN January 1970, Vol. 222, No. 1, pp. 114–121

OFFPRINT **890**

MODELS OF OCEANIC CIRCULATION

by D.˙James Baker, Jr.

Analysis of the ocean's response to the forces that move its waters
is aided by small models that isolate major features of the physics
of the ocean currents. One such model has given encouraging results.

How does the ocean respond to the forces that set its waters in motion? This response, the general circulation of the ocean, cannot be analyzed solely by direct observations; the system is too complex. One way to simplify it arbitrarily is to build a laboratory model of the ocean. Professor Allan Robinson and I have developed several such models in our laboratory at Harvard University. With the newest of them, which generates currents and makes them visible with a dye, we have obtained some significant results.

Our model is not a large tank of water, as one might expect it to be. Only 40 centimeters in diameter, it basically consists of two circular blocks of clear plastic, one above the other, with a space of one centimeter between them. The upper surface of the lower block is spherically convex and the lower surface of the upper block is correspondingly concave. The space is filled with the "ocean," one centimeter deep. The model therefore represents a segment of the surface of a sphere, as though a large circular portion of the North Atlantic had been lifted out of the earth, and it is operated in a nearly vertical position to correspond with the orientation of the North Atlantic as seen from a point in space above the Equator. The model is mounted on a turntable that simulates the rotation of the earth [see illustration on this page].

In studying ocean currents with models we are taking an approach that has been fruitful in the investigation of other large-scale phenomena of nature. This approach involves three closely interrelated enterprises: observation, theorizing and laboratory experiment. On the basis of careful observation of the natural phenomenon one can formulate a theory to explain what is happening. (The complexity of the phenomenon and inadequate knowledge of its basic physics make it necessary to start with a simplified theory that incorporates only the physics believed to be crucial to the phenomenon.) Armed with such a theory, one can construct simple laboratory models that also isolate the essential physics. Since a model can be controlled, the investigator can gain a good understanding of how it works. He uses the results from the model to frame better questions to ask during his observation of the original phenomenon, and the answers from the observation enable him to refine both the theory and the model. In this way he can hope at last to reach an adequate understanding of the natural system he set out to investigate.

Models can be dangerous. If the results from a simplified theory or experiment agree with observation, one is tempted to think that the true explanation is in hand. The fact is that another

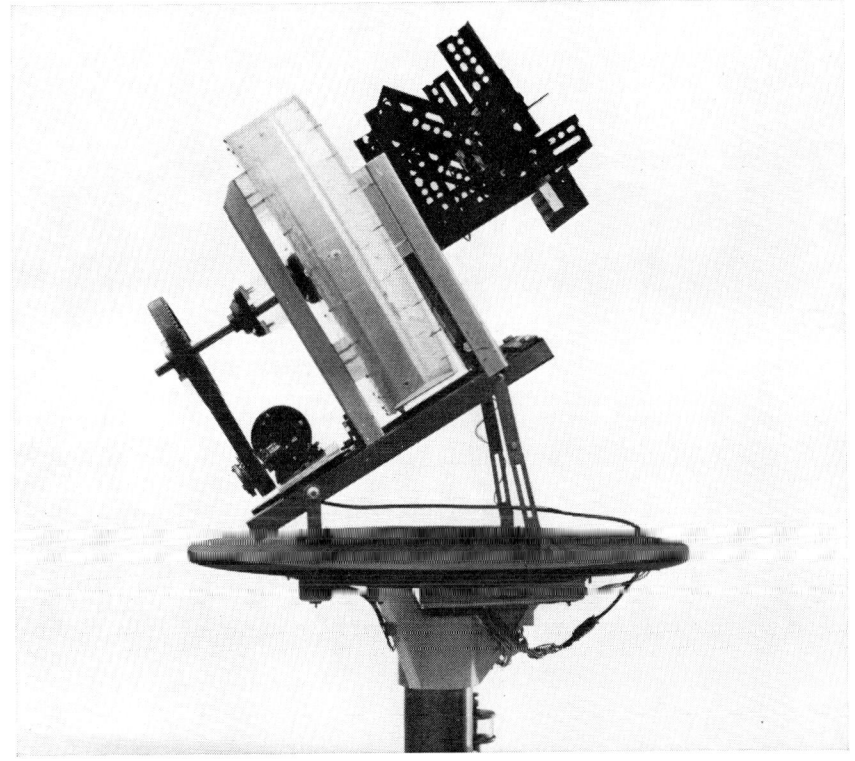

MODEL OF OCEAN is set up in the author's laboratory at Harvard University. The water is contained in the circular plastic blocks, where it is held in a spherical shape to reproduce the curvature of the North Atlantic Ocean as seen from a point in space above the Equator. The tilt of the model is based on the same perspective. Motion is imparted to the water by the lower plastic block, which is driven by the motor and belt at left. Framework at right is a camera mounting. Model is mounted on a turntable to simulate the rotation of the earth.

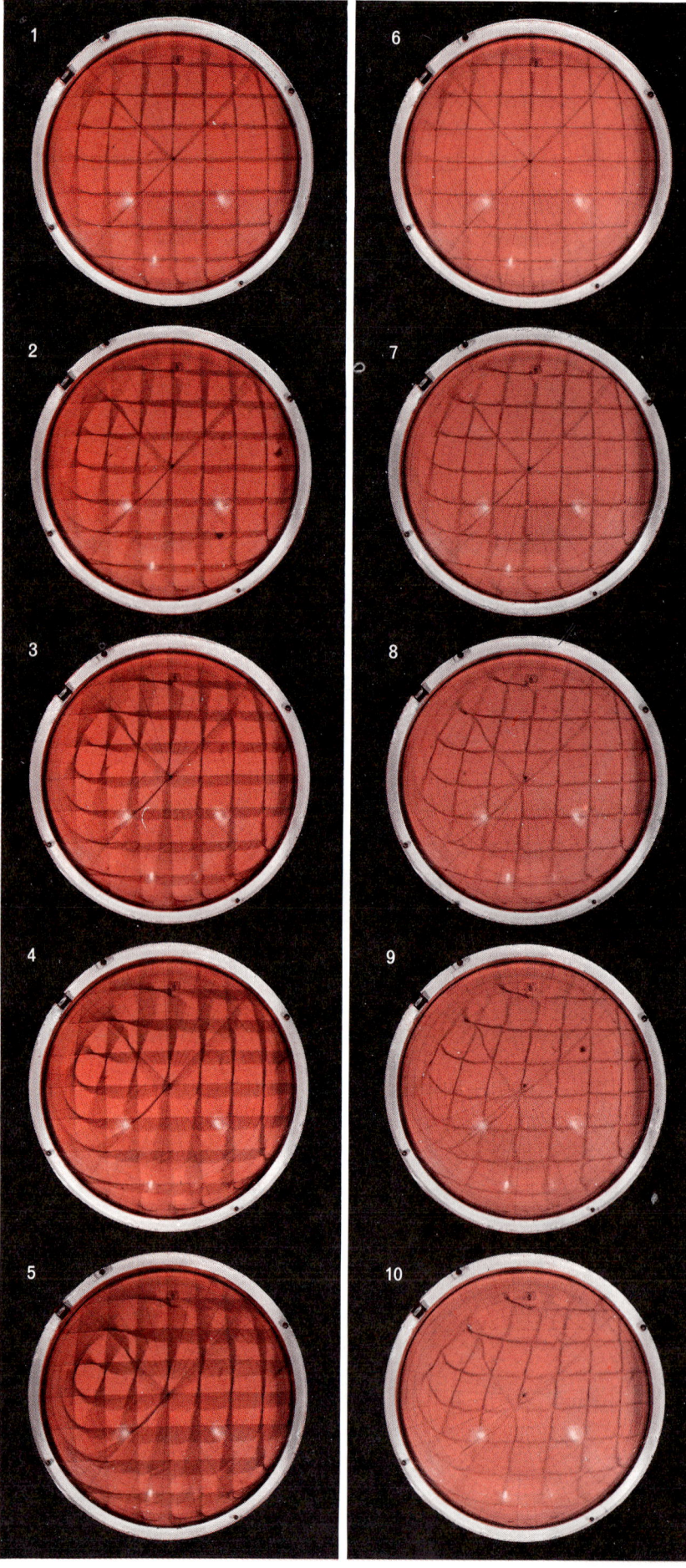

simplified theory might have yielded equally good agreement. One must be careful not to draw unwarranted conclusions. The more of the significant physics that is included in the model, the less likely one is to be led astray.

In the general circulation of the world ocean the input of heat from the sun is converted into large-scale motions: the broad, slow equatorial currents, the narrow oceanic jet streams and the abyssal flows. The currents are produced either directly by the unequal heating of the Equator and the poles or indirectly by the conversion of the sun's heat into winds [see "The Atmosphere and the Ocean," by R. W. Stewart; SCIENTIFIC AMERICAN Offprint 881]. However they are produced, the ocean currents move so slowly and over such great distances that both the curvature of the earth's surface and the daily rotation of the earth play a fundamental role in their dynamics. Other factors that influence the currents are the shape of the ocean floor and the basic vertical stratification of the water from least dense at the top to most dense at the bottom.

The wind-driven currents are the relatively rapid flows in the general oceanic circulation. They are usually strongest at the surface and at the western side of the oceans. The narrow, swift western currents, fed by the broad equatorial flows, include the Gulf Stream, the Japan Current (or Kuroshio) in the Pacific and the Somali Current in the Indian Ocean. Such currents have been known to man ever since he began sailing the high seas.

In 1947 Harald U. Sverdrup, who was then director of the Scripps Institution of Oceanography, showed that the broad equatorial currents in the Pacific could be explained by the action of the wind and the rotation and curvature of the earth. Henry M. Stommel of the Woods Hole Oceanographic Institution and Walter H. Munk of the Scripps Institu-

FLOW IN MODEL is made visible by the dark lines of dye. At left from top to bottom is a series of photographs (1–5) starting 15 seconds after the dye was created and made every 10 seconds thereafter. The driving block was being operated slowly, so that friction rather than inertia dominates the flow. A deep southward drift is evident, as observed in the Atlantic Ocean, and on the west side flows the mass-conserving western boundary stream that resembles the Gulf Stream. A similar series of photographs (6–10), made every 10 seconds after the beginning of flow and employing a somewhat different procedure with dye, appears at right.

tion of Oceanography later showed that the longitudinal barriers presented by the continents force the wind-driven equatorial currents in the Northern Hemisphere to turn northward to form the western boundary currents. A combination of wind and topography then turns the western boundary currents to the east, where they broaden, slow down and eventually turn south to join the equatorial flow again. Thus the circulation is closed in large gyres, or circular motions.

These early models of oceanic circulation were consistent with observations up to that time, but they left a number of features of the system unexplained or ambiguous. For example, the early models predicted the transport of water (its velocity averaged over depth) in terms of the effects of wind stress, the curvature of the earth and the rotation of the earth. The angular momentum put into the water by the wind was assumed to be dissipated by friction developed through contact between the water and the ocean bottom or continental boundaries. The inertia of moving water was neglected. The details of friction were left unspecified. In theoretical calculations the magnitude of the frictional effect could be determined by adjusting the width of a hypothetical western boundary current to equal the observed width of the Gulf Stream or the Japan Current. It was only by assuming a fairly large amount of friction that the theoretical equations could be solved. Nonetheless, agreement with observation, at least for the qualitative nature of the overall flow and the transports of the western boundary currents, was reasonably good.

Then subsequent studies by Stommel of the observational data showed that friction was small compared with inertia in the region of the Gulf Stream. Indeed, he found that a better approximation of the observed situation could be achieved by neglecting friction altogether in favor of inertia. Jule C. Charney of the Massachusetts Institute of Technology and George W. Morgan of Brown University were able to show that reasonably good theoretical models of the Gulf Stream could be made by including inertia instead of friction.

The inertia of the moving water, that is, its tendency to keep going in the original direction or at the original speed when it is forced to change its direction or speed, has thus replaced friction as a mechanism for redistributing momentum in the newer models. Hence the newer models are known as inertial mod-

SEPARATED PARTS of the model basin reveal the configuration whereby the water is held in the shape of a segment of the surface of a sphere, which is the ocean's configuration. Bottom block is convex and top one concave; their normal separation is one centimeter.

els. An understanding of this process, which is unique to fluid flow, requires a review of the simple physics of the oceanic circulation, so that the inertial models can be seen in their proper perspective.

The first point to be noted is that most ocean currents have a motion that is geostrophic, or affected by the turning of the earth. One can visualize the effect in terms of a large rectangular block in the North Atlantic Ocean [*see top illustration on page 806*]. If the water slopes slightly upward from south to north, a gradient of pressure is set up that would make the water flow toward the south in order to equalize the difference in height, provided that no other forces were present. On the rotating earth, however, particles of water in the Northern Hemisphere "feel" a thrust to the right of their motion because of the Coriolis effect. (Anyone can observe this effect by getting on a merry-go-round with a companion and attempting to toss a ball to his companion over a distance of a few yards. On a merry-go-round that is rotating counterclockwise as seen from above the ball will appear to curve to the thrower's right because the merry-go-round rotates out from under it. The ball will miss the companion unless the

thrower has compensated for the effect.) In the block of water an equilibrium would be reached when the pressure-gradient force balanced the Coriolis force, and the motion of the water would be at right angles to the slope, or to the west in the present example. This final, balanced motion is the geostrophic flow.

On the earth the Coriolis effect is greatest near the poles, where the earth turns out from underneath a moving object most rapidly. The effect decreases to zero at the Equator, since the curvature of the earth makes the surface of the earth parallel to the axis of rotation there. This change in the Coriolis factor with latitude, caused by the curvature of the earth, is the fundamental reason for the existence of the western boundary current, as will soon become apparent.

Now let us consider the average circulation and wind stress in the North Atlantic. The simplest model is a square basin with a wind that blows to the east in the northern part and to the west in the southern part; the wind is at a maximum at the northern and southern edges and is zero halfway between them. When the wind is turned on, its westerly direction in the southern part of the basin tends to push the water there toward the west, but the water is deflected toward the right, or north, by the Coriolis effect. The uppermost slab of water exerts a frictional force on the next-lower slab, which is thereby set in motion in a direction still farther to the right. The process is repeated until, at a depth of perhaps 100 meters, friction becomes negligible. The average flow of this entire layer, which is named the Ekman layer after the Swedish oceanographer V. Walfrid Ekman (who first described it nearly 70 years ago), is at a right angle to the wind direction.

Thus the wind in the southern half of our square basin representing the North Atlantic transports water to the north in the thin Ekman layer. Correspondingly the wind in the northern half of the basin transports water to the south. The water therefore tends to pile up in the middle [see bottom illustration on page 806]. The effect of this tendency, since water seeks a level, is a downward thrust that acts through the entire depth of the water. As a result of the thrust, water in the northern part of the basin tends to flow to the north and water in the southern part to the south. But again the Coriolis effect is felt, directing the average deep flow in the northern part of the basin to the east and in the southern part to the west.

Since the average deep flow in the southern part of the basin feels a weaker Coriolis effect but the same downward thrust, however, the amount of water transported in that part of the basin is larger than it is in the northern part. Hence there is a net transport of water to the west, which causes a net slope downward from west to east. As a result the net geostrophic flow over the entire basin is to the south (a basic eastward tendency deflected to its right by the Coriolis effect). Without the decrease in the Coriolis effect between the North Pole and the Equator there would be no net southward flow.

Obviously the water cannot move south over the entire basin; if it did, the basin would soon empty. There has to be a return flow. This requires a narrow northward stream with a transport equal to the southward transport that results (in a roundabout way) from the action of the wind. The northward stream is the western boundary current, which is seen in every ocean with such a wind system. If there were no southward transport, no western boundary current would be needed. Thus the curvature of the earth, as reflected in the changing Coriolis factor, is essential for the existence of the Gulf Stream and similar western boundary currents.

The net geostrophic flow to the south in a basin such as the North Atlantic also conforms to the requirement for the conservation of angular momentum or vorticity. (For fluids it is more convenient to speak of vorticity, which is proportional to the angular momentum of a spherical particle of fluid centered on the point of interest. The vorticity is equal to twice the angular velocity around an axis through the point.) Before the wind acts, every vertical column of water has the angular velocity of the earth, which as seen from above the North Pole is counterclockwise. Another way to put it would be to say that such a column of water in the Northern Hemisphere has positive vorticity. Since the direction of the wind over the North Atlantic is clockwise on the average, the wind puts in negative vorticity. The water must adjust to this condition either by spinning more slowly or by moving south, where the spin of the earth is slower. The southward movement, as we have seen, is what actually takes place.

The western boundary current also must conserve vorticity. At the outset its vorticity is positive, but small, since the current is near the Equator. As the current flows northward it moves into a region with larger positive vorticity. It therefore needs a source of positive vorticity to achieve the balance required by conservation.

In the older frictional models this source of vorticity was provided by a rubbing of the current on its western side, where it adjoins a continent, or on the ocean bottom. In the more realistic inertial models positive vorticity is carried in by the flow entering the stream from its right-hand edge. Thus where

WORKING FLUID of model is represented in color. The fluid is sealed between the upper and the lower plastic block when the model is in operation. By means of the linkage at lower right the tilt of the model can be changed to move the equator in relation to the fluid.

the deep geostrophic flow is toward the Gulf Stream, that is, westward, a continuous source of positive vorticity is available. It was here, in the lower half of the basin, that the calculations of Charney and Morgan neglecting friction in the Gulf Stream were successful in relating observation to theory.

In the upper part of the basin, however, the flow is away from the stream, and such a balance of vorticities is not possible. What actually happens? It is probably in this region that the dissipation necessary for a steady state takes place. Although the mechanism of dissipation can be ignored when one is studying part of the oceanic circulation, any model of the circulation that includes all its features must provide for dissipation, otherwise the system would never reach its observed steady state. Most of the observed streams break away from the coast and meander in this region. Theory has not yet come abreast of the problem. Although there are many speculations, there is still no consistent, unambiguous theory for a general circulation closed by an inertial stream.

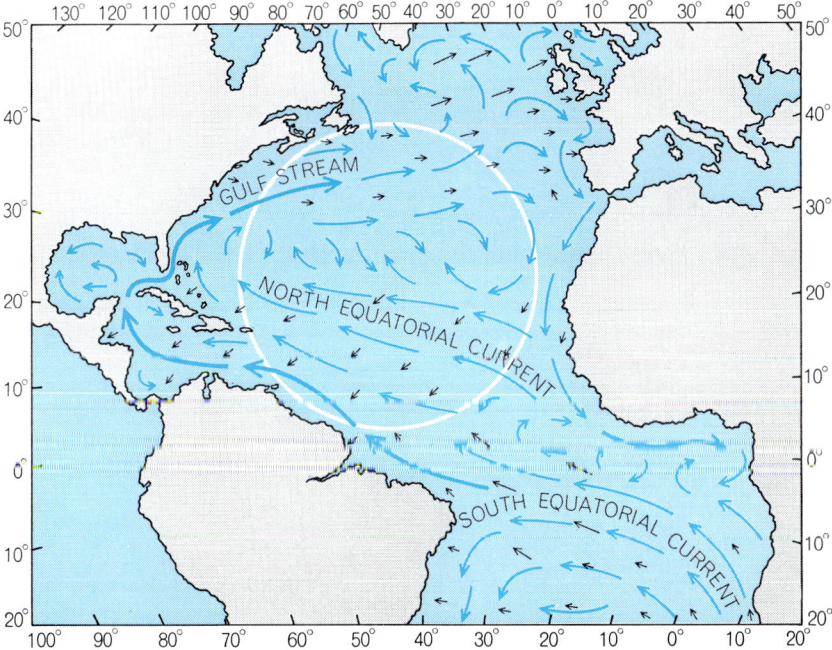

CURRENT AND WIND PATTERNS in the North Atlantic, which the model represents, are mapped. Currents are in color and winds are in black; in each case higher intensities are indicated by longer arrows. Large circle shows a typical area represented by the model.

It was the difficulty of studying fully inertial schemes either analytically or by computer that led us to try to construct a laboratory model of oceanic circulation. In the ideal model the variables could be manipulated so that we could first make a circulation dominated by friction (for comparison with the well-understood theory) and then move into the inertial range. We could also branch out into other problems, such as the effect of oscillating winds.

We cannot, however, make a perfect model of the ocean. There are two major problems. The first is a problem of scale. For example, the Reynolds number for the circulation (which may be defined here as the characteristic size of the basin times the characteristic velocity of the fluid divided by the viscosity) measures the relative importance of inertia and friction. For the ocean the number is large: typically 10^{11}. In the laboratory it is difficult to achieve a Reynolds number larger than about 10^4 because the scale is small and it is not possible to find useful fluids with a viscosity much lower than the viscosity of water.

The second problem is that the stratification of the ocean cannot be modeled satisfactorily. The ocean itself is held to the spherical surface of the earth by the earth's gravitational field. The equipotentials, or surfaces of constant density, are therefore essentially spherical, modified slightly by the earth's rotation. The vertical distance between any two is

essentially constant. The stratification tends to constrain the water to move along equipotentials, or horizontally rather than vertically.

It is unfortunately not possible to produce spherical equipotentials in the laboratory. When a fluid is rotated in the laboratory, the equipotentials are parabolas of revolution, or paraboloidal rather than spherical. The distance between any two of them, measured perpendicularly to the free surface, is not constant. In fact, the change in distance between equipotentials exactly cancels the effects due to the curvature of the free surface. Since, as I have pointed out, the curvature of the earth, represented by the changing Coriolis factor, is essential in producing western boundary currents, a stratified model rotated in the laboratory is not a good model of oceanic circulation. So far we have not been able to build a model of a stratified ocean and still retain the effects of curvature.

There is one way to avoid this problem: one can use a homogeneous fluid instead of a stratified one. A homogeneous fluid has a constant density, and so there is no tendency for the fluid to move along equipotential surfaces. Motion will be determined by rotation, wind stress and surface curvature. Still, the effect of the lack of stratification is difficult to assess. The homogeneous model should be considered as just the first step in a series of models, each one including more of the important effects.

When we came to design the model, we had to consider the important fact that the dynamics of the ocean are constrained by the overall geometric shape of the ocean. The first point is that the oceans are thin, so thin that the relative vertical and horizontal scales are in a ratio of about one to 1,000, much like the paper on which this article is printed. Hence accelerations in the vertical direction are small.

A second constraint is the importance of curvature in that the currents move over such distances that the change in the Coriolis factor is significant. There also have to be boundaries so that the broad equatorial flows can be turned northward or southward. A final restriction is that the total depth of the water must be much greater than the depth that is directly under the influence of the winds (the Ekman layer).

Within these constraints Robinson and I, with the financial support of the Office of Naval Research, designed and built our model ocean. We have tried to observe the constraints, but our model is not as thin as the ocean; the ratio of vertical to horizontal is about one to 40 rather than one to 1,000. The reason is that we have to keep the horizontal dimension within the bounds of what we can handle easily in the laboratory while still allowing a minimum depth of water to accommodate our devices for measuring fluid flow. Nonetheless, the geom-

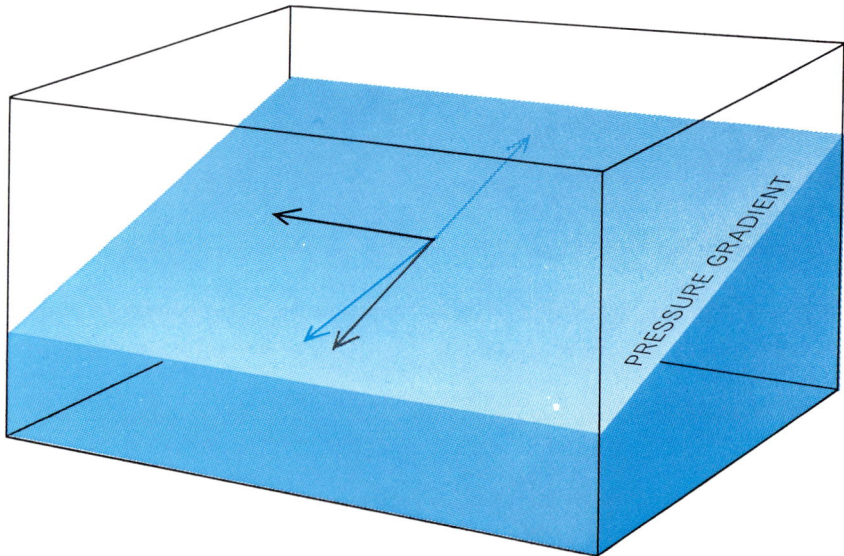

GEOSTROPHIC FLOW, which is westward here, results from a balance between the Coriolis effect (*light color*) and the pressure gradient (*gray*). Coriolis effect arises from the rotation of the earth, which is counterclockwise as seen from above the North Pole. Pressure gradient reflects slope of water. Because of the gradient, the water tends to flow south, but the Coriolis effect deflects it to the right (*dark color*), producing net westward flow.

etry of the model is close enough to the earth's so that the theory applicable to the ocean is also applicable to the model.

The fluid is held in the shape of a spherical shell by the two nested plastic blocks. As a result we cannot blow a wind over the free surface, because there is no free surface. We can achieve a velocity in the water, analogous to the velocity imparted to ocean water by wind, by rotating the lower block.

Our scheme for making the flow visible employs a solution of .04 percent thymol blue, which is often used as a pH indicator: it is yellow when the solution is acidic and brownish-blue when the solution is basic (alkaline). When two sets of electrodes are placed in the solution and an electric current is passed through the water from one electrode to the other, the solution becomes acidic near one of the electrodes and basic near

the other. If the solution used at the outset is acidic but almost at the point of becoming basic, the slight change of pH caused by the electric current will change the color of the solution near the basic electrode. The darkened fluid created in this way will move with the local velocity of the water, as can be seen on the cover of this issue and in several of the photographs accompanying this article.

For our model we made one of the electrodes a grid of wires that is located midway between the two plastic surfaces, that is, halfway down in the water. The second electrode, which is another grid of wires, is attached to the upper plastic block. With this system we can measure both longitudinal and latitudinal flow in any part of the basin at half the depth of the water. Therefore the patterns of flow that appear in the accompanying photographs represent the deep geostrophic flow in the ocean.

Our model is a segment of the surface of a sphere; if the entire sphere existed, it would have a radius of 53.5 centimeters. The lower block, which is our "wind," rotates typically at a speed of about .06 revolution per minute. The rotation of the entire model by the turntable, to simulate the rotation of the earth, is typically at a rate of about 60 revolutions per minute. At this rate the thickness of the Ekman layer is about one millimeter, which is in fact small

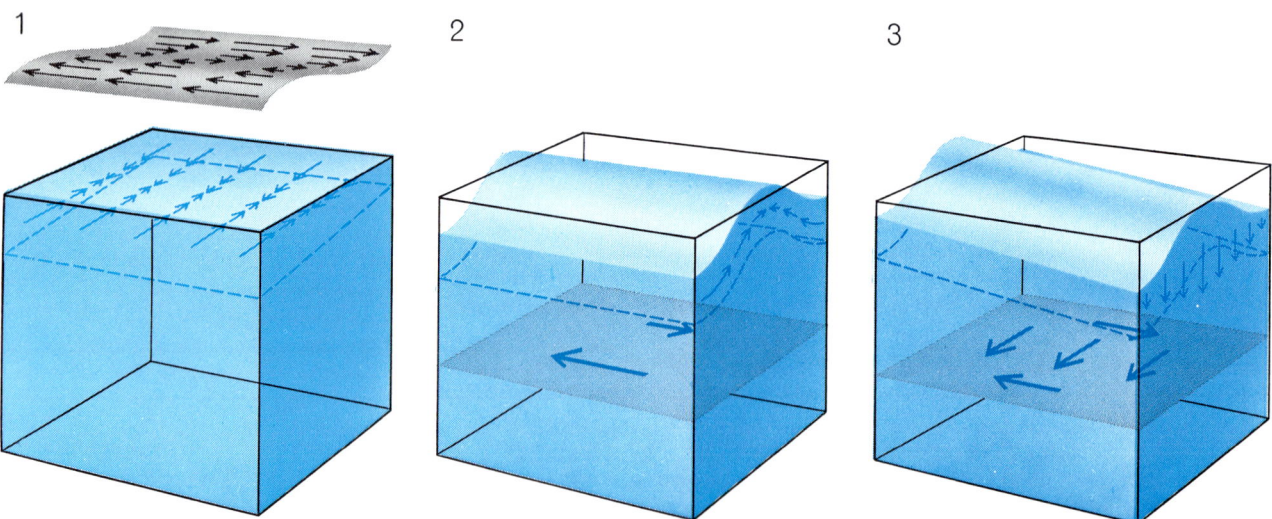

EFFECTS OF WIND are portrayed in a square block of water representing the Atlantic Ocean between about 15 degrees and 55 degrees north latitude. The typical wind pattern (*1*) is easterly in the northern part of the basin and westerly in the southern part; intensity is indicated by arrows. The winds push the water in the same directions, but the water is deflected to the right by the Coriolis effect, so that in an upper layer known as the Ekman layer the water is transported to the north in the southern part of the basin and to the south in the northern part. Hence the water tends to pile up in the middle of the basin (*2*), producing a downward thrust that tends to make the water flow northward in the northern part of the basin and southward in the southern part. The average deep flow, however, is to the east and west respectively, because of the Coriolis effect. Since the average deep flow in the southern part of the basin feels a weaker Coriolis force but the same downward thrust, more water is transported in that part of the basin than in the northern part. The net transport is therefore to the west, producing a slope downward from west to east (*3*). The resulting eastward flow is deflected to its right or southward by the Coriolis effect, so that the net geostrophic flow in the basin is toward the south.

compared with the total depth of the water (one centimeter).

Results from the model have been highly encouraging. By rotating the lower block very slowly we have been able to produce a circulation that is predominantly frictional, that is, the inertial effects are negligible compared with the frictional ones. Hence we have a basis for testing the frictional theory of oceanic circulation. Theory and experiment have shown good agreement in this range. One can see in the photograph on the cover, for example, the major features of a typical gyre. The predicted deep southward drift is evident, and on the left side appears the mass-conserving western boundary current or "Gulf Stream," which gains the required positive vorticity by rubbing on the bottom.

It is also possible to study more than isolated gyres with this model. On the second day of experimentation, when we adjusted the latitude so that the equator went through the lower half of the basin, a current appeared at the equator and moved in the direction opposite to the velocity of the rotating block. This was an equatorial undercurrent, a phenomenon that is also evident in the ocean. For example, the Cromwell Current in the Pacific moves eastward along the Equator below the surface in the direction opposite to the surface velocity.

Still, we must be careful here not to draw unjustified analogies or conclusions. The circulation we have achieved is dominated by friction and is a first step toward the ultimate model, where friction will not predominate. It is certain now that the Gulf Stream is not dominated by friction, and it is probable that equatorial currents are not either. Moreover, the stratification of the ocean plays an important role in the dynamics of equatorial currents. For these reasons the physics of equatorial undercurrents in our model that look like real ocean currents must be studied carefully before their relevance is established.

Even if our currents are not directly relevant to the earth, we gain a helpful insight into general equatorial dynamics. We have observed that in the model the equatorial undercurrent is always in the direction opposite to the velocity of the block, that is, the current can flow either east or west [see bottom illustration at left]. As the forcing velocity is increased, however, the inertial efforts become more important and we find that the behavior of our currents is not so symmetrical. As the inertial effects increase, the undercurrent that flows westward breaks

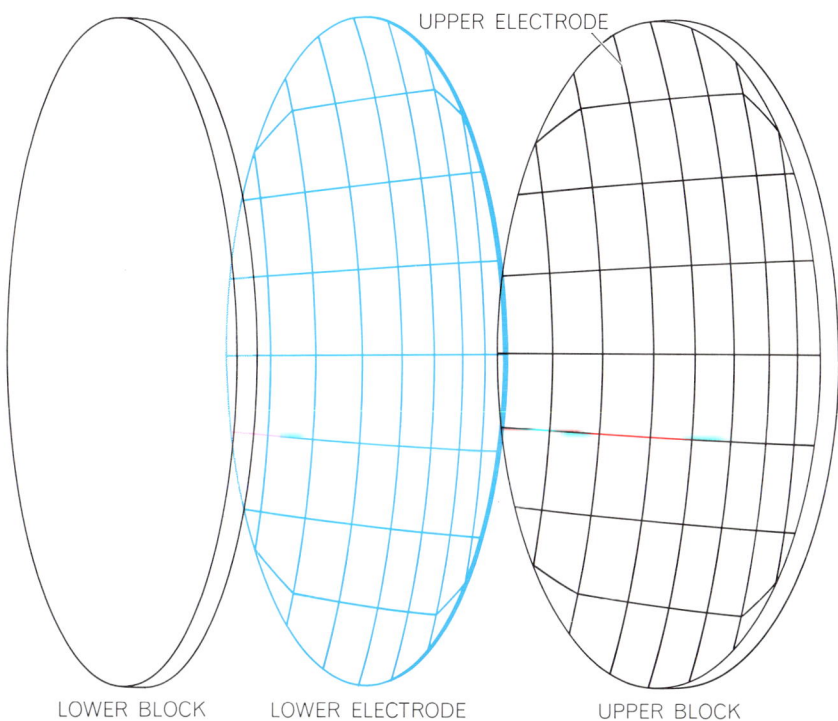

ELECTRODE SYSTEM for the model consists of two wire grids, one in the top plastic block and one halfway down in the water. Electric current in the grids darkens the color of the water, so that patterns of flow in the water can be seen and also readily photographed.

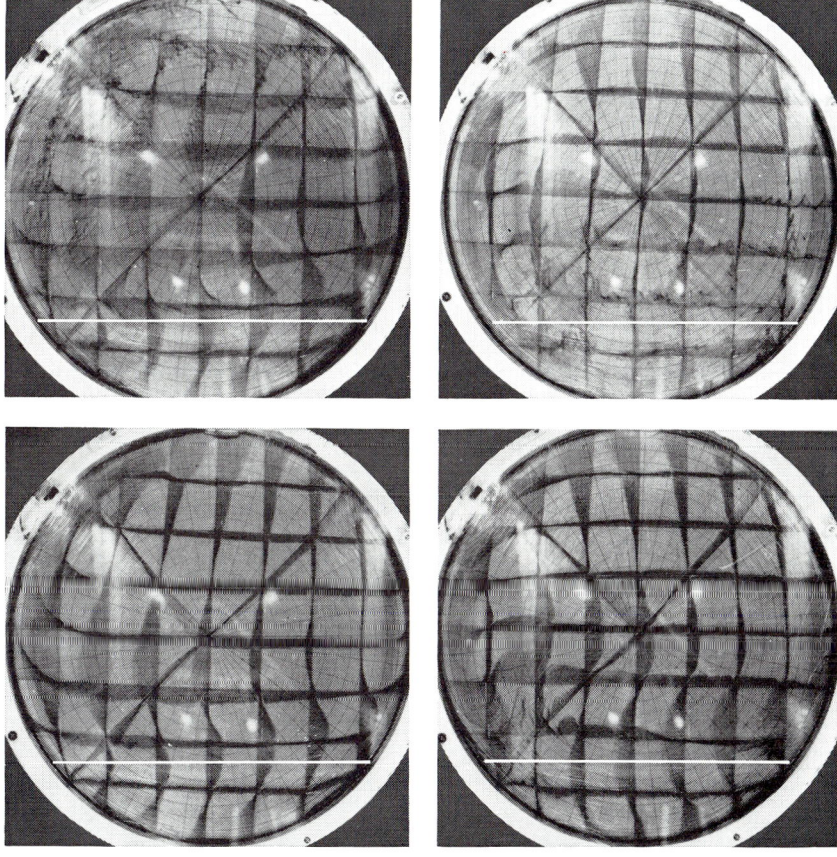

EQUATORIAL FLOW changes as inertial effects increase. The photographs at bottom have the minimum inertial effects and the ones at top have the maximum inertial effects. In the photographs at left the driving block was rotating clockwise; in the photographs at right, counterclockwise. The white lines show the equator. In each case the undercurrent at the equator moves in a direction opposite to the direction of the model's forcing velocity.

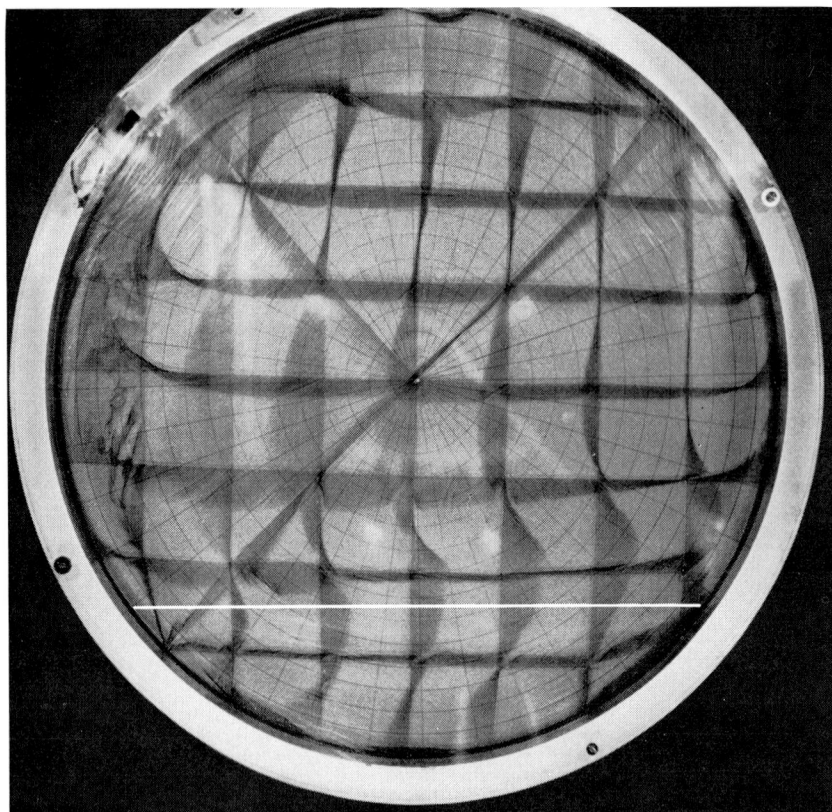

EASTERN BOUNDARY CURRENT, rather weak and flowing northward, gives evidence of inertial effects as the driving block is rotated more rapidly. A western boundary current, a southward interior flow and an equatorial undercurrent also appear. The white horizontal line represents the position of the equator; center of basin is 13 degrees north latitude.

up and disappears. At the moment it is not clear whether the current is flowing too fast to be clearly visualized by our technique or whether it becomes unstable. We plan to use more sophisticated methods of visualizing and measuring flow in order to clarify the matter.

It is particularly interesting that the eastward equatorial undercurrent remains stable in the model. Perhaps the explanation of the difference between the eastward and the westward current when the model is in the inertial range lies in the Coriolis effect. If the eastward current deviates to the north of the equator, the Coriolis effect, which is to the right, forces it back toward the equator. In the case of a westward current an instability could arise from the fact that if the current deviates slightly to the north, the Coriolis effect turns it farther to the north, and if the current deviates to the south, the Coriolis effect turns it farther in that direction.

What happens to equatorial undercurrents on the earth? In the Pacific and the Atlantic the equatorial undercurrent flows eastward and appears to be stable. In the Indian Ocean, however, the winds reverse direction every six months. Observations there show that an undercurrent exists during only one season; it is an eastward current. During the other months the equatorial flow is generally weak and variable and possibly unstable. We are studying our model carefully to see what light can be shed on the apparent instability of the westward undercurrent.

It is highly encouraging to see that several basic features of oceanic circulation appear in our model. Nonetheless, we still do not have a model of a fully inertial oceanic circulation. We have seen several qualitative changes in the circulation, however, as we have stepped up the velocity in order to increase the inertial effects. Three effects that can be attributed to inertia have appeared.

The first one was the migration of the center of the gyre toward the northeast. Such a migration is to be expected, because as the inertia of the water becomes important the fastest part of the western boundary current tends to move farther north. The center of the gyre thus is also shifted.

The second inertial effect that we have observed is the appearance of a relatively weak northward boundary current on the east coast [see top illustration on this page]. Although eastern boundary currents have been observed in the ocean and are probably inertial, the dynamics of such currents are poorly understood. The model apparently provides a framework for studying them.

An unsteadiness of the circulation is the third major inertial effect observed. From motion pictures we found that the pattern of flow oscillated irregularly when the speed of the rotating bottom block was high enough [see illustration below]. We found the estimated period to be in rough agreement with the free-oscillation period predicted by theory. A clearer picture of the oscillatory nature of this rapid, inertial flow will be possible with new procedures that we are now developing for the model.

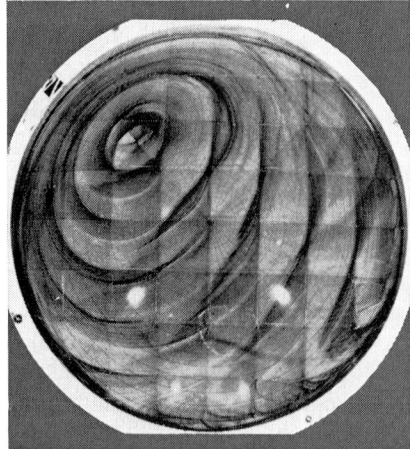
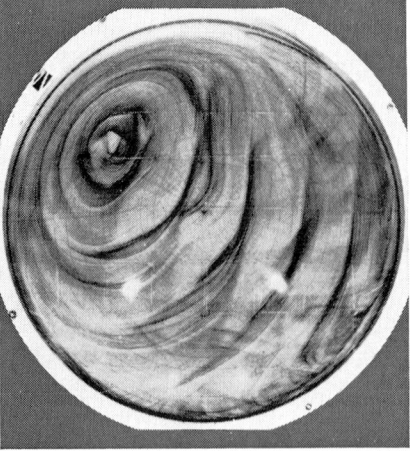

EFFECTS OF INERTIA appear when the driving block is rotated at high speed. The photograph at right was made .3 second after the one at left. In that time the configuration of the gyre, which is the closed and basically circular pattern of flow, has changed appreciably.

The Author

D. JAMES BAKER, JR. is assistant professor of oceanography at Harvard University. Following his graduation from Stanford University in 1958 he did graduate work in soft-X-ray physics at Cornell University, taking his Ph.D. there in 1962. He spent the next year measuring the equatorial current in the Indian Ocean as a postdoctoral fellow with the University of Rhode Island. The following year he studied photosynthesis at the University of California at Berkeley, beginning his association with Harvard in 1964 as a research fellow. His avocations include photography, travel and playing ragtime piano.

Bibliography

WIND-DRIVEN OCEAN CIRCULATION. Edited by Allan R. Robinson. Blaisdell Publishing Company, 1963.

A LABORATORY MODEL FOR THE GENERAL OCEAN CIRCULATION. D. J. Baker, Jr., and A. R. Robinson in *Philosophical Transactions of the Royal Society of London, Series A*, Vol. 263, in press.

SCIENTIFIC AMERICAN February 1970, Vol. 222, No. 2, pp. 32–40

OFFPRINT **891**

THE AFAR TRIANGLE

by Haroun Tazieff

A new ocean seems to be in the making where three great rift systems intersect near the junction of the Red Sea and the Gulf of Aden. The violence of the process has created a nightmarish desert landscape.

In the northeastern part of Ethiopia, at the juncture of the Red Sea and the Gulf of Aden, lies a region known as the Afar triangle. It is a wild and rugged country, featured by below-sea-level deserts, towering escarpments, fissures, volcanoes and craters. Few men have explored the region, and until recently it was terra incognita as far as its geology and even its exact geography were concerned. Now, however, the discovery of detailed evidence for the drift of continents and the growth of oceans has focused considerable interest on the Afar triangle [see "The Origin of the Oceans," by Sir Edward Bullard; SCIENTIFIC AMERICAN Offprint 880]. The triangle seems to be a focal point for new oceans in the making. What is more, whereas elsewhere the process that is producing continental separation is hidden in the depths of the ocean, here we can see it taking place in direct view on dry land.

The theory that the earth's present continents were once united in a great land mass and have gradually drifted apart is now generally accepted. Evidence collected over the past 10 years, mainly by exploration of the worldwide system of ridges running along the middle of the oceans, has outlined a convincing picture of how the continents were separated [see "The Confirmation of Continental Drift," by Patrick M. Hurley; SCIENTIFIC AMERICAN Offprint 874]. In brief, the process seems to be as follows. The material of the continents is a layer of comparatively light sialic rock, resting on a denser basaltic magma underneath. Stresses in the earth's crust may crack the sialic layer, producing faults and fissures that can be as much as 20 meters wide. Then molten magma wells up into the fissure, sometimes spilling out over the surface. The magma hardens into solid rock, thus holding apart the separated sialic blocks. Over long periods of time the same stresses create new fissures parallel to the old ones; these fissures too are filled with magma. Examination of the oceanic ridge has shown that it is composed of parallel strips of hardened magma, indicating that the crust was repeatedly fissured along the axis of the ridge and that the continental blocks thus moved farther and farther apart. The upwelled strips of basalt are distinguishable from one another by differences in the direction of their magnetic polarity, which

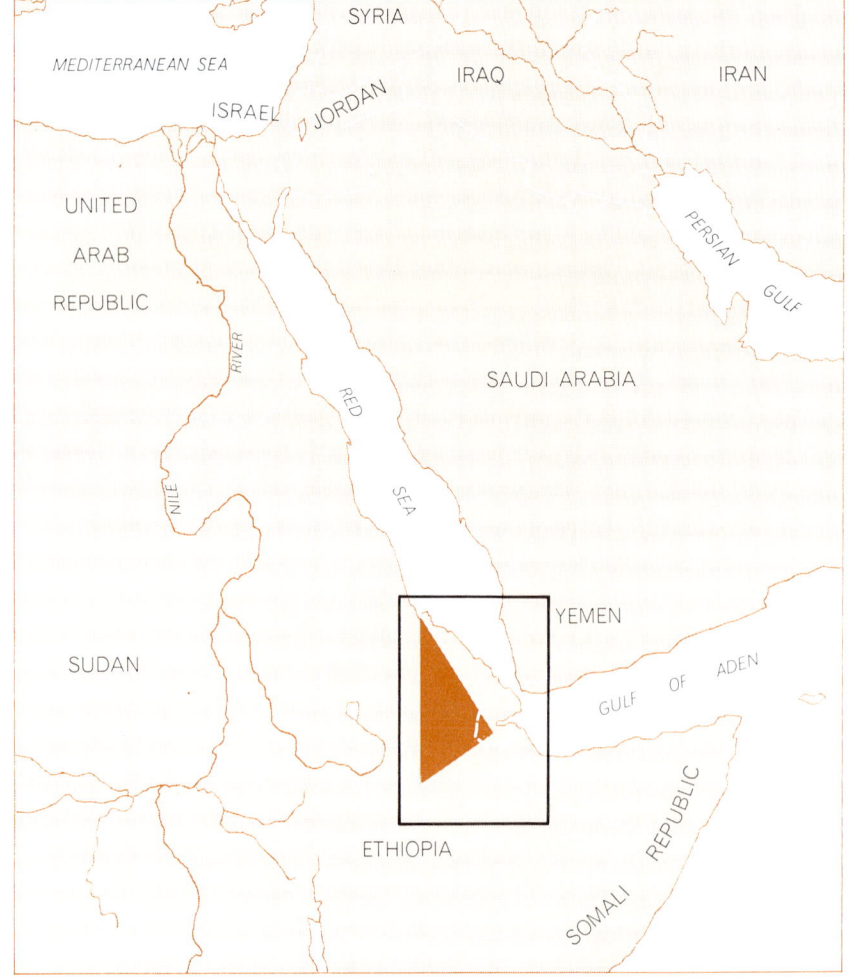

AFAR TRIANGLE (*color*) is in northeastern Ethiopia where the Red Sea rift, the Carlsberg Ridge of the Indian Ocean and the Rift Valley system of East Africa meet. Its northern section was once submerged, and part is still below sea level (*map, pages 812–13*).

MASSIVE FAULT BLOCKS characterize the terrain of the Afar triangle. The aerial photograph above shows a region near 12 degrees north latitude, where faulting has caused huge segments of surface to subside below the level of the surrounding plateau. The depressed structures are whitened in places by salt or gypsum.

STAIRSTEP LANDSCAPE of the eastern shore of Lake Giulietti (below) is further evidence of the extensive downfaulting that has shaped much of the Afar triangle. Such long, depressed blocks, bordered by fault zones, are called graben structures by geologists; the graben structures here are among the world's most spectacular.

reflect changes in the earth's magnetic field with time.

The overall conclusion from studies of the mid-oceanic ridges is that the continents have been moving apart at an average rate of a few centimeters per year. In the south Atlantic, for instance, it appears that the ocean has been widening at the rate of 2.5 centimeters per year, which indicates that the South American continent began to separate from Africa approximately 180 million to 200 million years ago.

Recent worldwide explorations of the ocean bottoms have shown that a rise running along the middle of the Indian Ocean, known as the Carlsberg Ridge, has a branch extending into the Gulf of Aden. The Red Sea bottom similarly has an axial ridge with physical properties like those of the oceanic ridges. The Gulf of Aden and Red Sea rifts, which are perpendicular to each other, meet in the Afar triangle. And the same region also lies at the northern end of the system of rift valleys that runs down the eastern side of the African continent. Once these facts were recognized, it was suddenly realized that the largely uncharted Afar triangle might offer an extraordinary opportunity for investingating the origin of new oceans.

The Afar triangle is one of the world's most forbidding regions. In addition to the fact that its terrain is all but impassable, the area is extremely hot; we were to find that the temperature rises to as high as 134 degrees Fahrenheit in the shade in summer and 123 degrees in winter. The region is inhabited only by nomadic tribes of fierce repute; the young warriors are said to mutilate male victims to offer trophies to their women, and they have been known to massacre armed parties for their weapons. Several exploring parties in the 19th century were slaughtered by the tribesmen.

Of the expeditions that had explored the Afar triangle before we began to survey it in 1967, the best-known was one carried out in the spring of 1928 by two Italians, Tullio Pastori and G. Rosina, a British mining engineer, L. M. Nesbitt, and half a dozen Ethiopians. The expedition's leader was Pastori, a hardy ore prospector who earlier had explored various parts of the region (and who, at the age of 60, in 1943 escaped from a British prisoner-of-war camp in Kenya and made his way on foot all the way to Alexandria on the Mediterranean coast). Nesbitt, who wrote a report on the 1928 expedition, vividly described the difficulties and dangers the expedi-

tion had encountered. It was apparently the first journey along the entire length of the Afar triangle, and Nesbitt's information provided a basis for the only detailed maps of the region that were available before our expeditions. The 1928 party was not equipped, how-

ever, to undertake a geological survey.

In 1967 we organized a team of specialists to study the geology of the Afar triangle with all possible thoroughness. Our group includes investigators from several universities in Italy, France and the U.S. It consists of three petrologists

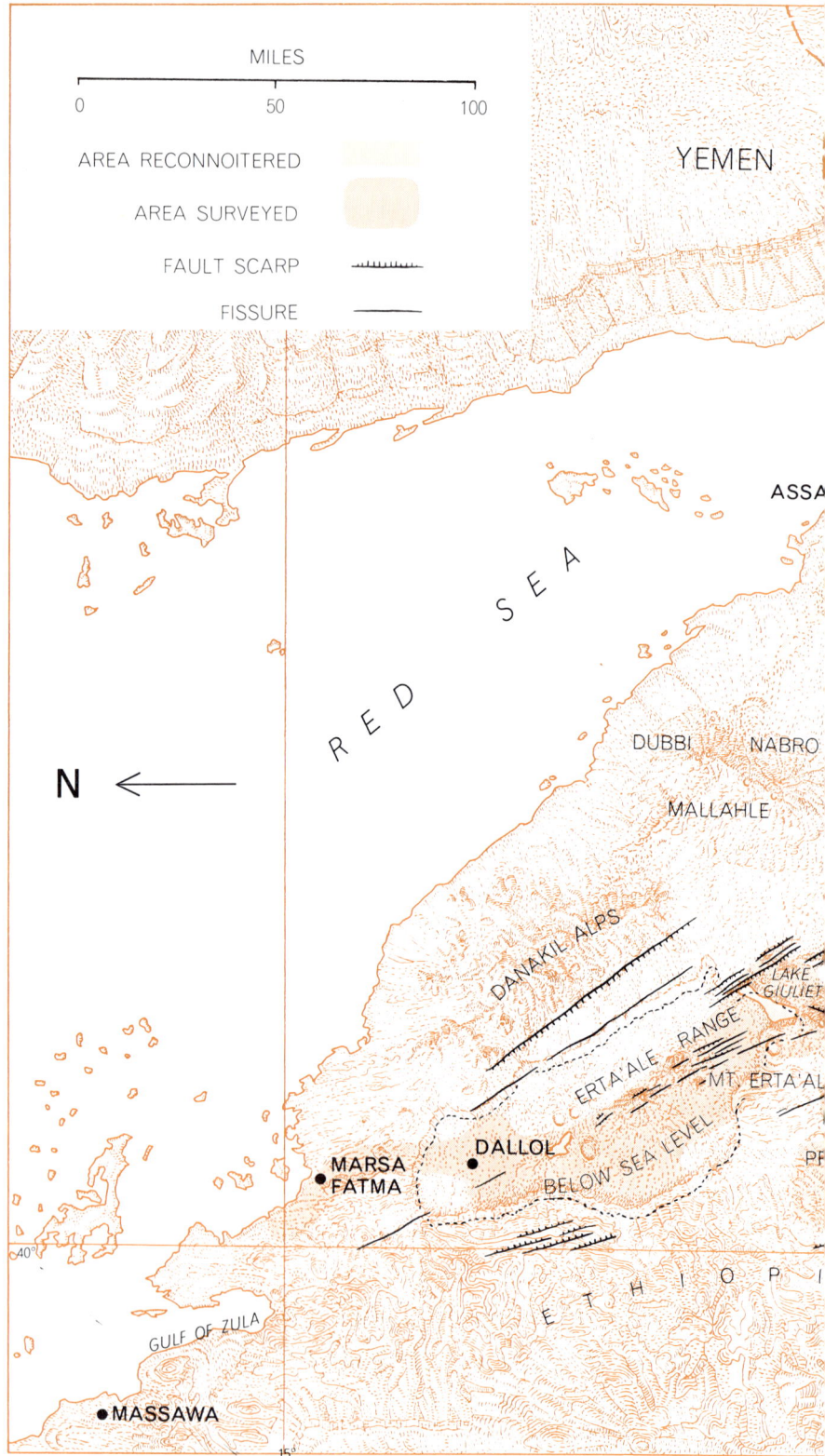

GEOLOGICAL SURVEY of the Afar triangle was conducted by the author and his associates beginning in 1967. In this map north is at the left and east at the top. The most inten-

(Giorgio Marinelli, Franco Barberi and Jacques Varet), four geochemists (Giorgio Ferrara, Sergio Borsi, J. L. Cheminée and Marino Martini), one tectonic geologist (Gaetano Giglia), two students of the geology of recent times (Hugues Faure and Colette Roubet), a geophysi-cist (Guy Bonnet), an oceanographer (Enrico Bonatti of the University of Miami) and a volcanologist (myself).

We have now completed three expeditions to the Afar triangle, all during the comparitively cool winter season (in 1967–1968, 1968–1969 and 1969–1970). Our party was unarmed and was not molested by the local people. For our explorations we have had the invaluable help of a helicopter, without which it would simply have been impossible to do serious, comprehensive fieldwork in that rugged country. So far we have

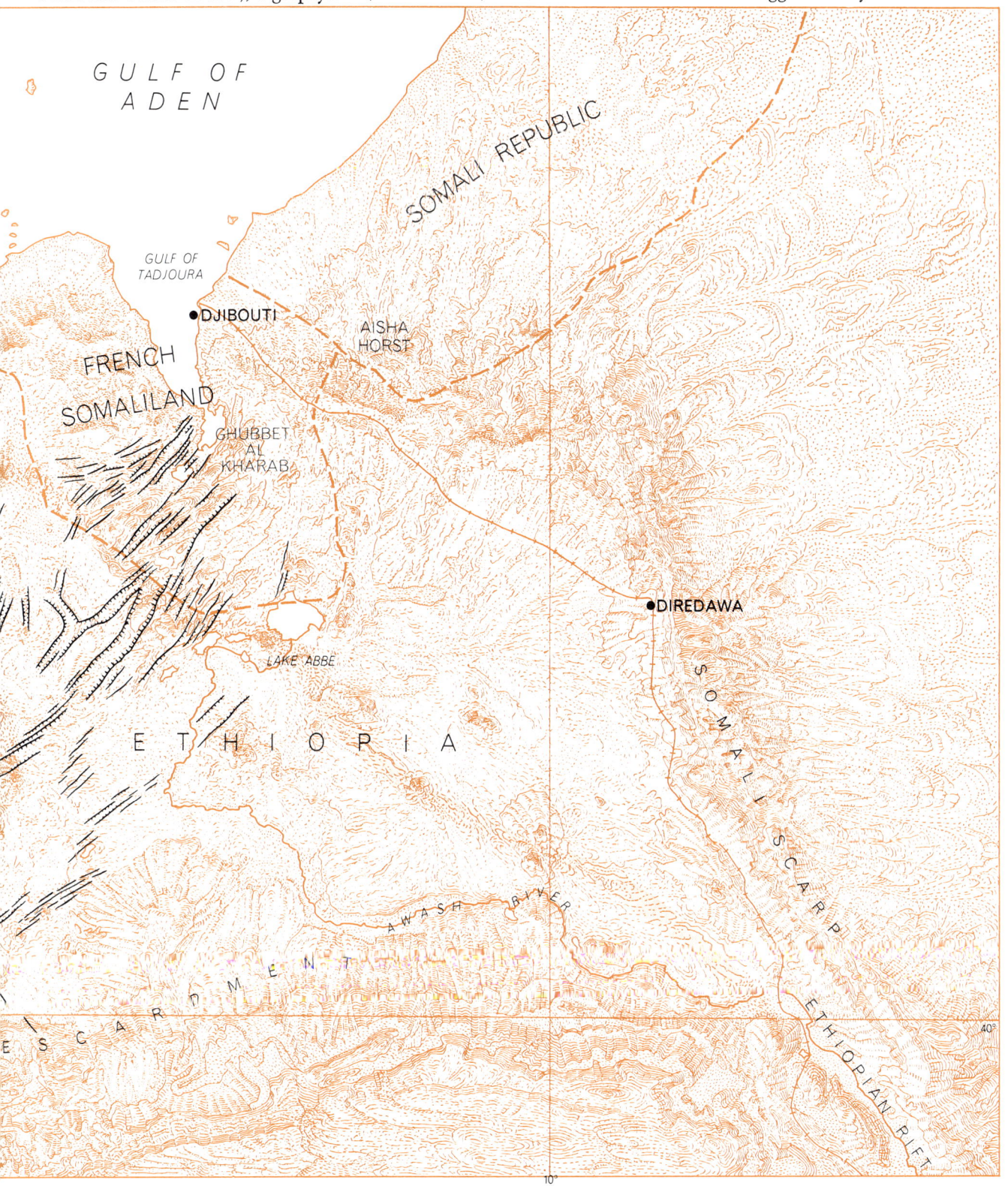

sive work was done in and near the area below sea level, between the Danakil Alps and the Ethiopian escarpment. In addition the group scouted widely by helicopter, landing more than 1,000 times to gather rock samples for later analysis and to study formations.

FLAT-TOPPED VOLCANO, Mount Asmara is composed of shards of volcanic glass such as are formed during underwater volcanic explosions. It resembles the numerous guyots, or submerged oceanic mountains, whose level summits are usually attributed to wave erosion. Because Mount Asmara was formed under water, it may be that a flat top is instead a feature common to all such volcanoes.

CINDER CONE near Lake Giulietti was built up at a point that overlies one of the innumerable fault lines found in the Afar triangle. A subsequent horizontal shift of one fault block has moved the far half of the cinder cone 100 meters ahead of the near half.

spent a total of 13 weeks in the field and have carefully mapped the geological structures and petrology of some 12,000 square miles in the northern half of the triangle. We have crisscrossed this entire area with the helicopter and have landed at more than 1,000 places to sample rocks and examine the tectonic structures on the ground.

On its northeast side the Afar triangle is separated from the Red Sea by a series of heights produced by deformations of the earth crust at faults and by volcanic action. Reading from north to south, these include a tectonic horst (upraised block) called the Danakil Alps, a mountain range formed by sedimentary and intrusive rocks, a volcanic massif composed of an active volcano and three volcanic piles crowned with calderas (craters of collapsed volcanoes) and, at the southern end, another mountainous horst. On the south side the triangle is bounded by the tall Somali scarp, from 4,900 to 6,500 feet high. On the west, the third side of the triangle, stands the huge Ethiopian escarpment, towering in some places to more than 13,000 feet. Here clifftops stand higher above the valley floor below them than anywhere else in the world. As for the floor of the triangle itself, it rises gently from about 400 feet below sea level near its northern end to more than 3,300 feet above sea level at the Somali end some 300 miles south.

What is one to make of this strange and spectacular landscape? Until very recently most geologists believed the Afar triangle was a funnel-shaped widening, produced in some unexplained manner, of the Great Rift Valley of East Africa. Our studies of the region's geology have led us to a completely different interpretation. The facts, as we have observed them on the scene, indicate that the floor of the triangle is actually a part of the Red Sea! In fact, the triangle from its northern apex down to the Ghubbet al Kharab at the western end of the Gulf of Aden and the Gulf of Tadjoura is a southwestern continuation of the central trough of the Red Sea that fades out close to 15 degrees south latitude. The tectonic trends of the Red Sea are evident and are geologically active throughout the area. There are none of the trans-rift structures or other formations that were once believed to account for the Afar triangle. It appears that as recently as some tens of thousands of years ago at least the northern half of this area was covered with seawater, with only the Danakil Alps and the high volcanoes standing above water as islands, and that most of the land has since been raised above sea level by tectonic uplifts through earthquakes, volcanic action and the rise of basaltic magma that filled the fissures.

The evidence that the Afar triangle is a part of the Red Sea floor, not some bizarre widening of the Ethiopian rift, can be seen on every hand. To begin with, we found that the observable facts contradicted other explanations of the region's topography. It had been suggested, for example, that the lofty escarpment on the west side of the triangle was produced by a downfolding of the high plateau, followed by erosion of the resulting hillside. Our observations turned up three important objections to this idea: (1) the blocks of faulted crust in the lower part of the escarpment are tilted westward—in the direction opposite to what would be expected in a downfold; (2) the supposed erosion should have deposited a vast amount of sediment in the triangle's closed basin, but almost none was found there; (3) the basin is filled with more than 3,000 feet of evaporites (salt formations deposited by the evaporation of seawater), and the basin's wall plunged down below this material at a steep angle, again indicating that the western boundary of the triangle was formed by slippage of the crust along a fault rather than by downwarping.

It had also been suggested that south of the Danakil Alps a belt of big calderas, apparently running north and south, was a continuation of the main Ethiopian rift and was a major active feature of the entire region. Field investigations show that no north-south belt exists, and that all the big calderas are located on a graben (a depressed section of crust) running north-northeast and south-southwest. This observation is an important one, because it again demonstrates that the Afar triangle is a part of the Red Sea and not an extension of the Ethiopian rift.

We found innumerable signs that the topography of the Afar triangle has been created by violent events that have occurred in very recent times, geologically speaking, and are still in progress. The entire northern half of the triangle shows clear evidence of extensive faulting of the crust and active crustal movement along the faults. North of latitude 13 degrees 10 minutes north all these faults are aligned along the axis of the triangle in the north-northwest to south-southeast direction [see illustration on pages 812–13]. South of 13 degrees 10 minutes north down to the Ghubbet al Kharab (11 degrees 30 minutes north) the same direction prevails, although there are also many faults and fissures running northwest-southeast and east-west. Along much of this part of the triangle the evidences of crustal movement are in plain view as wide-open fissures, horsts and grabens that form a classic graben structure of steps down the sides of a major depression. Even where the graben structure is now hidden under deposits of volcanic material and evaporites (but is still detectable by geomagnetic mapping), the fault axis is shown visibly by potash salt domes, explosion craters, boiling springs and other signs of volcanic activity below the surface. These eruptions follow the same line in the north-northwest to south-southeast direction.

This direction is precisely that of the Red Sea, so that the whole of the northern part of the Afar triangle can be regarded as part of the sea. All the evidence suggests that the waters of the Red Sea extended south-southeastward as far as the Somali scarp in the geologically recent past and that the present absence of water in the Afar triangle is only a temporary phase in the development of the ocean.

That the fissuring and displacements of the crust are going on actively at the present time is shown by several signs we were able to observe. Here and there we found fresh faults cutting through very young structures, such as alluvial fans and cones of active volcanoes. The volcano called Erta'Ale, the most active in all East Africa, has its main cone sliced by tectonic fissures that are parallel to the Red Sea axis. This is an exceptional phenomenon; volcanic fissures are usually not parallel but radial. We ourselves actually witnessed a significant event during our study: an earthquake in the middle of the depression on March 26, 1969, produced an appreciable slippage of the crust down a fault in the north-northwest to south-southeast direction.

Further evidence that the Afar triangle is actually a scene of oceans in the making came from the examination of the rocks themselves. Our samples gave every indication that the triangle's central trough contains no sialic (that is, continental) rock. The rocks, all very young, of the Erta'Ale volcanic range, which runs parallel to the Red Sea central trough, are preponderantly basaltic and typical of the rocks of oceanic ridges. From analysis of more than 100

BLACK RIBBONS of fresh basalt in the Erta'Ale mountains mark zones where molten rock has poured out of fissures in the floor of the Afar triangle. The basalt is chemically similar to the magmas that have welled up from the rifts in the earth's mid-oceanic ridges.

LOW VOLCANIC CONE that marks the northern end of the Erta'-Ale mountain range is further evidence that much of the area was submerged in the recent geologic past. The shards of glass that make up the cone are formed during an underwater eruption as lumps of hot lava turn the water about them into superheated steam. The steam explodes the lumps into hundreds of fragments.

specimens from this range we estimate that its composition is more than 90 percent basalts, the rest consisting of varieties of volcanic rock: dark trachytes (8 percent) and rhyolites (.5 percent). Chemically these rocks all show an evolutionary relationship, forming a practically continuous series from olivine basalts to the dark trachytes and rhyolites. The relationship has been confirmed by analysis of their strontium; the ratio of the rare isotope strontium 87 to the common one, strontium 88, in all these rocks is uniform and about the same as that in oceanic basalts. Another index also shows clearly that they were derived from oceanic magmas. The differentiation into the rock varieties in the series apparently came about mainly through gravity separation of components as the original magmas evolved in a series of steps, marked by a distinct iron enrichment in the middle stages. The end products are glassy trachytes and rhyolites, giving evidence that they were produced in a highly fluid state.

We found that a parallel volcanic range in the triangle, the Alayta range, has the same character, structurally and petrographically, as Erta'Ale. To sum up, the petrology of both ranges seems to indicate that they were born as upwellings through fissures in the crust, that their parent material was basaltic magma such as is characteristic of ocean floors, and hence that there is no sialic crust immediately under these ranges. All of this suggests that in the northern part of the Afar triangle continental blocks have already been severed from each other. Evidently this area is a part of the Red Sea ocean-forming system that is separating Arabia from the African continent. Similarly, our survey of the region indicates that the Danakil Alps and two smaller horsts we have detected south of latitude 13 degrees (one close to the Ethiopian escarpment and the other in the middle of the Afar triangle) are continental structures in the process of being split off and separated from the Ethiopian plateau.

Many signs that the northern region of the Afar triangle was covered with seawater in quite recent times turned up in our on-the-ground explorations. On a terrace at the foot of the Ethiopian scarp Faure found a stone axe that was encrusted with seashells, indicating that the sea had covered it after it was abandoned. This axe is from the Acheulean period, not more than 200,000 years ago. We found coral reefs of the Quaternary period (geologically the most recent) in the lava fields of the region. Scattered

about on the floor of the triangle are ash rings, resembling the one at Diamond Head in Hawaii, such as are known to be formed under water; these consist entirely of the shards of volcanic glass called hyaloclastites, which are typical of underwater basaltic eruptions. We even discovered a flat-topped cone, now standing on the dry beds of the triangle, that bears every resemblance to the famous guyots, or flat-topped seamounts, of ocean floors.

The ash rings were particularly interesting to me because I had previously witnessed the formation of two such structures from submarine eruptions, one in the Azores in 1957 and another south of Iceland in 1965. I found that the ash ring is formed as the result of secondary steam explosions from a volcanic eruption under water. The primary eruption tears the molten magma into pieces and hurls them into the water. Because of their large surface-to-volume ratio (a ratio much larger than it would be for a quiet lava flow) the lumps of hot lava transfer enough heat to the water im-

mediately around them to turn it into superheated steam. This steam generates secondary explosions that shatter the lumps into tiny pieces of glass (hyaloclastites) and throw them high into the air—from half a kilometer to more than a kilometer above sea level (which is a great deal higher than material is tossed by the usual volcanic explosions of the Hawaiian or Strombolian type on land). The tiny fragments falling back around the volcano's vent form a large rim that is frequently horizontal and highly regular, because the fragments are deposited under water. We found that the Afar ash rings looked very fresh and the most eroded one (presumably the oldest) was overlain with corals and shells of marine animals of the Pleistocene period, all of which adds to the evidence that the region was covered by the sea not long ago.

Our examination of the apparent guyot we found in the depression seems to cast a new light on the origin of seamounts. Hundreds of these flat-topped submarine mountains, with their tops in

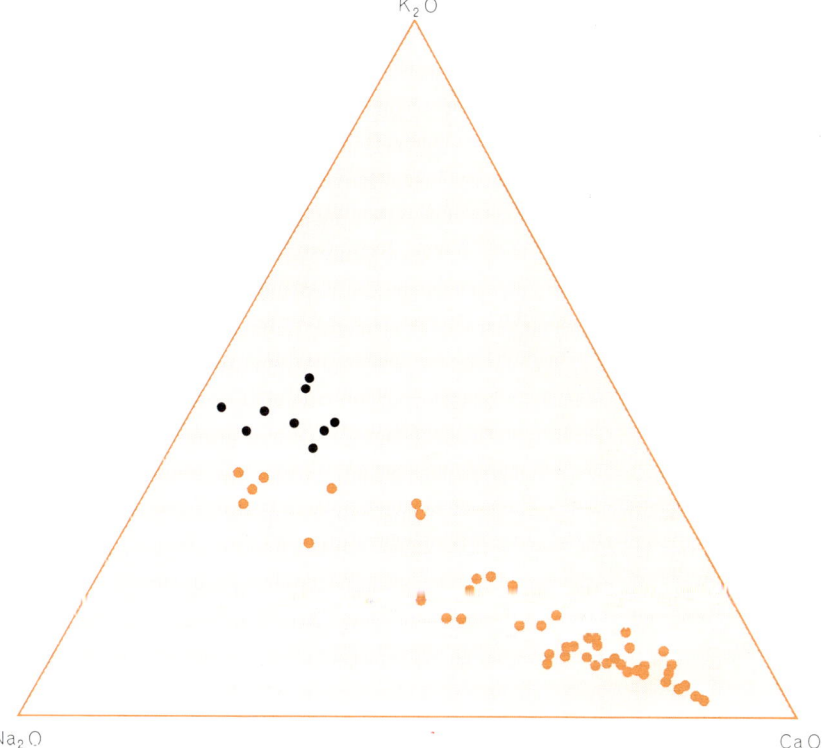

DIFFERENTIATION DIAGRAM shows that the chemical composition of volcanic rock from Mount Erta'Ale (*colored dots*), midway between the Ethiopian escarpment and the Danakil Alps, is different from the composition of the rock from the Pierre Pruvost volcano (*black dots*), which lies close to the escarpment. The Pruvost lavas result either from a "contamination" of deep magmas by the sialic material in the earth's crust or from a melting of the crust itself. The lack of such contamination in the Erta'Ale lavas suggests that, as the Alps and the escarpment have drifted apart, they have split open the earth's crust along a vast rift zone, with the result that no crust remains below the Erta'Ale range.

some cases as much as 1,000 fathoms or more below the sea surface, have been discovered in the world ocean. To explain their flat tops, it is generally supposed that the peaks formerly stood above the surface of the water and were eroded flat by the waves, and that the tops are now submerged because the sea bottom sank or the ocean level rose. We found that the truncated cone resembling a seamount in the Afar triangle was built in a way that would account for its flat top simply by the method of its construction. This pile, called Mount Asmara, is about 1,200 feet high and tapers from a width of a mile and a quarter at the base to two-thirds of a mile at the top. It was apparently formed by a buildup of layer on layer of beautiful golden hyaloclastites, which consist of palagonitized olivine basalt and are clearly of submarine origin. Such a process, produced by successive eruptions from a volcanic vent, can account for Mount Asmara's flat top. I am wondering if many of the seamounts in the oceans may not have been built in the same way under water. Perhaps their tops never have in fact been above water but may someday rise out of the ocean as the building process goes on. At all events, it appears that Mount Asmara in the not distant past was totally submerged, with its base more than 1,600 feet below sea level.

If we suppose that the Afar triangle is a part of the Red Sea, the coast of Arabia matches very well the contour of the "coastline" of the part of the African continent from which it is assumed to be separating; the match is at least as good as that between Africa and South America on opposite sides of the Atlantic.

We still have to explain why the axis of the Afar rift is displaced somewhat westward from that of the central ridge in the Red Sea and why the two troughs are now separated. That question will have to await our further explorations in the southern half of the depression. It seems possible that the Gulf of Aden ridge and rift system, thrusting into the depression at right angles to the Red Sea axis, may be exerting a powerful influence that could account for the displacement.

Meanwhile the information obtained so far about the Afar triangle raises an interesting economic question. Because of the absence of sialic crust below the axes of active volcanic ranges of basaltic composition and the probable closeness of the hot mantle to the surface in the northern part of the triangle, a great deal of heat flows into the underground rock strata there. These strata are highly porous and absorb a vast amount of fresh water that drains into the floor of the triangle from the surrounding highlands in the rainy season. Consequently it seems likely that subterranean fields of superheated water and steam underlie parts of this desert region where impermeable strata prevent them from escaping into the atmosphere. If they could be tapped, they might supply millions of kilowatt-hours of cheap electricity per year. This reservoir could supply power to the nearby seaports (Assab, Massawa, Djibouti) to support large new industries (aluminum and other metallurgies, petrochemistry, fertilizers, canneries) in which electricity is the main cost factor. With ores and other raw materials shipped to these ports at low cost, the price of finished products would also be low, and the area could be expected to have a tremendous economic growth. This almost desert region might be an industrial megalopolis in the future—a future far less remote than the geologic one, when the Gulf of Aden and the Red Sea will have expanded into new oceans.

SALT PLAIN lies near the northern apex of the Afar triangle. Each year, when rainfall in the highlands to the east and west drains into this low area, the plain is covered by pools of brine. These are too shallow, however, to hinder the passage of salt caravans.

The Author

HAROUN TAZIEFF is with the French Center for National Research. "My childhood was devoted to the idea of becoming a sailor and an arctic explorer," he writes. "Eventually I became a geologist and spent 99 percent of my field time between the Tropics. Once a geologist, I dreamed of high-mountain tectonics and of central Asia; I was confined to old cratons and Africa or Southeast Asia and western America." Tazieff was born in Warsaw and educated in Belgium; he was graduated from the Free University of Brussels in 1932 and has degrees in agronomy and geology from the State Agronomic Institute at Gembloux and the University of Liège respectively. "Besides my interest in investigating the mechanisms of volcanic eruptions and of rift genesis," he writes, "I am fond of Van Gogh's, Gauguin's and some others' painting, of rugby football (by now I am one of the oldest forwards in the world, I think), of knowledge theories, of Stephane Mallarmé's and Robert Vivier's poetry, of mountaineering, of international politics, of the geology of the moon, of architecture and of many other things."

Bibliography

A STUDY OF TECHNOLOGY ASSESSMENT: REPORT OF THE COMMITTEE ON PUBLIC ENGINEERING POLICY, NATIONAL ACADEMY OF ENGINEERING. Committee on Science and Astronautics, U.S. House of Representatives. U.S. Government Printing Office, July, 1969.

TECHNOLOGY: PROCESSES OF ASSESSMENT AND CHOICE. REPORT OF THE NATIONAL ACADEMY OF SCIENCES. Committee on Science and Astronautics, U.S. House of Representatives. U.S. Government Printing Office, July, 1969.

FORECASTS OF SOME TECHNOLOGICAL AND SCIENTIFIC DEVELOPMENTS AND THEIR SOCIETAL CONSEQUENCES. Theodore J. Gordon and Robert H. Ament. The Institute for the Future Report R-6; September, 1969.

SCIENTIFIC
AMERICAN   October 1970, Vol. 223, No. 4, pp. 30–41   OFFPRINT **892**

THE BREAKUP OF PANGAEA

by Robert S. Dietz and John C. Holden

Pangaea is the single land mass that is believed to have given
rise to the present continents. Its outline has now been plotted
and its further disruption has been projected into the future.

The history of science is replete with outrageous hypotheses. They are mostly forgotten, as best they should be, but from time to time one of them turns out to be true. So it was with the concept that the earth is a sphere spinning in space, supported by nothing at all. Now it also seems to be with the theory of continental drift, which in its extreme form holds that all the continents were once joined in a single great land mass. Named Pangaea, this universal continent was somehow disrupted, and its fragments—the continents of today—eventually drifted to their present locations.

Over the past three years geologists and geophysicists have been forced to abandon the old dogma that the crust of the earth is essentially fixed and to accept the new heresy that it is quite mobile. The notion that continents can drift thousands of kilometers in a few hundred million years is now generally accepted. Geology therefore finds itself in much the same position that astronomy was in at the time of Copernicus and Galileo. Textbooks are being rewritten to embrace the new mobilistic viewpoint.

Although the theory of continental drift has triumphed, many of its details remain uncertain. Advocates of drift are challenged to say exactly how the present continents fitted together to form Pangaea, or alternatively to reconstruct the two later supercontinents Laurasia and Gondwana, which some theorists prefer to a single all-embracing land mass. The original concept of Pangaea ("all lands") was proposed in the 1920's by Alfred Wegener. Most attempts to improve on his reconstruction have been rather generalized sketches showing how the continents might have been joined. A few workers have made jigsaw fits with considerable care but without taking advantage of the latest concepts in geotectonics. Recently British theorists have presented detailed reconstructions showing how land masses were juxtaposed before the opening of either the Atlantic or the Indian Ocean, but their solutions show only the relative motions of the masses involved.

In this article we present a reconstruction of Pangaea in which the continents are assembled with cartographic precision. For the first time Pangaea is positioned on the globe in absolute coordinates. This reconstruction is accompanied by four maps that show the breakup and subsequent dispersion of the continents by the end of the four major geologic periods covering the past 180 million years: the Triassic, the Jurassic, the Cretaceous and the Cenozoic.

The guiding rationale for our reconstruction is the drift mechanism associated with plate tectonics and seafloor spreading [see illustration on page 4]. According to this concept the earth has a strong lithosphere, or outer shell of rock, about 100 kilometers thick. Presumably in response to forces generated in the asthenosphere, the weak upper mantle of rock underlying the lithosphere, the shell was broken up into a number of separate plates. There are now some 10 major plates, plus numerous additional subplates. The continents resting on these plates were rafted across the surface of the globe.

The mechanism of plate movement is not yet clear. The plates may be pushed, carried by convection cells in the mantle, driven by gravitational forces or pulled. We prefer a model based on pulling; we suspect that plates are colder and heavier at one boundary than elsewhere and thus dive down into the earth's mantle along "subduction" zones. These zones usually show themselves as deep trenches, which are disposed principally around the periphery of the Pacific. As a result a tear, or rift, widens along the opposite boundary of the plate; this rift is filled by a solid flow of viscous mantle rock and by dikes of molten tholeiitic basalt (a differentiated partial melt of the mantle). Because the mantle rock and its basaltic derivative are both heavier than the granitoid rock of the continents they assume a level about four kilometers below sea level. Consequently such a pulled-apart region always becomes new ocean floor. As two adjacent plates continue to pull apart, basaltic dikes continue to pour into the suboceanic rift, which remains midway between the two plates. This highly symmetrical process, which creates new ocean basins or continuously repaves old ocean floors, is termed seafloor spreading. The rate of spreading, measured from the mid-ocean rift to either plate, is from one centimeter per year (10 kilometers per million years) to several times that figure. This is remarkably rapid by geological standards, being many times faster than mountains are elevated by tectonism or leveled by erosion. For example, the North American plate is moving westward the length of one's body in a lifetime.

The discovery of a mid-ocean ridge system some 40,000 kilometers long, winding through all the ocean basins, was an important prelude to the seafloor-spreading hypothesis. It was soon recognized that the ridge has a fossa, or axial depression, into which dikes of basalt are continuously being injected. This linear depression in the ridge marks the location of the rift. The term "mid-ocean," although appropriate for the part of the ridge system in the Atlantic

SUBCONTINENT OF INDIA, originally attached to what is now Antarctica, made the longest migration of all the drifting land masses: approximately 9,000 kilometers in 200 million years. This picture, taken at an altitude of 650 kilometers from *Gemini XI* in September, 1966, shows all of the subcontinent. The Himalayan mountains, 3,700 kilometers away, are just visible on the horizon.

822

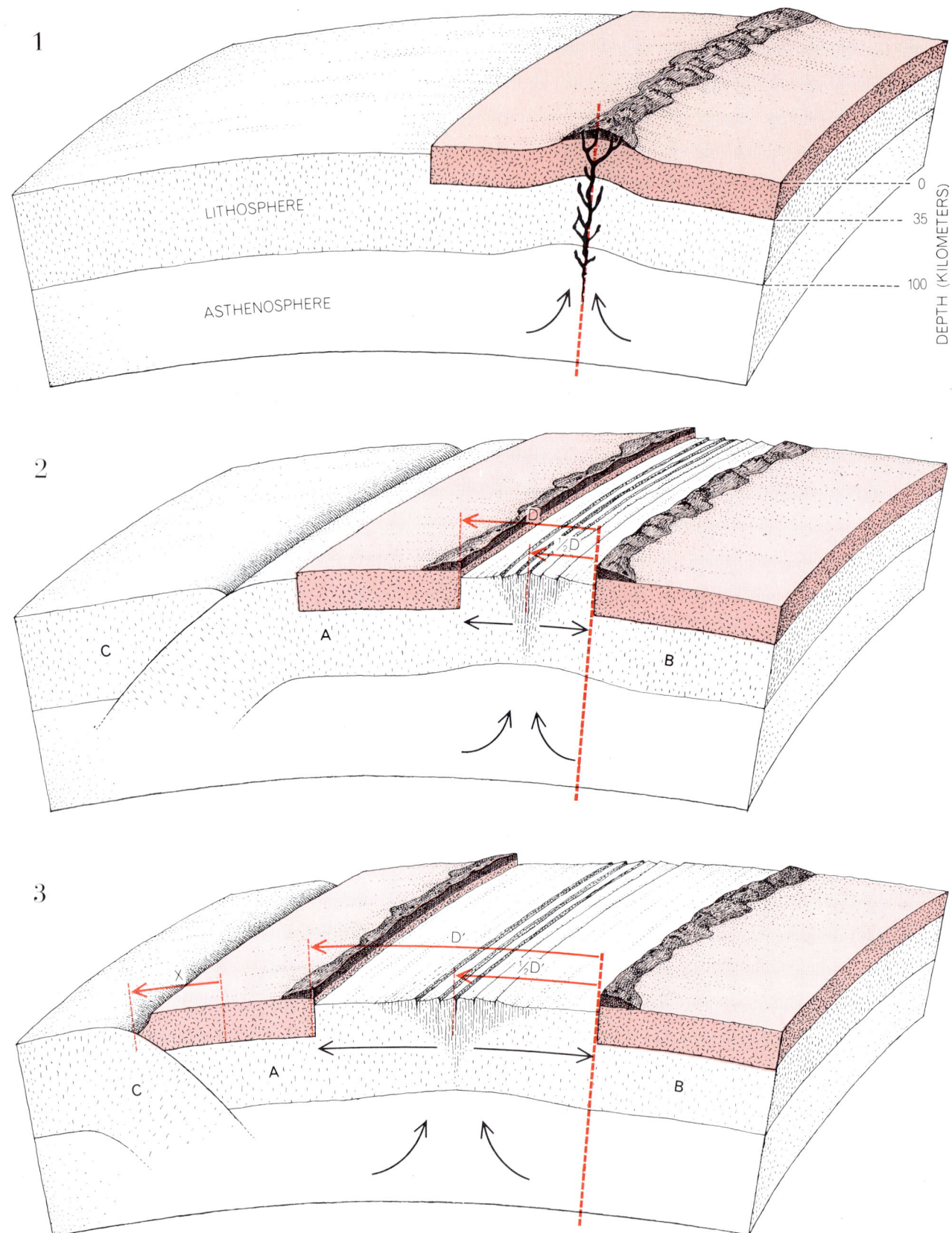

THEORY OF PLATE TECTONICS provides a mechanism for continental drift. The process begins (1) when a spreading rift develops under a continent (*color*) that is resting on a single crustal plate. Molten basalt from the asthenosphere spills out. The second simultaneous requirement for continental drift is the formation of a zone of subduction, or trench, into which oceanic crust of the new moving plate (*A*) is pulled and "consumed" (2). As the new continent carried by plate *A* is rafted to the left, a new ocean basin is created between the two land masses. In the third stage (3) the continent on plate *A* encounters and overrides the trench for some distance (*X*) and eventually reverses, or flips, its direction from west-dipping to east-dipping. Because the continent on plate *B* is here arbitrarily fixed, the mid-ocean rift migrates to the left, remaining in the center of the new ocean basin, whose width is *D'*.

and the Indian Ocean, is a misnomer for the ridge in the Pacific. The Atlantic and the Indian Ocean are rift oceans, formed where continents were once split apart; therefore it is natural for the axis of spreading, marked by the ridge system, to remain in the center of these two oceans. The Pacific, on the other hand, is not a rift ocean; it is clearly the ancestral ocean, and it is becoming smaller as new ocean basins grow. Although the Pacific also has a ridge, it runs north-south well to the east of the ocean's center.

In reality the crustal motions are considerably more complex than the ones we have just outlined. The trenches and rifts apparently migrate, and the opposing plates are also subject to displacements produced by internal shears. The "megashears," the large zones of slippage along plate boundaries, also seem able to accommodate minor amounts of crustal extension or compression. Few of the plates are "ideal" in the sense of being rectilinear, of having a rift matched by an opposing trench and of having these two antithetical zones connected by a megashear. The Antarctic plate, for example, has no trench at all. Perhaps this anomaly is partly explained by the fact that a sphere cannot be covered with rectangles.

We can visualize the continents as being passively rafted over the surface of the globe as embedded plateaus of sialic (granite-like) rock resting on the even larger and thicker crustal plates. The continents have generally maintained their size and shape since the breakup of Pangaea. There have been some accretions with the formation of mountain belts, but these have been mostly confined to the sides of continents facing the Pacific. The sides of continents facing rift oceans (the Atlantic and the Indian Ocean) show little change; hence they can be fitted together almost as neatly as pieces of a jigsaw puzzle.

In contrast, the crustal plates can change in size or shape either by the addition of new ocean floor along the rifts or by the resorption of oceanic crust in trenches. Thus it has been possible for the North American and South American plates moving toward the Pacific to grow larger at first and then smaller as they passed over the great circle of the earth and now converge toward the central Pacific. An even more tortured history is reflected in the complex evolution of the Caribbean Sea region, caught as a "gore" between the North American and South Ameri-

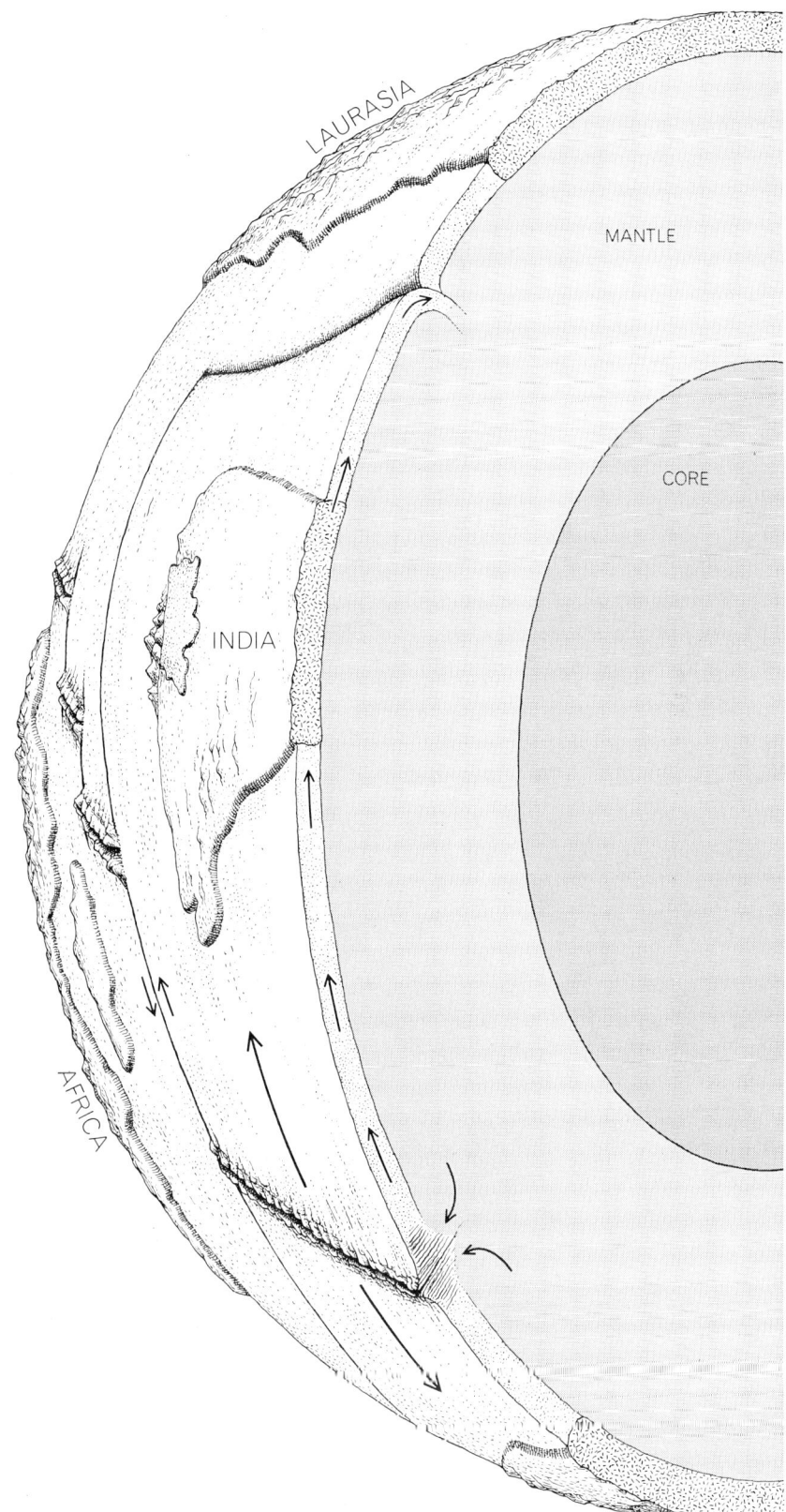

NORTHWARD DRIFT OF INDIA exemplifies how far a land mass can be carried when tectonic conditions are favorable. The plate carrying the Indian land mass is nearly a perfect rectangle, which was sliced away from Antarctica within the primitive universal continent of Pangaea. The plate that rafted India then migrated northward toward and subducted into the Tethyan trench, which ran east-west near the Equator. The plate evidently glided freely along parallel "megashears" on its eastern and western boundaries without interacting with the other crustal plates of the world. India finally collided with and underthrust the southeast margin of Asia, creating the Himalayas, which are thus two plates thick.

can plates, and the Scotia Sea region, similarly trapped between the South American and Antarctica plates. As we shall see, in at least one case two plates evidently collided, producing a mid-continent mountain range: the Himalayas.

In making our reconstruction of Pangaea we selected for fitting not the present coastlines of continents but the contour lines where the continental slope reaches a depth of 1,000 fathoms, or about 2,000 meters [see illustration below]. This isobath was selected because it is approximately halfway down the continental slope and thus marks roughly half the height of the vertical walls created when the continents first rifted. On the assumption that these walls subsequently slumped to a condition of stable repose, the 1,000-fathom isobath closely delineates the location of the original break.

For joining the two sides of the Atlantic we have followed, with some modification, the reconstruction proposed by Sir Edward Bullard, J. E. Everett and A. G. Smith of the University of Cambridge. For closing the Indian Ocean we have used the best-fit computer solutions of Walter P. Sproll, a colleague of ours in the Marine Geology and Geophysics Laboratory of the Environmental Science Services Administration. His studies provide precise fits between Australia and Antarctica and between Antarctica and Africa. The three continents together constitute most of Gondwana. Presumably India was also part of the Gondwana complex, but where it was attached remains unclear. Fortunately the pattern of fracture zones in the ocean floor provides crude but useful dead-reckoning tracks showing how the continents drifted. Using such tracks, we have placed the west coast of India against Antarctica rather than against

western Australia, the fit that is often proposed.

Another difficult fit is presented by the bulge of Africa and the bight of North America. The areas of mismatch, particularly that caused by the Florida-Bahamas platform, are sufficiently large for one to reasonably argue that Africa and North America were never joined. On this assumption instead of Pangaea one obtains two unconnected supercontinents as the antecedent land masses: Laurasia in the Northern Hemisphere and Gondwana in the Southern. This version of the continental-drift theory has important adherents.

We nevertheless prefer the Pangaea reconstruction; in our view the areas of mismatch can reasonably be regarded as modifications that arose after Africa and North America began drifting apart. We regard the Florida-Bahamas platform as a sedimentary infilling of a small ocean basin that appeared when Africa

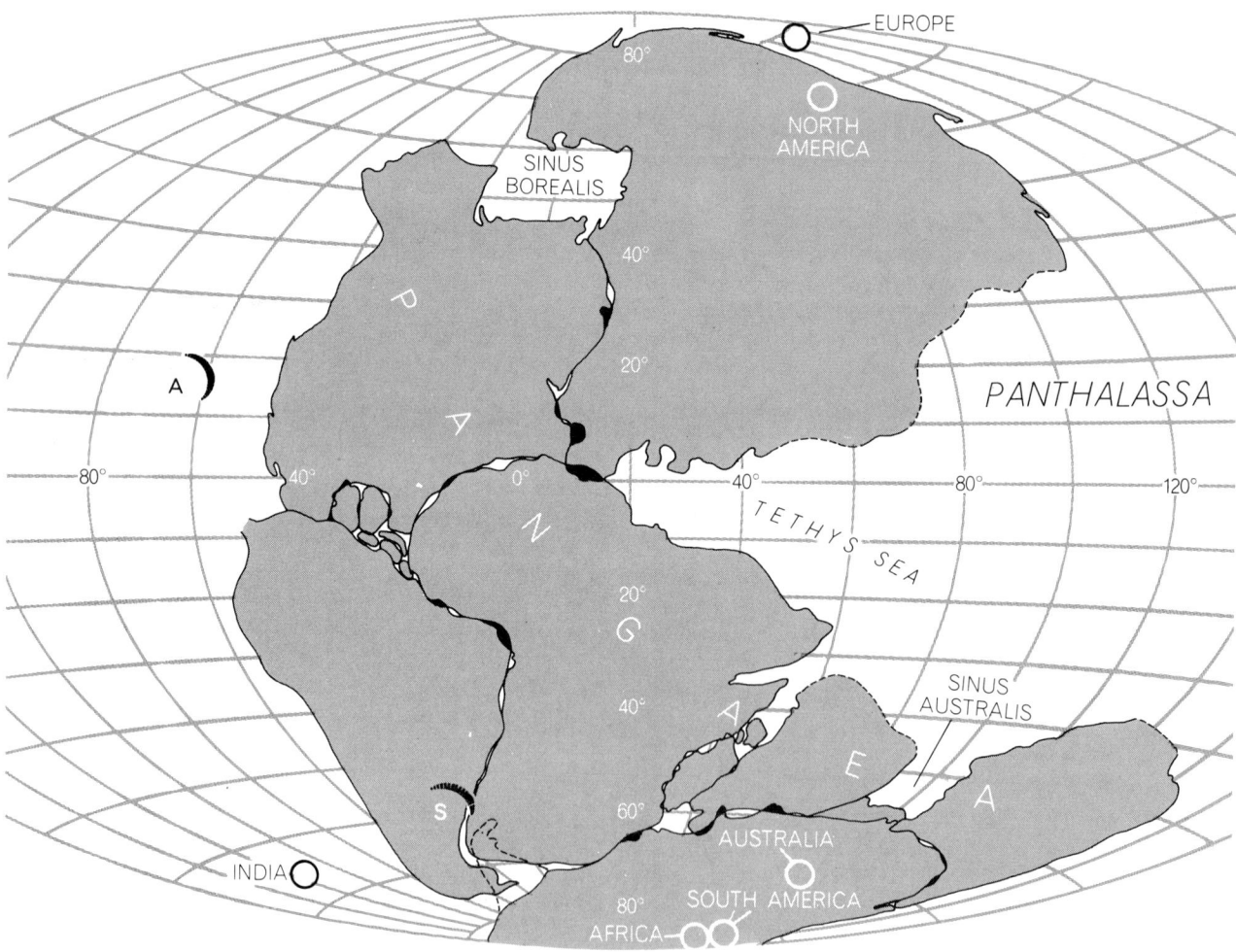

UNIVERSAL LAND MASS PANGAEA may have looked like this 200 million years ago. Panthalassa was the ancestral Pacific Ocean. The Tethys Sea (the ancestral Mediterranean) formed a large bay separating Africa and Eurasia. The relative positions of the continents, except for India, are based on best fits made by computer, using the 1,000-fathom isobath to define continental boundaries. When the continents are arranged as shown, the relative locations of the magnetic poles in Permian times are displaced to the positions marked by circles. Ideally these positions should cluster near the geographic poles. The hatched crescents (A and S) serve as modern geographic reference points; they represent the Antilles arc in the West Indies and Scotia arc in the extreme South Atlantic.

and North America first began to pull apart. Without this assumption the platform unaccountably overlays a large portion of the bulge of Africa [see illustration on page 830].

According to our reconstruction, Pangaea was a land mass of irregular outline surrounded by the universal ocean of Panthalassa: the ancestral Pacific. The fit between North America and Africa provides the principal connection between the future block of northern continents and the future group of southern ones. On the east the Tethys Sea, a large triangular bight, separated Eurasia from Africa; the present Mediterranean Sea is a remnant of the Tethys. Other major indentations in the outline of Pangaea (adapting terminology from the moon) can be named Sinus Borealis, the ancestral Arctic Ocean, and Sinus Australis, a southern bay off the Tethys separating India from Australia. Our fully closed reconstruction of the Central American

region is problematical. An alternate possibility is that the Gulf of Mexico is the remnant of an oceanic arm extending into the Americas from Panthalassa—a Sinus Occidentalis.

When measured down to the 1,000-fathom isobath, the total area of Pangaea was 200,000 square kilometers, or 40 percent of the earth's surface—equal to the area of the present continents measured to the same isobath. When the future continents were still part of Pangaea, they were generally to the south and east of their present location, so that the amount of land in the two hemispheres was almost equally balanced. (Today two-thirds of all the land lies north of the Equator.) The Y-shaped junction connecting North America, South America and Africa was located in the South Atlantic not far from the present position of Ascension Island. If New York had been in existence at the time, it would have been on the Equator

and at longitude 10 degrees east (rather than 74 degrees west). Spain would also have been on the Equator, but it would have been near its present longitude. Japan would have been in the Arctic, well north of its position today. India and Australia would have bordered the Antarctic, far to the south of where they are now.

The great event that broke up Pangaea and set its fragments adrift evidently began no more than 200 million years ago, or in the last few percent of geologic time. There may have been—indeed, there probably was—"predrift drift" that assembled Pangaea from two or more smaller land masses. The evidence is still scanty, however, and does not bear directly on this discussion.

We take the immediate prelude to the breakup of Pangaea to be the first large outpourings of basaltic rock along the continental margins being es-

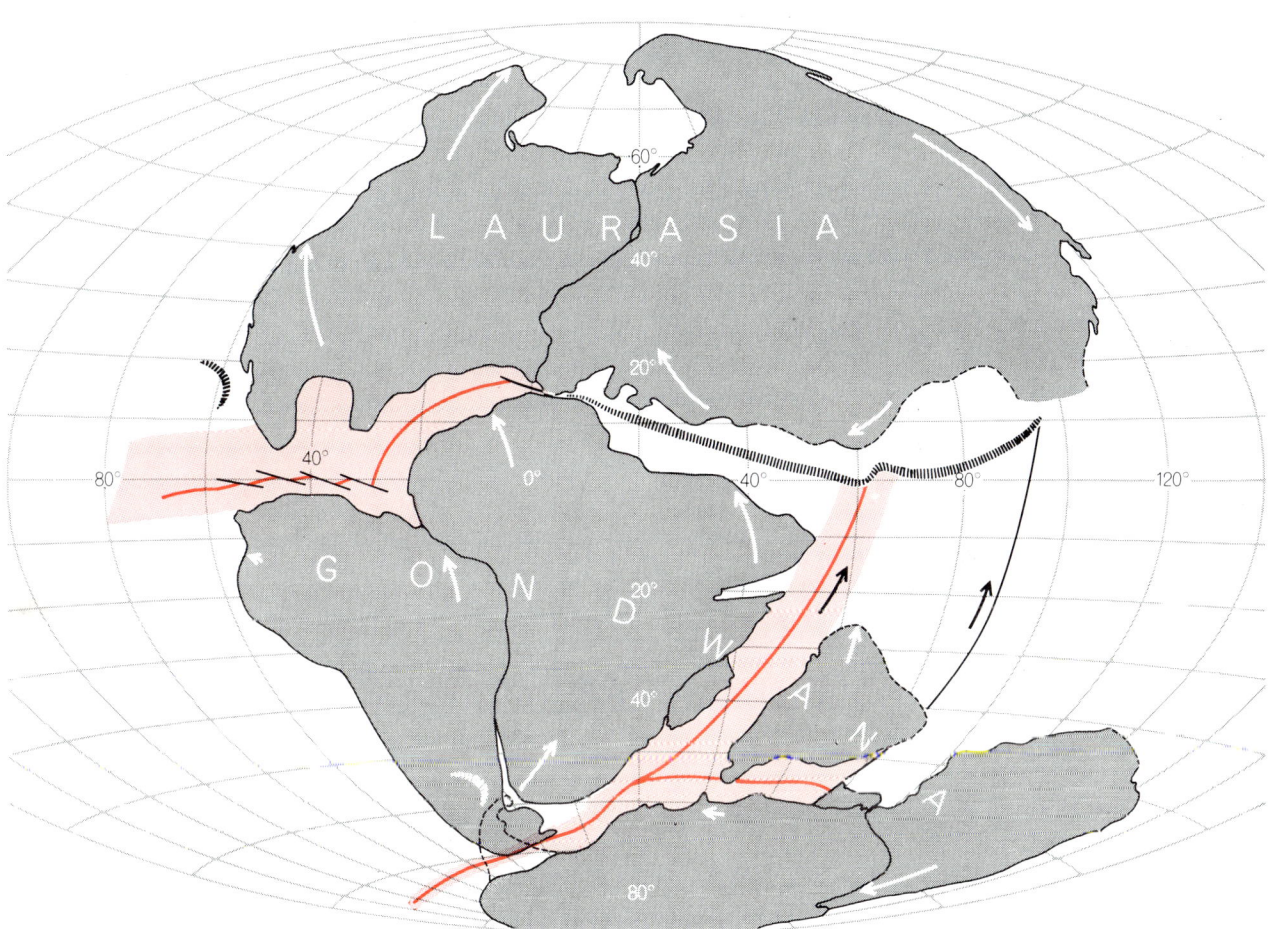

AFTER 20 MILLION YEARS OF DRIFT, at the end of the Triassic period 180 million years ago, the northern group of continents, known as Laurasia, has split away from the southern group, known as Gondwana. The latter has started to break up: India has been set free by a Y-shaped rift (heavy line in color), which has also begun to isolate the Africa–South America land mass from Antarctica-Australia. The Tethyan trench (hatched lines in black), a zone of crustal uptake, runs from Gibralter to the general area of Borneo. Black lines and black arrows denote megashears, zones of slippage along plate boundaries. The white arrows indicate the vector motions of the continents since drift began. Oceanic areas tinted in color represent new ocean floor created by sea-floor spreading.

tablished by rifting. The Triassic New-ark series of basaltic flows along the east coast of the U.S. is a good example. Measurements of radioactivity indicate that the most ancient of these rocks are about 200 million years old, yielding a date that coincides with the middle of the Triassic period. As we interpret the evidence, two extensive rifts were initiated in Pangaea about 200 million years ago, which resulted in the opening of the Atlantic and the Indian Ocean by the end of the Triassic period 180 million years ago [see illustration on preceding page]. The northern rift split Pangaea from east to west along a line slightly to the north of the Equator and created Laurasia, composed of North America and Eurasia. The Laurasian land mass evidently rotated clockwise as a single plate around a pole of rotation that is now in Spain, creating a western "Mediterranean" that ultimately became part of the Gulf of Mexico and the Caribbean Sea. The southern rift split South America and Africa as a single land mass away from the remainder of Gondwana, consisting of Antarctica, Australia and India. Soon afterward (if not simultaneously) India was severed from Antarctica by a smaller rift to begin its rapid drift northward.

During the Jurassic period, from 180 to 135 million years ago, the direction of drift established by the Triassic rifts continued, further opening up the Atlantic and the Indian Ocean [see illustration below]. As North America drifted to the northwest, the Atlantic became more than 1,000 kilometers wide and probably remained fully connected to the Pacific. The east coast of the present U.S. ran almost east and west at a latitude of about 25 degrees north, so that coral reefs were able to grow all along the edge of the Atlantic continental shelf to the present Grand Banks, off Nova Scotia.

During the 45-million-year Jurassic period the Atlantic rift extended northward, blocking out the Labrador coast-line and possibly initiating the opening up of the Labrador Sea between North America and Greenland. The interaction between the African and Eurasian plates forced the region of Spain to rotate counterclockwise 35 degrees, opening up the Bay of Biscay. The Tethys Sea, forerunner of the Mediterranean, continued to close at its eastern end. The Tethys was not only a zone of crustal subduction, or trench, but also a zone of shear along which Eurasia slid westward with respect to Africa. The compression associated with the Tethys trench raised bordering mountains composed of deep-water sediments.

At the close of the Jurassic an incipient rift began splitting South America away from Africa, entering from the south and working only as far north as where Nigeria is today. The tectonic situation first resembled the one now found in the rift zone along the back-bone of high Africa (the region from Ethiopia to Tanzania) and then gradually opened farther to form a body of water resembling the Red Sea of today.

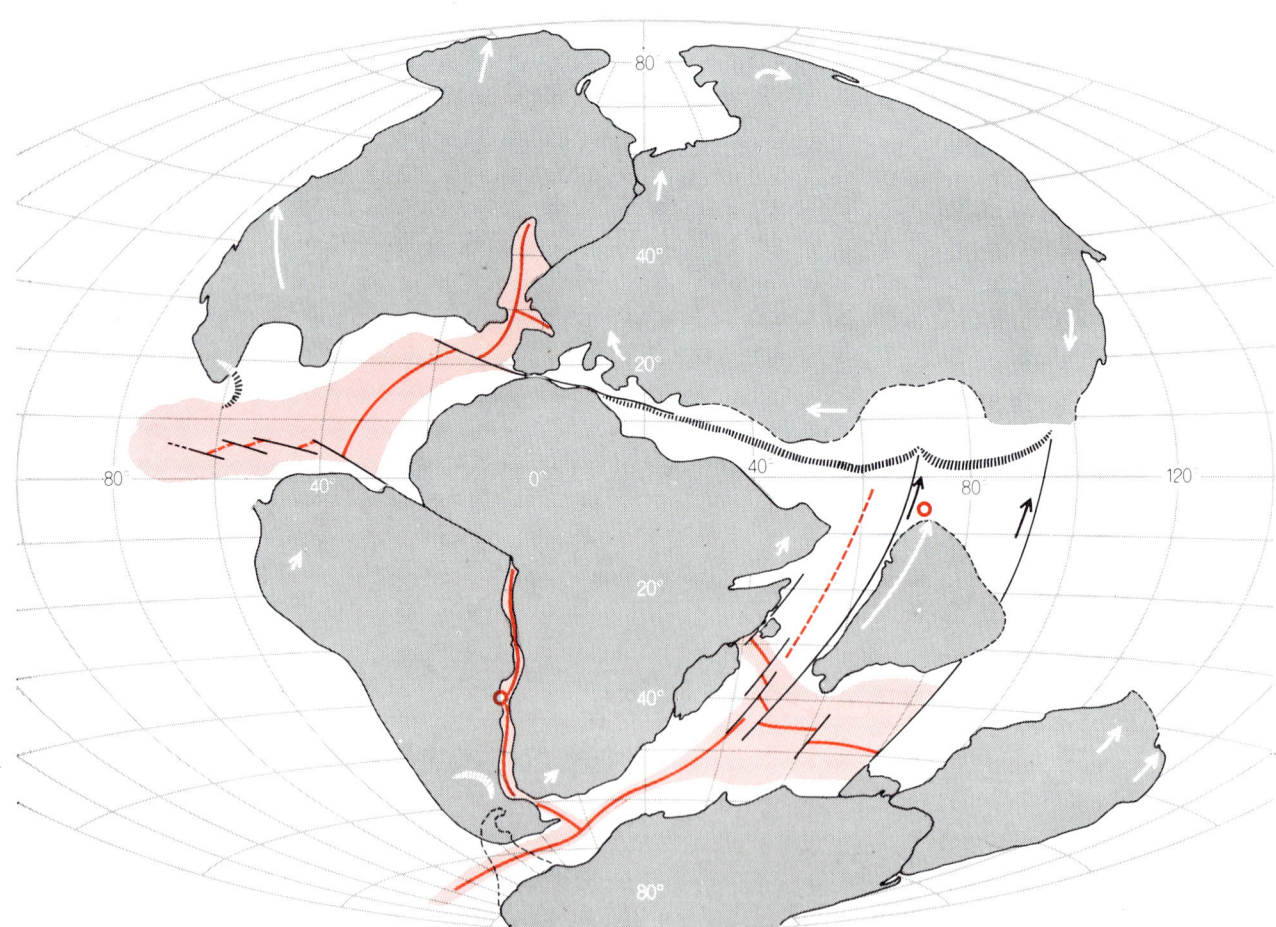

AFTER 65 MILLION YEARS OF DRIFT, at the end of the Jurassic period 135 million years ago, the North Atlantic and the Indian Ocean have opened considerably. The birth of the South Atlantic has been initiated by a rift. The rotation of the Eurasian land mass has begun to close the eastern end of the Tethys Sea. The Indian plate is about to pass over a thermal center (*colored dot*) that will soon pour out basalt to form the Deccan plateau. Later the hot spot will create the Chagos-Laccadive ridge in the Indian Ocean. Similarly, in the South Atlantic the Walvis thermal center (*colored dot*) will create the Walvis and Rio Grande "thread ridges."

At first freshwater sediments created thick deposits in pockets opened by faults; these sediments were overlain by deposits of salt.

By the end of the Cretaceous period, some 70 million years later (and 65 million years ago), the rupture of South America and Africa was complete, and the South Atlantic had widened rapidly to at least 3,000 kilometers [see illustration below]. Meanwhile the rift in the North Atlantic had switched from the west side of Greenland to the east side, blocking out its eastern margin (without, however, penetrating to the Arctic Ocean). Africa had drifted northward about 10 degrees and continued its counterclockwise rotation as the Eurasian plate rotated slowly clockwise. These two opposed motions nearly closed the eastern end of the Tethys Sea. The slow westward rotation of Antarctica continued. All the continents were now blocked out except for the remaining connection between Greenland and northern Europe and between Australia and Antarctica.

Although it is not shown on our maps, an extensive north-south trench system must have existed in the ancient Pacific to consume by subduction the rapid westward drift of the two plates carrying North and South America. North America presumably encountered this trench in the late Jurassic and early Cretaceous, with the result that the Franciscan fold belt, the predecessor of the California Coast Ranges, was accreted to the western margin of the continent. It appears that the trench was eventually overridden and "stifled" by North America's continued westward drift. Such trenches have the capacity to resorb ocean crust but not the lighter granitic crust of continents.

At about the same time, or soon afterward, South America first encountered the Andean trench and began to displace the trench westward, without ever overriding it. The early Andean fold belt resulted from this encounter. It seems likely that the trench originally dipped toward the west but was flipped over to its present eastward dip.

In the Cenozoic period (from 65 million years ago to the present) the continents drifted to the positions we observe today. The mid-Atlantic rift propagated into the Arctic basin, finally detaching Greenland from Europe [see top illustration on next two pages]. There were three other major developments during the Cretaceous: (1) the two Americas were rejoined by the Isthmus of Panama, created by volcanism and the arching upward of the earth's mantle, (2) the Indian land mass completed its remarkable journey northward by colliding with the underbelly of Asia and (3) Australia was rifted away from Antarctica and drifted northward to its present position.

In the collision of India with Asia the northern margin of the Indian plate was subducted below the Asiatic plate, creating the Himalayas. On India's passage to the north early in the Cenozoic its western margin crossed a fixed source of

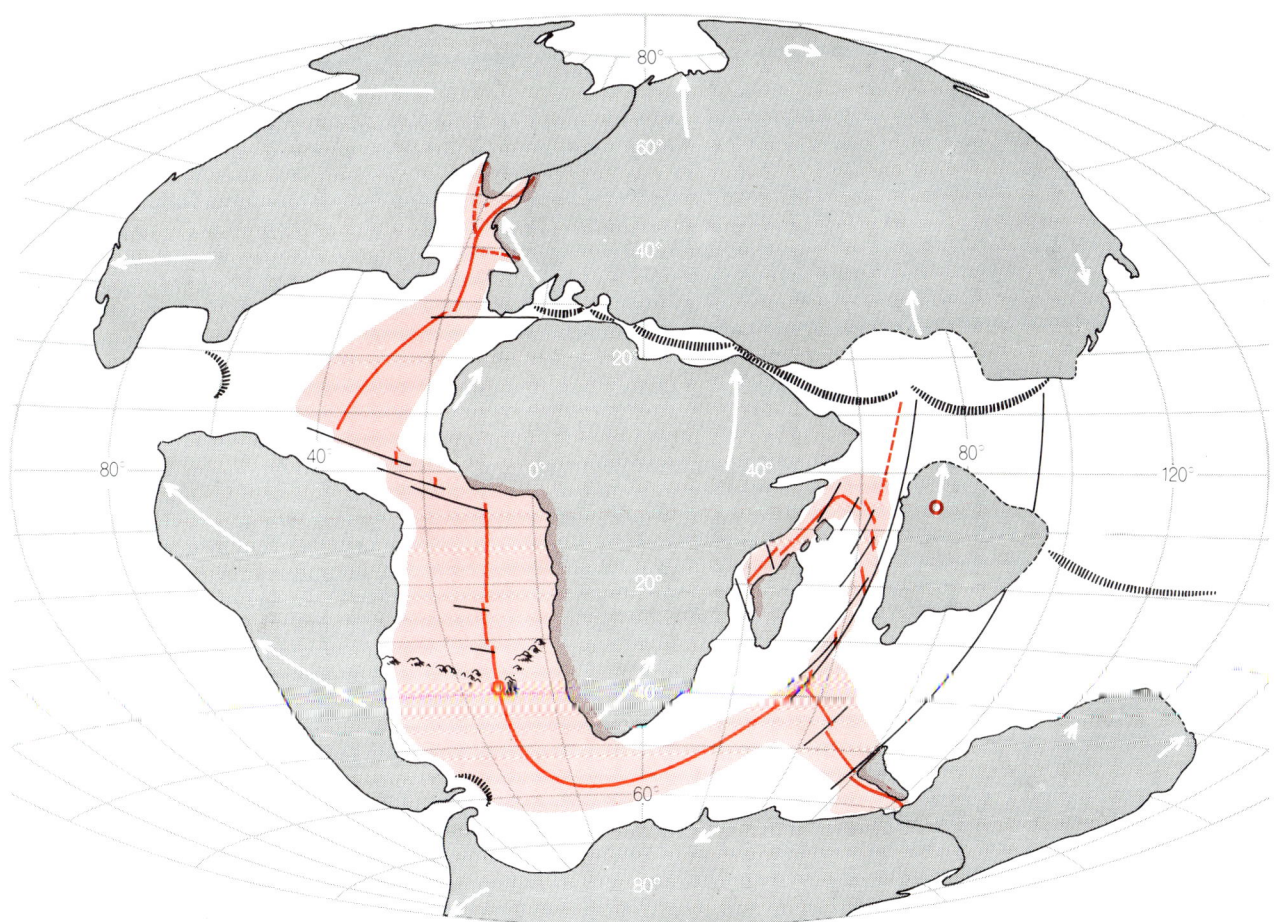

AFTER 135 MILLION YEARS OF DRIFT, 65 million years ago at the end of the Cretaceous period, the South Atlantic has widened into a major ocean. A new rift has carved Madagascar away from Africa. The rift in the North Atlantic has switched from the west side to the east side of Greenland. The Mediterranean Sea is clearly recognizable. Australia still remains attached to Antarctica. An extensive north-south trench (not shown) must also have existed in the Pacific to absorb the westward drift of the North American and South American plates. Note that the central meridian in all these reconstructions is 20 degrees east of the Greenwich meridian.

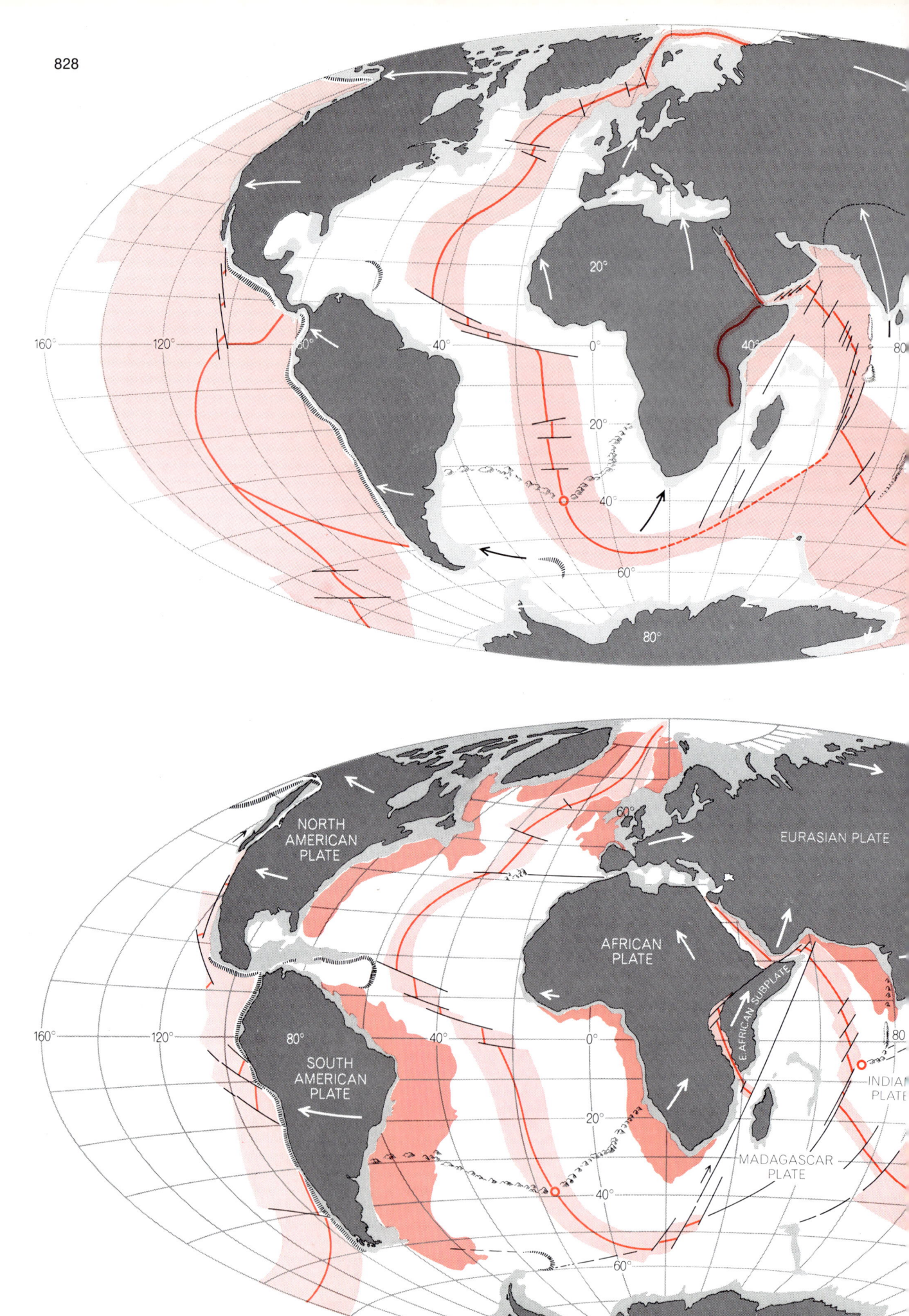

NORTH AMERICAN PLATE

EURASIAN PLATE

AFRICAN PLATE

SOUTH AMERICAN PLATE

E. AFRICAN SUBPLATE

INDIAN PLATE

MADAGASCAR PLATE

ANTARCTIC PLATE

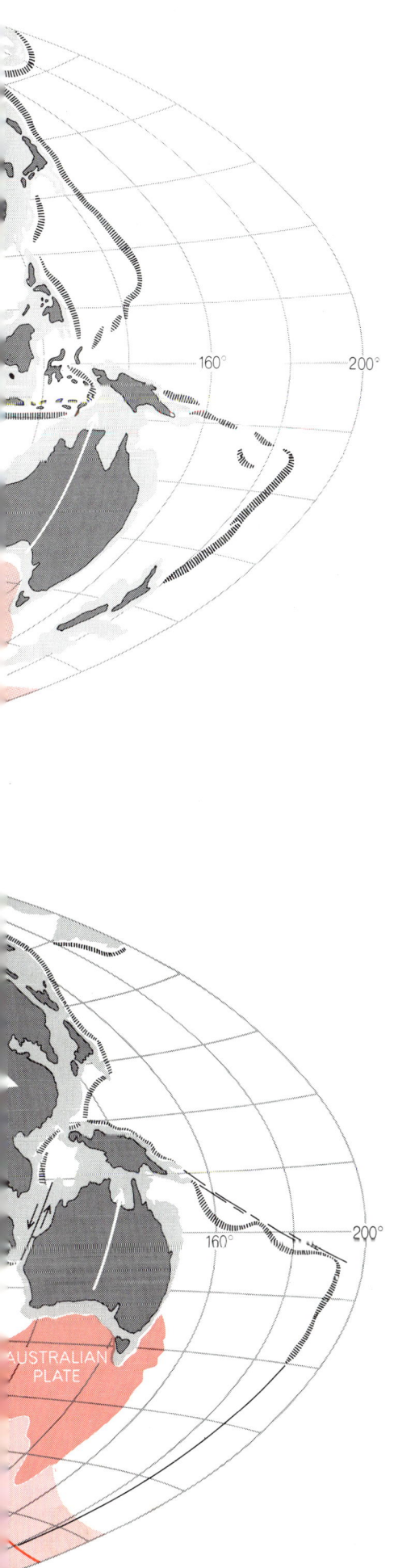

WORLD AS IT LOOKS TODAY was produced in the past 65 million years in the Cenozoic period. Nearly half of the ocean floor was created in this geologically brief period, as shown by the areas stippled in color. India completed its flight northward by colliding with Asia and a rift has separated Australia from Antarctica. The North Atlantic rift finally entered the Arctic Ocean, fissioning Laurasia. The widening gap between South America and Africa is closely traced by the thread ridges produced by the Walvis thermal center. The Antilles and Scotia arcs now occupy their proper positions with respect to neighboring land masses.

basaltic magma rising from the earth's upper mantle near the Equator. Molten rock erupted through the crust and poured onto the Indian subcontinent, laying down the basalts of the Deccan plateau. Even after India had left the hot spot behind, magma continued to stream out on the ocean floor, producing the Chagos-Laccadive ridge, which became covered with coral as it subsided into the Indian Ocean. Finally, a branch of the Indian Ocean rift split Arabia away from Africa, creating the Gulf of Aden and the Red Sea, and a spur of this rift meandered west and south into Africa.

Less pronounced changes during the Cenozoic period included the partial closing of the Caribbean region and the continued widening of the South Atlantic as new ocean crust was emplaced by sea-floor spreading. As the Atlantic continued to open in the far north the northwestward movement of the Eurasian land mass was halted and reversed, simultaneously reversing its sense of slippage with respect to Africa. The new direction of shear has been strongly impressed on the tectonic character of the Mediterranean and the Near East. The major north-south rift in the Indian Ocean largely ceased spreading and became instead a megashear that accommodated the counterclockwise and northward rotation of the African plate.

The reader will have observed that our maps of continental drift show more than relative positions and motions; the land masses, beginning with Pangaea itself, are assigned absolute geographic coordinates. Since this has not been attempted before we shall

briefly describe how we arrived at our results. In the mobile world of plate tectonics one must assume that all parts of the crust are capable of moving and almost surely have moved.

After an extensive search for some absolute reference point, we finally concluded that the Walvis thermal center, or hot spot, might provide what we sought. In reaching this conclusion we accepted a hypothesis put forward by J. Tuzo Wilson of the University of Toronto. He had suggested that the Walvis ridge and the Rio Grande ridge in the South Atlantic are nemataths, or "thread ridges" of basalt, that had been poured onto the spreading ocean floor from a fixed lava orifice rising from a deep, stagnant region of the mantle. As new floor was carried past the orifice, lava would periodically pour out and form a small volcanic cone. By observing the location of succeeding cones as they merged into a ridge one can establish the absolute direction taken by the crust in that region. A study of the Walvis and Rio Grande ridges enabled us to establish not only the drift of the South American plate with respect to the African plate but also any motion the two plates may have had in some other direction [*see illustration on page 831*].

Unfortunately the Walvis hot spot did not exist earlier than about 140 million years ago, so that its usefulness as a fixed point does not go back earlier than the end of the Jurassic period. To trace crustal motions during the first 60 million years after the breakup of Pangaea one has to rely on dead reckoning. We have made the assumption that Antarctica has moved very little from its original location when it was part of Pan-

WORLD 50 MILLION YEARS FROM NOW may look something like this. The authors have extrapolated present-day plate movements to indicate how the continents will have drifted by the end of what they propose to call the Psychozoic era (the age of awareness). The Antarctic remains essentially fixed but may rotate slightly clockwise. The Atlantic (particularly the South Atlantic) and the Indian Ocean continue to grow at the expense of the Pacific. Australia drifts northward and begins rubbing against the Eurasian plate. The eastern portion of Africa is split off, while its northward drift closes the Bay of Biscay and virtually collapses the Mediterranean. New land area is created in the Caribbean by compressional uplift. Baja California and a sliver of California west of the San Andreas fault are severed from North America and begin drifting to the northwest. In about 10 million years Los Angeles will be abreast of San Francisco, still fixed to the mainland. In about 60 million years Los Angeles will start sliding into the Aleutian trench.

FIT OF AFRICA AGAINST NORTH AMERICA was made by the authors' colleague Walter P. Sproll with the aid of a computer. As in the reconstruction of Pangaea, it is assumed that each continent actually extends out into the ocean and halfway down the continental slope, where the ocean reaches a depth of 1,000 fathoms. The North American "coast" between A and A' was matched for best fit to the African "coast" between B and B'. White areas are gaps in the fit; black areas are overlaps. The overlap produced by the Bahamas platform, an enormous area half the size of Texas, is specially depicted in dark color. The authors propose that the platform represents an accumulation of sediments followed by coral growth after the two continents became separated. The largest gap in the proposed fit between the two continents is found off the Spanish enclave of Ifni. The Ifni gap may have been created when a small section of Africa split off and was translated 190 kilometers to the southwest, forming the eastern group of the Canary Islands.

gaea. This seems reasonable because the Antarctica plate is entirely surrounded by a system of rifts and megashears; there is no associated trench toward which the plate would tend to move away from its polar position.

Independent support for this assumption is obtained by plotting the position of the North and South poles before the dispersal of Pangaea. These positions are obtained by studying the direction of magnetization in rocks of the Permian period, as obtained by E. Irving of the Dominion Observatory in Canada and by other workers. We plotted the Permian pole positions with respect to each continent as it exists today and then rotated these pole positions as needed to assemble the continents into our version of Pangaea. By this method the pole positions should ideally cluster at one of the geographic poles. Actually there is some scatter, as can be seen in our reconstruction of Pangaea [*see illustration on page 824*], but all the positions do fall within either the Arctic Circle or the Antarctic Circle.

We can now summarize how the continents have moved in time and space. The two Americas have drifted a long way, generally westward. North America has drifted more than 8,000 kilometers west northwest; the tip of Florida once lay in the South Atlantic near the present position of Ascension Island. Moving toward the Tethyan trench system, India and Australia were carried far to the north. Africa rotated counterclockwise perhaps 20 degrees as the Eurasian land mass, similarly moving toward the Tethyan trench, rotated clockwise a roughly equal amount. India's remarkable flight is probably attributable to its being rafted on an "ideal" plate. Approximately rectangular, the Indian plate was sliced away from Pangaea by a rift along what is now India's east coast and then was free to move northward toward a major trench. This northward movement was facilitated by two parallel megashears.

Decades ago Wegener proposed that the drift of the continents was vectored by forces he termed *Westwanderung* (westward drift) and *Polarfluchtkraft* (flight from the poles). Although real, these forces are minuscule and not likely to be the underlying cause of drift. Our solution, however, does support Wegener's hypothesis of a westward flight, which, like the slip of the atmosphere, directly opposes the earth's rotation. We have also inferred a latitudinal drift, but from the South Pole only, or, paraphrasing Wegener's terminology, a *Sudpolarfluchtkraft*.

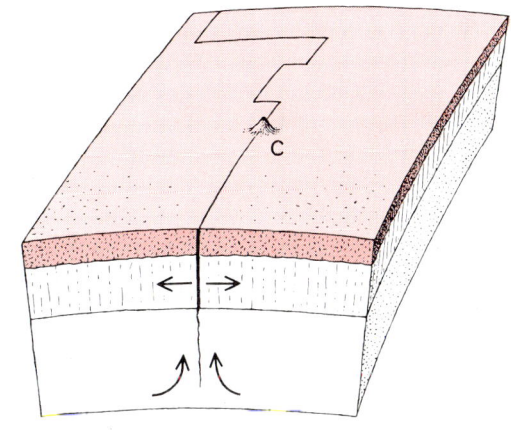

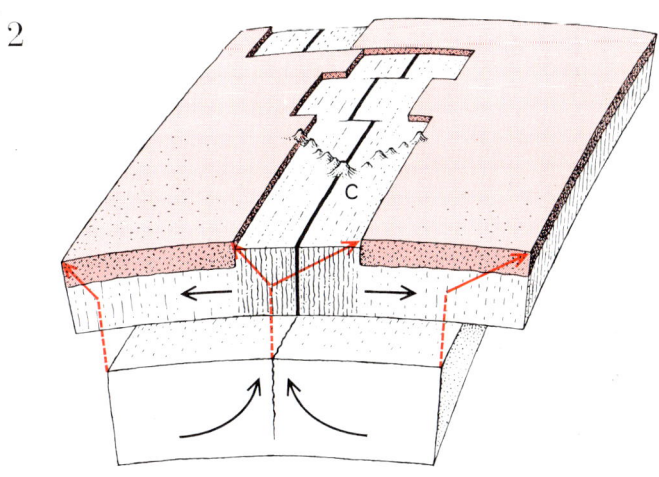

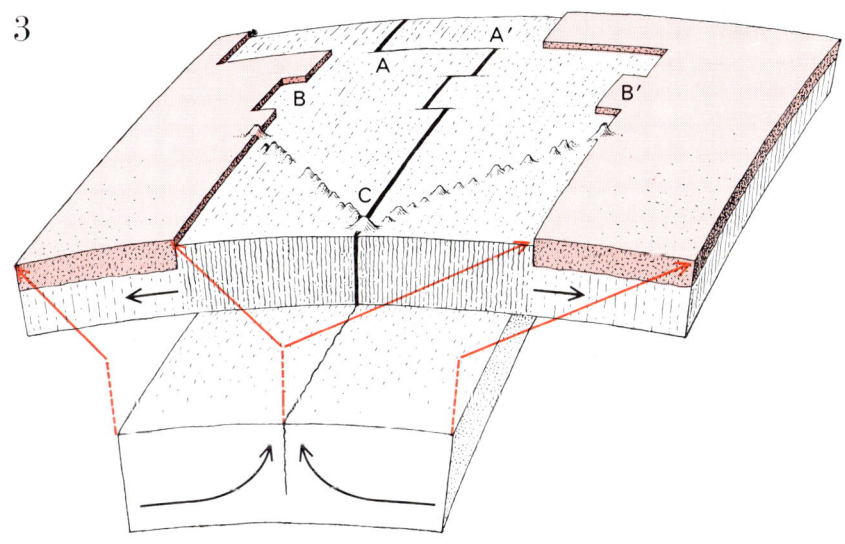

SEPARATION OF SOUTH AMERICAN AND AFRICAN PLATES can be traced in absolute geographic coordinates by observing the orientation of the thread ridges, a *V*-shaped stream of volcanoes, produced by the Walvis thermal center (*C*). The hot spot has evidently been pouring out magma from a source deep in the mantle for the past 140 million years. The three-part diagram illustrates a hypothesis first proposed by J. Tuzo Wilson of the University of Toronto. The thread ridges show that the South American and African plates have been not only drifting rapidly apart but also migrating northward. Features such as the strike of the ridge-ridge transform faults (*A–A'*) and matching indentations on opposing continents (*B–B'*) can do more than indicate the relative motion of two plates.

The Authors

ROBERT S. DIETZ and JOHN C. HOLDEN are marine geologists with the Environmental Science Services Administration, working at the Atlantic Oceanographic and Meteorological Laboratories in Miami. Dietz, who obtained his Ph.D. from the University of Illinois in 1941, has written extensively on geotectonics, geosynclines, continental drift (with the late Harry H. Hess of Princeton he was originator of the sea-floor-spreading concept that is now widely accepted as the underlying mechanism of continental drift), the morphology and structure of the ocean floor, the history of ocean basins, the evolution of continental shelves and slopes, marine mineral resources and astroblemes. Holden is completing work on his doctorate in micropaleontology at the University of California at Berkeley. "As a switch from the megathinking world of plate tectonics and continental drift," he writes, "I find solace in investigating the microcosmic world revealed by the microscope."

Bibliography

THE ORIGIN OF CONTINENTS AND OCEANS. Alfred Wegener. Methuen & Co. Ltd., 1924.

CONTINENT AND OCEAN BASIN EVOLUTION BY SPREADING OF THE SEA FLOOR. Robert S. Dietz in *Nature*, Vol. 190, No. 4779, pages 854–857; June 3, 1961.

HISTORY OF OCEAN BASINS. H. H. Hess in *Petrologic Studies: A Volume in Honor of A. F. Buddington*, edited by A. E. J. Engel, Harold L. James and B. F. Leonard. The Geological Society of America, 1962.

CONTINENTAL DRIFT. J. Tuzo Wilson in *Scientific American*, Vol. 208, No. 4, pages 86–100; April, 1963.

RISES, TRENCHES, GREAT FAULTS AND CRUSTAL BLOCKS. W. Jason Morgan in *Journal of Geophysical Research*, Vol. 73, No. 6, pages 1959–1982; March 15, 1968.

GEOTECTONIC EVOLUTION AND SUBSIDENCE OF BAHAMA PLATFORM. Robert S. Dietz, John C. Holden and Walter P. Sproll in *Geological Society of America Bulletin*, Vol. 81, No. 7, pages 1915–1927; July, 1970.

SCIENTIFIC
AMERICAN November 1970, Vol. 223, No. 5, pp. 104–115   OFFPRINT **893**

WHY THE SEA IS SALT

by Ferren MacIntyre

The sea contains more than 70 elements in addition to sodium
and chlorine. The global cycles that remove and replenish them
involve rainfall, volcanoes and the spreading of the ocean floor.

According to an old Norse folktale the sea is salt because somewhere at the bottom of the ocean a magic salt mill is steadily grinding away. The tale is perfectly true. Only the details need to be worked out. The "mill," as it is visualized in current geophysical theory, is the "mid-ocean" rift that meanders for 40,000 miles through all the major ocean basins. Fresh basalt flows up into the rift from the earth's plastic mantle in regions where the sea floor is spreading apart at the rate of several centimeters per year. Accompanying this mantle rock is "juvenile" water—water never before in the liquid phase—containing in solution many of the components of seawater, including chlorine, bromine, iodine, carbon, boron, nitrogen and various trace elements. Additional juvenile water, equally salty but of somewhat different composition, is released by volcanoes that rim certain continental margins, such as those bordering the Pacific, where the sea floor seems to be disappearing into deep trenches [see illustration on these two pages].

The elements most abundant in juvenile water are precisely those that cannot be accounted for if the solids dissolved in the sea were simply those provided by the weathering of rocks on the earth's surface. The "missing" elements, such as chlorine, bromine and iodine, were once called "excess volatiles" and were attributed solely to volcanic emanations. It is now recognized that juvenile water may have nearly the same chlorinity as seawater but is much more acid due to the presence of one hydrogen ion (H^+) for every chloride ion (Cl^-). In due course, as I shall explain later, the hydrogen ions are removed and replaced by sodium ions (Na^+), yielding the concentration of ordinary salt (NaCl) that constitutes 90-odd percent of all the "salt" in the sea.

The chemistry of the sea is largely the chemistry of obscure reactions at extreme dilution in a strong salt solution, where all the classical chemist's "distilled water" theories and procedures break down. The father of oceanographic chemistry was Robert Boyle, who demonstrated in the 1670's that fresh waters on the way to the sea carry small amounts of salt with them. He also made the first attempt to quantify saltiness by drying seawater and weighing the residue, but his results were erratic because some of the constituents of sea salt are volatile. Boyle found that a better method was simply to measure the specific gravity of seawater and from this estimate the amount of salt present. Since the distribution of density in the sea is important to oceanographers, the same calculation is routinely performed today in reverse: the salinity is deduced by measuring the electrical conductivity of a sample of seawater, and from this and the original temperature of the sample one can compute the density of the seawater at the point the sample was taken.

In 1715 Edmund Halley suggested that the age of the ocean and thus of the world might be estimated from the rate of salt transport by rivers. When this proposal was finally acted on by John Joly in 1899, it gave an age of some 90 million years. The quantity that Joly measured (total amount of x in ocean divided by annual river input of x) is now recognized as the "residence time" of the constituent x, which is an index of an element's relative chemical activity in the ocean. Joly's value is about right for the residence time of sodium; for a more reactive element (in the ocean environment) such as aluminum the residence time is as brief as 100 years.

Not quite 200 years ago Antoine Laurent Lavoisier conducted the first analysis of seawater by evaporating it slowly and obtaining a series of compounds by fractional crystallization. The first compound to settle out is calcium carbonate ($CaCO_3$), followed by gypsum ($CaSO_4 \cdot 2H_2O$), common salt (NaCl), Glauber's salt ($Na_2SO_4 \cdot 10H_2O$), Epsom salts ($MgSO_4 \cdot 7H_2O$) and finally the chlorides of calcium ($CaCl_2$) and magnesium ($MgCl_2$). Lavoisier noted that slight changes in experimental conditions gave rise to large shifts in the relative amounts of the various salts crystallized. (In fact, some 54 salts, double salts and hydrated salts can be obtained by evaporating seawater.) To get reproducible results for even the total weight of salt one must remove all organic matter, convert bromides and iodides to chlorides, and carbonates to oxides, before evaporating. The resulting weight, in grams of salt per kilogram of seawater, is the salinity, $S^0/_{00}$. (The symbol $^0/_{00}$ is read "per mil.")

In actual practice the total weight of salt in seawater is nowadays never determined. Instead the amount of chloride ion is carefully measured and a total for all other ions is computed by applying the "constancy of relative proportions." This concept dates back to the middle of the 19th century, when John Murray eliminated confusion about the multiplicity of salts by observing that individual ions are the important thing to talk about when analyzing seawater. Independently A. M. Marcet concluded from many measurements that various ions in the world ocean were present in nearly constant proportions, and that only the absolute amount of salt was variable. This constancy of relative proportions was confirmed by Johann Forchhammer and again more thoroughly by Wilhelm Dittmar's analysis of 77 samples of seawater collected by H.M.S. *Challenger* on the first worldwide oceanographic cruise. These 77 samples are probably the last ever analyzed for all the major constituents. Their average salinity was close to 35⁰/₀₀, with a normal variation of only ±2⁰/₀₀.

In the 86 years since Dittmar reported eight elements, 65 more elements have been detected in seawater. It was recognized more than a century ago that elements present in minute amounts in seawater might be concentrated by sea organisms and thereby raised to the threshold of detectability. Iodine, for example, was discovered in algae 14 years before it was found in seawater. Subsequently barium, cobalt, copper, lead, nickel, silver and zinc were all detected

first in sea organisms. More recently the isotope silicon 32, apparently produced by the cosmic ray bombardment of argon, has been discovered in marine sponges.

There are also inorganic processes in the ocean that concentrate trace elements. Manganese nodules (of which more below) are able to concentrate elements such as thallium and platinum to detectable levels. The cosmic ray isotope beryllium 10 was recently discovered in a marine clay that concentrates beryllium. In all, 73 elements (including 13 of the rare-earth group) apart from hydrogen and oxygen have now been detected directly in seawater [*see illustration on page 837*].

It is only in the past 40 years that geochemists have become interested in

the chemical processes of the sea for what they can tell us about the history of the earth. Conversely, only as geophysicists have pieced together a comprehensive picture of the earth's history has it been possible to bring order into marine chemistry.

The earth's present atmosphere and ocean are not primordial but have been liberated from chemical and mechanical entrapment in solid rock. Perhaps four billion years ago, or a little less, there was (according to many geophysicists) a "grand catastrophe" in which the earth's core, mantle, crust, ocean and atmosphere were differentiated from an original homogeneous accumulation of material. Estimates of water released during the catastrophe range from a third to 90 percent of the present volume of the ocean. The catastrophe is not finished even yet, since differentiation of the mantle continues in regions of volcanic activity. Most of the exhalations of volcanoes and hot springs are simply recycled ground water, but if only half of 1 percent of the water released is juvenile, the present production rate is sufficient to have filled the entire ocean in four billion years.

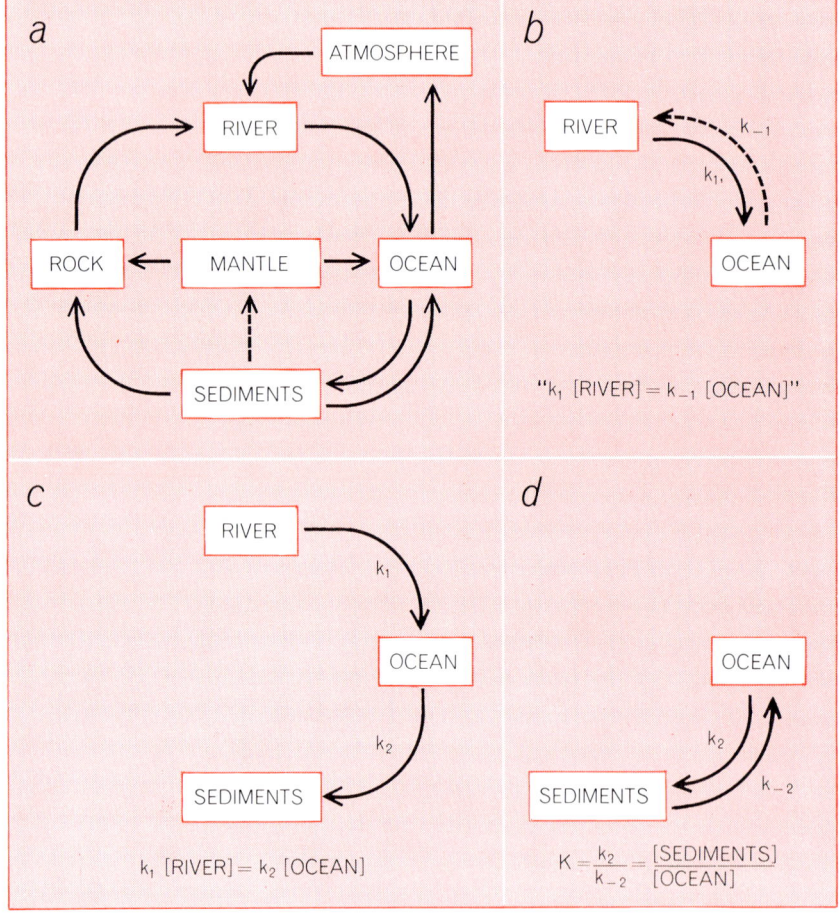

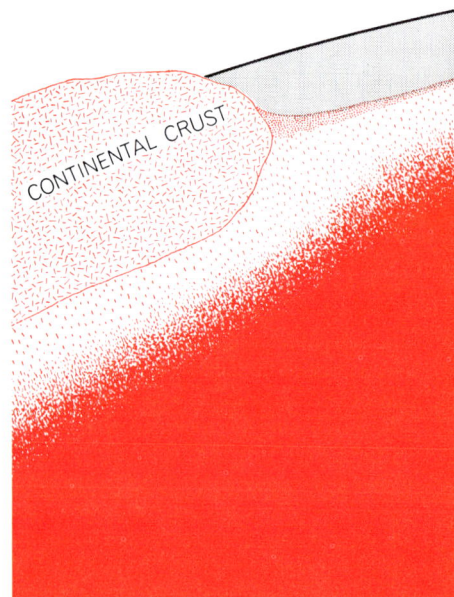

GRAND GEOCHEMICAL CYCLE (*a*) summarizes the global pathways taken sooner or later by the three-score elements that pass through the ocean and maintain its saltiness. The three "thalassochemical" models (*b, c, d*) abstracted from it are more helpful when trying to understand the rate laws governing the transport of specific elements. The rate constants, k, are expressed as a fraction: one over some number of years. The brackets enclose concentrations of the element being studied, specified according to its environment. The "cyclic" model (*b*) accounts for 90 percent of the chloride in river water. Its rate law is in quotation marks because extra factors, such as the area of the ocean, must be incorporated. The "steady state" model (*c*) works well for reactive trace metals; the reciprocal of k_2 is simply the residence time in the ocean. The "equilibrium" model (*d*) seems the most appropriate for the hydrogen ion (H^+) and the ions of the major metals, such as sodium.

MAGIC SALT MILL at the bottom of the sea, imagined in the old Norse folktale, turns out to be not so fanciful after all. The modern explanation of why the sea is salt invokes the concept of the "mid-ocean" rift and sea-floor spreading, as depicted here in cross section. The rift is a weak point be-

There is evidence that the salinity of the ocean has not changed greatly since the ocean was formed; in any event the salinity has been nearly constant for the past 200 million years (5 percent of geologic time). The composition of ancient sediments suggests that the ratio of sodium to potassium in seawater has risen from about 1 : 1 to its present value of about 28 : 1. Over the same period the ratio of magnesium to calcium has risen from roughly 1 : 1 to 3 : 1 as organisms removed calcium by building shells of calcium carbonate. It is significant, however, that the total amount of each pair of ions varied much less than the relative amounts.

If we look at rain as it reaches the sea in rivers, we find a distinctly nonmarine mix to its ions. If we catch it even earlier as it tumbles down young mountains, the differences are even more pronounced. This continual input of water of nonmarine composition would eventually overwhelm the original composition of the ocean unless there were corrective reactions at work.

The overall geochemical cycle that keeps the marine ions closely in balance involves a complex interchange of material over decades, centuries and millenniums among the atmosphere, the ocean, the rivers, the crustal rocks, the oceanic sediments and ultimately the mantle [see "a" in illustration on page 834]. Because this overall picture is too general to be of much use, we abstract bits from it and call them thalassochemical models (thalassa is the Greek word for "sea"). One model involves simply the cyclic exchange of sea salt between the rivers and the sea; the cycle includes the transport of salt from the sea surface into the atmosphere, where salt particles act as condensation nuclei on which raindrops grow [see "b" in illustration on page 834]. This process accounts for more than 90 percent of the chloride and about 50 percent of the sodium carried to the sea by rivers.

Another useful abstraction is the "steady state" thalassochemical model. If the ocean composition does not change with time, it must be rigorously true that whatever is added by the rivers must be precipitated in marine sediments [see "c" in illustration on page 834]. Oceanic residence times computed from sedimentation rates, particularly for reactive trace metals, agree well with the input rates from rivers. Unfortunately residence times do not reveal the mechanism by which an element is removed from seawater. For residence times greater than a million years it is often helpful to invoke the "equilibrium" model, which deals only with the rate of exchange between the ocean and its sediments [see "d" in illustration on page 834].

To understand how the earth maintains its geochemical poise over a billion-year time scale we must return to the circle of arrows—the weathering and "unweathering" processes—of the geochemical cycle. This circle starts with primordial igneous rock, squeezed from the mantle. Ignoring relatively minor heavy metals such as iron, we can assume that the rock consists of aluminum, silicon and oxygen combined with the alkali metals: potassium, sodium and calcium. The resulting minerals are feldspars (for example $KAlSi_3O_8$). Rainwater picks up carbon dioxide from the air and falls on the feldspar. The reaction of water, carbon dioxide and feldspar typically yields a solution of alkali ions and bicarbonate ions (HCO_3^-) in which is suspended hydrated silica (SiO_2). The

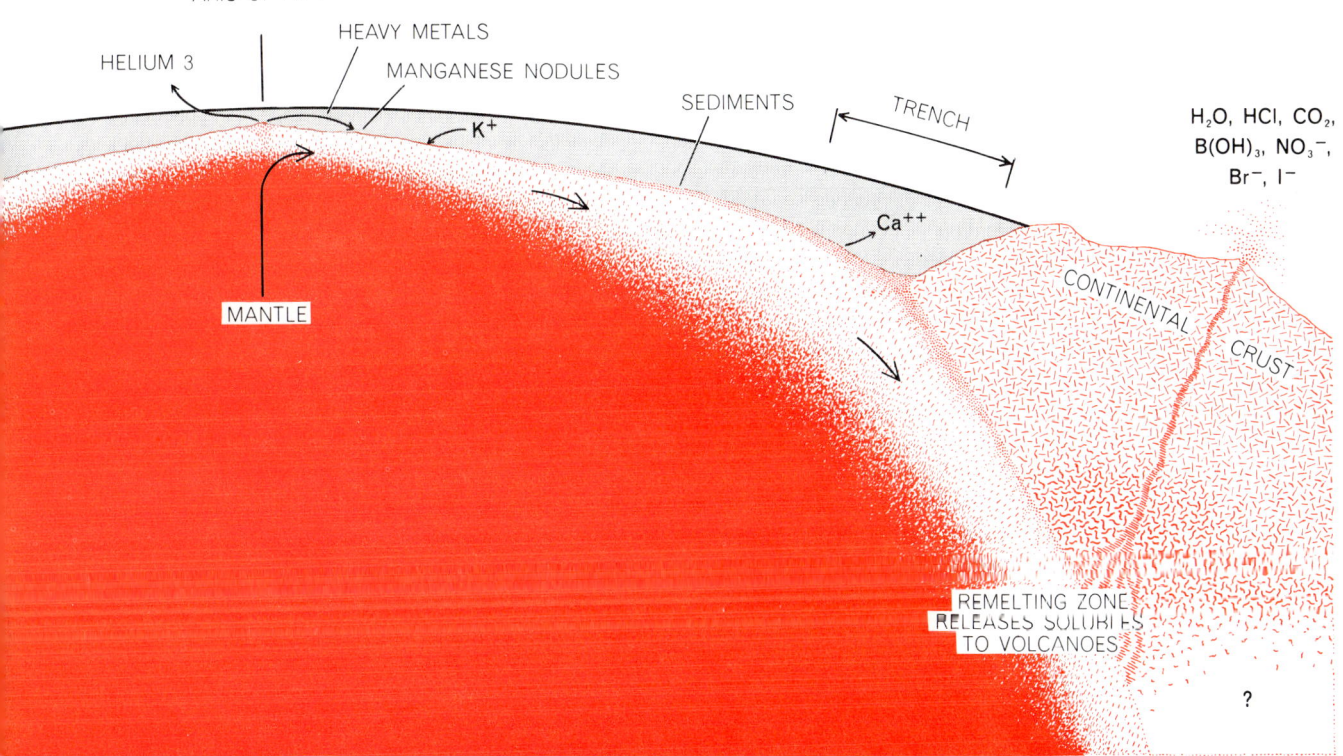

tween rigid plates, or segments, in the earth's crust. Although the driving mechanism is not yet understood, the plates move apart a few centimeters a year as fresh basalt from the plastic mantle flows up between them. The new basalt releases "juvenile" water (water never before in liquid form) and a variety of elements, including heavy metals that become incorporated in manganese nodules and the rare isotope helium 3, which escapes finally into space. At the continental margin (right) the lithospheric plate is subducted, forming a trench and carrying accumulated sediments with it. (The plate apparently thickens en route as plastic basalt "freezes" to its underside.) As it descends the plate remelts and releases soluble elements and ions that are ejected into the atmosphere by volcanoes. They maintain the saltiness of the sea and together with weathered crustal rock, such as granite, provide the stuff of sediments.

residual detrital aluminosilicate can be approximated by the clay kaolinite: $Al_2Si_2O_5(OH)_4$ [see Step 1 in illustration on page 838]. A mountain stream carries off the ions and the silica. The kaolinite fraction lags behind, first as a friable surface on weathering rock, then as soil material and finally as alluvial deposits in river valleys. If the stream evaporates in a closed basin, such as one finds in the western U.S., the result is a "soda lake" containing high concentrations of carbonates and amorphous silica.

In mature river systems the kaolinite fraction reaches the sea as suspended sediment. Encountering an ion-rich environment for the first time, the aluminosilicate must reorganize itself into new minerals. One such mineral, which seems to be forming in the ocean today, is the potassium-containing clay illite [see Step 2 in illustration on page 838]. These "clay cation" reactions may take decades or centuries. They are poorly understood because graduate students who study them invariably leave before the reactions are complete. The net effect of such reactions is to tie up and remove some of the potassium and bicarbonate ions, along with aluminum, silicon and oxygen.

A biologically important reaction, usually confined to shallow water, allows marine organisms to build shells of calcium carbonate, which precipitates when calcium (Ca^{++}) and bicarbonate ions react. If dilute hydrochloric acid is present (it is released by volcanoes), it reacts even more rapidly with bicarbonate, forming water and carbon dioxide and leaving free the chloride ion. When marine organisms die and sink to about 4,000 meters, they cross the "lysocline," below which calcium carbonate redissolves because of the high pressure. We have now traced the three metallic ions removed from igneous rock to three separate niches in the ocean. Sodium remains dissolved, potassium precipitates in clays on the deep-sea floor and calcium precipitates in shallow water as biogenic limestone: coral reefs and calcareous oozes.

Ages pass and the geochemical cycle rolls on, converting ocean-bottom clay into hard rock such as granite. When old sea floor finally reaches a region of high pressure and temperature under a continental block, it still contains some free ions that can react with the clay to reconstitute hard rock. A score of reaction schemes are possible. In Step 3 in the illustration on the opposite page I have chosen to build a "granite" from equal parts of potassium feldspar, sodium feldspar, potassium mica and quartz. (Notice that calcium is missing because it has dissolved from the sediments during their descent into the deep-ocean trenches that carry the sediments under the continental blocks.) The reaction written in Step 3 uses up all the silica formed in Step 1.

The goal of this geochemical exercise has now been reached. First, we have shown that of all the substances that enter the ocean, only sodium and chlorine remain abundantly in solution. Of the other elements, the amount remaining in solution is less than a hundredth of the amount delivered to the ocean and precipitated from it. Second, we have made a start at explaining the observed sodium-potassium ratios: in basalt this ratio is about 1 : 1, in seawater 28 : 1 and in granite 1 : 1.2. If the weight of sodium tied up in granite were about 140 times as great as the weight of sodium dissolved in the sea, the slight excess of potassium over sodium in granite would explain the sea's deficiency in potassium.

We now have working models for thinking about the circulation of the major elements, but we have barely scratched the true complexity and subtlety of seawater. The sources and sinks of the minor elements are now being explored. In many cases we can only guess at what the natural marine form of an element is because our detection techniques either convert all forms to a common form for analysis or miss some forms completely. Moreover, certain ions seem to behave capriciously in the ocean. For example, at the pH (hydrogen-ion concentration) of seawater, vanadium should appear as $VO_2(OH)_3^{--}$, an ion with a double negative charge; instead it seems to exist in positively charged form, perhaps as VO_2^+.

Much of what is known about elements in the sea can be summarized in an oceanographer's periodic table [see illustration on page 839]. The usefulness of the usual kind of periodic table to the chemist is that it arranges chemically similar elements in vertical columns and presents behavioral trends in horizontal rows. The oceanographer's table shows how these regularities are disrupted in the ocean environment.

First of all, more than a dozen elements have never been detected in seawater, although two of them (palladium and iridium) exist in parts per billion in marine sediments and another (platinum) is present in manganese nodules. The second interesting feature of the oceanographer's table is the tendency for the "upper" and "outer" elements, those in the raised wings, so to speak, to be the most plentiful in the sea. The "upper" tendency simply reflects the greater cosmic abundance of light elements. (Lithium, beryllium and boron, however, are fairly scarce even cosmically.)

The "outer" trend can be explained in quantum-mechanical terms by the presence or absence of electrons in d orbitals, the electron shells principally involved in forming complexes. Elements in the first three columns at the left have no d orbitals; those in the last four columns at the right have full d orbitals. Both characteristics favor weak chemical bonds, with the result that these two groups of elements tend to ionize readily and remain in solution, either by themselves or in simple combination with oxygen and hydrogen. In contrast, the elements in the center of the table with partially filled d orbitals form strong chemical bonds and compounds that precipitate readily; thus they can exist only at low concentration in solution. For silver and the surrounding group of metals the most stable complexes are formed with the most abundant seawater ion: chloride. Most of the other elements that are hungry for d electrons form their complexes with oxygen, or oxygen plus some protons (hydrogen nuclei).

Ordinarily the oxidation state of metals

CURRENTLY RECOVERED FROM SEAWATER

ELEMENTS IN SHORT SUPPLY

RANGE OF BIOLOGICALLY CAUSED CHANGE

RANGE OF ANALYSES

METALS CONCENTRATED IN MANGANESE NODULES

COMPOSITION OF SEAWATER has been a challenge to chemists since Antoine Laurent Lavoisier made the first analyses. The logarithmic chart on the opposite page shows in moles per kilogram the concentration of 40 of the 73 elements that have been identified in seawater. A mole is equivalent to the element's atomic weight in grams; thus a mole of chlorine is 35 grams, a mole of uranium 238 grams. Only four elements are now recovered from the sea commercially: chlorine, sodium, magnesium and bromine. Recovery of other scarce elements is not promising unless biological concentrating techniques can be developed. Manganese nodules are a potential source of scarce metals but gathering them from the deep-sea floor may not be profitable in this century.

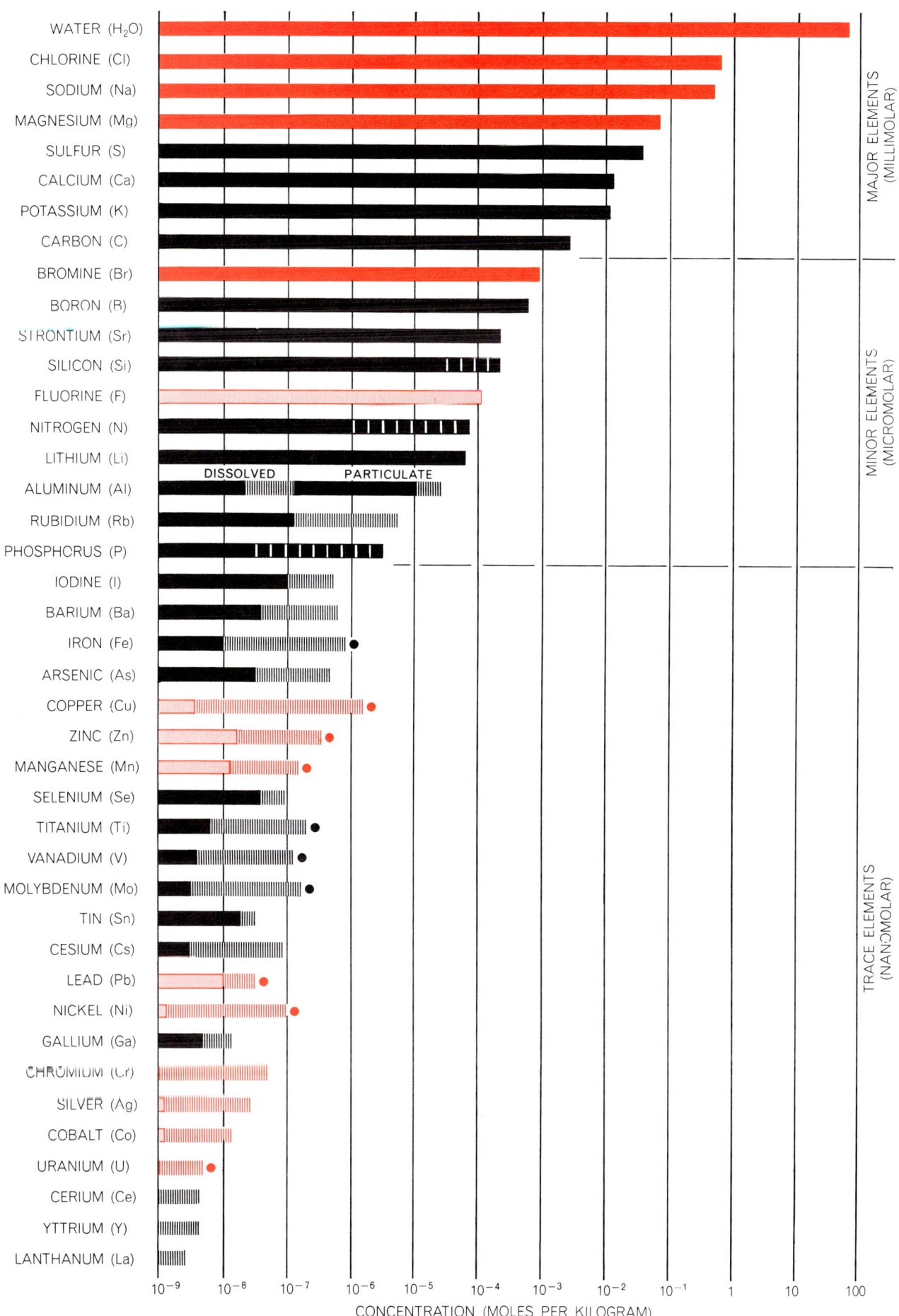

STEP 1: WEATHERING OF IGNEOUS ROCK

$$\left\{ \begin{array}{c} CaAl_2Si_2O_8 \\ \text{ANORTHITE} \\ 2KAlSi_3O_8 \\ \text{POTASSIUM FELDSPAR} \\ 2NaAlSi_3O_8 \\ \text{SODIUM FELDSPAR} \end{array} \right\} + 9H_2O + 6CO_2 \longrightarrow \left\{ \begin{array}{c} Ca^{++} \\ 2K^+ \\ 2Na^+ \\ 6HCO_3^- \end{array} \right\} + 8SiO_2(aq) + 3Al_2Si_2O_5(OH)_4 \\ \text{"KAOLINITE"}$$

IGNEOUS ROCK   +   RAINWATER $\longrightarrow$ STREAM WATER   +   DETRITUS

STEP 2: EQUILIBRATION IN OCEAN

$$3Al_2Si_2O_5(OH)_4 + 2K^+ + 2HCO_3^- \longrightarrow 2K(AlSiO_4)Al_2(OH)_2O_2(Si_2O_4) + 5H_2O + 2CO_2 \uparrow \text{ (DEEP WATER)}$$

"KAOLINITE"   +   SEAWATER $\longrightarrow$ CLAY (ILLITE)

$$Ca^{++} + 2HCO_3^- \xrightarrow{\text{ORGANISMS}} CaCO_3 \downarrow + H_2O + CO_2 \uparrow \text{ (SHALLOW WATER)}$$

$$2HCl + 2HCO_3^- \xrightarrow{\text{VULCANISM}} 2Cl^- + 2H_2O + 2CO_2 \uparrow$$

STEP 3: METAMORPHOSIS OF SHALE (CLAY)

$$2K(AlSiO_4)Al_2(OH)_2O_2(Si_2O_4) + Na^+ + Cl^- + 8SiO_2 \xrightarrow[\text{PRESSURE}]{\text{HEAT}} \left\{ \begin{array}{c} KAlSi_3O_8 \\ \text{POTASSIUM FELDSPAR} \\ NaAlSi_3O_8 \\ \text{SODIUM FELDSPAR} \\ KAl_2(AlSi_3O_{10})(OH)_2 \\ \text{POTASSIUM MICA} \\ SiO_2 \end{array} \right\} + HCl + 2SiO_2 + AlSi_2O_5(OH)$$

CLAY   +   INTERSTITIAL WATER $\longrightarrow$ "GRANITE" + VOLCANIC GAS + QUARTZ + PYROPHYLLITE

STEP 4: LEFT BEHIND IN OCEAN

$$Na^+ + Cl^-$$

ONLY SALT REMAINS after the ocean "laboratory" has finished processing the complex of chemicals removed from igneous rock by rainwater containing dissolved carbon dioxide. Step 1 yields a solution of alkali ions and bicarbonate (HCO_3^-) ions in which hydrated silica (SiO_2) and aluminosilicate detritus are suspended. In crystalline form the aluminosilicate would be kaolinite. In the ocean (*Step 2*) the "kaolinite" is complexed with potassium ions (K^+) to form illite clay. Marine organisms use the calcium ion (Ca^{++}) to make calcium carbonate shells, which form sediments in shallow water. Hydrochloric acid (HCl), injected by undersea volcanoes, reacts with bicarbonate ions, returning some carbon dioxide to the atmosphere. In Step 3 clay is metamorphosed into "granite." Sodium chloride (*Step 4*) remains. Although some of this sequence is hypothetical, something very similar seems to take place.

avid for *d* electrons would be determined by the oxidation potential of seawater, which is a measure of its ability to extract electrons from a substance just as its *p*H is a measure of its ability to extract protons. The oxidation potential of seawater has the high value of .75 volt, enabling it to extract the maximum possible number of electrons from nearly all elements except the noble metals (platinum group) and the halogens (fluorine family).

Surprisingly, however, the oxidation potential of seawater does not seem to control the oxidation states of many metals that have partially filled *d* shells. One reason is that most reactions proceed by a mechanism in which only a single electron is transferred at a time. Such transfers occur most readily when the reactants are adsorbed on surfaces where atomic geometry and electric-charge distribution are able to expedite

the redistribution of electrons (hence the utility of catalysts, which provide such surfaces). But surfaces of any kind are few and far between in the ocean, and (with the exception of manganese nodules) those that do exist are poor catalysts. A second reason for the failure of the sea's oxidation potential to control valence states is that organisms sometimes excrete electron-rich substances, which then remain in that reduced state in spite of seawater's apparent capacity to oxidize them.

Manganese nodules are porous chunks of metallic oxides up to several centimeters in diameter, widely distributed over the ocean floor. They evidently exist because they are autocatalytic for the reaction that produces them. Because of their porous structure, nodules have a surface area of as much as 100 square meters per gram. The autocatalytic property seems to extend to an

entire suite of metals that coprecipitate with manganese: iron, cobalt, nickel, copper, zinc, chromium, vanadium, tungsten and lead. Nodules found on the flanks of oceanic ridges contain significant concentrations of metals, such as nickel, that are scarce in seawater itself. This suggests that the nodules are collecting juvenile metals as the metals leak from the mantle at the fissure of the ridge. One would like to know why the nodule metals are present in oxide form rather than, as one would expect, in carbonate form.

The level of the discussion so far might best be called thalassopoetry. The discussion can be made more serious in two ways. One approach—the "geochemical balance"—has employed a computer to follow in detail as many as 60 elements as they move through the geochemical cycle, from igneous rock back

|←——NO d-ORBITALS——→|←———————— PARTIALLY FILLED d-ORBITALS ————————→|←———FULL d-ORBITALS———→|

																	He
OH^-																	
Li^+	Be	$B(OH)_3$										HCO_3^-	NO_3^-	O_2	F^-		Ne
Na^+	Mg^{2+}	$Al(OH)_3$										$Si(OH)_4$	HPO_4^{2-}	SO_4^{2-}	Cl^-		Ar
K^+	Ca^{2+}	Sc	$Ti(OH)_4$	VO_2^+	CrO_4^{2-}	$Mn(OH)_2$	$Fe(OH)_3$	$CoCl^+$	Ni^{2+}	$CuCl^+$	Zn^{2+}	Ga	$Ge(OH)_4$	$HAsO_4^{2-}$	SeO_4^{2-}	Br^-	Kr
Rb^+	Sr^{2+}	Y	Zr	Nb	MoO_4^{2-}	Tc	Ru	Rh	–	Pd	$AgCl_3^{2-}$	$CdCl_2$	In	Sn	$Sb(OH)_6^-$	Te	I^- / IO_3^-	Xe
Cs^+	Ba^{2+}	RARE EARTHS 3+	Hf	Ta	WO_4^{2-}	Re	Os	Ir	Pt	$AuCl_2^-$	$HgCl_3^-$	Tl^+	$Pb(OH)^+$	BiO^+	Po	At	Rn
Fr^+	Ra^{2+}	Ac	Th	Pa	$UO_2(CO_3)_3^{4-}$												

MAJOR ELEMENTS   MINOR ELEMENTS   TRACE ELEMENTS   DETECTED   UNDETECTED

PERIODIC TABLE, as prepared by the "thalassochemist," shows the form in which the detectable elements appear in seawater. In each box the element normally found in that place in the usual periodic table is shown in color; the elements associated with it are in black. Thus carbon appears predominantly as HCO_3^-, arsenic as $HAsO_4^{2-}$ and so on. The superscripts show the number of positive or negative charges carried by each ion. Iodine's two forms, I^- and IO_3^-, are about equally common. Except for the noble gases (*last column at right*), all the elements dissolved in the sea must be present as ions. When an element (other than a noble gas) is shown by itself, without a plus or minus charge, it means that its preferred ionic form in seawater is not yet established.

to metamorphosed sediments. In the second approach the actual chemistry of each element is followed by applying the thermodynamic methods of Josiah Willard Gibbs to systems regarded as being near equilibrium. This effort was launched by Lars Gunnar Sillén of Sweden and has been pursued by Robert M. Garrels of Northwestern University and by Heinrich D. Holland of Princeton University.

Of course no chemist in his right mind would talk seriously about equilibria in a system of variable temperature, pressure and composition that was poorly stirred, had variable inputs and contained living creatures. On the other hand, the observed uniformity of the ocean and the long periods available for reacting suggest that at least the major components are sufficiently close to equilibrium to make an investigation worthwhile. (We *know* the minor constituents are not in equilibrium.)

The equilibrium approach is based on Gibbs's phase rule, which states that the number of phases (P) possible in a system of C components at equilibrium is given by the equation $P = C + 2 - F$, where F is the number of "degrees of freedom," or quantities that may be independently varied without changing the number of phases or their composition (although F may change their relative proportions). The 2 enters the equation because only two variables, temperature and pressure, are important in most chemical reactions.

One of Sillén's most comprehensive ocean models has nine components: water, hydrochloric acid, silica, three hydroxides (aluminum, sodium and potassium), carbon dioxide and the oxides of magnesium and calcium. Observation of sea-floor sediments, aided by laboratory studies, suggests that a nine-phase ocean will result [*see illustrations on page 840*]. If C and P both equal nine, the phase rule states that the number of degrees of freedom (F) must equal two. Logically these are temperature (which can vary over the oceanic range from −2 degrees Celsius to 30 degrees) and the chloride ion concentration (which can shift over the normal oceanic range without changing the composition of the stable phases).

A diagrammatic view of how the nine components sort themselves into phases is shown in the bottom illustration on the opposite page. Note that the liquid phase contains ions not listed either as components or phases (for example H^+ and OH^-). Thermodynamics need not consider them explicitly because they do not vary independently; their concentrations are fixed by the equilibrium constants that connect the observed phases. Thus $H_2O = H^+ + OH^-$. Moreover, one knows that the product of H^+ and OH^- is a thermodynamic constant, which equals 10^{-14} mole per liter. Similar relations tie the entire system into a comprehensible whole, so that when all the calculations are performed one has discovered the equilibrium concentra-

tions of five cations (H^+, Na^+, K^+, Mg^{++} and Ca^{++}) and four anions (Cl^-, OH^-, HCO_3^- and CO_3^{--}).

It may seem peculiar to discuss an "atmosphere" containing only water vapor and carbon dioxide. One could easily add oxygen and nitrogen to the list of components. Since they would add no new phases, they would raise the number of degrees of freedom from two to four ($9 = 11 + 2 - 4$). The two new F's would be the total atmospheric pressure and the ratio of oxygen to nitrogen. In the study of the ocean, however, the partial pressure due to carbon dioxide is more significant than the total pressure of the atmosphere. Moreover, the presence of gaseous oxygen and nitrogen has little importance for the inorganic environment of the ocean, so that it is simpler to omit them and just as "real."

Suppose now we perturb the equilibrium of the model ocean by assuming that a submerged volcano has suddenly released enough hydrochloric acid (HCl) to double the amount of chloride ion (Cl^-). The dissociation of hydrochloric acid releases enough H^+ ions to raise the total number of hydrogen ions in the ocean from the former equilibrium value of 10^{-8} mole per liter to $10^{+.3}$. This excess of hydrogen ions almost immediately pushes all the available carbonate ions (CO_3^{--}) to bicarbonate ions (HCO_3^-) and the latter to carbonic acid (H_2CO_3). These shifts, however, only slightly depress the pH, which remains

COMPONENTS (C)	PHASES (P)	VARIABLES (F)
H₂O	1 GAS	TEMPERATURE
HCl	2 LIQUID	Cl⁻
SiO₂	3 QUARTZ (SiO₂)	
Al(OH)₃	4 KAOLINITE (t·o CLAY)	
NaOH	5 MONTMORILLONITE (Na·t·o·t CLAY)	
KOH	6 ILLITE (K·t·o·t CLAY)	
MgO	7 CHLORITE (Mg·t·o·t CLAY)	
CO₂	8 CALCITE (CaCO₃)	
CaO	9 PHILLIPSITE (Na·K FELDSPAR)	

NINE MAJOR COMPONENTS IN SEA can, to a first approximation, be combined into nine distinctive phases to satisfy the "phase rule" that governs systems in equilibrium. The rule, formulated in the 19th century by Josiah Willard Gibbs, prescribes the number of phases P, components C and degrees of freedom F in such a system: $P = C + 2 - F$. When the number of phases and components are equal, the number of degrees of freedom, F, must be two, which allows both the temperature and the chloride-ion concentration to vary without altering the number of phases. In the clay-containing phases (4, 5, 6, 7) the letter "t" stands for a tetrahedral crystal structure; the letter "o" stands for an octahedral structure.

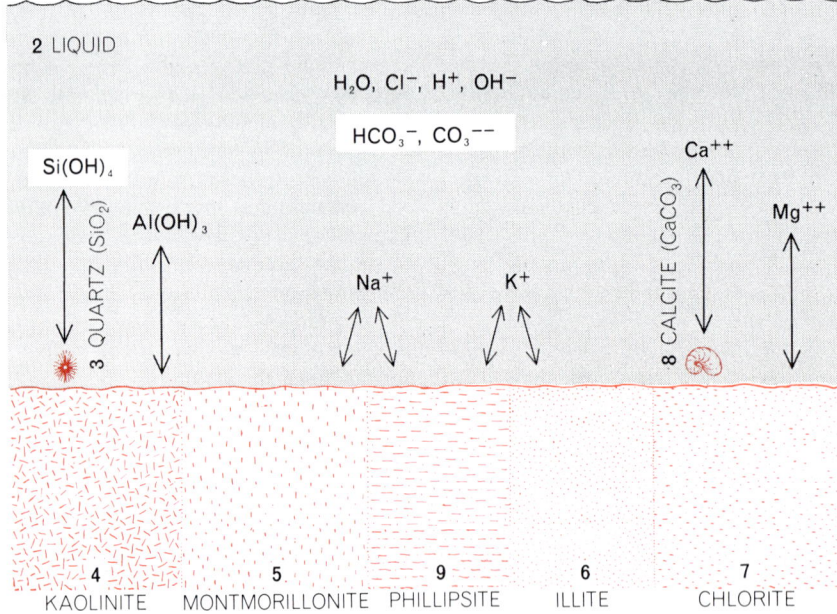

1 GAS

H₂O VAPOR, CO₂

2 LIQUID

H₂O, Cl⁻, H⁺, OH⁻

HCO₃⁻, CO₃⁻⁻

Si(OH)₄   Ca⁺⁺   Mg⁺⁺

Al(OH)₃

Na⁺   K⁺

3 QUARTZ (SiO₂)   8 CALCITE (CaCO₃)

4   5   9   6   7
KAOLINITE   MONTMORILLONITE   PHILLIPSITE   ILLITE   CHLORITE

EQUILIBRIUM OCEAN MODEL, consisting of nine phases and nine components, shows how the principal constituents of the ocean distribute themselves among the atmosphere, the ocean and the sediments. Three of the constituents (HCO₃⁻, CO₃⁻⁻ and Si(OH)₄) are not included among nine listed components but appear as equilibrium products of those that are listed, as do seven ions (H⁺, K⁺, Na⁺, Ca⁺⁺, Mg⁺⁺, Cl⁻, OH⁻). Two of the solids are shown as biological "precipitates": "quartz" (3) in the form of silicate structures built by radiolarians and "calcite" (8) in the form of calcium carbonate chambers built by foraminifera. The method of precipitation is unimportant as long as the product is stable. The equilibrium model goes far to explain why the ocean has the composition it does.

high until the slow circulation of the ocean brings the hydrogen ions in direct contact with the clay sediments on the sea floor.

The structure of clay is such that oxygen atoms at the free corners of polyhedrons carry unsatisfied negative charges, which attract positive ions [*see top illustration on page 844*]. Because the ocean is so rich in sodium ions (Na⁺), they occupy most of the corners of clay polyhedrons. When the excess hydrogen ions come in contact with the clay, they quickly replace the sodium ions and set them adrift. This fast reaction is limited in scope because the surface and interlayer ion-exchange capacity of clay is not very great. Much more capacity is provided when the structure of the clay is rearranged; for example, the conversion of montmorillonite to kaolinite also consumes hydrogen atoms and releases sodium. Given sufficient time—centuries—such rearrangements inexorably take place, and the pH of the ocean slowly drifts back to its equilibrium value. The charge on the excess chloride introduced by the volcano will then be balanced not by H⁺ but by Na⁺. This slow equilibration mechanism can be regarded as the ocean's "pH-stat" (in analogy with "thermostat"). This clay-cation model suggests that the pH of the ocean has been constant over the span of geologic time and that hence the carbon dioxide content of the atmosphere has been held within narrow limits.

If only the pH-stat were available for leveling surges in pH, the ocean might be subjected to violent local fluctuations. For fast response pH control is taken over by a corbonate buffer system [*see bottom illustration on page 844*]. In fact, until recently oceanographers neglected the clay-cation reactions and assumed that the carbonate-buffer system almost completely determined the pH of the ocean.

One might think that if the carbon dioxide content of the atmosphere were to decrease, carbon dioxide would flow from the sea into the atmosphere, leading to a general depletion of all carbonate species in the ocean and eventually to the dissolution of some carbonate sediments. In actuality something quite different happens because the carbonate system is its own source of hydrogen ions. Removal of carbon dioxide from water reduces the concentration of carbonic acid (H₂CO₃), the hydrated form of carbon dioxide. Replacement of this acid from bicarbonate ions requires a hydrogen ion, which can only be obtained by converting another bicar-

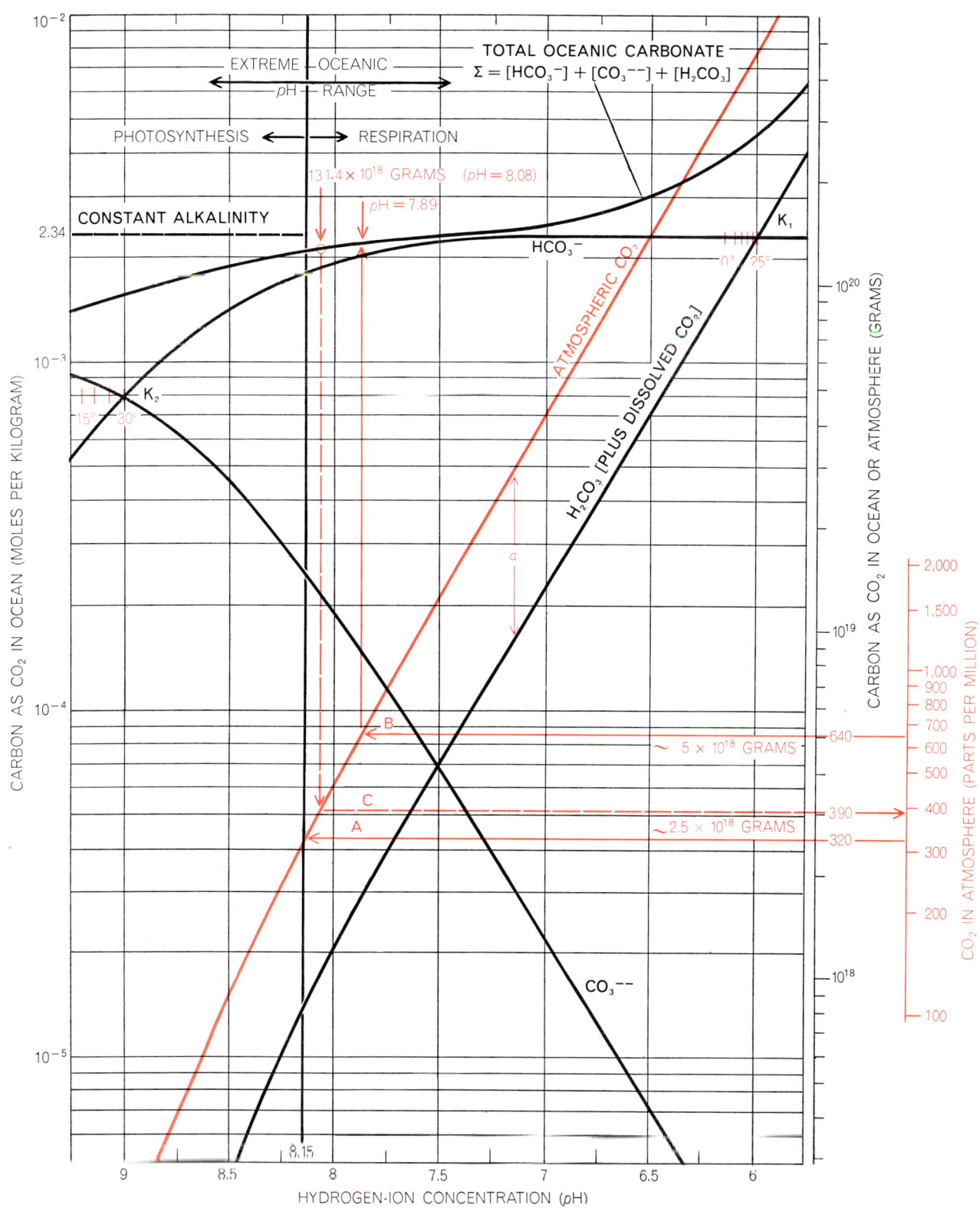

OCEANIC CARBONATE SYSTEM can be represented by a "Bjerrum diagram" that shows how carbonate in its several forms varies with the ocean's pH, or hydrogen-ion concentration. The diagram is plotted for a constant "carbonate alkalinity" of 2.34×10^{-3} moles of carbonate per kilogram of seawater (*scale at left*). "System point" K_1 shows where the concentrations of bicarbonate ion (HCO_3^-) and carbonic acid (H_2CO_3) are equal. At K_2 the concentrations of bicarbonate and carbonate (CO_3^{--}) are equal. The exact locations of K_1 and K_2 are shown for a range of temperatures (in degrees Celsius) at constant conditions of salinity and pressure. The top curve, Σ, is the sum of oceanic carbonate in all its forms. The normal pH of the ocean is 8.15. The two short arrows at top mark the normal biological limits: at 7.95 the available oxygen has been consumed by respiration; at 8.35 photosynthesis has removed so much carbon dioxide that absorption from the atmosphere rises sharply. The limits of oceanic pH lie between 7.45 and 8.6. The amount of carbon dioxide in the atmosphere (*colored curve and scale at far right*) is related to the amount of carbon dioxide dissolved in the ocean by alpha (α), the average worldwide solubility of carbon dioxide in seawater. The consequences of doubling the carbon dioxide in the atmosphere from 320 parts per million (A) to 640 parts (B) are discussed in the text of the article, as is line C.

bonate ion to carbonate. The overall reaction is $2HCO_3^- \rightarrow H_2CO_3 + CO_3^{--}$. Thus instead of dissolving existing sediments, removing carbon dioxide from the sea may actually precipitate carbonate. This reaction can be seen in the "whitings" of the sea over the Bahama Banks, where cold deep water, rich in dissolved carbon dioxide and calcium, is forced to the surface and warmed. As carbon dioxide escapes into the air, the pH drops and aragonite ($CaCO_3$) precipitates, turning large areas of the ocean white with a myriad of small crystals.

The reaction above conserves charge, which means that the "alkalinity"—the traditional name for the concentration of sodium ion ("alkali") needed to bal-

ance this negative charge—is also conserved. The "carbonate alkalinity," defined as the bicarbonate concentration plus twice the carbonate concentration, is useful because it remains fixed even when the relative amounts of the two species vary.

The system can be visualized with the help of the illustration on the opposite page, which is the "Bjerrum plot" for carbonic acid at constant alkalinity. It takes its name from Niels Bjerrum, who introduced such plots in 1914; it shows the interrelations between the various compounds in the world carbonate system as a function of pH. Although the diagram ignores variations of pressure, temperature and salinity, it displays the

essential features of the system.

The Bjerrum plot facilitates a semi-quantitative discussion of the relation of atmospheric carbon dioxide to oceanic carbon dioxide. Over the next 20 years we shall burn enough fossil fuel to double the amount of carbon dioxide in the atmosphere from 320 parts per million to 640. On the plot this is indicated by shifting the line A, corresponding to 320 parts per million, to position B, 640 parts per million.

To produce this shift some 2.5×10^{18} grams of carbon dioxide must be added to the atmosphere. If the altered atmosphere were to come to equilibrium with the ocean, the pH of the ocean would drop from its present value of 8.15 to

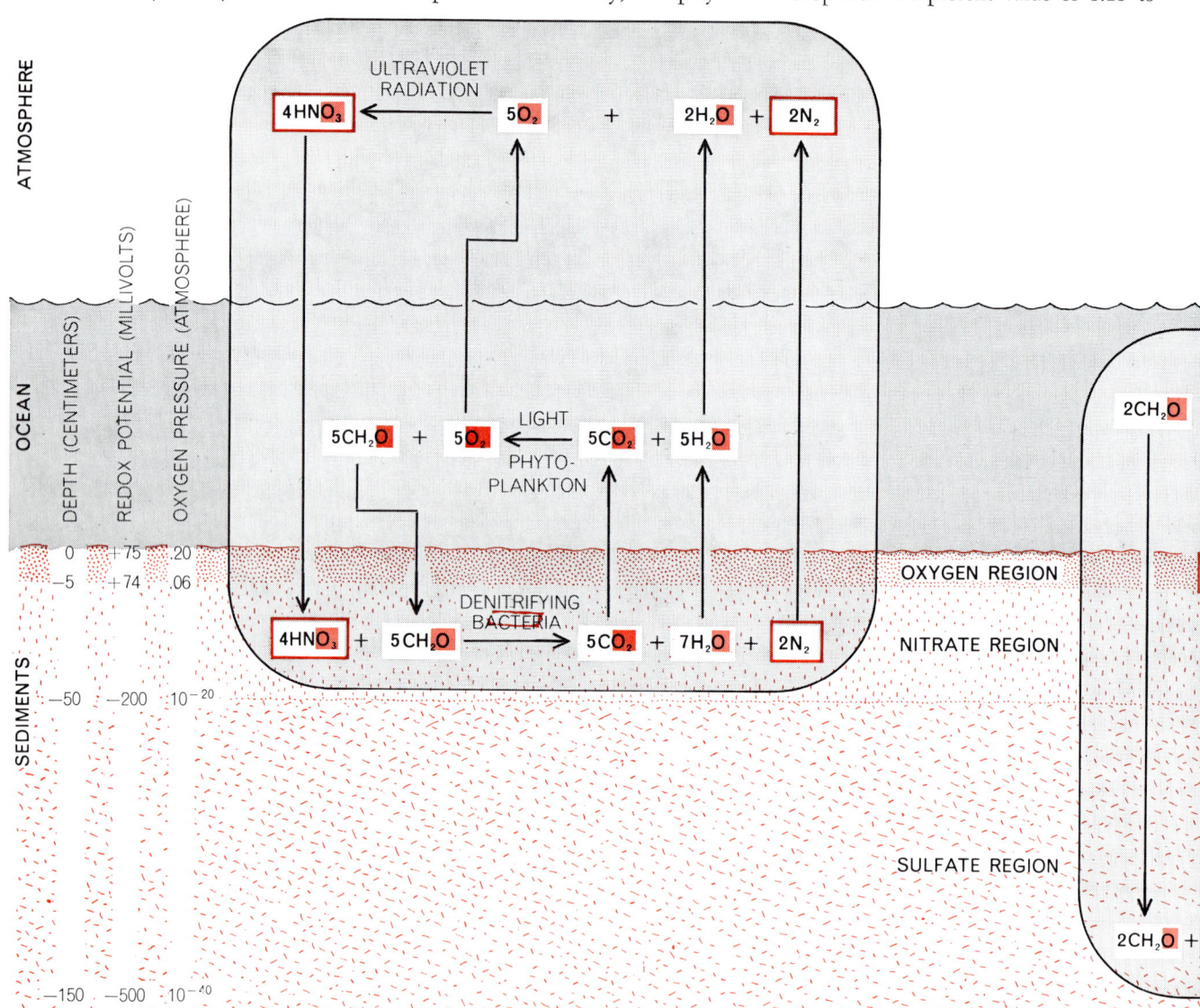

BACTERIA IN MARINE SEDIMENTS, although scarce by terrestrial soil standards, play a major role in replenishing the oxygen of the atmosphere and in limiting the accumulation of organic sediments. The bacteria concerned are buried in fine-grained sediments from several centimeters to several tens of centimeters below the ocean floor, with limited access to free oxygen for respiration. Thus deprived, they use the oxygen in nitrates and sulfates to oxidize organic compounds, represented by CH_2O. The actual reactions are far more complex than indicated here. The net result, however, is that denitrifying bacteria (*left*) release free nitrogen and convert carbon to a form (carbon dioxide) that can be reutilized by phytoplankton. These organisms, in turn, release free oxy-

7.89—still well within the range tolerated by marine organisms. This cannot happen, however, because the total mass of carbon dioxide in the ocean (Σ in the Bjerrum plot) plus the carbon dioxide in the atmosphere would have to increase from its present value, 128.9×10^{18} grams, to 138.3×10^{18} grams. The difference, 9.4×10^{18} grams, is nearly four times the amount added to the atmosphere.

The long-term equilibration process for such an atmospheric doubling can be broken down into two steps. First the pH-buffer system operates: 2.5×10^{18} grams, or 2 percent of the total mass, is added to the world system at constant alkalinity. The result of this step is the

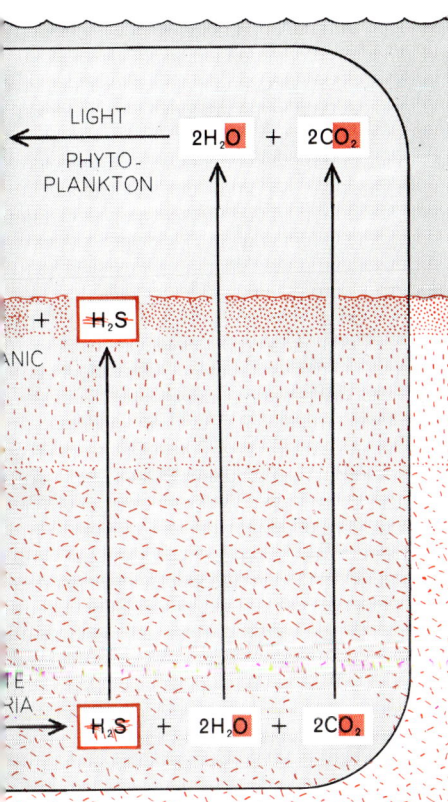

gen. Without the cooperative effort of these two groups of organisms the oxygen in the atmosphere might all be fixed by high-energy processes within some 10 million years. The sulfate bacteria (*right*) play a role in the recycling of sulfur and oxygen.

line C in the diagram, corresponding to a total mass of 131.4×10^{18} grams, an atmospheric carbon dioxide content of 390 parts per million and an oceanic pH of 8.08. Next, if the ocean has time to equilibrate with its sediments, the pH-stat will operate, returning the system to pH 8.15 at a constant total mass. The result of this step is that the alkalinity rises by 2 percent, which in terms of the Bjerrum plot means that the system will return to normal except that all the numbers on the concentration axes will be multiplied by 1.02. The long-range effect of a sudden doubling of the atmosphere's carbon dioxide, therefore, is to increase the ultimate value 2 percent, from 320 parts per million to 326, and some of that increase will ultimately find its way into vegetation and humus.

It is obvious that rates are crucial in the global distribution of carbon dioxide. The wind-stirred surface layer of the sea exchanges carbon dioxide rapidly with the atmosphere, requiring less than a decade for equilibration. Because this layer is only about 100 meters deep it contains only a tiny fraction of the ocean's total volume. Large-scale disposal of atmospheric carbon dioxide therefore requires that the gas be dissolved and transported to deep water.

Such vertical transport takes place almost exclusively in the Weddell Sea off the coast of Antarctica. Every winter, when the Weddell ice shelf freezes, the salt excluded from the newly formed ice increases the salinity and hence the density of the water below. This ice-cold water, capable of containing more dissolved gas than an equal volume of tropical water, cascades gently down the slope of Antarctica to begin a 5,000-year journey northward across the bottom of the ocean. The carbon dioxide in this "antarctic bottom water" has plenty of time to come to equilibrium with clay sediments.

Enough fossil fuel has been burned in the past century to have raised the carbon dioxide content of the atmosphere from about 290 parts per million to 350 parts. Since the actual level is now 320 parts per million, about half of the carbon dioxide put into the air has been removed. Although proof is lacking, a principal removal agent is undoubtedly antarctic bottom water. The process is so slow, however, that the carbon dioxide content of the atmosphere may reach 480 parts per million before the end of the century. By then it should be clear if man's inadvertent global experiment (altering the atmosphere's carbon dioxide content) will have the predicted ef-

fect of changing the earth's climate. In principle an increase in atmospheric carbon dioxide should reduce the amount of long-wavelength radiation sent back into space by the earth and thus produce a greenhouse effect, slightly raising the average world temperature.

Having described an equilibrium model of the ocean that neglected the atmosphere's content of nitrogen and oxygen, I should not leave the reader with the impression that the continued presence of these two gases in the atmosphere is independent of the ocean. If the ocean were truly in equilibrium with the atmosphere, it would long since have captured all the atmospheric oxygen in the form of nitrates, both in solution and in sediments. This catastrophe has apparently been averted by the intervention of certain marine bacteria that have the happy faculty of releasing nitrogen gas from nitrate compounds and of converting the oxygen to a form that can later be liberated by phytoplankton.

The story is this. A variety of high-energy processes in the atmosphere continuously break the triple chemical bond that holds two nitrogen atoms together in a nitrogen molecule (N_2). The bonds can be broken by ultraviolet photons, by cosmic rays, by lightning and by the explosions in internal-combustion engines. Once dissociated, nitrogen atoms can react with oxygen to form various oxides, which are then carried to the ground by rainfall. In the soil these oxides are useful as fertilizer. Ultimately large amounts of them reach the sea. They do not, however, accumulate there and no one is really sure why.

The best guess is that denitrifying bacteria in oceanic sediments use the oxygen of nitrate to oxidize organic molecules when they run out of free oxygen [*see left half of illustration on these two pages*]. The nitrogen is released directly as a gas, which goes into solution but is available for return to the atmosphere. The oxygen emerges in molecules of water and carbon dioxide. The carbon dioxide is assimilated by phytoplankton, which build the carbon into organic compounds and release the oxygen as dissolved gas, also available for return to the atmosphere. Without these coupled biological processes the atmospheric fixation of nitrogen would probably exhaust the world's oxygen supply in less than 10 million years. Nevertheless, the amount of nitrogen returned to the atmosphere from the sediments is so small that we may never be able to measure it directly: the yearly return is less than one two-thousandth of the total nitrogen dissolved in the sea.

Another little-known epicycle in the global oxygen cycle probably has the effect of limiting the net accumulation of carbon in the form of oil-bearing shale, tar sands and petroleum. After denitrifying bacteria have consumed the nitrate

in young sediments, sulfate bacteria begin oxidizing organic matter with the oxygen contained in sulfates [*see right half of illustration on pages 843–844*]. The product, in addition to water and carbon dioxide, is hydrogen sulfide, the foul-smelling compound that characterizes environments deficient in oxygen.

In undisturbed mud the hydrogen sulfide never reaches the surface because it is inorganically reoxidized to sulfate as soon as it comes in contact with free oxygen. It seems likely that the bacterial turnover of oxygen in sulfate is so rapid that half of the world's oxygen passes through this epicycle in about 50,000 years.

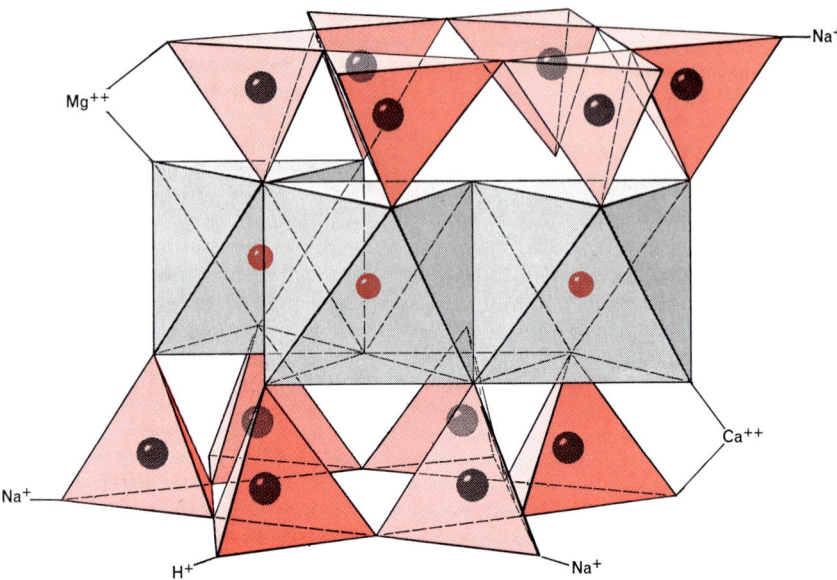

THREE-LAYER CLAY PARTICLE has a layer of octahedrons sandwiched between two layers of tetrahedrons. Each octahedron consists of an atom of aluminum surrounded by six closely packed atoms of oxygen. Each tetrahedron consists of a silicon atom surrounded by four atoms of oxygen. The polyhedrons are tied into layers at shared corners where a single oxygen atom is bonded to a silicon atom on one side and to an aluminum atom on the other. At the free corners the oxygen atoms bear unsatisfied negative charges that attract cations such as sodium ($Na+$) and potassium ($K+$). If the hydrogen-ion concentration should rise in the vicinity of clay, free hydrogen ions tend to be exchanged for sodium ions, which are released. In addition, many doubly charged metal ions can replace Si^{4+} at the centers of tetrahedrons and Si^{4+} can replace Al^{3+} in the octahedrons. Whenever this occurs, another cation is bound to the structure to conserve charge. Such reactions apparently exert considerable control over the ocean's composition and hydrogen-ion concentration.

The global activities of man have now reached such a scale that they are beginning to have a profound effect on marine chemistry and biology. We are learning that even the ocean is not large enough to absorb all the waste products of industrial society. The experiment involving the release of carbon dioxide is now in progress. DDT, only 25 years on the scene, is now found in the tissues of animals from pole to pole and has pushed several species of birds close to extinction. The concentration of lead in plants, animals and man has increased tenfold since tetraethyl lead was first used as an antiknock agent in motor fuels. And high levels of mercury in fish have forced the abandonment of some commercial fisheries. (Lead and mercury are systemic enzyme poisons.) Of the total petroleum production some .2 percent gets slopped into the sea in half a dozen major accidents each year. (At least six of the rare gray whales died last year after migrating through the oil slick off Santa Barbara caused by the blowout of a well casing belonging to the Union Oil Company.) Conceivably a persistent oil film could change the surface reflectivity of the ocean enough to alter the world's energy balance. The rapid increase in the use of nitrogen fertilizers leaves a nitrate excess that runs into rivers, lakes and ultimately reaches the sea. The sea can probably tolerate the runoff indefinitely but along the way the nitrogen creates algal "blooms" that are hastening the dystrophication of lakes and estuaries.

It is fashionable today to view the ocean as the last global frontier, waiting only technological "development." Thermodynamically it is easier to extract fresh water from sewage than from seawater. Ecologically it is wiser to keep our concentrated nutrients on land than to dilute them beyond recall in the ocean. Sociologically, and probably economically, it makes more sense to process our junkyards for usable metals than to mine the deep-sea floor. The task is to persuade our engineers and business companies that working with sewage and junk is just as challenging as oceanography and thalassochemistry.

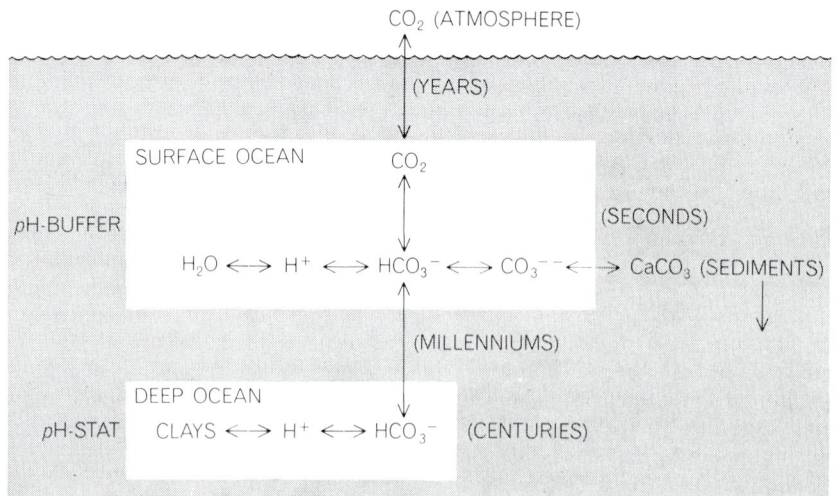

HYDROGEN-ION CONCENTRATION, or *p*H, of the ocean is controlled by two mechanisms, one that responds swiftly and one that takes centuries. The first, the "*p*H-buffer," operates near the surface and maintains equilibrium among carbon dioxide, bicarbonate ion (HCO_3^-), carbonate ion (CO_3^{--}) and sediments. The slower mechanism, the "*p*H-stat," seems to exert ultimate control over *p*H; it involves the interaction of bicarbonate ions and protons (H^+) with clays. Clay will accept protons in exchange for sodium ions (primarily).

The Author

FERREN MACINTYRE is research associate in the Marine Science Institute of the University of California at Santa Barbara. He writes: "I worked for eight years in the real world—as lumberman, machinist, machine designer, loftsman— before retiring to college. Turned chemist because Ambrose Nichols, then at San Diego State, was the best teacher I've ever met. B.A. in chemistry from the University of California at Riverside, 1960; Ph.D. in physical chemistry from the Massachusetts Institute of Technology in 1965. Spent three years at the Scripps Institution of Oceanography, where my specialty was the 'top half' of the ocean (on a logarithmic scale), examining the physicochemical hydrodynamics of the top few millimeters of the sea and the interchange of chemicals between sea and air. But anyone can be a specialist; the problem is to generalize sufficiently to find our way back to a human habitat that reflects our claim to intelligence. My other interests include folk music, building musical instruments and rock-climbing."

Bibliography

THE OCEANS: THEIR PHYSICS, CHEMISTRY, AND GENERAL BIOLOGY. H. U. Sverdrup, Martin W. Johnson and Richard H. Fleming. Prentice-Hall, Inc., 1961.

THE COMPOSITION OF SEA-WATER, SECTION I: CHEMISTRY, in *The Sea: Ideas and Observations on Progress in the Study of the Seas, Vol. II,* edited by M. N. Hill. Interscience Publishers, 1963.

CHEMICAL OCEANOGRAPHY. Edited by J. P. Riley and G. Skirrow. Academic Press, 1965.

THE OCEAN AS A CHEMICAL SYSTEM. Lars Gunnar Sillén in *Science,* Vol. 156, No. 3779, pages 1189–1197; June 2, 1967.

MARINE CHEMISTRY: THE STRUCTURE OF WATER AND THE CHEMISTRY OF THE HYDROSPHERE. R. A. Horne. Wiley-Interscience, 1969.

SCIENTIFIC
AMERICAN January 1971, Vol. 224, No. 1, pp. 32–42

OFFPRINT **894**

THE GLOBAL CIRCULATION
OF ATMOSPHERIC POLLUTANTS

by Reginald E. Newell

Worldwide wind and temperature patterns and the behavior of trace
substances are studied in an effort to learn what effect changes
in the atmosphere caused by man may have on the earth's climate.

Pollution is more than a plume of smoke rising above a factory or a yellowish haze hanging over a city. The foreign substances man introduces into the air spread all over the globe and rise into the upper atmosphere. It is therefore important to learn how each of the major pollutants enters the atmosphere, the speed and extent of its spread and the ways in which it may alter the atmosphere and thus affect temperature and precipitation both locally and worldwide. In this article I shall discuss the movement of the various pollutants and review what is known about the effect on temperature of foreign substances in the atmosphere.

Of the total of 164 million metric tons of pollutants emitted each year in the U.S., about half comes from automobiles. Of the main component, carbon monoxide, 77 percent is from automobiles; so are most of the hydrocarbons and much of the oxides of nitrogen. The oxides of sulfur come mainly from electric power plants, small particles mainly from power plants and industry. In addition to these more obvious pollutants, vast quantities of water and carbon dioxide are produced by the burning of fossil fuels. Other pollutants are lead from automotive gasoline and ozone, which is produced by the action of sunlight on automobile exhaust. Radioactive substances introduced by man's activity include fission products from weapons tests, such as strontium 90, cesium 137 and iodine 131, and neutron-activated casing materials, such as tungsten 185, manganese 54, iron 55, rhodium 102 and cadmium 109. A satellite carrying a portable power plant using plutonium 238 as fuel accidentally burned up in the atmosphere over the southern Indian Ocean in April, 1964, at a height of from 40 to 50 kilometers instead of going into orbit, and the radioactivity from that point source has been tracked over the globe. The reprocessing of nuclear-fuel elements from power plants releases krypton 85, a radioactive gas with a half-life of about 10 years, which is gradually accumulating in the atmosphere.

Trace substances are, of course, present in nature. Carbon dioxide is taken up by plants during their growth cycle and released by the decay of plant material. Sulfur is also involved in plant processes and is abundant in the ocean, from which it is released by sea spray; it is sometimes injected high into the atmosphere in large amounts during volcanic eruptions. Ozone is produced in the upper atmosphere and carried downward toward the earth's surface, where it is destroyed [see "The Circulation of the Upper Atmosphere," by Reginald E. Newell; SCIENTIFIC AMERICAN, March, 1964]. As for radioactivity, radon and thoron gas emanate from the soil and decay in the atmosphere, giving rise to a chain of radioactive substances; some of the end products, such as lead 210, are transported up into the stratosphere. Cosmic rays entering the atmosphere collide with air molecules, usually in the stratosphere, to form radioactive nuclides such as beryllium 7, sodium 22 and carbon 14, some of which live long enough to find their way down to the surface.

Whether natural or man-made, some of these trace substances occur as gases, others as aerosols: finely divided liquid droplets or solid particles. Many are involved in phase transformations. For example, water vapor (gas) may be cooled to the point where it changes to water droplets or to ice crystals, forming clouds; sulfur dioxide gas may change in moist air to droplets of sulfuric acid. The gases diffuse and mix quite easily but the aerosols are governed by a number of factors that limit their spread. Large particles with radii of 10 microns (thou-

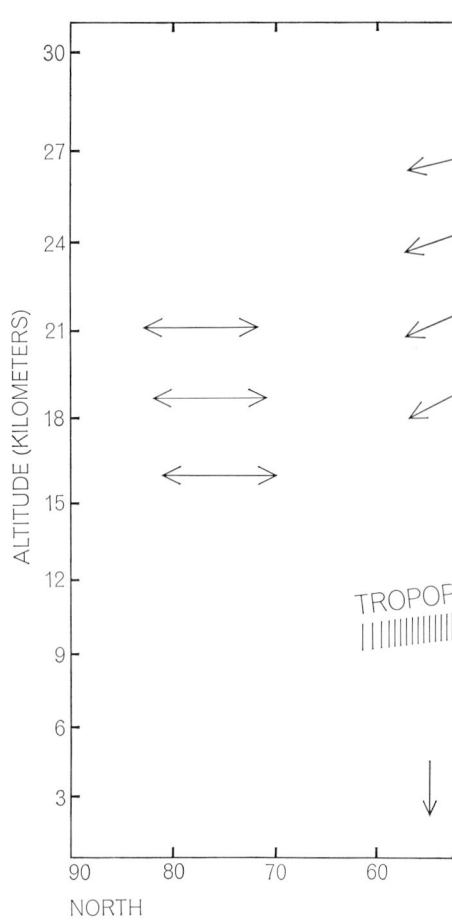

DISTRIBUTION AND MOVEMENTS of
ozone (*black*), water (*gray*) and aerosols, or
small solid and liquid particles (*color*), are

sandths of a millimeter) or more can be washed out of the air by raindrops or fall out directly. Very small particles can grow by coagulation until they too can be trapped in clouds and be washed out. Particles larger than about .3 micron cannot reach the upper atmosphere under normal circumstances because their fall velocities are greater than the average updraft speeds. Larger particles are nevertheless found in the upper atmosphere, some being introduced directly by volcanic eruptions, others growing there from smaller particles and gases introduced by the air motion. (Incidentally, the electrostatic precipitators on smokestacks are good for trapping particles with radii larger than about a micron; so is the human nose. Smaller particles can penetrate into the lungs, however, and so it is not always the larger particles one sees pouring from chimneys that are the most hazardous to health.)

Trace substances are moved over the globe by wind systems that fluctuate in strength and direction from day to day as cyclones and anticyclones move around the globe. The lowest 10 to 15

kilometers (six to nine miles) of the atmosphere, where the temperature decreases with height, is called the troposphere; above it to about 50 kilometers is the stratosphere, which in some respects resembles a series of stratified layers. Air parcels in both the troposphere and the stratosphere can be tabbed and tracked from one day to another [see illustration on next page]. The prevailing, or mean, wind, which is revealed by averaging over a long time period, blows from west to east over much of the middle latitudes in both hemispheres, but at low latitudes (and in the upper regions in general in summer) the prevailing wind is from the east. The mean wind transfers trace substances fairly rapidly round the globe; for example, the 35-meter-per-second west to east flow at 30 degrees north latitude gives a transit time of about 12 days. Clouds of debris from a nuclear explosion or volcanic eruption can often be identified as they make several circuits of the globe.

While the prevailing wind is along lines of latitude, north-south oscillations occur at the same time; the resulting north-south drift and an accompanying up-and-down motion give rise to an

overturning pattern at low latitudes called the Hadley-cell circulation [see illustration on page 850]. In the Tropics this large cell is a dominant feature of the circulation, whereas at middle latitudes north-south eddies in the prevailing-wind systems overshadow the mean north-south drifts. Air can be exchanged between the hemispheres by both the mean Hadley circulation and eddies in the upper tropical troposphere. As for vertical exchange, air passes into the stratosphere from the troposphere at low latitudes in the Hadley-cell circulation. Most of the transfer back into the troposphere is thought to occur close to the tropopause level near the middle-latitude jet stream [see illustration on these two pages], and some transfer back into the stratosphere may also occur there.

In the troposphere clouds, rain and thunderstorms are evidence of considerable vertical motion. Vertical velocities may reach 10 to 20 meters per second in thunderstorms but are generally no more than 10 centimeters per second in normal middle-latitude cyclones and anticyclones. In the stratosphere it is much harder to push an air parcel up or down

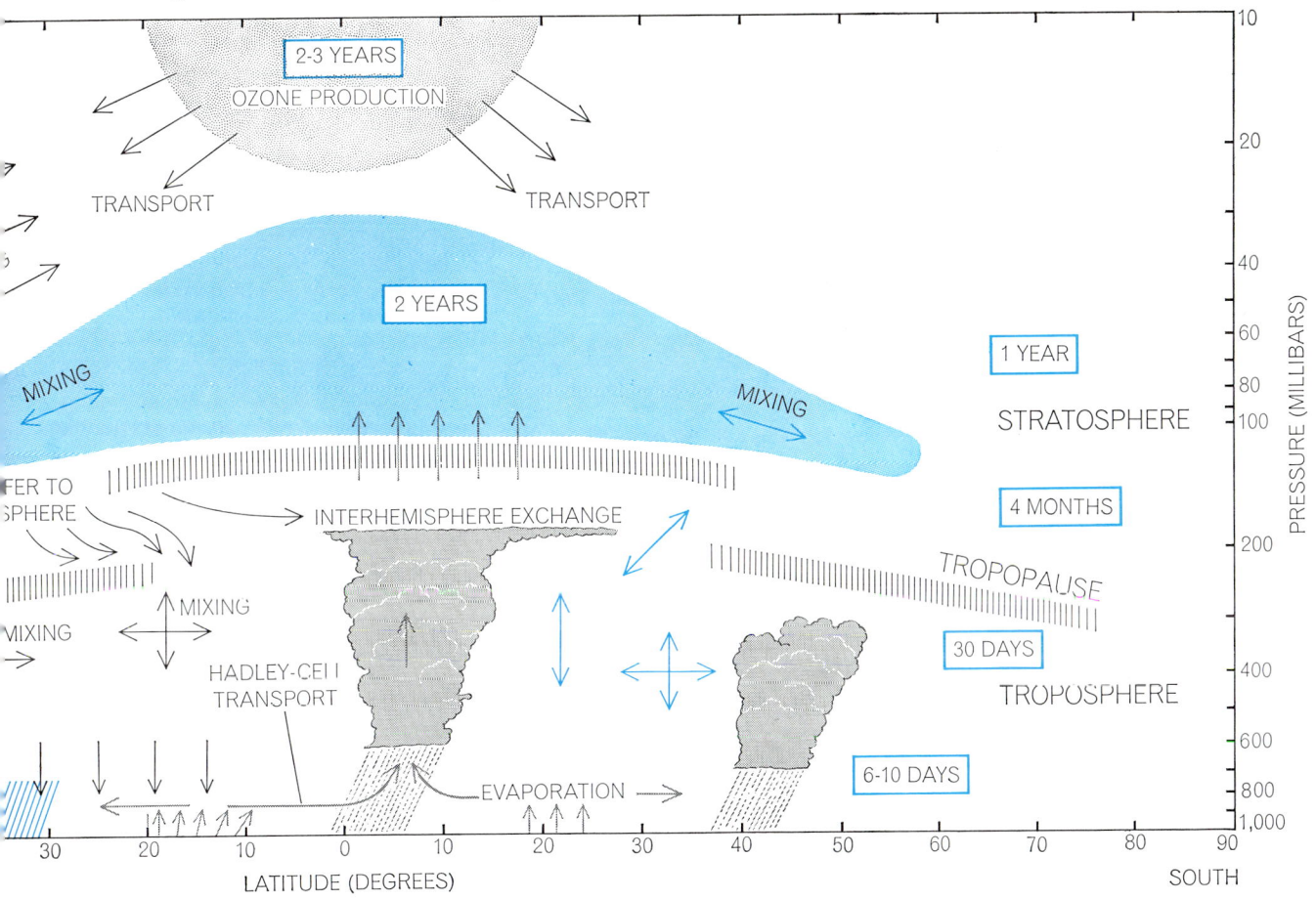

indicated on a schematic diagram drawn along a line of longitude. Effects illustrated for one hemisphere occur in both hemispheres. Boxed figures are residence times for aerosols. Pressure, measured in millibars, is often used by meteorologists as a measure of altitude. The tropopause is the boundary between the troposphere and the stratosphere; its altitude varies with latitude as indicated.

because the temperature increases with height, and so vertical motions rarely exceed a few centimeters per second and are often much smaller. The vertical spread of a trace substance in the stratosphere is therefore rather slow, like the downward migration of a card in a pack of cards that is being shuffled. There is much less shuffling at low latitudes than there is over the polar regions, so that trace constituents can stay in the equatorial stratosphere for several years.

Once the material reaches the bottom of the pack and enters the troposphere it can be mixed vertically rather rapidly. Small particles spend about 30 days in the troposphere before being washed out by rain. Gases spend varying periods there depending on the "sinks" by which each is removed from the atmosphere: incorporation into cloud droplets, reactions with other gases, loss to finely divided liquid or solid particles or the earth's surface and so on. Generally the tropospheric residence times of gases are from about two to four months—provided that there *is* a sink. (Krypton 85, for example, has no known sink and disappears only by radioactive decay.)

The global temperature pattern shows that the coldest air is over the Equator near the tropical tropopause at all seasons and over the winter pole in the stratosphere. The lowest temperatures over the Equator occur in January. Notice that temperature increases with latitude in the lower stratosphere. The temperature pattern is maintained by the liberation of latent heat, by radiation and by the motion of air masses, the sum of which is roughly in balance. When more water is rained out of a given air column than is evaporated into it from the surface, the column gains the latent heat that is liberated. The net effect of latent-heat liberation and radiative processes is that tropospheric air is heated at low latitudes and cooled at high latitudes, generating energy. In the troposphere the net effect of the radiative processes by themselves is to cool the air at all latitudes. Contributions come from absorption of incoming solar radiation, and from absorption and reemission of terrestrial long-wave radiation, by carbon dioxide, water vapor and ozone. Water vapor dominates in the lower layers, producing cooling of up to two degrees per day, with carbon dioxide of secondary importance; in the strato-

sphere carbon dioxide and ozone dominate [*see top illustration on page* 9]. Horizontal and vertical motions can also produce temperature changes. If air is forced to rise, it cools as it expands; if it is forced to sink, it contracts and becomes warmer, as in a bicycle pump. The former effect is thought to maintain the low temperatures over the tropical tropopause, the upward motion being forced from below, where the latent heat is liberated. On the other hand, compression is responsible for the inversion over the Los Angeles basin during a good fraction of the year—the increase of temperature with altitude that inhibits vertical mixing and traps pollutants near the surface.

Carbon dioxide is quite well mixed vertically, whereas water-vapor concentrations decrease with distance from the source (the earth's surface) and ozone decreases away from its source in the middle stratosphere [*see bottom illustration, page* 853]. Aerosols show two regions of high concentration. One is at the source (ground level). The other, in the lower stratosphere, is due to the direct injection of particles and the injection of gases from which particles form, together with a very slow removal rate.

It is fairly clear that if the present atmospheric temperature structure is maintained by a combination of air motions and effects that involve trace substances, it can be altered by changes in the concentration of trace substances. One therefore needs to know the natural cycles of the atmospheric trace constituents that are important in this context and the changes in the cycles that are being or may in the future be brought about by man.

Ozone (O_3) is produced by the photodissociation of molecular oxygen and the recombination of molecular with atomic oxygen, primarily above about 22 kilometers and at low latitudes. Ozone is transported toward the poles and downward by atmospheric motions between the subtropics and the high latitudes. From the lower stratosphere in the middle latitudes the ozone seeps into the troposphere, and it is eventually destroyed at the earth's surface or in reactions with aerosols in the surface layer. There is a maximum of ozone in the lower stratosphere in the spring and in the troposphere in the late spring; it is caused by an increase in the supply of energy to the stratosphere from the troposphere in this season and a concomitant increase in the large-scale mixing motions. Ozone stays for about four months in the middle-latitude lower

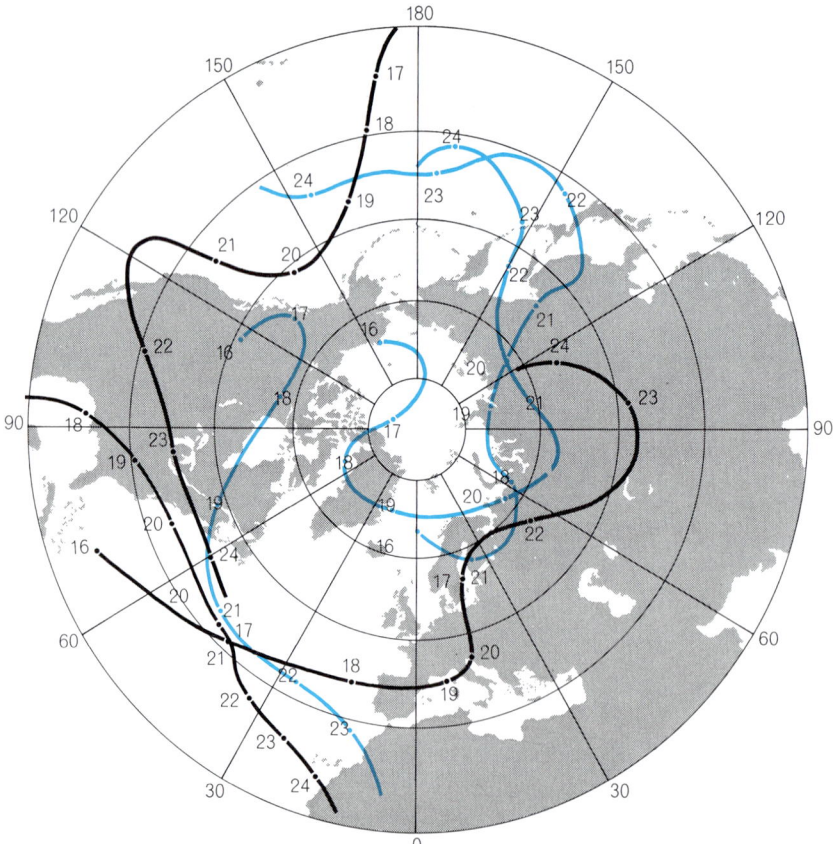

MOVEMENT OF AIR around the world is determined by identifying specific air parcels and tracking them. These tracks were worked out by Edwin Danielsen of the National Center for Atmospheric Research, who identified parcels in the troposphere (*black*) and stratosphere (*color*) and tracked them. The numbers represent successive days in April, 1964.

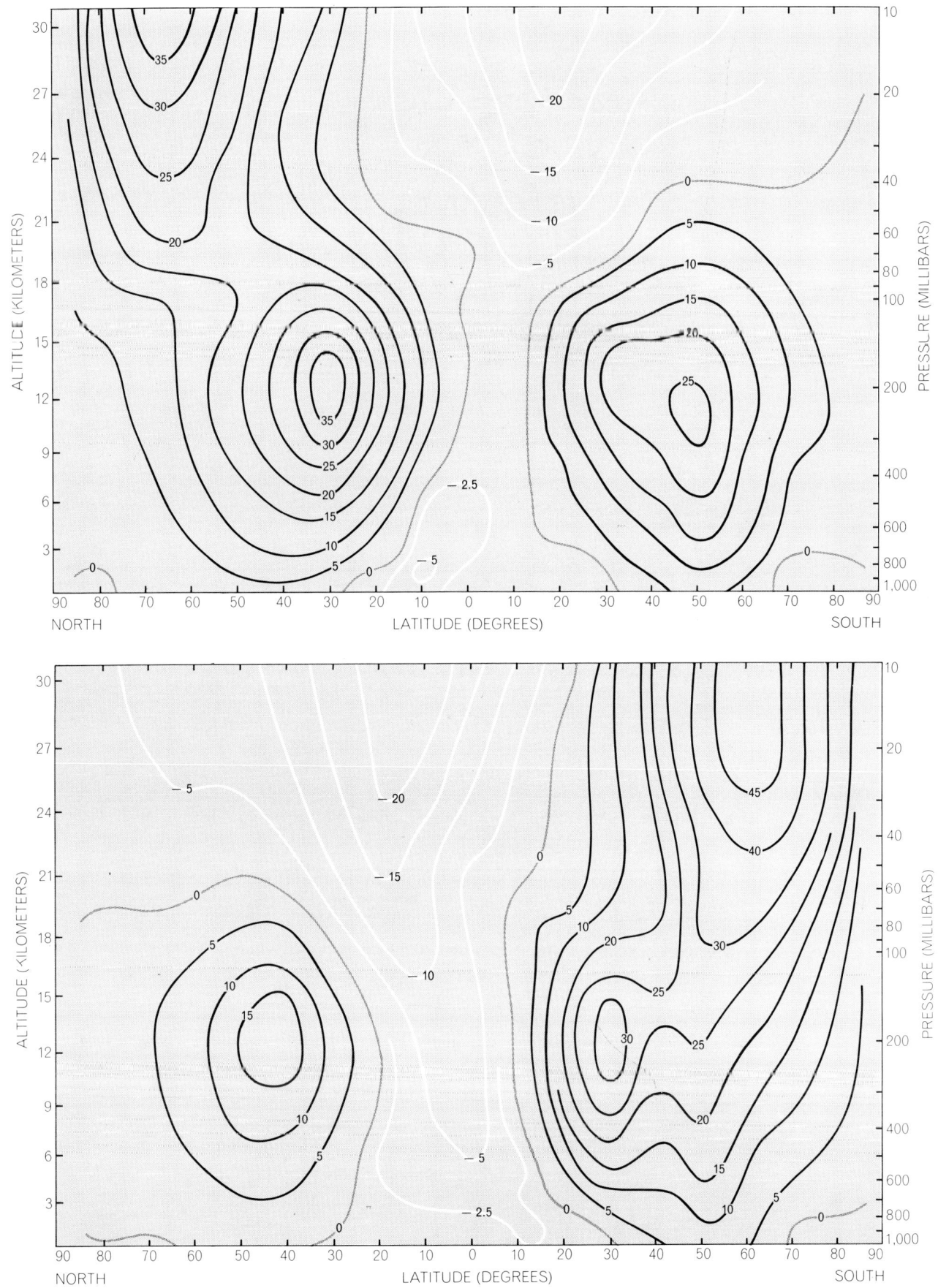

PREVAILING WINDS are revealed by averaging observations over a period of time. Here the mean east-west wind speed is shown as computed for two three-month periods: December–February (*top*) and June–August (*bottom*). The speed is given in meters per second, with positive numbers indicating winds blowing from west to east (*black contour lines*) and negative numbers for east-to-west winds (*white lines*). Note that seasonal changes in the lower atmosphere are larger in Northern Hemisphere than in Southern.

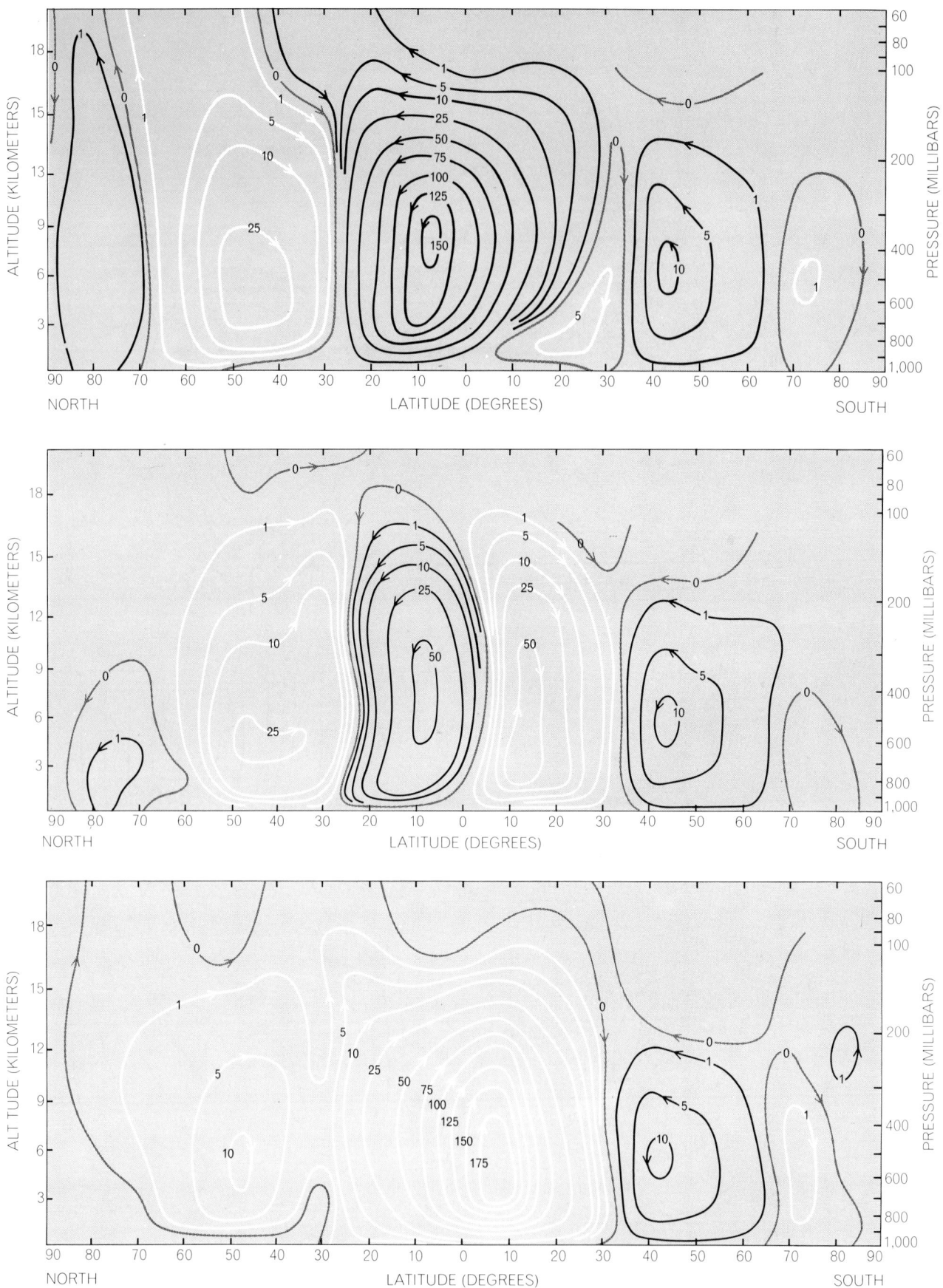

MASS FLUX IN THE TROPICS takes the form of Hadley-cell circulation, the result of a combination of north-south and up-and-down motions. It is shown here for three seasons; December–February (*top*), March–May (*middle*) and June–August (*bottom*). The difference between adjacent contour values gives the flux in millions of metric tons per second based on average values from all longitudes; actually the pattern varies with longitude. The charts are based on work of Dayton G. Vincent and John W. Kidson.

stratosphere and from two to three months in the troposphere.

Most of the steps of the cycle have been verified in detail by a "dynamical-numerical" model that was developed at the Environmental Science Services Administration's Geophysical Fluid Dynamics Laboratory at Princeton University and has been applied to the ozone problem by Syukuro Manabe and B. G. Hunt. In such a model equations governing temperature, wind speeds and ozone concentration are solved with a computer to trace the evolution of the fields over a period of time. The models have been developed to help in weather forecasting and in understanding the general circulation of the global atmosphere and theories of climatic change. Radiative effects, the earth's surface properties, clouds and even ocean temperatures all have to be included to achieve a good representation of the atmosphere, and the necessary degree of resolution in space and time can only be obtained with the largest computers. As far as can be seen, the influence of man-made ozone (in city smog) is small compared with the natural cycle of production, transport and loss, which totals about two billion tons per year. (This is on a global scale, of course, and is of small comfort to people whose smarting eyes make them quite aware of the ozone in the smog they live with.)

The water-vapor cycle in the troposphere depends on the difference between precipitation and evaporation in a given air column [see illustration on page 854]. In the subtropics there is an excess of evaporation over precipitation, so that these latitudes can be regarded as source regions; the opposite situation occurs over the middle and low latitudes. Water vapor is therefore transported from subtropical latitudes toward the Pole and toward the Equator. Large-scale eddy processes govern the movement toward the Pole and the mean Hadley-cell circulation governs the movement toward the Equator. The average rainfall at a point on the globe is about 100 centimeters per year and the average residence time of a water-vapor molecule in the troposphere is about 10 days.

The water vapor produced by man—by fuel combustion, for example—is not a significant fraction of the natural tropospheric cycle. (Again, people who live near power-plant cooling towers will think otherwise.) Nevertheless, more subtle effects could well be produced by interference with the evaporation-precipitation cycle. (Efforts have already been made to do this on a small scale by

spreading a thin film on the water surface of some reservoirs in Australia to prevent excessive evaporation.) One would need a complete dynamical-numerical model to see what effect a given change would have if extensive regions were altered. Less water vapor evaporated would lead to less liberation of latent heat—but also to less cooling by the radiative effect; only a full model can give a proper idea of the interlocking feedbacks and their net effect.

The water-vapor balance of the stratosphere is much more delicate. Alan W. Brewer of the University of Toronto has suggested that most of the water vapor in the stratosphere enters through the region near the tropical tropopause in the rising branch of the Hadley-cell circulation. The temperature is close to −80 degrees Celsius in that region, and so the air can hold only minute amounts of moisture; in the process of passing through, most of the moisture from the troposphere is frozen out and precipitates, staying in the troposphere as cirrus clouds. (The fact that the frost point throughout the lower atmosphere even at middle latitudes is close to the temperature near the tropical tropopause forms the observational basis for Brewer's suggestion.) Henry Mastenbrook of the Naval Research Laboratory has been monitoring the water vapor in the stratosphere since 1964, flying a frost-point device (developed by Brewer) on high-altitude balloons. He finds that the content in the lower stratosphere varies sea-

sonally in phase with the temperature at the tropical tropopause, whereas the content at 30 kilometers varies only slightly, with average values throughout of only two to three micrograms of water vapor per gram of air.

From the mean rising motion and the water-vapor content near the tropical tropopause one can calculate the mass of water entering the stratosphere; it is about seven million grams per second. Now, 500 supersonic transport planes flying at 21 kilometers (70,000 feet) would inject about two million grams of water vapor per second directly into the stratosphere. Since this water is introduced above the cold trap, and since it is introduced at a rate that is of the same order of magnitude as the natural rate, it is clearly going to lead to a significant increase in the water-vapor content at high levels. Wherever rising motion and low temperatures exist together in the stratosphere, clouds may form as expansion and concomitant cooling allow the air to reach the local frost point. Such clouds are occasionally observed near 25 kilometers over Norway and Iceland and over the Antarctic in winter, and also near 80 kilometers at high latitudes in summer and near the tropical tropopause —regions in which temperatures become very low. Increased water-vapor content in the stratosphere, then, would be expected to produce increased cloudiness and therefore a change in the albedo, or reflectivity, of the earth. Again, however, a full dynamical-numerical model

	PARTIC-ULATES	SULFUR OXIDES	NITROGEN OXIDES	CARBON MONOXIDE	HYDRO-CARBONS
POWER AND HEATING	8.1	22.1	9.1	1.7	.6
VEHICLES	1.1	.7	7.3	57.9	15.1
REFUSE DISPOSAL	.9	.1	.5	7.1	1.5
INDUSTRY	6.8	6.6	.2	8.8	4.2
SOLVENT EVAPORATION					3.9
TOTAL	16.9	29.5	17.1	75.5	25.3

POLLUTANTS emitted in U.S. in 1968 are shown. Estimates, in millions of metric tons, are by George B. Morgan and colleagues of National Air Pollution Control Administration.

	NATURAL	MAN-MADE
OZONE	2×10^9	SMALL
CARBON DIOXIDE	7×10^{10}	1.5×10^{10}
WATER	5×10^{14}	1×10^{10}
CARBON MONOXIDE	?	2×10^8
SULFUR	1.42×10^8	7.3×10^7
NITROGEN	1.4×10^9	1.5×10^7

NATURAL AND MAN-MADE trace-gas cycles are compared (in metric tons). Sulfur and nitrogen data are from E. Robinson and R. C. Robbins of the Stanford Research Institute. (The man-made carbon dioxide input is not really a cycle; some remains in the atmosphere.)

is required in order to study the possible changes.

The largest pollutant by mass is carbon dioxide, and here there is evidence that man's activities are indeed altering the concentrations formerly controlled by nature. Some 70,000 million tons per year are involved in the natural cycle, corresponding to a fluctuation of nine parts per million by volume in the carbon dioxide concentration of 320 parts per million. Plants and trees start taking up carbon dioxide in the spring and continue to do so until the fall; then the leaves drop, vegetable matter begins to decay and the direction of the net transfer is from matter to air. This seasonal cycle can be monitored on the ground or in the lower troposphere. Superimposed on the seasonal cycle is a long-term increase, produced by the burning of fossil fuels. The carbon dioxide increase expected from the fuel that has been burned is about 1.8 parts per million per year, yet the observations show an increase of only about .7 part per million per year [see "The Carbon Cycle," by Bert Bolin; SCIENTIFIC AMERICAN Offprint 1193]. Where does the rest of the carbon dioxide go? Some of it may be incorporated in the biosphere, but it is thought that the largest fraction is dissolved in the oceans, which all together contain about 60 times as much carbon

dioxide as the atmosphere does. The solubility diminishes as the temperature of the water increases. There has been some concern that as carbon dioxide in the air increases, presumably raising the air temperature through the "greenhouse effect," the water temperature will rise also, releasing some of the carbon dioxide from the ocean back into the air. Proponents of this view argue that a runaway effect will occur, with the additional carbon dioxide producing still more atmospheric heating.

It is well to bear in mind, however, that the net radiative contribution of carbon dioxide is to cool the atmosphere. With computer programs developed by Thomas G. Dopplick, my colleagues at the Massachusetts Institute of Technology and I have rerun the radiative-heating computations, assuming a tripled carbon dioxide concentration of 1,000 parts per million and the same temperature and cloudiness distributions. We find that the cooling rate diminishes by only a small fraction of a degree per day in the lower levels and actually increases in the stratosphere. Other things being equal, one could interpret a smaller cooling rate as effective heating in the troposphere. The outgoing infrared flux emitted by carbon dioxide is smaller for higher concentrations; for the atmosphere to radiate the original amount of infrared radiation back to space it would have to

radiate at a higher effective temperature. Other things are not equal, however. If the temperature distribution changed, it is likely that the clouds, albedo and therefore the radiation-stream balance would change also. For example, a slightly higher temperature would mean more evaporation and hence more water-vapor radiative cooling. Although it is tempting to argue from our results that little net temperature change should be expected, it is again clear that the proper way to proceed is to run a complete dynamical-numerical model with higher carbon dioxide levels and see what happens.

Another substance that is copiously produced by combustion is carbon monoxide. Levels of 50 parts per million are not uncommon in city streets, with values up to several hundred parts per million in traffic tunnels and underground garages; weekly average levels of 20 parts per million are sometimes found in the Los Angeles basin. The toxic effect is proportional to the ambient-air concentration and the time of exposure. Carbon monoxide that enters the bloodstream combines with hemoglobin, forming carboxyhemoglobin, and thus reduces its capacity to carry oxygen. Impairment of mental function, as measured by visual performance and ability to discriminate time intervals, occurs when carboxyhemoglobin in the blood

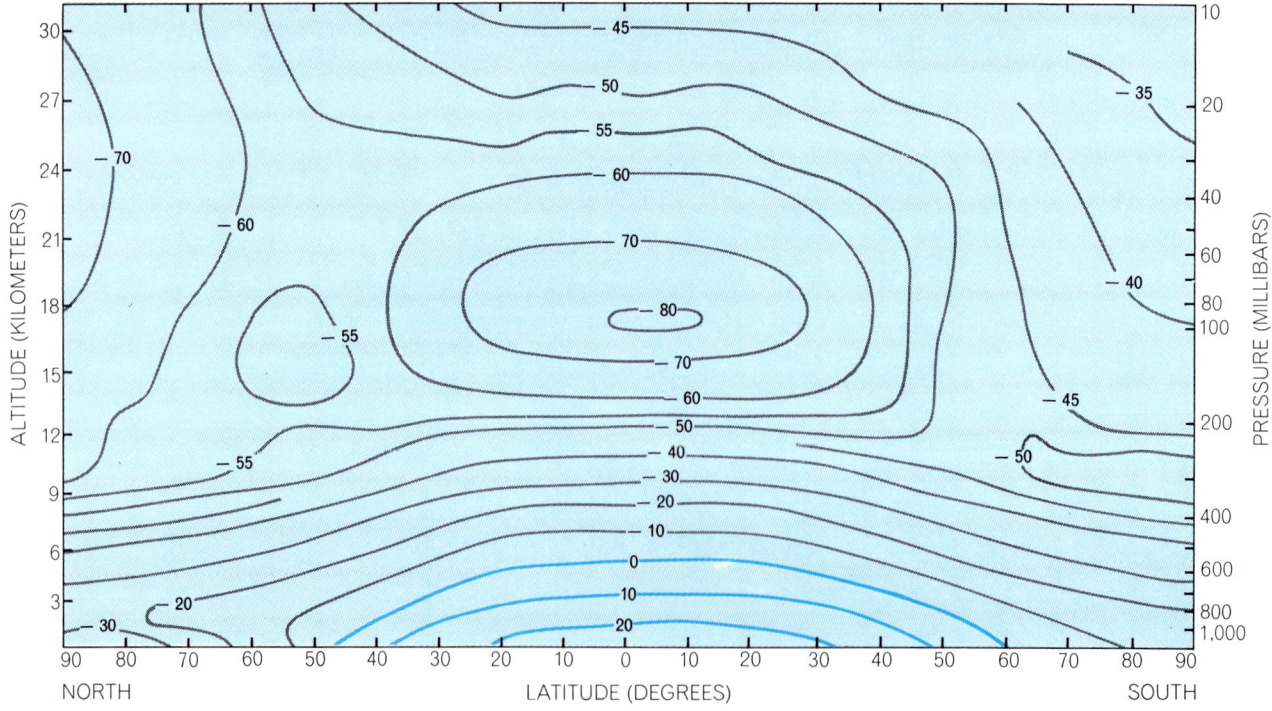

TEMPERATURE PATTERN is given for a north-south cross section of the atmosphere for the three-month period December–February. The isotherms, or contour lines of equal temperature, indicate the temperature in degrees Celsius. The coldest air is over the Equator and near the winter pole. In the latter region the temperature sometimes reaches −85 degrees C. at certain longitudes.

goes above about 2.5 percent (compared with a normal level of about .5 percent). Such values accompany exposure to 200 parts per million for about 15 minutes or 50 parts per million for about two hours.

The 200 million tons injected each year would correspond to an increase of .03 part per million per year. No such increase has been observed, and so a search for sinks of carbon monoxide is under way.

Christian E. Junge of the University of Mainz and his student Walter Seiler have reported that near sea level there is a boundary close to the Equator with high carbon monoxide values to the north and low values to the south; in the upper troposphere aircraft measurements show no such boundary [see bottom illustration on page 855]. Such a carbon monoxide discontinuity is compatible with the fact that there are more automobiles in the Northern Hemisphere and with what is known of the Hadley-cell circulation. At low levels the air just north of the discontinuity is streaming toward the Equator, whereas the air to the south has come from the Southern Hemisphere, with a much smaller population of automobiles. (A similar boundary was frequently revealed by surface-air measurements of strontium 90 when nuclear tests were being held in the Northern Hemisphere.) Higher up in the troposphere eddy mixing eliminates the interhemisphere gradients.

There is still much to be learned about the carbon monoxide sink. Junge's data show a decrease in the stratosphere, and it has been suggested that carbon monoxide combines above the tropopause with hydroxyl radicals (OH) to produce carbon dioxide and hydrogen. There are no direct measurements of the hydroxyl radical in the lower stratosphere and the computed amounts, together with the reaction rate, are just barely sufficient to account for the loss of 200 million tons per year. Furthermore, the rate at which air is transferred from the upper troposphere to the lower stratosphere also seems a shade too low to ascribe the entire loss to this path. The ocean, which has been found to contain carbon monoxide, has been suggested as both a sink and a source; so little is known that it is difficult to say which! Clearly many more observations are needed all over the atmosphere and in the ocean at different latitudes.

Nitrogen constitutes the largest fraction of the atmosphere and is involved in a variety of reactions, including plant growth. The total atmospheric turnover in all forms is about 10 billion tons per

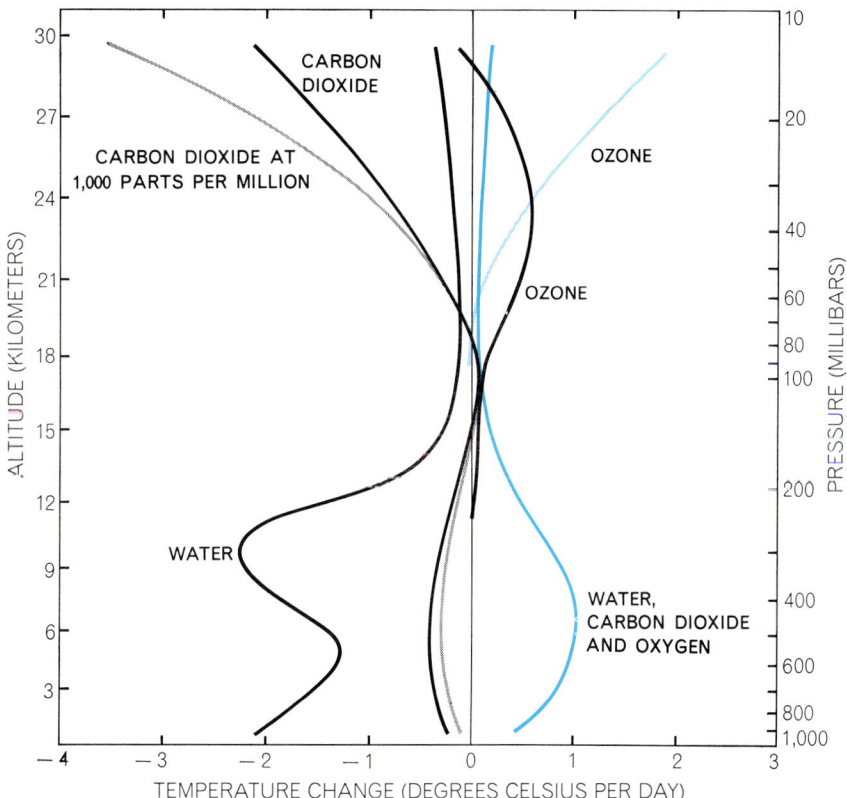

RADIATIVE TEMPERATURE CHANGE is brought about by the absorption of solar visible and ultraviolet (*light color*) and near-infrared (*dark color*) radiation and the absorption and reemission of thermal radiation from the earth (*black*) by various atmospheric gases. The direction and magnitude of each effect vary with altitude, as shown for at the Equator in January. If the carbon dioxide concentration were tripled to 1,000 parts per million, its effects would change as shown (*gray*), according to computations by author's group.

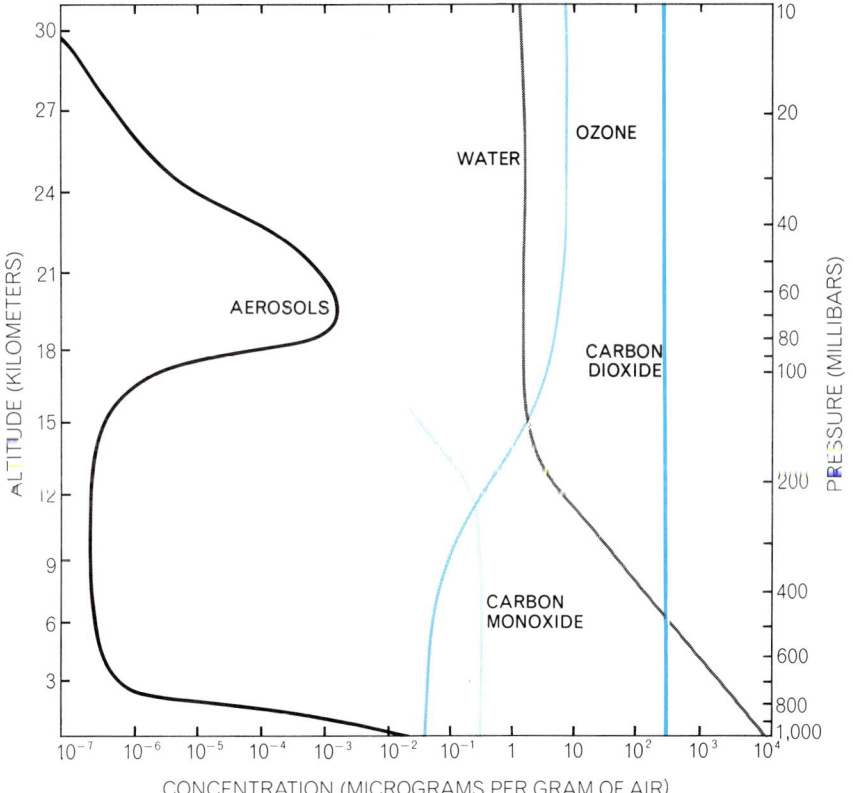

MIXING RATIO, or concentration, of the trace constituents of the atmosphere varies with altitude, decreasing with distance from their sources, except in the case of carbon dioxide.

year, whereas the amount introduced by man is only about 50 million tons. While this much pollution is very important in some regions (city smogs are produced by the action of sunlight on oxides of nitrogen from automobiles), on a global scale it seems that nature's contributions dominate.

There is no such clear distinction for the sulfur compounds. Sulfur is an abundant constituent of ocean water and is released to the atmosphere by sea spray and by biological decay; it is removed by precipitation, intake by vegetation and direct deposition. The total amount involved in the natural cycle is about 142 million tons per year; the 73 million tons injected by man as a pollutant is therefore a very significant amount. This additional sulfur is thought to end up in the ocean and, as Erik Eriksson of the University of Stockholm has stressed, increases the acidity of terrestrial waters. (Fish cannot live in waters of high acidity, and they already shun some inland waterways in Sweden; man has been given a warning.) Sulfur dioxide and particles together in the air seem to produce respiratory ailments.

A considerable amount—no one knows exactly how much—of sulfur dioxide emitted to the atmosphere ends up as sulfate ions or as ammonium sulfate par-

ticles. The time necessary for the sulfur dioxide to disappear as a gas and become incorporated into particles varies from about half an hour to a few days, depending on the air's moisture content and other factors. Measurements made by Junge show that the very small particles called Aitken nuclei (radius about .03 micron) decrease in concentration above the tropopause in a manner consistent with the view that their source is the troposphere. Their composition is unknown. In addition there is a layer of somewhat larger particles (mean radius about .3 micron, with some radii of a micron or more) in the lower stratosphere, composed of ammonium sulfate or sulfuric acid. James P. Friend of New York University and Richard D. Cadle of the National Center for Atmospheric Research have independently verified Junge's finding of this layer of larger particles. Junge has suggested that the large particles in the lower stratosphere may grow on the Aitken nuclei from the gases injected there, and gradually coagulate to form larger particles before they fall, or are transported by exchange, into the troposphere.

When the distribution of sizes of particles of various kinds in the lower troposphere is measured, it is found that most of the mass is accounted for by particles

whose radius is between .1 micron and 10 microns. Observations over land away from major cities have shown an increase in worldwide particle mass in the past 10 years, even though over some city areas there has been a decrease. There has been considerable speculation that an increase in the concentration of particles in the troposphere would cause more solar radiation to be scattered back to space and thus contribute to the lowering of terrestrial temperatures. There are no measurements to support this speculation. In fact, George Robinson of the British Meteorological Office has found that solar radiation is absorbed, rather than scattered, by tropospheric aerosols.

The efforts now being made to reduce particulate pollution, coupled with the relatively short washout time for aerosols, provide reason to hope the long-term effect of large man-made aerosols may be small. If particulate material is not removed but is simply more finely divided, however, more will find its way into the stratospheric regions where the residence times are long and the influence of aerosols generated at the surface could be appreciable.

In the case of hydrocarbons, very little is known about the natural cycle and consequently about the worldwide effect of man's interference. Methane (marsh gas) is abundantly produced by nature and finds its way to the stratosphere, but there are no global-scale measurements. Yet hydrocarbons do appear to play a role in smog formation and cannot be ignored at the local level.

Great volcanic eruptions, such as that of Krakatoa in 1883 or Mount Agung on Bali in 1963, increase the layer of stratospheric aerosols, so that colorful sunsets are produced all around the world. After Krakatoa, investigators suggested that the sunsets were due to the injection of sulfuric acid as well as volcanic dust, basing this suggestion on the fact that there is a strong odor of sulfur near active volcanoes. Samples taken in the lower stratosphere over Australia after the Bali eruption showed that this was indeed the case.

As the concentration of particles (and perhaps their size too) increases, solar radiation is intercepted and less arrives at the ground. Such blocking has often been proposed as a major cause of climatic change [see "Volcanoes and World Climate," by Harry Wexler; SCIENTIFIC AMERICAN Offprint 843]. We examined stratospheric temperature patterns after the Bali eruption in March, 1963, and found there was an immediate rise in stratospheric temperatures, with values

DIFFERENCE between precipitation and evaporation at each latitude is given in terms of the amount of heat lost through evaporation or gained through precipitation in each column of air. (About 600 calories are required to evaporate a column of water one centimeter high covering one square centimeter.) The curve, for December–February, is based on three sets of data. Imbalances of precipitation (top) and evaporation (bottom) are redressed by air motions that transport water toward Equator and toward poles (arrows).

as high as eight degrees C. above average being recorded; the rise was global in extent [*see illustrations on next page*]. It was mainly concentrated in the region above 15 kilometers and there were concomitant changes in the wind field. Presumably the small particles intercepted and absorbed solar radiation. (Their size is of the same order as the wavelength of light.)

The variation of the temperature change with latitude and height was very similar to the patterns that have been found for clouds of radioactive debris in the stratosphere. Such clouds persist longest over the Equator and slope downward toward the poles. The volcanic cloud from Bali did likewise—and the same slope was discussed in a report on Krakatoa published in 1888. The 1963 temperature increase is the largest climatic change ever observed by man. (There were no balloon observations of the stratosphere after Krakatoa.) It serves to warn us that we should watch the sulfur cycle carefully and try to learn soon whether the increased amounts injected into the troposphere by man can find their way into the stratosphere.

Investigators are able to base deductions about atmospheric motions on a network of 800 balloon sounding stations, at many of which temperature is also measured; global maps of temperature constructed from satellite observations are also becoming available. Yet the pollutants in the air we breathe are measured at very few places, and there are no systematic measurements above the surface layers. Moreover, most of the surface measurements are made near cities. It is perfectly well established that if a city has oil-fired or coal-fired power plants, there will be sulfur in the air; if it has automobiles on crowded streets, there will be carbon monoxide and oxides of nitrogen. The most intensive monitoring efforts near cities cannot provide information concerning the global buildup of pollutants.

Water vapor and ozone have been measured above the surface because of their interest as natural trace substances. Moisture sensors are included on the same balloon flights that collect wind and temperature information, but the sensors only operate up to about six to eight kilometers. Mastenbrook's special frost-point hygrometer, which records moisture content up to about 30 kilometers, makes only one flight a month from Washington, D.C.—pitifully inadequate for a study of the global distribution of stratospheric moisture. About six balloon stations measure ozone up to about

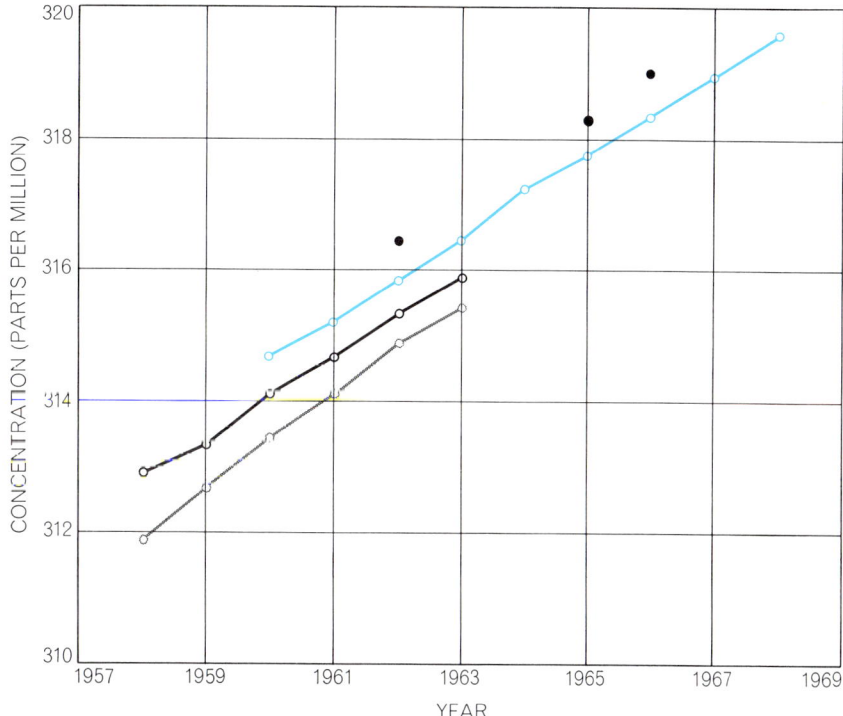

LONG-TERM INCREASE in atmospheric carbon dioxide is revealed by four investigations. The data are from measurements at Barrow, Alaska, by John Kelley of the University of Washington (*black dots*), aircraft observations by Bert Bolin and Walter Bischof of the University of Stockholm (*color*) and measurements at Mauna Loa in Hawaii (*black curve*) and in the Antarctic (*gray*) by Charles Keeling of the Scripps Institution of Oceanography.

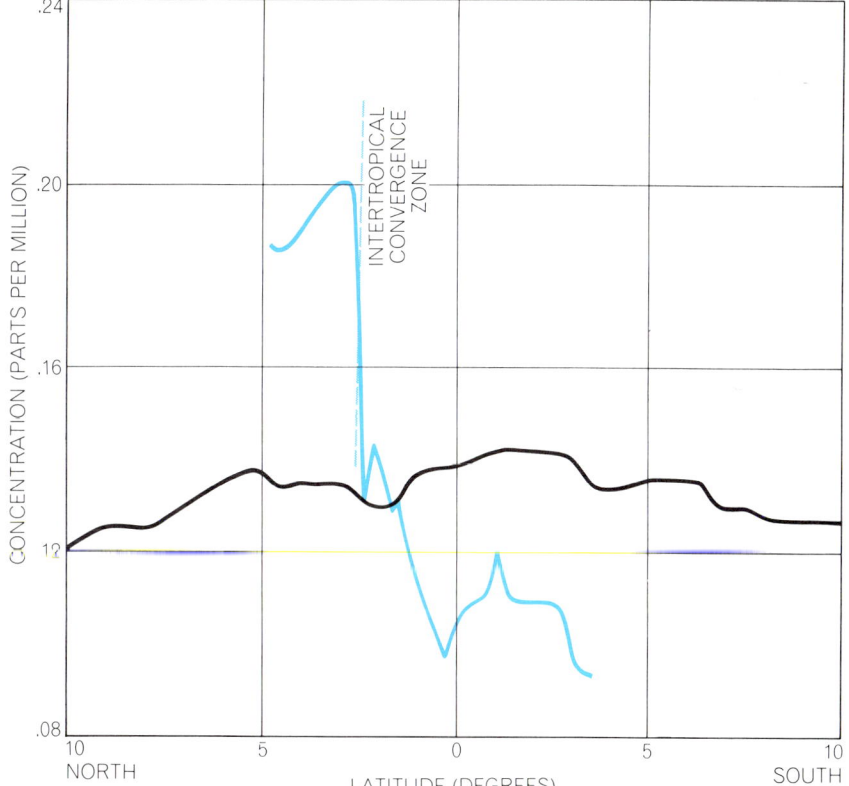

CARBON MONOXIDE measurements made from a ship at sea (*color*) and from an aircraft at 10 kilometers (*black*) by Christian E. Junge and Walter Seiler of the University of Mainz reveal a change in concentration at sea level but not in the upper troposphere. The boundary is explained by the larger number of automobiles in the Northern Hemisphere and the rising motion of the Hadley-cell circulation at the Intertropical Convergence Zone (*broken line*). Higher up eddy currents mix the air, eliminating the interhemisphere gradient.

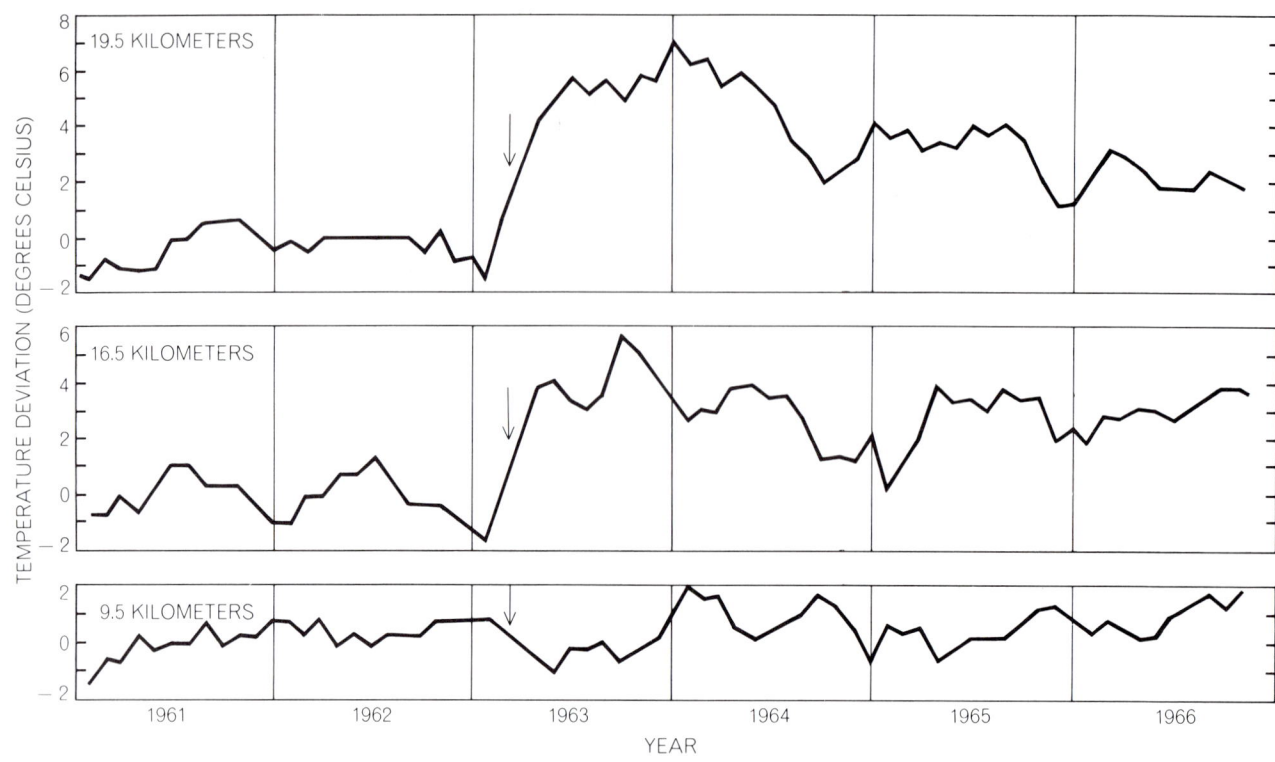

TEMPERATURE CHANGES above Port Hedland, Australia, show the heating effect (first noted by James G. Sparrow of the University of Adelaide) of particles from the 1963 eruption of Mount Agung on Bali. Monthly means were calculated for five years before the eruption (*arrows*); deviations from the means were computed and three-month running averages were plotted.

30 kilometers once a week. The Atomic Energy Commission launches balloons from four sites to obtain samples from up to 35 kilometers, mainly for analysis of radionuclides, although carbon dioxide has sometimes been included. High-altitude sampling aircraft such as the U-2 have also been used, but as nuclear tests in the atmosphere have decreased these sampling programs have naturally been pared. Practically all the published data on carbon monoxide measured away from the surface is represented in Junge's results. (The instrument he used was made portable and carried on a commercial jet flight; it occupied one seat, the observer another.) A few spot measurements of sulfur dioxide and sulfate above the surface layers have been made by Hans Georgii of the University of Frankfurt; again no extensive coverage is available.

There are many opportunities to collect data on atmospheric trace substances by taking advantage of commercial airline flights and regular ocean voyages as well as special oceanographic cruises, the unique AEC balloon network, the surface-air sampling networks established to monitor global radioactivity levels, mountaintop observatories and so on. Now that some important trace substances are being introduced into the atmosphere by man at rates comparable to those in the natural cycle, it seems appropriate to start making these measurements.

I must mention in closing that one cannot take the global view of pollutants without feeling some concern about the rate of use of natural resources and the rate of generation of pollutants. Both could be slowed considerably by a serious effort to use every pound of fuel in the most efficient manner possible.

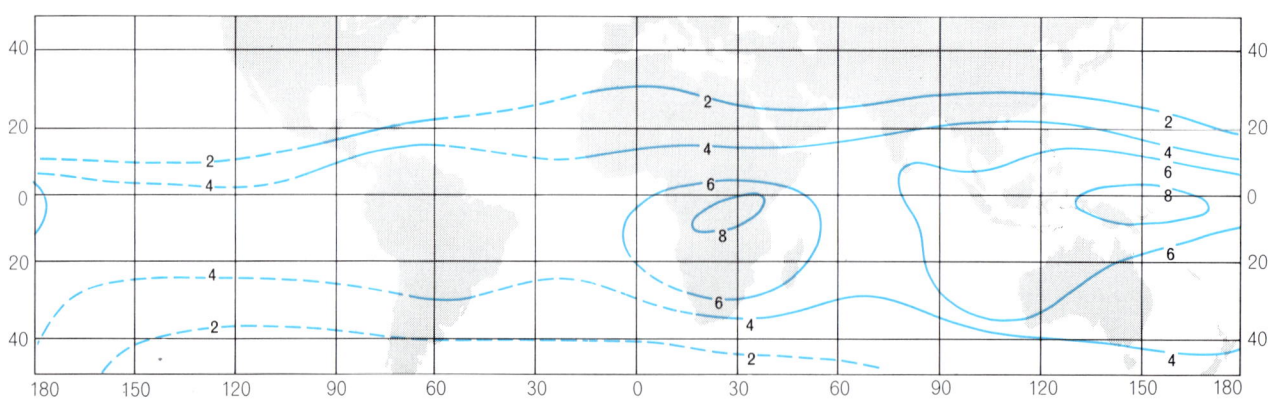

HEATING EFFECT after Mount Agung is mapped. The isotherms give the increase in mean temperature at about 19 kilometers, in degrees Celsius, from January before the eruption to one year later. (Broken lines indicate uncertainty due to scarcity of observations.)

The Author

REGINALD E. NEWELL is professor of meteorology at the Massachusetts Institute of Technology. He was born in England and received a bachelor's degree in physics from the University of Birmingham in 1954. He then moved to M.I.T. to study meteorology, receiving his master's degree in 1956 and his Sc.D. in 1960. Since 1961 he has taught a course on upper-atmosphere physics. Newell writes that he enjoys visits to nearby beaches with his four children, "preferably when the power-plant plumes and the smoke from the trash burned off Boston harbor are blowing away from the beach." He also enjoys debating pollution and political problems with his wife and others. Aside from the atmosphere his professional interests include the circulation of planetary atmospheres, auroras and theories of climatic change.

Bibliography

INTERNATIONAL SYMPOSIUM ON TRACE GASES AND NATURAL AND ARTIFICIAL RADIOACTIVITY IN THE ATMOSPHERE. *Journal of Geophysical Research*, Vol. 68, No. 13, pages 3745–4016; July 1, 1963.

PROCEEDINGS OF THE CACR SYMPOSIUM: ATMOSPHERIC CHEMISTRY, CIRCULATION AND AEROSOLS, AUGUST 15–25, 1965, VISBY, SWEDEN. *Tellus*, Vol. 18, No. 2–3, pages 149–684; 1966.

MAN'S IMPACT ON THE GLOBAL ENVIRONMENT: ASSESSMENT AND RECOMMENDATIONS FOR ACTION. Report of the Study of Critical Environment Problems (SCEP). The M.I.T. Press, 1970.

SCIENTIFIC
AMERICAN May 1971, Vol. 224, No. 5, pp. 30–42

THE OLDEST FOSSILS

by Elso S. Barghoorn

The remains of ancient bacteria and algae, some of them more than three billion years old, have been found in Africa, Australia and Canada. They provide evidence on the earliest stages of evolution.

How did life originate on the earth? It was not so long ago that attempts to answer this question defined more areas of uncertainty than of agreement. Today a fossil record that was virtually unknown before the 1950's has been found to bear witness to three of the key events in the earliest stages of organic evolution. The fossils come from widely separated parts of the earth. All are preserved in unusual rocks of the Precambrian, the first and by far the longest interval in geologic history. The earliest of the fossils are more than three billion years old.

All that is known or conjectured about the terrestrial origin of life suggests that the first appearance of living organisms was preceded by the gradual development of a complex chemical environment. This environment is usually pictured as a kind of primordial broth, filled with such "organic" molecules as amino acids, sugars and other biologically important substances that came into existence through nonorganic processes. Millions of years must have been required for the accumulation, elaboration and differentiation of the broth. That period may be called the time of chemical evolution, a concept that owes much to the work of a leading student of abiotic synthesis, Cyril A. Ponnamperuma of the Ames Research Center of the National Aeronautics and Space Administration. As a preliminary stage in the total process of organic evolution, chemical evolution of course reaches its climax when lifeless organic molecules are assembled by chance into a living organism. This first form of life is what the Russian biochemist A. I. Oparin calls a "protobiont."

Judging from the various forms of life we know today, the first protobionts were probably microscopic in size and single-celled in structure, perhaps resembling the modern coccoid, or spherical, bacteria. Rather than imagining some specific organism, however, let us consider this first form of life more abstractly and give it a name that describes how it lives rather than how it looks. Call it a heterotroph, which is to say an organism that cannot manufacture all its own nutrients but must feed on organic molecules in the broth that surrounds it. (This assumes that the organism is immersed in an aqueous medium or at least rests on a wet surface; a supply of water is essential to protoplasmic life.) It seems reasonable to suppose that the organism was a heterotroph; to have at this stage a full-fledged autotroph, an organism that can manufacture its organic nutrients out of inorganic substances, would be asking too much of chance.

We can, however, demand that heterotrophs rather promptly give rise to autotrophs. Otherwise, as Preston Cloud of the University of California at Santa Barbara puts it, once the heterotrophs had "gobbled up all the goodies" within reach they would die off, making another accident of biosynthesis necessary in order to get things going again.

One way or another, then, autotrophs must evolve from an original heterotrophic population. This event in the development of life marks the time when the organic nutrients of the primordial broth are depleted and photosynthesis—the most plausible form of organic self-nutrition—must have been invented. Such an assumption is not based solely on logical considerations of thermodynamics and physiology. It is also supported by geological evidence that early organisms depended on photosynthesis to sustain themselves. Among the most ancient formations in the geological record are some showing signs that small amounts of free oxygen—the gaseous by-product of photosynthesis—were present early in the history of the earth. This chemical evidence coincides with indications in the fossil record that some of the most primitive forms of terrestrial life, organisms that resembled modern bacteria and blue-green algae, were then increasing in numbers and diversity.

Many modern bacteria are photosynthetic, although they do not produce free oxygen; all modern blue-green algae are photosynthetic and all produce free oxygen. The evidence suggesting that the first autotrophs were organisms of the same primitive kind is significant for an additional reason. Bacteria and blue-green algae are alone among living things in the simplicity of their cells. They have neither a membrane-enclosed nucleus nor such specialized cellular organelles as mitochondria. Their genetic material is diffused throughout the cell, and they are incapable of either mitosis (body-cell division) or meiosis (germ-cell division). Both kinds of cell division require that the genetic material be organized into chromosomes.

Bacteria and blue-green algae are thus fundamentally different from other organisms: all other plants, all animals and the many forms that are neither entirely plant nor entirely animal. These other organisms have cells with nuclei and specialized organelles or specialized intracellular structures; they are called eukaryotic (truly nucleated), whereas bacteria and blue-green algae are prokaryotic (prenuclear). It would be surprising if the autotrophs on the lowest rungs of the evolutionary ladder were anything but prokaryotic.

The prokaryotic cell is notable for still another reason. Any organism whose genetic material is diffused throughout the

OLDEST KNOWN BACTERIUM, one of two primitive forms of life preserved in a Precambrian rock formation in South Africa, appears as a raised rectangular shape in this electron micrograph. What is seen is a carbon replica of the polished surface of a rock sample, shadowed with heavy metal. The fossil bacteria come from cherts of the Fig Tree formation; they are from .5 to .75 micron long and about .25 micron wide. The organisms are some 3.2 billion years old and have been given the name *Eobacterium isolatum.*

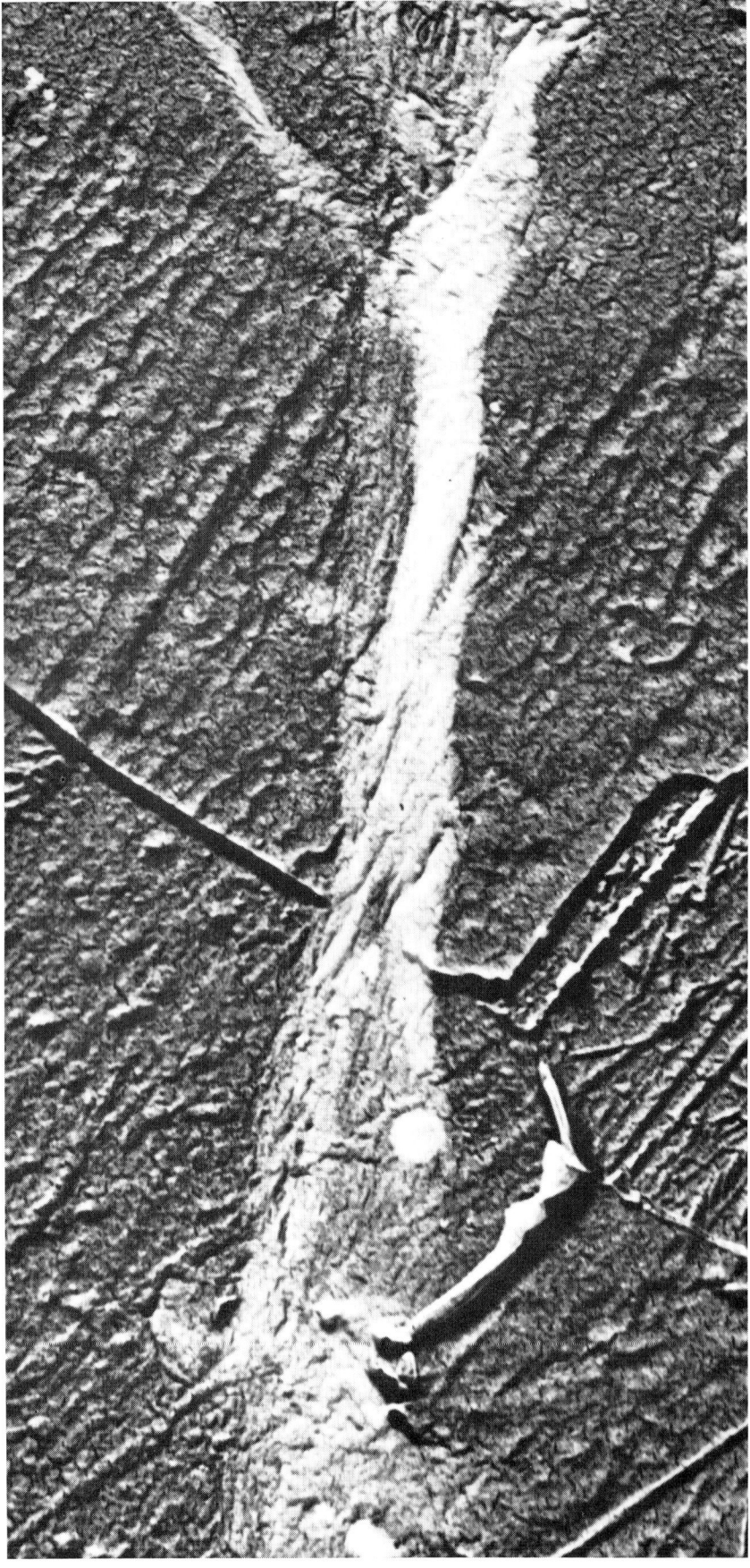

THREADLIKE FILAMENT of organic matter resembling decomposed plant tissue is another kind of fossil that appears in electron micrographs of the Fig Tree cherts. Some specimens are nine microns long. Not identifiable with any known organism, the filaments might conceivably be polymerized abiotic molecules from the "primordial broth."

cell and whose reproduction does not involve a recombination of parental genes is genetically conservative. In such an organism random mutations, instead of being preserved when they benefit the organism, tend to be damped out in a few generations. The blue-green algae are an outstanding example of such genetic conservatism. Many living species are almost indistinguishable in structure from species that flourished a billion or more years ago.

Against this background of fact and conjecture, how many of the events in the early stages of organic evolution can be labeled as being outstanding? There appear to be three such events, each of them a kind of threshold-crossing. The first, obviously the *sine qua non*, is successful biosynthesis: a crossing of the threshold separating the initial period of abiotic chemical evolution from the subsequent organic period. Perhaps someday fossil evidence of the earth's earliest organisms, the heterotrophs that crossed this first threshold, will be discovered. In the meantime the remains of their successors, the early photosynthetic autotrophs, provide the necessary proof that the first threshold had been passed.

The next threshold can be characterized as the threshold of diversification. As evolution progresses the first photosynthetic organism should not be limited to a few similar forms. Instead they should develop differences in shape and structure indicative of roles in a variety of ecological niches.

The third threshold divides prokaryotic organisms from all others. A world populated solely by bacteria and blue-green algae is conceivable; indeed, that is apparently the way it was. Viewed from our present vantage such a world seems poor in possibilities for further evolution. The extraordinary variety of plant and animal life that has arisen on the earth over the past 600 million years is due entirely to the invention of the eukaryotic cell, with its potential for genetic diversity.

Thanks to a lucky accident of fossil preservation, we now have evidence that each of these thresholds was successfully crossed during the vast span of Precambrian times. The Precambrian represents nearly four billion of the earth's first 4.5 billion years. It began with the formation of the earth and ended some 600 million years ago with the dawn of the Paleozoic era [*see illustration on page 869*]. Nothing is known of the first billion years or so; the world's oldest-known rocks, found in Africa, are not much more than three billion years old.

Precambrian rocks are found not only in Africa but also on every other continent. The best-known areas are the Canadian Shield in North America and the Fenno-Scandian Shield in Europe, but more than a third of Australia is also a Precambrian shield and there are sizable Precambrian areas in South America and Asia. Some Precambrian rocks are of igneous origin and some are of sedimentary origin. Most of the sediments have been heavily metamorphosed: changed in form and chemical composition by heat and pressure. In these metamorphic rocks not only fossils but also the faintest traces of organic matter have been obliterated.

A few Precambrian sediments have escaped substantial alteration. Extensive deposits of black shale, black chert and other stratified sediments are scattered through the major shield areas in virtually unmetamorphosed condition. These carbon-rich rocks—for example the formations in the Lake Superior region of North America, in the Transvaal of South Africa and in parts of western Australia—look even to the experienced eye very much like certain sediments of the Carboniferous epoch that are a mere 300 million years old.

The oldest-known group of Precambrian sediments is located in the border region between the Republic of South Africa and Swaziland. The formation is called the Swaziland Sequence; its stratified rocks are thousands of feet thick. One series of strata in the middle of the sequence, known as the Fig Tree formation, is well exposed in the Barberton Mountain Land, a gold-mining district near the town of Barberton in the eastern Transvaal. The Fig Tree rocks consist of black, gray and greenish cherts, interbedded with jaspers, ironstones, slates, shales and graywackes. In places the chert beds are 400 feet thick. The chert is usually fractured, but it is cemented together and the fractures are filled with quartz; parts of the formation show little evidence of metamorphism. The Fig Tree cherts contain traces of organic matter and a few microfossils.

The Barberton Mountain Land has long been an active mining area, and its geology has been studied in considerable detail. The age of the graywackes and shales has been determined independently by several laboratories using a method based on the decay of radioactive strontium and rubidium. This particular radioactive clock started to run 3.1 billion years ago, but there is good reason to believe the sediments were deposited

somewhat earlier. Recently rocks lying close to the base of the Swaziland Sequence (and thus well below the level of the Fig Tree cherts) were shown to be 3.36 billion years old. In light of these age determinations it seems probable that the age of the Fig Tree cherts is in excess of 3.2 billion years.

In 1965 I collected cherts from several localities in the Fig Tree series, and samples were prepared for examination in my laboratory at Harvard University that summer. Two techniques were employed: thin sections of chert were cut for examination by reflected and transmitted white light under the microscope, and carbon replicas of etched and polished chert surfaces were made. The carbon replicas were then "shadowed" with metal and examined by transmission electron microscopy. J. William Schopf joined me in the study of the specimens.

When Schopf and I examined the thin sections under the light microscope, we could see that the rock matrix contained abundant laminations of dark-colored and virtually opaque organic matter. The

laminations were irregular but were usually aligned parallel to the strata of the chert, suggesting that they had originally been formed as part of an aqueous sedimentary deposit. No distortion was evident where the organic matter crossed the boundaries of the individual grains of chalcedony comprising the chert, which suggested to us that the process of deposition had emplaced the organic material within a silica-rich matrix before the silica was crystallized into chert. There was no evidence whatever that the silica was of secondary origin.

It was difficult to discern distinct bodies within the layers of organic matter under the light microscope. Our first success in isolating a Fig Tree organism was achieved with the carbon-replica technique; the electron microscope revealed a number of rod-shaped structures, preserved both in profile and in cross section. The rods are very small. They range in length from slightly under .5 micron to a little less than .7 micron, and in diameter from about .2 micron to a little more than .3 micron. In cross section the cell wall is sometimes seen to

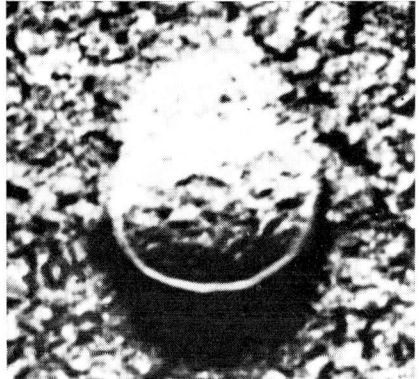

TWO CROSS SECTIONS of *Eobacterium* are seen in electron micrographs of metal-shadowed carbon replicas. In the fossil at left the outer and inner layers of the cell wall are visible. The wall, .015 micron thick, resembles the wall of living bacteria of the bacillus type.

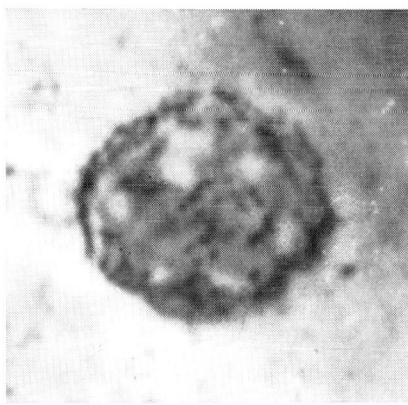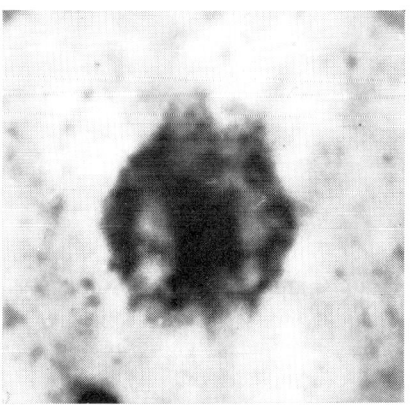

ALGA-LIKE SPHERES, seen in photomicrographs of thin sections of rock, are the other fossil organisms that are found in the Fig Tree formation. The diameter of the spheres is usually less than 20 microns. The organism is named *Archaeosphaeroides barbertonensis*.

consist of an inner and an outer layer, with a total wall thickness of .015 micron [*see top illustration on preceding page*]. This is comparable to the cell wall of many modern bacteria in both structure and dimension.

Electron microscopy also revealed the presence of organic material in the form of irregular, threadlike filaments lacking discernible structural detail. The threads are clearly native to the chert and not the products of contamination. They are as much as nine microns long and resemble decomposed plant material. Although these filaments are almost certainly of biological origin, they cannot be identified with any known type of organism. It has been suggested, probably as a result of wishful thinking, that they may be polymerized strands of abiotic organic matter from the primordial broth.

Schopf and I later succeeded in resolving larger microfossils in thin sections of Fig Tree chert under the light microscope. These fossils are spheroidal; measurements of 28 particularly well-defined specimens show that the majority are between 17 and 20 microns in diameter [*see bottom illustration on preceding page*]. Some have a darkened interior, as if cytoplasm within the spheroid had coalesced and become "coalified." Just as the rod-shaped organisms revealed by the electron microscope resemble certain modern bacteria, so the spheroids are not unlike some modern blue-green algae of the coccoid group. They may even be among the evolutionary precursors of such algae.

We have named the rods *Eobacterium isolatum,* a new genus and species. The generic name (*eo-* is the Greek root for "dawn") points to the great antiquity of the organism and to its bacterium-like appearance; the specific name defines its noncolonial, single-cell habit of growth. The spheroids we have named *Archaeosphaeroides barbertonensis,* also a new genus and species. Again the generic name refers to the organism's great age and its appearance; the specific name identifies its place of discovery. The existence of these two organisms, successful inhabitants of an aquatic environment more than three billion years ago, is evidence that the first evolutionary threshold—the transition from chemical evolution to organic evolution—had been safely crossed at some even earlier date. We now know that at least two living organisms appeared well before the first third of earth history had passed. If we accept the evidence (to which we shall return) that the alga-like Fig Tree organisms were photosynthetic, an important

geochemical event must also have occurred. With the onset of photosynthesis free oxygen would have begun to appear among the other constituents of the environment. The appearance of free oxygen was an event destined to have profound influence, both biological and geological, on the subsequent history of the earth.

Evidence of the second evolutionary threshold-crossing comes from North America. A remarkable outcropping of Precambrian rocks along the shore of Lake Superior in western Ontario shows a sequence of sediments known as the Gunflint Iron formation. The rocks at the base of the Gunflint formation include beds of black chert three to nine inches thick. The beds are exposed—in some places more or less continuously and in others as isolated outcrops—over a distance of some 115 miles in Ontario, from the vicinity of Schreiber on the east to Gunflint Lake on the west. Like the much earlier Fig Tree series, the Gunflint formation has been the subject of detailed geological investigation.

Granite underlies the Gunflint formation unconformably, that is, the basement rock had been eroded before the Gunflint sediments were deposited on its surface. Radioactive clocks have yielded two independent dates for the granite. A concentrate of its biotite contents indicates an age, in terms of the argon-potassium ratio, of 2.5 ± .75 billion years. The age of a whole-rock sample, in terms of its rubidium-strontium ratio, is 2.36 ± .70 billion years. The age of the granite thus provides a "floor" of maximum age under the Gunflint formation. The Gunflint cherts cannot be any older than these dates.

Micas separated from rocks in the upper levels of the Gunflint formation, collected near Thunder Bay, indicate an age in terms of the argon-potassium ratio of 1.60 ± .05 billion years. For technical reasons this figure is only 80 percent of the true age; it must therefore be adjusted to 1.90 ± .20 billion years. The micas thus provide a "ceiling" of minimum age for the cherts that lie at the base of the Gunflint formation. It seems reasonable to set the age of the cherts at approximately two billion years, which makes them a billion years or so younger than the Fig Tree cherts.

The only rocks in the Gunflint formation that contain microfossils are the cherts. Like the Fig Tree cherts, they are evidently the product of deposition in an aqueous environment that was rich in silica. Most of the Gunflint fossils are three-dimensional, and many show ex-

quisite anatomical details. It has been argued that the structure of these fossils has been preserved by the infiltration of silica from the surrounding sediments. In my opinion the organisms were preserved without distortion by being deposited in a siliceous solution that later crystallized into chert, much as a modern biological specimen is preserved by being embedded in plastic. The soft structure of the organism owes its preservation to the almost complete incompressibility of the silica matrix. This is as unusual as it is fortunate. In most instances of fossil preservation the matrix is composed of relatively plastic sediments that are much compressed during consolidation, with the result that any preservation of soft tissues, let alone preservation in three-dimensional form, is a rarity.

How were the Gunflint cherts deposited? The picture in Ontario is rather clearer than it is in the Transvaal. The Gunflint cherts were apparently precipitated and consolidated around an underlying complex of basement rocks, consisting of greenstone boulders and finer conglomerates, that seems to have been continuously submerged at the time. "Domes" of algae, which are visible to the unaided eye in samples of Gunflint rock, grew on the surface of the boulders. Algal "pillars" grew perpendicularly to the domes; their fossil remains consist of alternate layers of coarsely crystalline quartz and fine-grained black chert [*see bottom illustration on opposite page*].

In the 1950's I was privileged to be associated with Stanley A. Tyler of the University of Wisconsin in collecting and analyzing specimens of Gunflint chert from the Schreiber area. Only a few preliminary studies of these and other Gunflint specimens were published before Tyler died, although by then we both knew that his work had added to the fossil record an entire new group of very ancient, primitive photosynthetic organisms whose existence had not been suspected. Indeed, even though many years of study have now been devoted to the Gunflint organisms, the formation continues to yield new finds. Eight genera of primitive Gunflint plants, comprising 12 species, have been described so far, yet accompanying this article are illustrations of new forms of undetermined taxonomic status.

The most abundant of the Gunflint microfossils are filamentous structures. The majority of them are between .6 micron and 1.6 microns in diameter but a few are more than five microns across. They vary in length up to several hundred microns. Some of the filaments have inter-

SAMPLE OF GUNFLINT CHERT of Ontario (*above*) has a knobby surface. The knobs, exposed by weathering, are tops of pillars formed by algae in shallow water where Gunflint organisms lived.

VERTICAL SECTION through a chert sample (*below*) shows the structure of the algal pillars. Layers of quartz and of fossil-bearing black chert alternate in each pillar like sets of nesting thimbles.

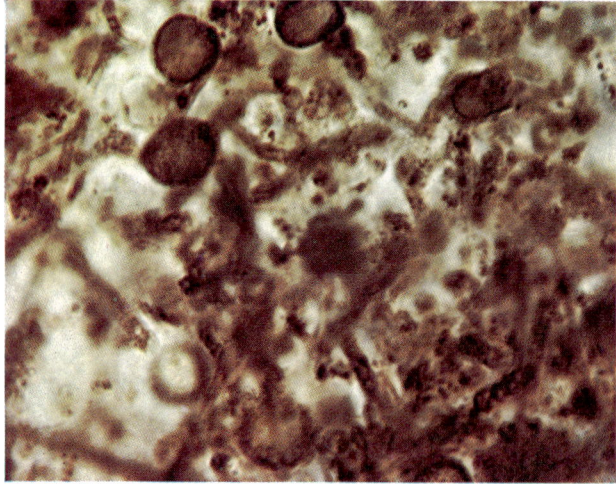

DENSE MIXTURE of organic detritus, spherical bodies and filaments is enlarged 250 diameters in this photomicrograph. All the micrographs on this page show thin sections of the Gunflint cherts.

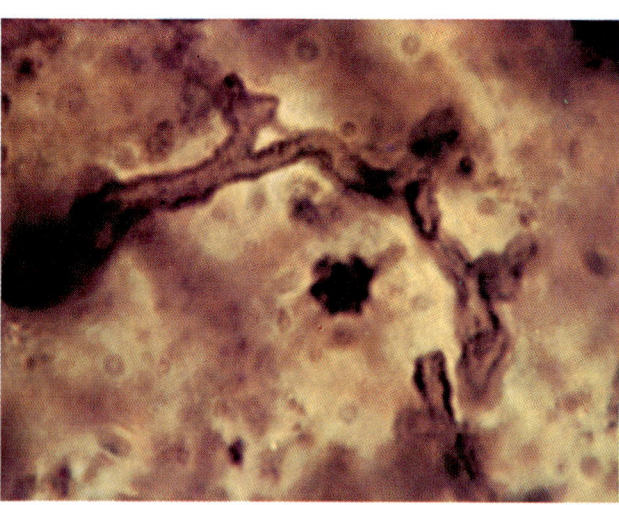

TUBULAR FILAMENT with branches and swellings is not obviously related to any living organism. Where it is not swollen it is about two microns across. It is called *Archaeorestis schreiberensis*.

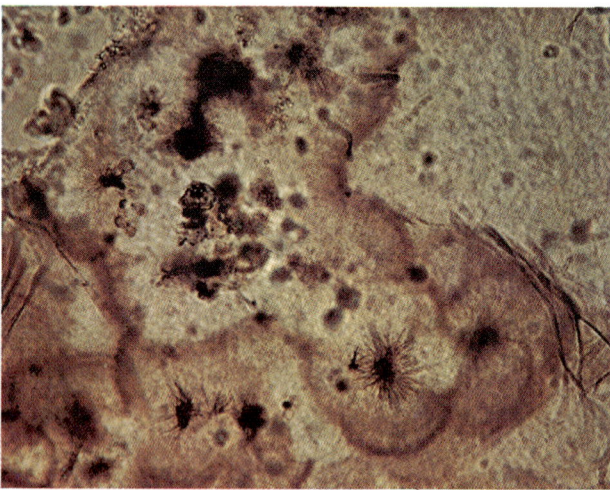

COLONY OF ALGAE is enlarged 200 diameters. The organisms, each a cluster of spikelets, have been named *Paleorivularia ontarica* because of their resemblance to the living genus *Rivularia*.

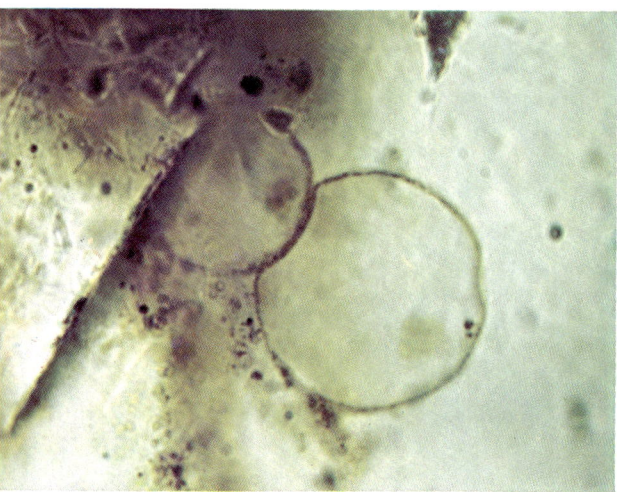

CLOSE CONTACT between a larger and a smaller cell is apparent in a chert section enlarged 1,000 diameters. Interrelations of this kind may represent a stage in the evolution of the eukaryotic cell.

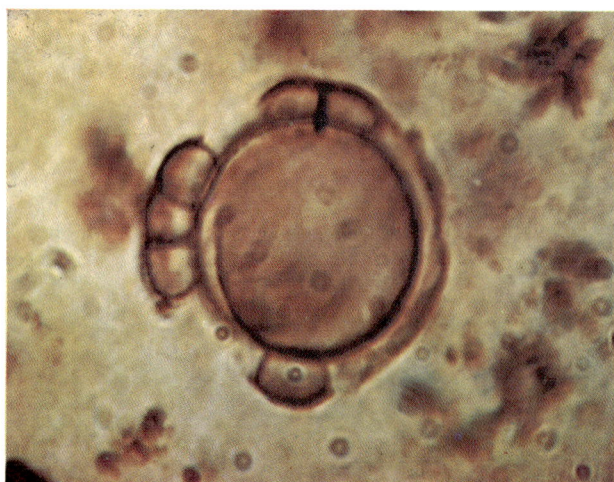

SPHERE WITHIN SPHERE, the inner one held away from the outer one by a number of smaller spheroids, is about 30 microns in diameter. The spheres comprise the extinct species *Eosphaera tyleri*.

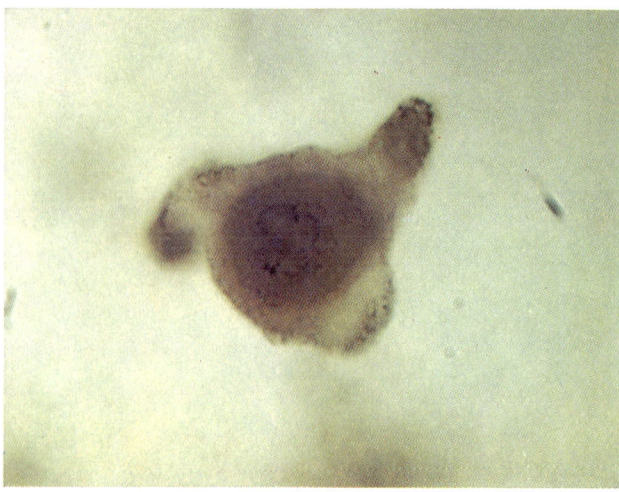

ASYMMETRICAL CELL, found in a recently prepared chert section, has not yet been formally classified. The Gunflint cherts have been studied for nearly 20 years, but they still yield new organisms.

nal walls at right angles to their length; others lack such walls. Where the walls are present, they can be broad or narrow. These primitive plants have been assigned on the basis of their morphology to four genera including five species. Among the present-day blue-green algae they resemble are the filamentous *Oscillatoria* and related forms; one of the four also shows resemblances to the modern iron-oxidizing bacterium *Crenothrix*.

Another genus of Gunflint organisms is represented by small spheroids, ranging from one micron in diameter to more than 16 microns. The walls of the spheroids vary both in thickness and in structural detail. Because of these variations they have been assigned to three separate species and placed in the genus *Huroniospora*. The spherical organisms are found in chert collected from many parts of the Gunflint formation but they are not equally abundant everywhere.

What kinds of organism are represented by the spheroids? We have only their simple morphology to judge by. Like their counterparts in the much older Fig Tree cherts, they might be noncolonial blue-green algae of the coccoid group. They might also, however, be the reproductive spores produced by the filamentous plants mentioned above, and some might even be the spores of iron bacteria. Still another possibility is that they are the fossilized bodies of free-swimming organisms whose flagella have not been preserved. Further study may lead to a choice among these alternatives.

The remaining Gunflint genera that have been intensively studied and described so far all show unusual characteristics. The organisms in one of the three genera are star-shaped and made up of radially arrayed filaments. The diameter of the "star" ranges from eight to 25 microns; in rare cases some of the filaments are branched. Although representatives of the genus are few in number and often poorly preserved, they are found throughout the Gunflint cherts.

The genus has been named *Eoastrion*—"dawn star"—to indicate the great antiquity and distinctive shape of its members. Two species have been established, one to accommodate the fossils with nonbranched filaments and the other the fossils with branched filaments. There are no clear analogies between the fossils and modern organisms, although in some respects *Eoastrion* resembles the curious iron- and manganese-oxidizing organism *Metallogenium personatum*.

A most peculiar organism comprises the next of these genera. Its fossils are

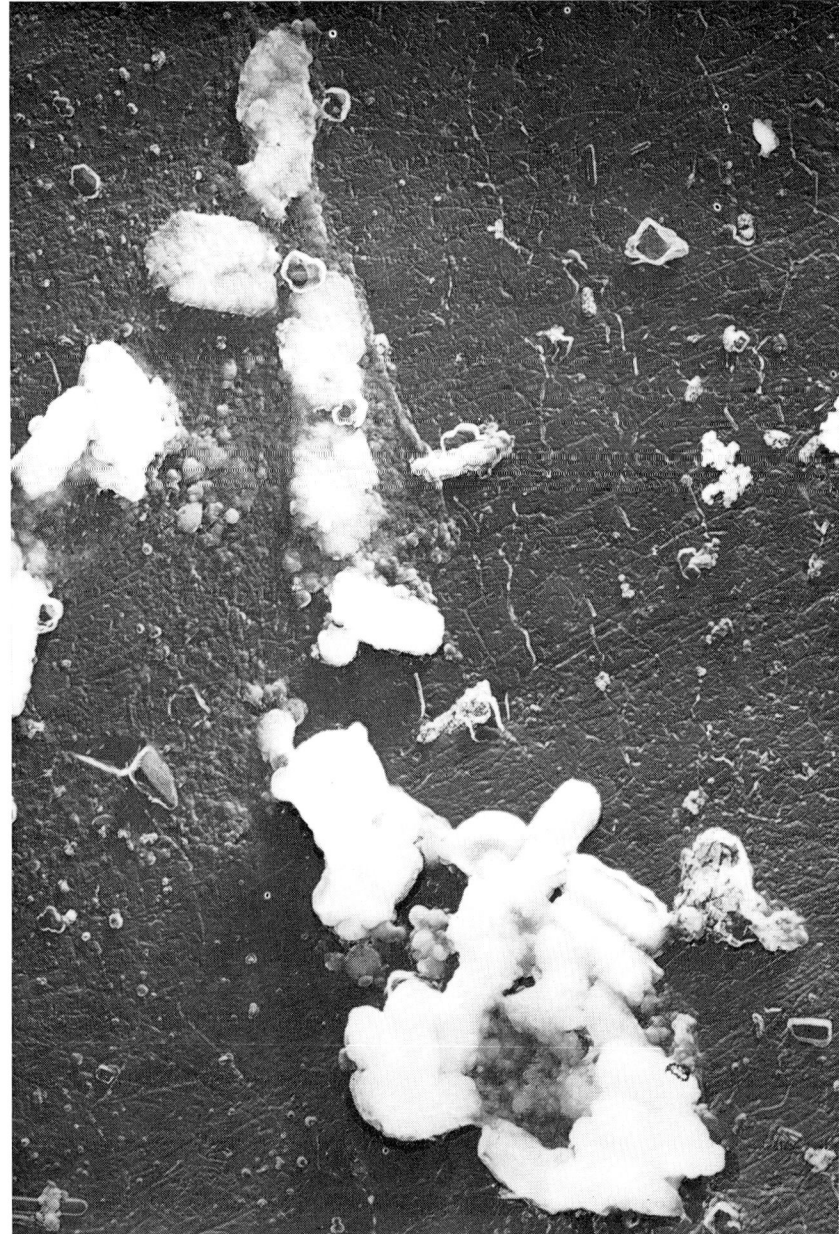

FOSSIL BACTERIA found in the Gunflint cherts are enlarged 30,000 diameters in this electron micrograph. Some two billion years old, they look like living rod-shaped species.

abundant in the cherts exposed near Kakabeka Falls, some 20 miles west of Thunder Bay. It consists of a bulb with a narrow stalk, surmounted by a structure that in some cases resembles an umbrella. The relative size of the three parts varies widely from specimen to specimen; the size of the bulb and the stalk together or of the umbrella alone ranges between 10 and 30 microns.

This odd plant has been assigned to the new genus *Kakabekia* and the species *umbellata* (which refers to the umbrella). Living organisms that superficially resemble it in form are certain multicellular polyps. All modern polyps, however, are many times larger.

The story of *Kakabekia* has had some interesting and continuing overtones. Late in 1964, quite independently (in fact, unaware) of paleontological investigations of the Gunflint fossils, Sanford M. Siegel discovered a strange new form while he was studying soil microorganisms that can survive extreme atmospheric conditions. The organism defied identification or assignment to any known taxonomic category. Siegel set it aside as being an enigma, although he kept his preparations, drawings and photomicrographs. A few months later a description of *Kakabekia* was published, and Siegel immediately noted a striking resemblance between his soil organism

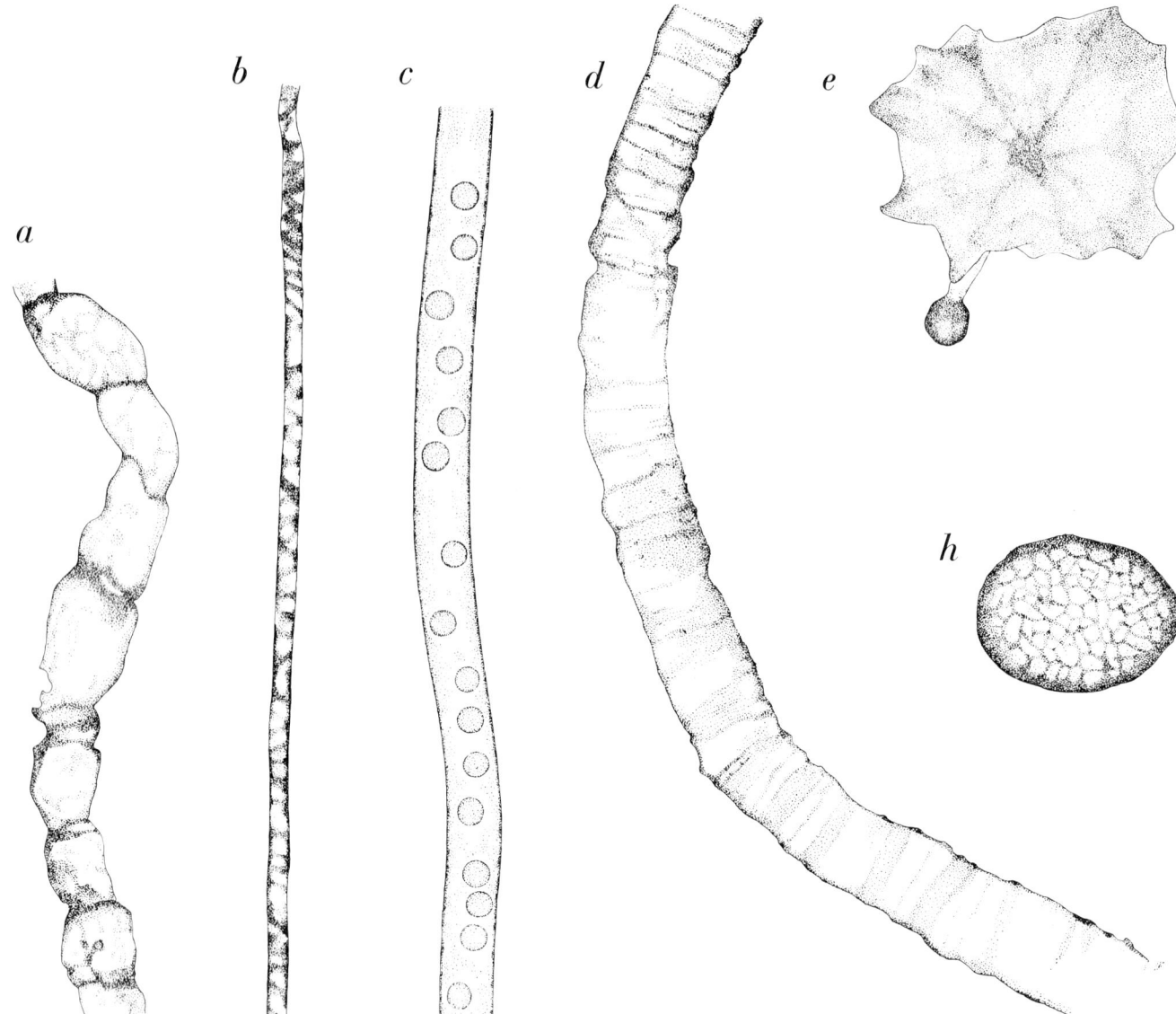

DIVERSITY OF FORM, evident among the plants fossilized in the Gunflint cherts, is indicative of evolutionary progress during the billion years that separate Fig Tree from Gunflint times. These composite drawings are based on a study of several specimens of each species and include three of the most unusual ones found. All are shown at the same scale: 2,500 times actual size. Filamentous organisms include two of the genus *Gunflintia*: *G. grandis* (*a*) and *G. minuta* (*b*). The other two are *Entosphaeroides amplus* (*c*) and *Animikiea septata* (*d*). The hydra-like *Kakabekia umbellata* (*e*) has a modern counterpart, if not a descendant, in a newfound soil

and some of the Gunflint specimens. Siegel's organism is very slow-growing, contains no chlorophyll and apparently has no nucleus; it may therefore be representative of a new group of prokaryotic microorganisms. It was first found in ammonia-rich soil collected at Harlech Castle in Wales, and it has since been recognized in soils from Alaska and Iceland and recently in soil from the slopes of the volcano Haleakala in Hawaii. Whether or not it is related to the two-billion-year-old *Kakabekia* is questionable, but the existence of two such bizarre forms is at least a remarkable evolutionary coincidence.

The eighth genus of Gunflint organisms comes from a single area near the easternmost outcropping of the chert, in the vicinity of Schreiber Beach. The organism consists of two concentric spheres; its outside diameter ranges from 28 to 32 microns. In most of the fossils the inner sphere is kept from contact with the outer one by as many as a dozen "spacers" in the form of small flattened spheroids. I have assigned this distinctive organism to the new genus *Eosphaera* ("dawn sphere") and the species *tyleri* (in honor of Tyler). No analogous living organism is known, nor has an organism resembling *Eosphaera* been found in any other Precambrian rocks. *Eosphaera* may be somewhat fancifully regarded as a "mistake" in evolution that did not survive the middle Precambrian.

Anyone who wanted to question that the organisms preserved in the Fig Tree cherts three billion years ago were photosynthetic could defend the negative position quite eloquently. Where the billion-years-younger Gunflint organisms are concerned, however, the affirmative evidence seems overwhelming. For one thing, the chemical analysis of organic material from the Gunflint cherts in several laboratories reveals the presence of the hydrocarbons pristane and phytane: two "chemical fossils" that can most reasonably be regarded as being breakdown products of chlorophyll. A second datum is the striking resemblance between the filamentous fossils that are the most abundant Gunflint forms and modern

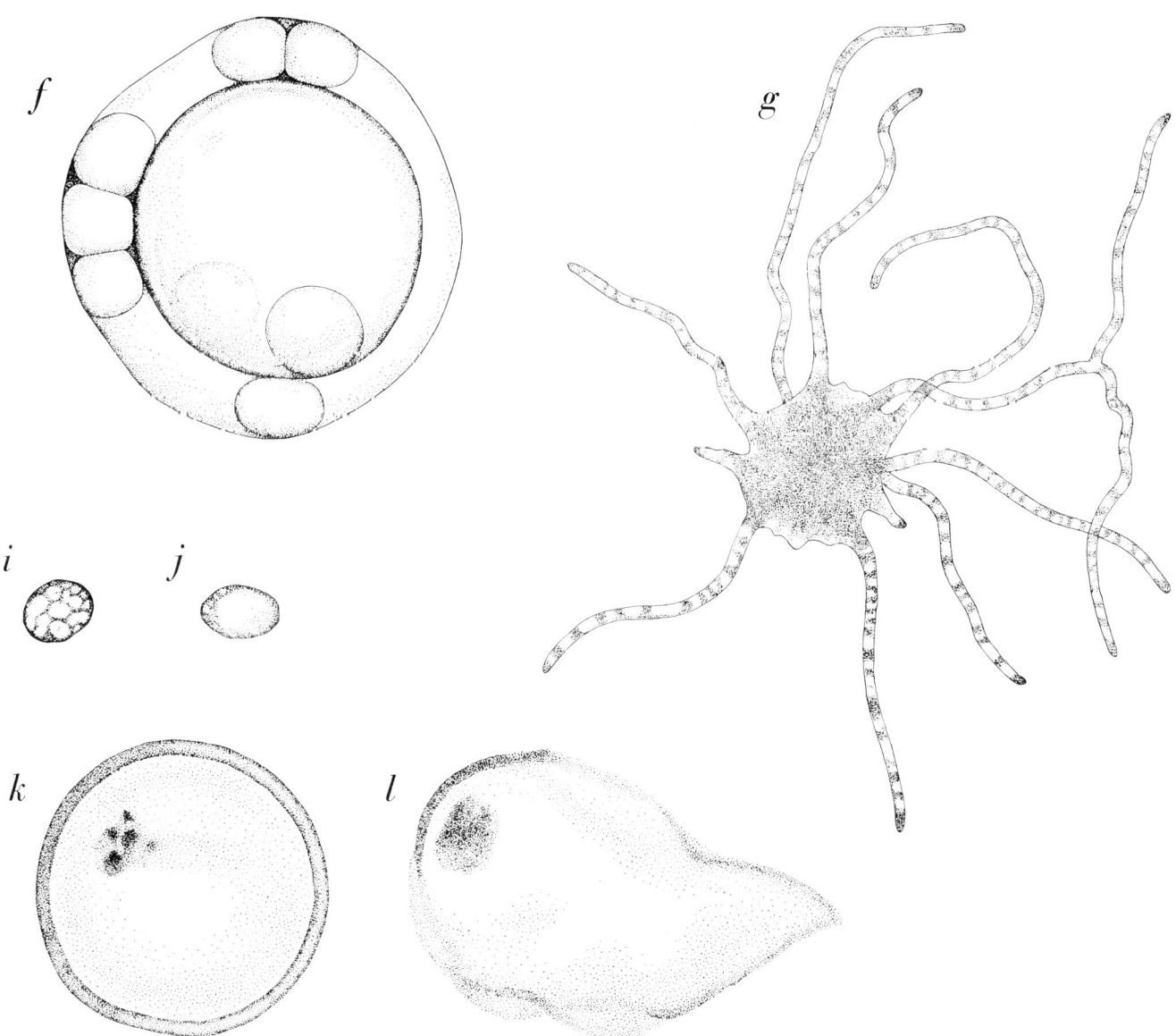

organism. The globular *Eosphaera tyleri* (*f*) evidently failed to survive Precambrian times. *Eoastrion bifurcatum* (*g*), with its array of filaments, is one of two such species. Three spherical organisms that differ chiefly in surface markings are assigned to the single genus *Huroniospora*; they are *H. microreticulata* (*h*), *H. macroreticulata* (*i*) and *H. psilata* (*j*). A one-celled organism that has not yet been classified (*k and l*) has internal features suggestive of a nucleus; the second specimen may have been fossilized as the process of cell division began. If these are eukaryotic cells, a key evolutionary event took place far earlier than is now thought.

photosynthetic blue-green algae. A third is the small domes and pillars in the Gunflint chert that resemble the structures formed in shallow water today by dome-building photosynthetic algae.

One may add the evidence of the relative abundance in Gunflint organic matter of the two nonradioactive carbon isotopes carbon 12 and carbon 13. The carbon in the carbon dioxide of the earth's atmosphere normally consists of about 99 percent carbon 12 and 1 percent carbon 13. In the process of photosynthesis, however, plants tend to fix slightly more carbon 12 than carbon 13, so that plant tissues are even poorer in the heavier isotope. Measurements of the ratio of the two isotopes in Gunflint organic material were made by Thomas C. Hoering of the Geophysical Laboratory of the Carnegie Institution of Washington. The results indicate that the Gunflint material too is poor in carbon 13, and to a degree that is almost identical with the depletion in modern algae and other photosynthetic plants. The billion-years-older Fig Tree organic material shows a carbon-12-to-carbon-13 ratio that is about the same as the ratio in the Gunflint material. This result lends credence to the view that the alga-like Fig Tree organisms probably were photosynthetic too.

It seems reasonable to conclude that even if oxygen was only a trivial component of the environment in Fig Tree times, the Gunflint organisms represent the intermediate agency that brought about the oxygen-rich environment at the end of the Precambrian. This is a development of major importance. It is not, however, the only development or even the most important development of Gunflint times. The variety of form (and therefore presumably of function) represented by the eight genera of Gunflint plants demonstrates that terrestrial life had crossed the second evolutionary threshold—the threshold of diversification—no less than two billion years ago.

The crossing of the third threshold is clearly documented in a series of late Precambrian formations, consisting primarily of limestones, sandstones and dolomites, found along the northern rim of the Amadeus basin in the Northern Territory of Australia. One member of the series is the Bitter Springs formation; a ridge in the Ross River area, about 40 miles from Alice Springs, consists of strata from its lower and middle levels. The exposed formations include isolated beds of dense black chert and laminated rocks that are associated with fossil structures built by colonial algae.

No absolute age is known for the Bitter Springs formation. Its top strata, however, lie some 4,000 feet below the lowest of the rocks that in the Ross River area are the boundary between formations of Precambrian age and those of the succeeding Cambrian period. The generally accepted date for the beginning of the Cambrian period is 600 million years ago, so that the lower position of the Bitter Springs cherts makes them considerably older. The Bitter Springs formation also underlies Precambrian sediments that are known to be some 820 million years old on the basis of their rubidium-strontium ratio. I think it is reasonable to assume that the Bitter Springs cherts are roughly a billion years old. This makes them only half as old as the Gunflint cherts and less than a third as old as the Fig Tree cherts. I collected samples of the Bitter Springs cherts in the Ross River area in April, 1965, and I added them to the Fig Tree samples for study with Schopf that summer.

A preliminary analysis of the abundant microfossils in the Bitter Springs cherts indicates that at least four general groups of lower plants lived in the shallow seas or embayments that covered this part of central Australia in late Precambrian times. As one would anticipate, the plants included filamentous blue-green algae akin to such living forms as *Oscillatoria* and *Nostoc*. Some of these filaments are more than 75 microns long; they are thickest (about 1.4 microns) at the center and taper toward the ends to less than one micron.

Our most exciting identifications during the preliminary analysis concerned the other three groups. On the basis of the internal structures that have been preserved, all three appear to represent various green algae. The green algae, unlike the blue-green algae, are eukaryotic! Thus the Bitter Springs cherts apparently contain the earliest-known fossil evidence that an organism potentially capable of sexual reproduction, or at least possessing a nucleus, had finally evolved.

By the time Schopf had finished his detailed analysis of the Bitter Springs specimens in 1968, he had concluded that they represent a total of three bacterium-like species, some 20 certain or probable representatives of blue-green algae, two certain genera of green algae, two possible species of fungi and two problematical forms. For one of the green algae, *Glenobotrydion aenigmatis*, an unusual coincidence of fossilization has preserved several specimens at various stages of mitotic division. By arranging the specimens in order one can recreate almost the entire sequence [*see illustration on page 870*].

With the discovery of a late Precambrian population of eukaryotic plants we approach the end of our story. The unequivocal appearance of the eukaryotic level of cellular organization at this point in earth history provides a welcome explanation for one of the principal puzzles of terrestrial evolution: Why do so many groups of higher organisms not appear in the fossil record until the span of geologic time is seven-eighths past? It is a good question, particularly in the light of what we know about the multicellular animals and higher algae early in the Paleozoic era.

Until recently the commonest explanation of the rarity of advanced organisms, multicellular animals in particular, before Cambrian times was that the supply of free oxygen in both the atmosphere and the hydrosphere had up to that point been extremely limited. Time was required to let the first prokaryotic organisms adjust to existence in an environment that had begun to contain this highly reactive element. Time was also needed for enough oxygen to leave the hydrosphere (where it was being produced by algal photosynthesis) and enter the atmosphere to establish a protective shield of ozone (O_3) between the earth's surface and the sun's harsh ultraviolet radiation. Until this shield existed neither shallow water nor bare ground would have been habitable.

An oxygen-poor environment was certainly a major obstacle in the path of

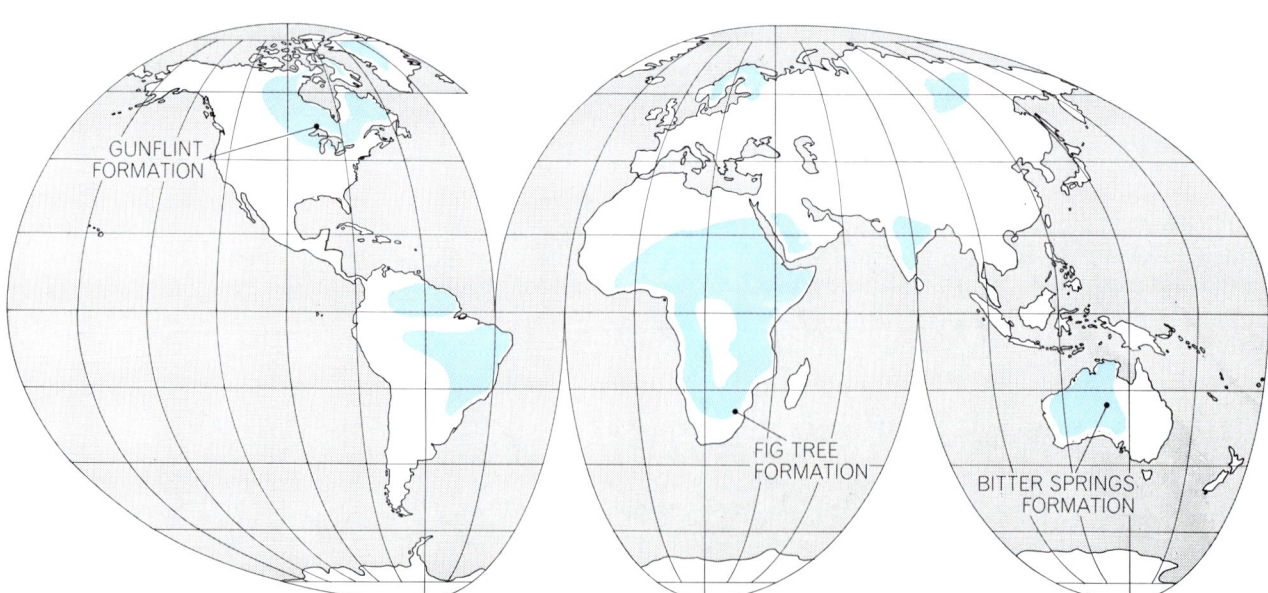

PRECAMBRIAN ROCKS occupy the old, low-lying "core" areas of continents throughout the world. Many, formed from accumulated sediments, are so altered by heat and pressure that any fossils have been obliterated. Some relatively unaltered exposures of Precambrian sediments, however, are rich in organic remains. Three such fossil-bearing Precambrian formations are shown on this map.

oxygen-dependent heterotrophs. In addition to the evidence provided by oxidized Precambrian formations, however, a number of recent studies indicate that the earth's atmosphere had actually accumulated enough oxygen to establish some kind of ozone shield considerably before the end of the Precambrian. The appearance of eukaryotic organisms late in the Precambrian, as indicated by the green algae of the Bitter Springs formation, provides a better explanation for the failure of higher organisms to appear until an even later time. The fundamental key to evolutionary progress is genetic variability. Sexual reproduction, which involves the recombination of heritable characteristics, is the highway to genetic variability and all its consequences, including the increased complexity of form and function at all levels of organization, that are thereafter apparent in the course of evolution.

This last of three thresholds, which separates the primitive world of cells without a nucleus from a world where sexual reproduction is possible, must have been crossed a long time before the formation of the Bitter Springs cherts. Only half a billion years or so later the Paleozoic seas were swarming with highly differentiated aquatic plants and animals, evolved from primitive forebears that had managed to cross all three Precambrian thresholds. Half a billion years does not seem to be evolutionary "room" enough to account for such epic progress. Moreover, there is evidence to suggest that developments in this direction may have begun in Gunflint times.

The Fig Tree, Gunflint and Bitter Springs cherts are not the only sources of Precambrian fossils, nor have Tyler, Schopf and I been the only investigators of the Precambrian fossil record. Indeed, work in the field has gone on for nearly a century. The emphasis has changed in recent years from an earlier concern with the curious "boundary" that has been traditionally accepted as separating the Precambrian from the beginning of the Paleozoic. We and our colleagues in many countries are primarily interested today in evidence that we hope will reveal even finer details of fossil cellular organization. It is a field that requires all levels of observation, from the macroscopic down to the electron microscopic. Workers in the field agree that the search for other stages and thresholds in the evolution of life must be focused on fine structures wherever these have been fortuitously preserved. The Precambrian fossil record is still meager, but significant gaps in the record are steadily

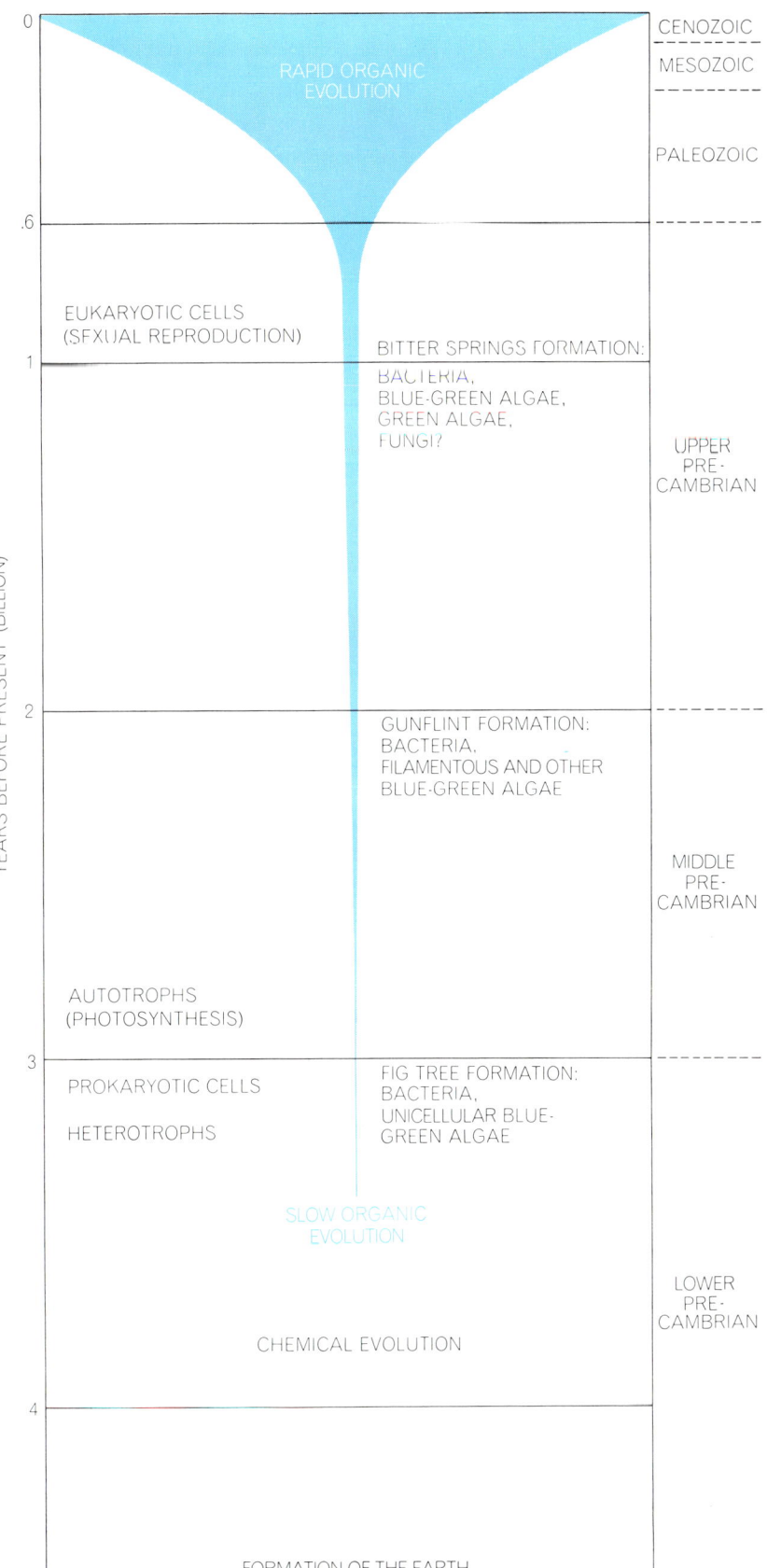

ORGANIC EVOLUTION is presented in terms of successively briefer episodes of biological advance. The Precambrian interval, the earliest and by far the longest, began when the earth was formed 4.5 billion years ago and ended 600 million years ago with the beginning of the Paleozoic era. The increasing abundance of species with the passage of time is indicated in color. Once organisms with eukaryotic, or truly nucleated, cells evolved, hastening evolutionary progress in late Precambrian times, the number of species multiplied explosively.

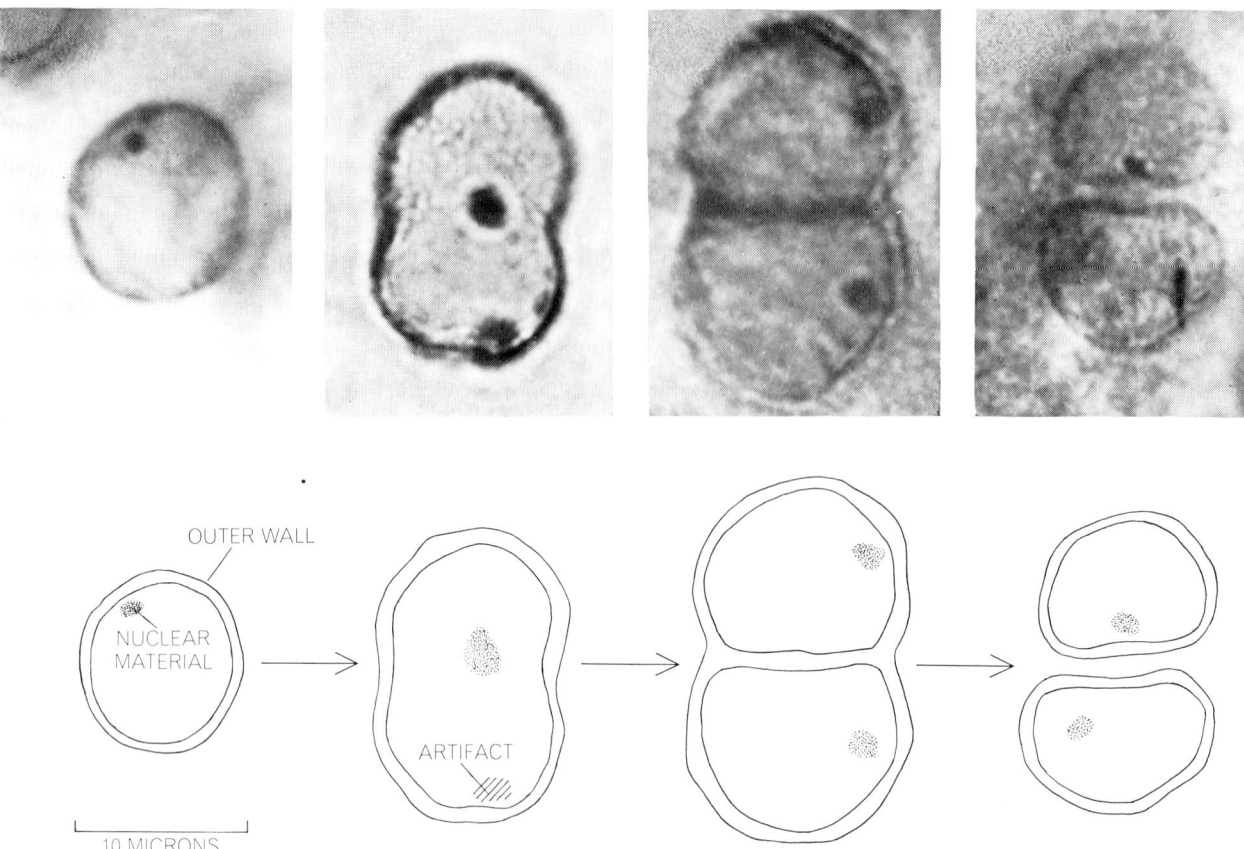

CELL DIVISION of a eukaryotic organism, a green alga of the genus *Glenobotrydion*, appears in the billion-year-old cherts of the Bitter Springs formation of Australia. J. William Schopf reconstructed the event (*drawings*), working with fossils that preserved individual organisms at different stages of division (*micrographs*). Existence of such biologically advanced algae in Bitter Springs times is proof that the evolutionary threshold leading to sexual reproduction and genetic variability had been crossed even earlier.

being filled. Indeed, the traditional concept of a clearly delimited Precambrian boundary may soon disappear from reconstructions of the history of terrestrial life.

Two final comments may be of interest to those who, like myself, are stimulated by the search for first causes no matter how often the search goes unrewarded. Some biologists now suggest that the organelles characteristic of the eukaryotic cell may once have been independent organisms that somehow came to live symbiotically inside larger host cells. It is not clear whether or not one of the host cell's responses to the presence of such a guest was a regrouping into a nucleus of the genetic material formerly scattered throughout the host's cytoplasm. If symbiosis was indeed the first step toward evolution of the eukaryotic cell, it may be that certain of the Gunflint organisms currently being studied show initial steps in this direction [see middle illustration at right on page 864]. A fascinating account of this transformation of various bacterial and algal organisms into components of the eukaryotic cell is given by Lynn Margulis of Boston University in a recent book. Mrs. Margulis' argument rests on biological grounds; the increasingly detailed Precambrian fossil record supports her thesis.

Astronomers and physicists have found evidence in recent years that molecules such as the hydroxyl radical (OH), carbon monoxide (CO), ammonia (NH_3), hydrogen cyanide (HCN) and formaldehyde (HCHO) are formed in the "empty" reaches of space. So far the ultimate in interstellar chemical complexity is represented by organic material that some investigators maintain they find in a peculiar class of meteorites known as carbonaceous chondrites. Until recently no carbonaceous chondrite had been recovered under circumstances that completely ruled out the possibility of accidental contamination by terrestrial organic material. As a result nagging questions about the chondrites have remained, particularly with regard to such biologically important molecules as amino acids (which are a *sine qua non* of terrestrial life). Recently, however, using stringent laboratory procedures to analyze carefully documented samples of the Murchison chondrite (which fell in southern Australia in September, 1969), Ponnamperuma and his associates at the Ames Research Center (in collaboration with Carleton B. Moore of Arizona State University and Ian R. Kaplan of the University of California at Los Angeles) seem to have proved the existence of extraterrestrial amino acids. Not only is the quantity of amino acids in the Murchison meteorite surprisingly high but also certain of them are unknown in terrestrial organisms and hence cannot be contaminants from the soil.

This finding opens up a new world of chemical evolution: a world of random synthetic processes not on the earth but in space, including the extraterrestrial formation of bodies (of which the carbonaceous chondrites seem to be fragments) that are rich in organic materials. The finding brings us back to the discussion at the beginning of this article. The chemical evolution of organic matter, the prelude to biogenesis on the earth, seems to have occurred elsewhere in the solar system or outside it. For knowledge of the earliest stages of organic evolution the biologist and paleontologist must rely on the chemist and astrophysicist.

The Author

ELSO S. BARGHOORN is professor of botany at Harvard University, curator of paleobotanical collections at the Harvard Botanical Museum and a member of the department of geology at the university. During the past three years he has in addition been a principal investigator in the Apollo space program. "My early interests in science," he writes, "were in astronomy, then chemistry, botany and finally paleontology. In college [Miami University of Ohio] I had a conflict between my yen for the track team and for the laboratory. At one point I quit college to work as a deckhand. I have long had photography as a hobby and still have, but I suppose I can also call myself an amateur farmer and gardener, since I enjoy raising my own vegetables and hay crop and maintaining an Angus per year for beef." Barghoorn has been associated with Harvard most of the time since 1937, when he entered as a graduate student. He received his master's degree there in 1938 and his Ph.D. in 1941.

Bibliography

MICROORGANISMS FROM THE GUNFLINT CHERT. Elso S. Barghoorn and Stanley A. Tyler in *Science*, Vol. 147, No. 3658, pages 563–577; February 5, 1965.

MICROORGANISMS THREE BILLION YEARS OLD FROM THE PRECAMBRIAN OF SOUTH AFRICA. Elso S. Barghoorn and J. William Schopf in *Science*, Vol. 152, No. 3723, pages 758–763; May 6, 1966.

PRECAMBRIAN MICRO-ORGANISMS AND EVOLUTIONARY EVENTS PRIOR TO THE ORIGIN OF VASCULAR PLANTS. J. William Schopf in *Biological Reviews*, Vol. 45, No. 3, pages 319–352; August, 1970.

CHEMICAL EVOLUTION AND THE ORIGIN OF LIFE: A COMPREHENSIVE BIBLIOGRAPHY. Compiled by Martha W. West and Cyril Ponnamperuma in *Space Life Sciences*, Vol. 2, No. 2, pages 225–295; September, 1970.

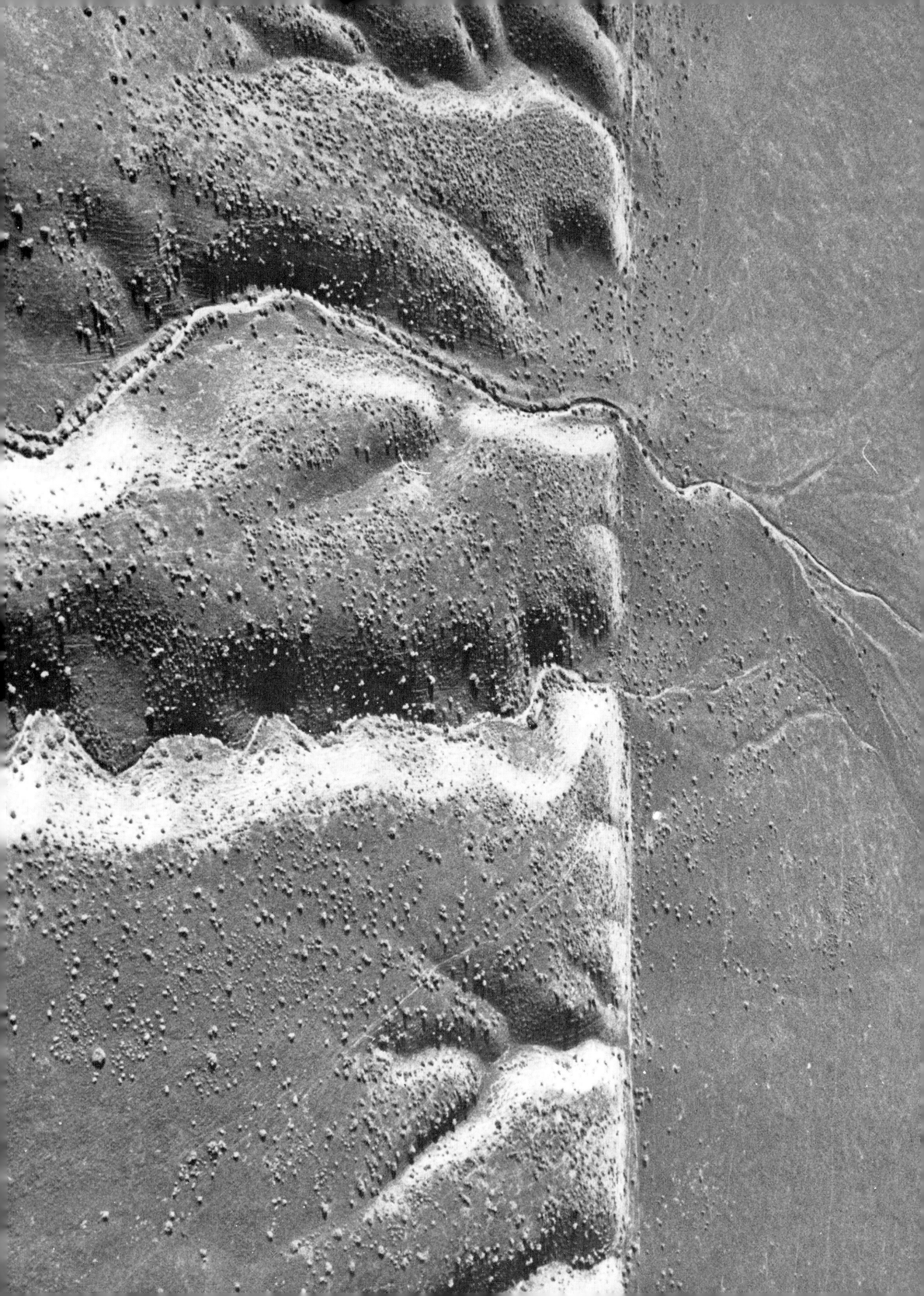

SCIENTIFIC AMERICAN November 1971, Vol. 225, No. 5, pp. 52–66

OFFPRINT **896**

THE SAN ANDREAS FAULT

by Don L. Anderson

This well-known break in the earth's crust is actually not one fault but a system of faults. The break separates a northward-moving wedge of California, including Los Angeles, from the rest of North America.

The San Fernando earthquake that occurred at sunrise on February 9, 1971, jolted many southern Californians into acute awareness that California is earthquake country. Although it was only a moderate earthquake (6.6 on the Richter scale), it was felt in Mexico, Arizona, Nevada and as far north as Yosemite National Park, more than 250 miles from San Fernando. It was recorded at seismic stations around the world. In spite of its relatively small size the San Fernando earthquake was extremely significant because it happened near a major metropolis and because its effects were recorded on a wide variety of seismic instruments. Within hours the affected region was aswarm with geologists mapping faults and seismologists installing portable instruments to monitor aftershocks and the deformation of the ground. It was immediately clear from data telemetered to the Seismological Laboratory of the California Institute of Technology in Pasadena from the Caltech Seismic Network that the earthquake was not centered on the much feared San Andreas fault or, for that matter, on any fault geologists had labeled as active. The faults in the area, however, are all part of the San Andreas fault system that covers much of California.

The San Andreas fault system (and its attendant earthquakes) is part of a global grid of faults, chains of volcanoes

and mountains, rifts in the ocean floor and deep oceanic trenches that represent the boundaries between the huge shifting plates that make up the earth's lithosphere. The concept of moving plates is now fundamental to the theory of continental drift, which was long disputed but is now generally accepted in modified form on the basis of voluminous geological, geophysical and geochemical evidence. The theory had received strong support from the discovery that the floors of the oceans have a central rise or ridge, often with a rift along the axis, that can be traced around the globe. Within the rift new crustal material is continuously being injected from the plastic mantle below, forming a rise or ridge on each side of the rift. The newly formed crustal material slides away from the ridge axis. Since the magnetic field of the earth periodically reverses polarity, the newly injected material "freezes" in stripes parallel to the ridge axis, whose north-south polarity likewise alternates. By dating these stripes one can estimate the rate of sea-floor spreading.

The San Andreas fault system forms the boundary between the North American plate and the North Pacific plate and separates the southwestern part of California from the rest of North America. In general the Pacific Ocean and that part of California to the west of the San Andreas fault are moving northwest

with respect to the rest of the continent, although the continent inland at least as far as Utah feels the effects of the interactions of these plates.

The relative motion between North America and the North Pacific has been estimated in a variety of ways. Seismic techniques yield values between 1½ and 2½ inches per year. The ages of the magnetic stripes on the ocean floor indicate a rate of about 2⅓ inches per year. Geodetic measurements in California give rates between two and three inches per year. The ages of the magnetic anomalies off the coast of California indicate that the oceanic rise came to intersect the continent at least 30 million years ago. Geologists and geophysicists at a number of institutions (notably the University of Cambridge, Princeton University and the Scripps Institution of Oceanography) have proposed that geologic processes on a continent are profoundly affected when a continental plate is intersected by an oceanic rise. At the rates given above the total displacement along the San Andreas fault amounts to at least 720 miles if motion started when the rise hit the continent and if all the relative motion was taken up on the fault. Displacements this large have not been proposed by geologists, but the critical tests would involve correlation of geology in northern California with geology on the west coast of Baja California, an area that has only recently been studied in detail. One can visualize how the west coast of North America may have looked 32 million years ago by closing up the Gulf of California and moving central and northern California back along the San Andreas fault to fit into the pocket formed by the coastline of the northern half of Baja California. This places all of California west of the San Andreas fault south of

DISPLACEMENT ALONG SAN ANDREAS FAULT is clearly visible in this aerial photograph of a region a few miles north of Frazier Park, Calif., itself 65 miles northwest of Pasadena, where the fault runs almost due east and west. This east-west section of the San Andreas fault is part of the "big bend," where the fault appears to be locked. The photograph is reproduced with north at the right. The hilly region to the left (south) of the fault line is moving upward (westward) with respect to the flat terrain at the right, causing clearly visible offsets in the two largest watercourses as they flow onto the alluvial plain.

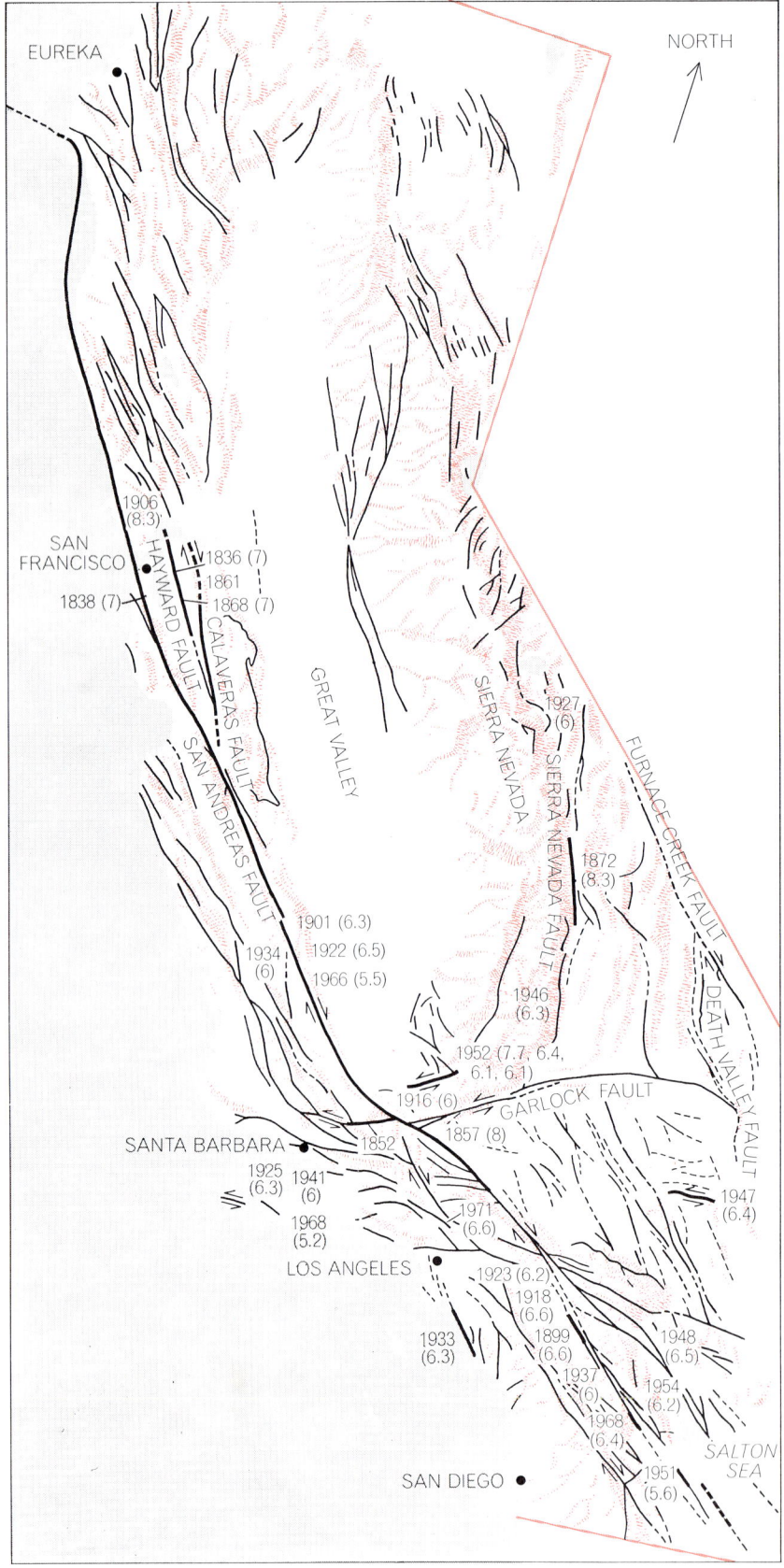

SIMPLIFIED FAULT MAP of California identifies in heavy black lines the faults that have given rise to major earthquakes since 1836. The magnitude of all but two of the earthquakes is given in parentheses next to the year of occurrence. For events that predated the introduction of seismological instruments the magnitudes are estimated from historical accounts. For two major events, the earthquakes of 1852 and 1861, information is too sparse to allow a magnitude estimate. Arrows parallel to the faults show relative motion.

the present Mexican border [*see illustration on page 880*].

California is riddled with faults, most of which trend roughly northwest-southeast, like most of the other tectonic and geologic features of California (such as the Sierra Nevada and the Coast Ranges). The prominent exceptions are the east-west-trending transverse ranges and faults that make up a band some 100 miles wide extending inland from between Los Angeles and Santa Barbara. The San Gabriel Mountains, which form the rugged backdrop to Los Angeles, are part of this complex geologic region, and it was here that the San Fernando earthquake struck. The northeast-trending Garlock fault and the Tehachapi Mountains, which separate the Sierra Nevada and the Mojave Desert, also cut across the general grain of California. The area to the west of most of the northwest-trending faults is moving northwest with respect to the eastern side. This is called right-lateral motion. If one looks across the fault from either side, the other side is moving to the right.

Motion on the Garlock fault is left-lateral, which, combined with the right-lateral motion on the San Andreas fault, means that the Mojave Desert is moving eastward with respect to the rest of California. Parts of the faults that have been observed or inferred to move as a result of earthquakes in historic times are shown in the illustration at the left. Also shown are the dates of the earthquakes and the magnitude of some of the more important ones. In general both the length of rupture and the total displacement are greater for the larger earthquakes. Horizontal displacements as great as 21 feet were observed along the San Andreas fault after the San Francisco earthquake of 1906, which had a magnitude of 8.3 on the Richter scale. (The Richter scale, devised by Charles F. Richter of Cal Tech, is logarithmic. Although each unit denotes a factor of 10 in ground amplitude, or displacement, the actual energy radiated by an earthquake is subject to various modifications.) The San Fernando earthquake produced displacements of six feet, whose direction was almost equally divided between the horizontal and the vertical.

The trend of the San Andreas fault system is roughly northwest-southeast from San Francisco to the south end of the Great Central Valley (the San Joaquin Valley) and again from the north of the Salton Sea depression to the Mexican border. The motion along the faults in these areas is parallel to the fault and is mainly strike-slip, or horizontal. Between these two regions, from the south end of the San Bernardino Mountains to the Garlock fault, the faults bend abruptly and run nearly east and west, producing a region of overthrusting and crustal shortening [*see illustration below*]. The attempt of the southern California plate to "get around the corner" as it moves to the northwest is responsible for the complex geology in the transverse ranges, for the abrupt change in the configuration of the coastline north of Los Angeles and ultimately for the recent San Fernando earthquake. The big bend of the San Andreas fault is commonly regarded by seismologists as being locked and possibly as being the location of the next major earthquake. Much of the motion in this region, however, is being taken up by strike-slip motion along faults parallel to the San Andreas fault and by overthrusting on both

sides of the fault. The displacements associated with the larger earthquakes in southern California in the vicinity of the big bend have averaged out to about 2½ inches per year since 1800. The Kern County earthquake of 1952 (magnitude 7.7) apparently took care of most of the accumulated strain, at least at the north end of the big bend, that had built up since the Fort Tejon earthquake of 1857 (magnitude 8).

The San Andreas fault system cannot be completely understood independently of the tectonics and geology of most of the western part of North America and the northeastern part of the Pacific Ocean. This vast region is itself only a part of the global tectonic pattern, all parts of which seem to be interrelated. The earthquake, tectonic and mountain-building activities of western North America are intimately related to the relative motions of the Pacific and North American plates. Just as it is misleading

to think of the San Andreas fault as an isolated mechanical system, so it is misleading to think of the entire San Andreas fault as a single system. The part of the fault that lies in northern California was activated earlier and has moved farther than the southern California section. The northern portion is less active seismically than the southern section and seems to have been created in a different way. It is also moving in a slightly different direction.

Measuring Displacements

There are several ways to measure displacements on major faults. Fairly recent displacements are reflected in offset stream channels [*see illustration on page 872*]. Many such offsets measured in thousands of feet are apparent across the San Andreas fault in central California, some of which can be directly related to earthquakes of historic times.

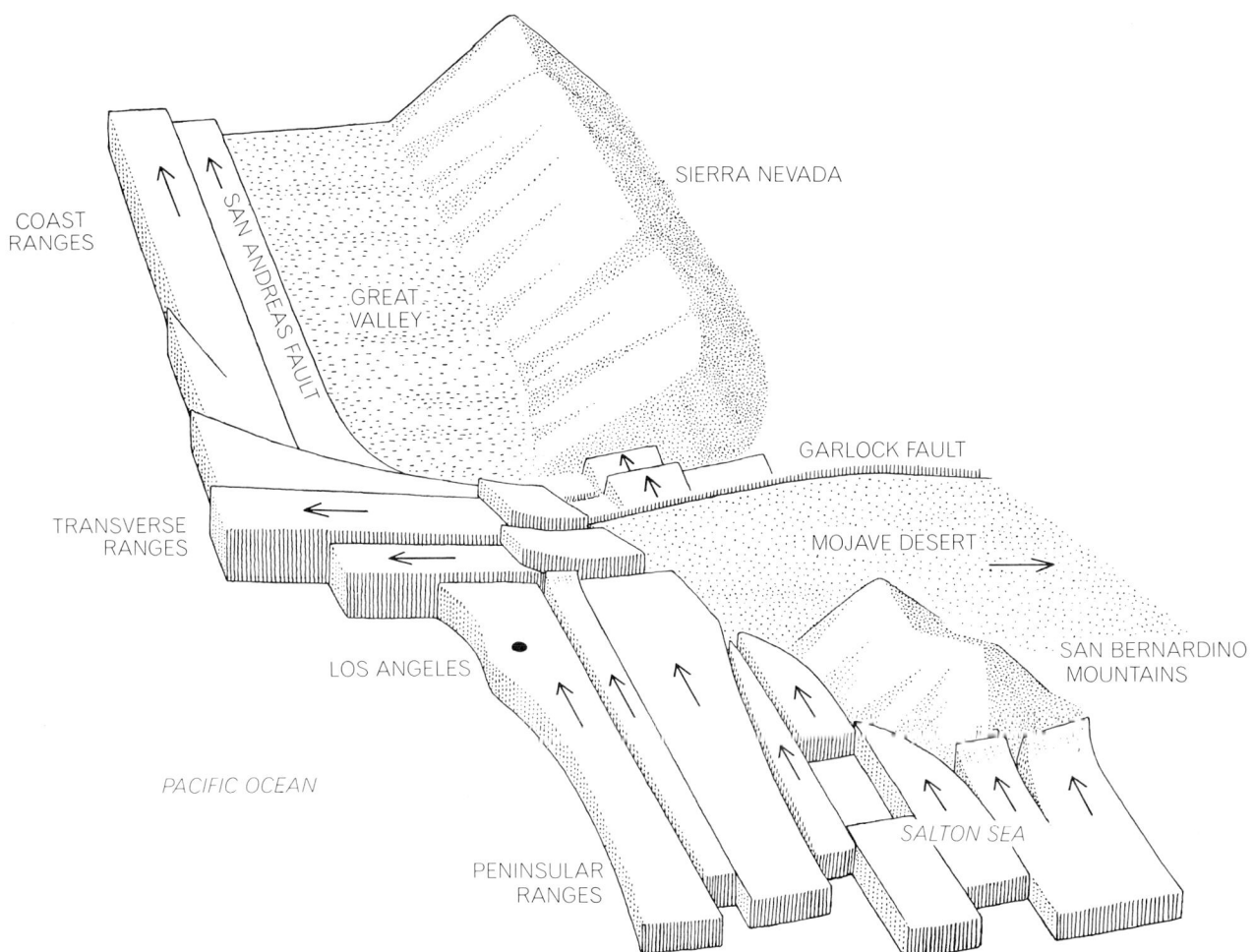

MOTION OF EARTH'S CRUST in southern California is generally northwest except where the lower group of blocks encounters the deep roots of the Sierra Nevada. At this point the blocks are diverted to the left (west), creating the transverse ranges and a big bend in the San Andreas fault system. Above the bend the blocks continue their northwesterly march, carrying the Coast Ranges with them. The Salton Sea trough at the lower right evidently represents a rift that has developed between two blocks.

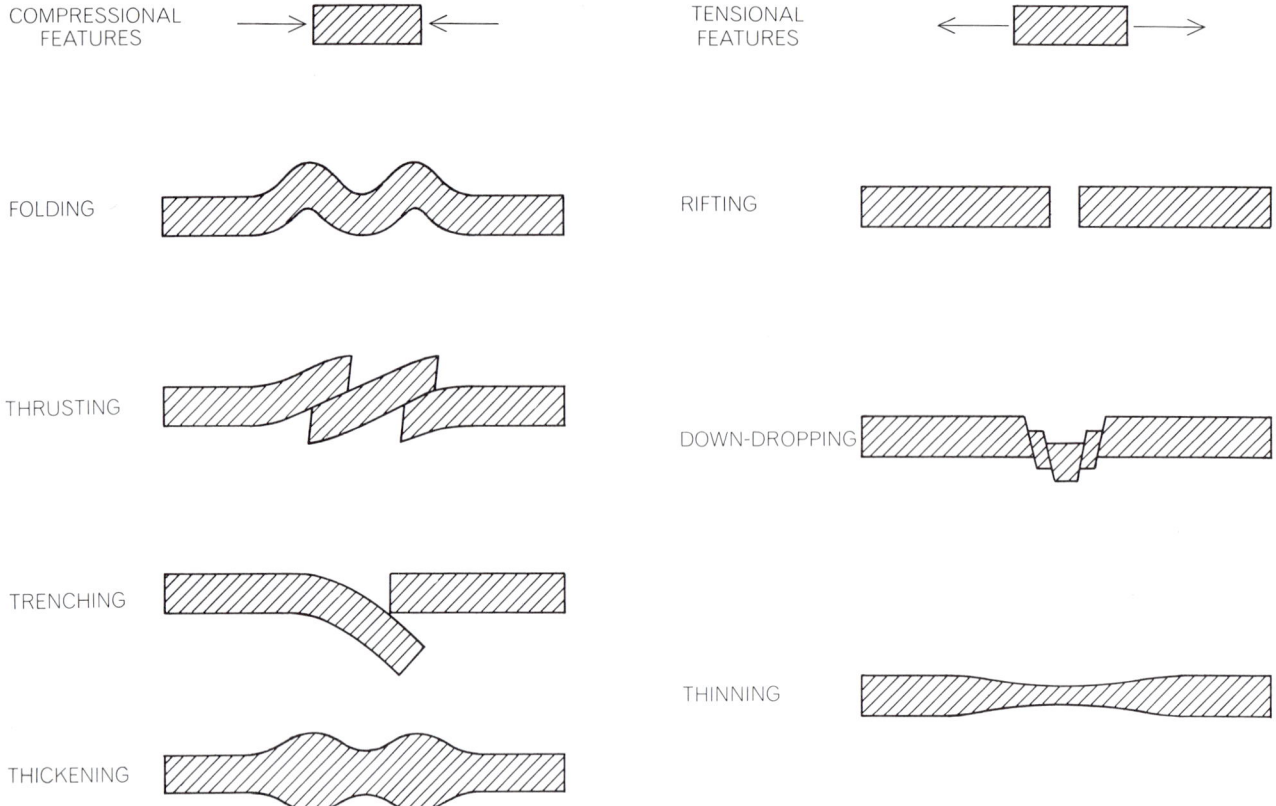

COMPRESSIONAL FEATURES

FOLDING

THRUSTING

TRENCHING

THICKENING

TENSIONAL FEATURES

RIFTING

DOWN-DROPPING

THINNING

RESPONSE OF CRUSTAL PLATES to compression (*left*) and tension (*right*) accounts for most geologic features. According to the recently developed concept of plate tectonics, the earth's mantle is covered by huge, rigid plates that can be colliding, sliding past one another or rifting apart. The rifting usually occurs in the ocean floor. The San Andreas fault marks the location where two plates are sliding past each other. Plate tectonics helps to explain how the continents have drifted into their present locations.

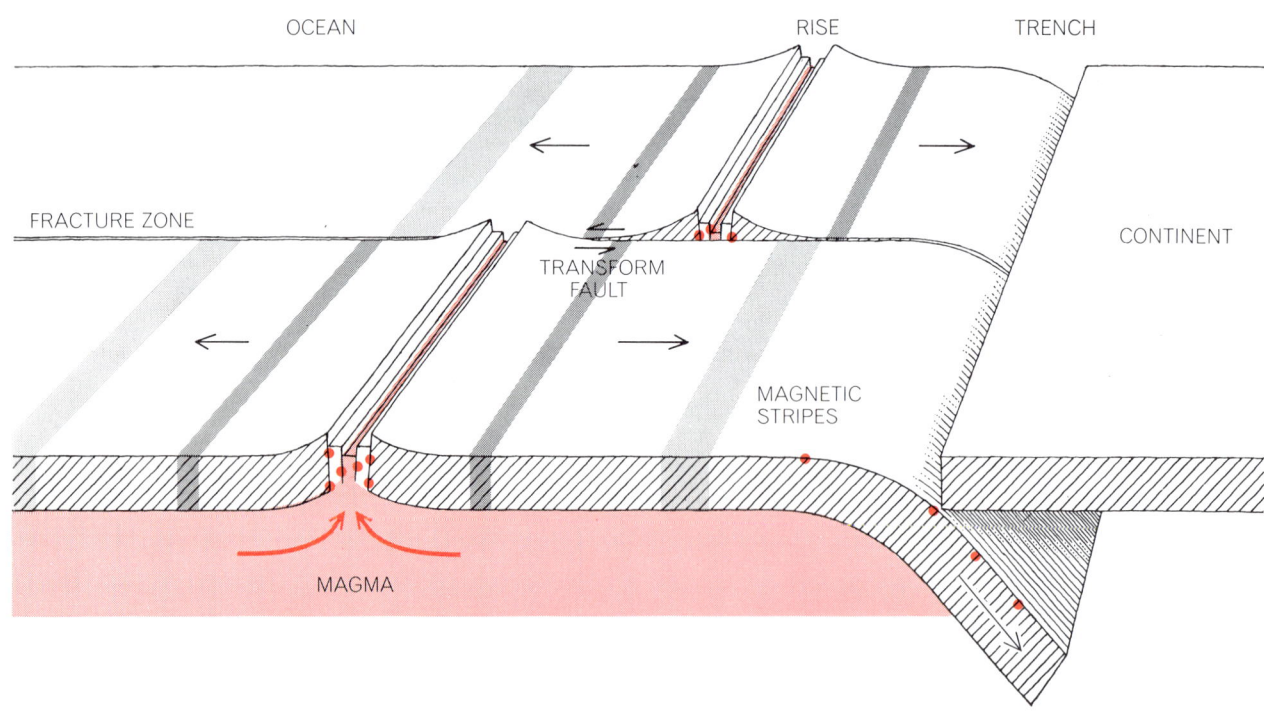

OCEAN   RISE   TRENCH

FRACTURE ZONE   CONTINENT

TRANSFORM FAULT

MAGNETIC STRIPES

MAGMA

RIFT IN OCEAN FLOOR (*color*) initiates three major features of oceanic plate tectonics. The rift is bordered by a rise or ridge created by magma pushed up from the mantle below. The magma solidifies with a magnetic polarity corresponding to that of the earth. When, at long intervals, the earth's polarity reverses, the polarity of newly formed crust reverses too, resulting in a sequence of magnetic "stripes." A trench results when an oceanic plate meets a continental plate. A fracture zone and transform fault result when two plates move past each other. Earthquakes (*dots*) accompany these tectonic processes. The earthquakes in the vicinity of a rise and along a transform fault are shallow. Deep-focus earthquakes occur where a diving oceanic plate forms a trench.

Erosion destroys this kind of evidence very quickly. By matching up distinctive rock units that have been broken up and moved with respect to each other it is possible to document offsets of tens to hundreds of miles. A sedimentary basin often holds debris that could not possibly have been derived from any of the local mountains; matching up these basins with the appropriate source region on the other side of the fault can provide evidence of still larger displacements. When these various kinds of information are combined, one obtains a rate of about half an inch per year for motion on the San Andreas fault in northern and central California over the past several tens of millions of years.

This is much less than the 2½ inches per year that is inferred for the rate of separation of Baja California and mainland Mexico and the rate that is inferred from seismological studies in southern California. There are several possible explanations for the discrepancy. Northern and southern California may be moving at different rates; this seems unlikely since they are both attached to the same Pacific plate. On the other hand, part of the compression in the transverse ranges may result from a differential motion between the two parts of the state. Another possibility is that all of the relative motion between the North American plate and the Pacific plate is not being taken up by the San Andreas fault or even by the San Andreas fault system but extends well inland. The fracture zones of the Pacific seem to affect the geology of the continent for a distance of at least several hundred miles.

The Great Central Valley and the Sierra Nevada lie between two major fracture zones that abut the California coast: the Mendocino fracture zone and the Murray fracture zone. The transverse ranges, the Mojave Desert and the Garlock fault are all in line with the Murray fracture zone. Recent volcanism lines up with the extensions of the Clarion fracture zone and the Mendocino fracture zone. The basins and range geological province of the western U.S., a region of crustal tension and much volcanism, may represent a broad zone of deformation between the Pacific plate and the North American plate proper. Seismic activity is certainly spread over a large, diffuse region of the western U.S.

Although the subject has been quite controversial, most geologists are now willing to accept large horizontal displacements on the faults in California, particularly the San Andreas. Displacements as large as 450 miles of right-lat-

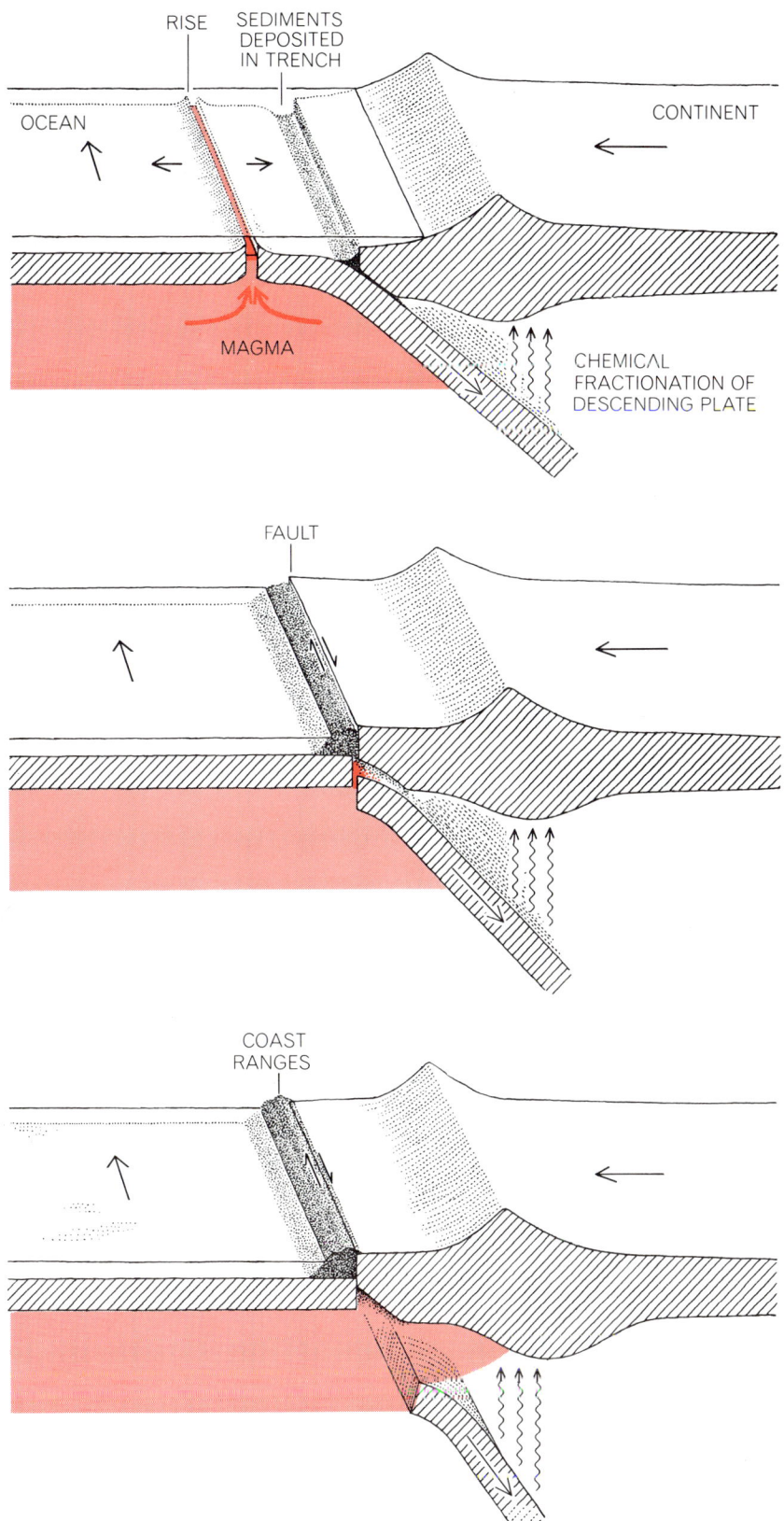

INTERACTION BETWEEN RISE AND TRENCH leads to mutual annihilation. The trench, formed as the oceanic plate dives under the continental plate, slowly fills with sediments carried by rivers and streams (*top*). Meanwhile the melting of the descending slab adds new material to the continent from below. When the axis of the rise reaches the edge of the continent, the flow of magma into the rift is cut off and trench sediments are scraped onto the western (that is, left) part of the oceanic plate (*middle*). The descending plate disappears under the continent and the sediments travel with the oceanic plate (*bottom*). The northern part of the San Andreas fault may have been formed in this way.

eral slip have been proposed for the northern segment of the fault. Displacements on the southern San Andreas fault are put at no more than 300 miles. This discrepancy has been puzzling to geologists. My own conclusion is that the part of northern and central California west of the San Andreas fault has moved northwest more than 700 miles and that the southern San Andreas fault has slipped about 300 miles, which makes the apparent discrepancy even worse. The discrepancy disappears if one drops the concept of a single San Andreas fault and admits the possibility that the two segments of the fault were initiated at different times.

The two-fault hypothesis is supported by straightforward extrapolation of the record on the ocean floor. The two San Andreas faults formed at different times, in different ways and may be moving at different rates. The record indicates that the western part of North America caught up with a section of the East Pacific rise somewhere between 25 million and 30 million years ago. Before the collision a deep oceanic trench existed off the coast such as now exists farther to the south off Central America and South America. The trench had existed for many millions of years, receiving

sediments from the continent; subsequently the sediments were carried down into the mantle by the descending oceanic plate, which was diving under the continent. Based on what we know of trench areas that are active today one can assume that the plate sank to 700 kilometers and that the process was accompanied by earthquakes with shallow, intermediate and deep foci.

The Origin of Continents

Let us examine a little more closely what happens when an oceanic rise, the source of new oceanic crust, approaches a trench, which acts as a sink, or consumer of crust. Evidently the rise and the trench annihilate each other. The oceanic crust and its load of continental debris, which was formerly in the trench, rise because the crust is no longer connected to the plate that was plunging under the continent. The trench deposits are so thick they eventually rise above sea level and become part of the continent. The deposits are still attached to the oceanic plate, however, and travel with it [*see illustration on preceding page*].

In the case of the Pacific plate off California the deposits move northwest

with respect to the continent. This is the stuff of coastal California north of Santa Barbara, particularly the Coast Ranges. According to this view, the northern segment of the San Andreas fault was born at the same time as northern California. The rise and the trench initially interacted near San Francisco, which then was near Ensenada in Baja California. Ensenada in turn was near the northern end of the Gulf of California, which was then closed.

The tectonics and geologic history of California, and in fact much of the western U.S., are now beginning to be understood in terms of the new ideas developed in the theories of sea-floor spreading, continental drift and plate tectonics. Many of the basic concepts were laid down by the late Harry H. Hess of Princeton and Robert S. Dietz of the Environmental Science Services Administration. Tanya Atwater of the University of California at San Diego and Warren Hamilton of the U.S. Geological Survey and their colleagues have made particularly important contributions by applying the concepts of plate tectonics to continental geology. We now know that the outer layer of the earth is immensely mobile. This layer, the lithosphere, is relatively cold and

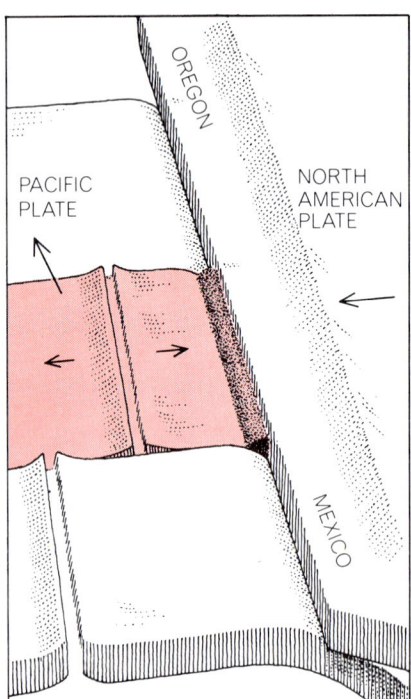 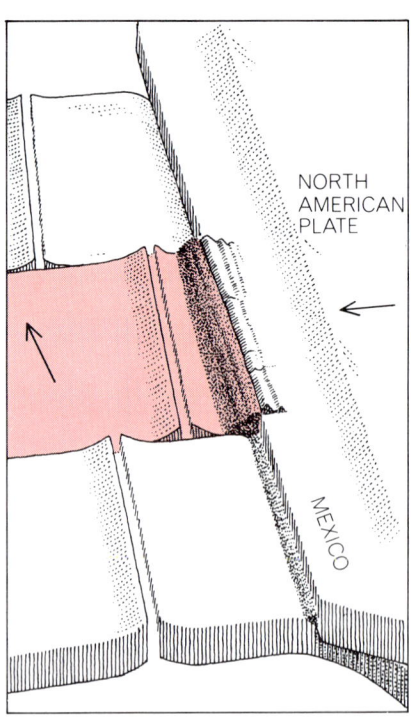 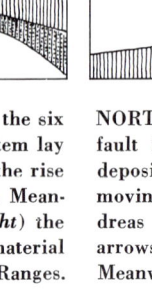 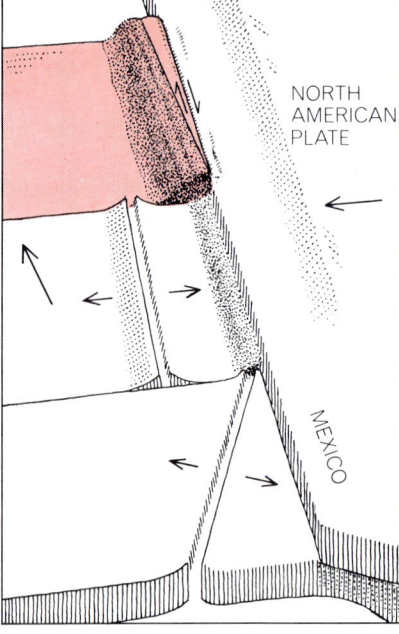

FORMATION OF SAN ANDREAS FAULT SYSTEM is depicted schematically in the six diagrams on these two pages. Some 30 million years ago (*left*) an oceanic rise system lay off the west coast of North America, which was carried by a plate moving toward the rise crests. The continental plate overrides the Pacific plate, producing a long trench. Meanwhile the entire Pacific plate is moving northwest. After a few million years (*right*) the rise nearest the continent is shut off. The trench by now has been filled with material eroded from the continent. These deposits will later become the California Coast Ranges.

NORTHERN SECTION of San Andreas fault is created when the former trench deposits become attached to the northward-moving Pacific plate (*left*). The San Andreas fault lies between the two opposed arrows indicating relative plate motions. Meanwhile to the south a tilted rise crest

rigid and slides around with little resistance on the hot, partially molten asthenosphere.

Where the crust is thick, as it is in continental regions, the temperatures become high enough in the crust itself to cause certain types of crustal rocks to lose their strength and to offer little resistance to sliding. There is thus the possibility that the upper crust can slide over the lower crust and that the moving plate can be much thinner than is commonly assumed in plate-tectonic theory. The molten fraction of the asthenosphere, called magma, rises to the surface at zones of tension such as the mid-oceanic rifts to freeze and form new oceanic crust. The new crust is exposed to the same tensional forces (presumably gravitational) that caused the rift in the first place; therefore it rifts in turn and subsequently slides away from the axis of the rise. In addition to providing the magma for the formation of new crust, the melt in the asthenosphere serves to lubricate the boundary between the lithosphere and the asthenosphere and effectively decouples the two. The rise is one of the types of boundary that exist between lithospheric plates and is the site of small, shallow tensional earthquakes.

When two thin oceanic lithospheric plates collide, one tends to ride over the other, the bottom plate being pushed into the hot asthenosphere. The boundary becomes a trench. When the lower plate starts to melt, it yields a low-density magma that rises to become part of the upper plate; this magma becomes the rock andesite, which builds an island arc on what is to become the landward, or continental, side of the trench. (The rock takes its name from the Andes of South America. Mount Shasta in California is primarily andesite, as are the island arcs behind the trenches that surround the Pacific.) The thickness of the crust is essentially doubled as a result of the underthrusting. The material remaining in the lower plate is now denser than the surrounding material in the asthenosphere, both because it has lost a low-density fraction and because it is colder; thus it sinks farther into the mantle. In some parts of the world the downgoing slab can be tracked by seismic means to 700 kilometers, where it seems to bottom out. By this process new light material is added to the crust and new dense material is added to the lower mantle. A large part of what comes up stays up; a large part of what goes down stays down.

The introduction of chemical fractionation and a mechanism for "unmixing" makes the process different from the one customarily visualized, in which gigantic convection cells carry essentially the same material in a continuous cycle. The new process is able to explain in a convincing way how continents are formed and thickened. As the continent thickens and rises higher because of buoyancy, erosional forces become more effective and dump large volumes of continental sediments into the coastal trenches. A portion of the sediments is ultimately dragged under the continent to melt and form granite. The light granitic magma rises to form huge granitic batholiths such as the Sierra Nevada. A batholith is a large mass of granitic rock formed when magma cools slowly at great depth in the earth's crust. It is carried to the surface by uplifting forces and exposed by erosion.

The concept of rigid plates moving around on the earth's surface and interacting at their boundaries has been remarkably successful in explaining the evolution of oceanic geology and tectonics. The oceanic plates seem to behave rather simply. Tension results in a rise, compression results in a trench and lateral motion results in a transform fault

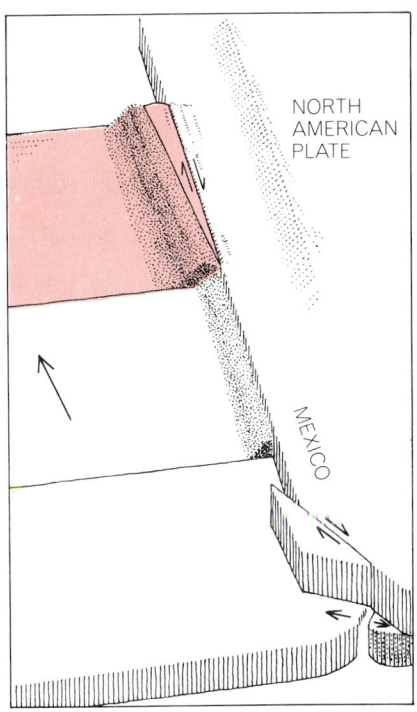

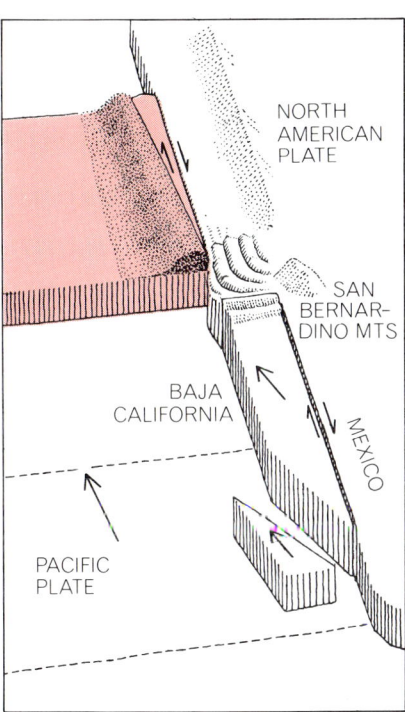

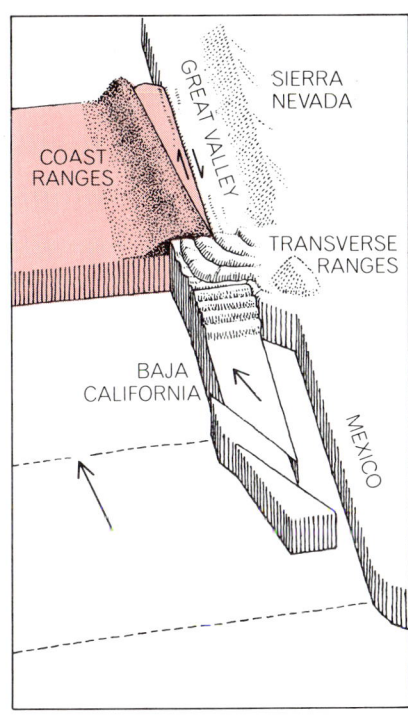

(not yet visible in the first pair of diagrams) is ready to encounter the continent end on at a break in the coastline south of Baja California. The collision (*right*) breaks off a part of the Baja California peninsula, which becomes attached to the Pacific plate and starts its journey to the northwest.

SOUTHERN SECTION of San Andreas fault is now fully activated (*left*) as the Baja California block begins sliding past the North American plate and collides with deeply rooted structures to the north, the Sierra Nevada and San Bernardino Mountains, which deflect the block to the west. More of Baja California breaks loose, opening up the Gulf of California. As Baja California continues to move northwestward (*right*) the Gulf of California steadily widens. The compression at the north end of the Baja California block creates the transverse ranges, which extend inland from the vicinity of present-day Santa Barbara.

880

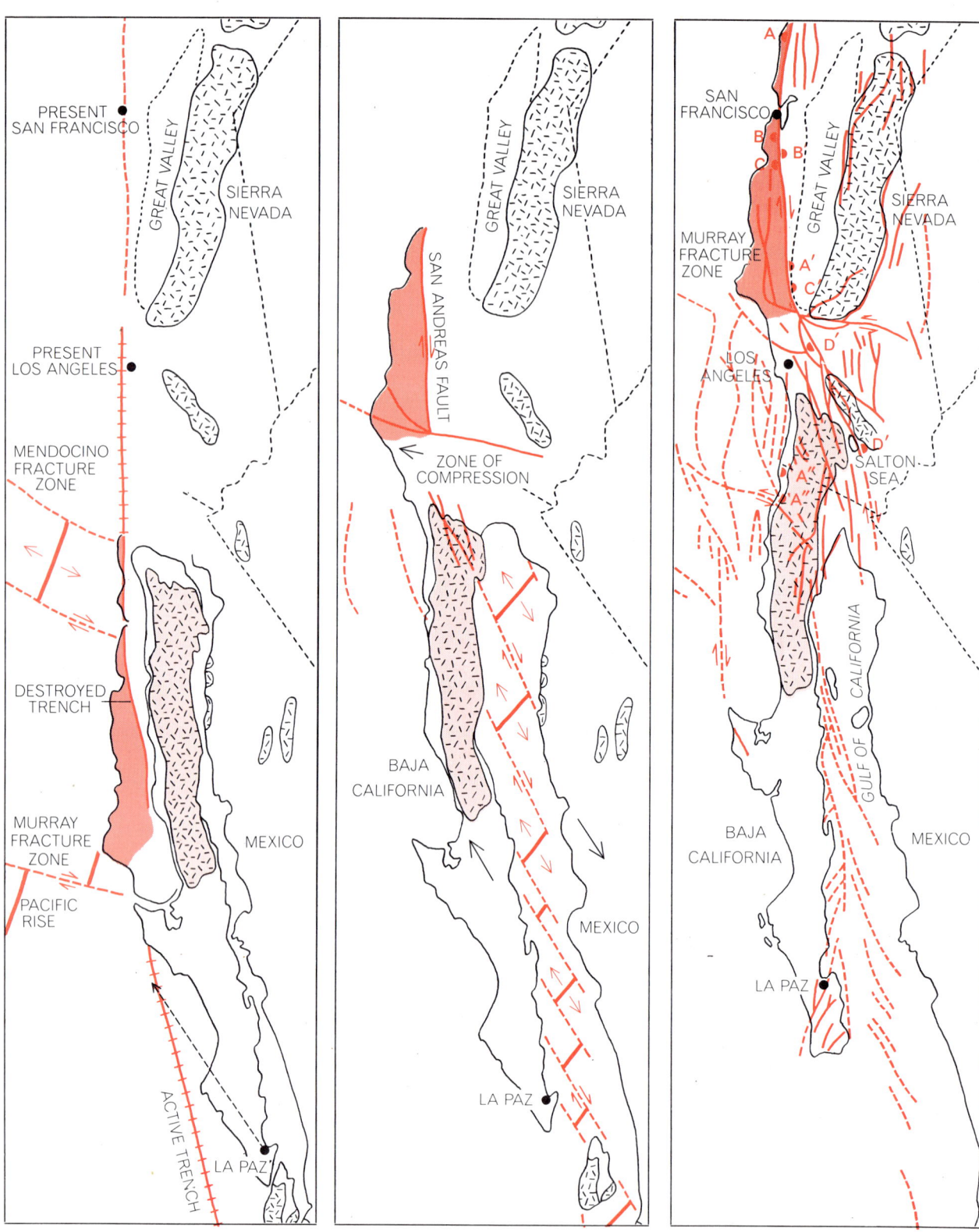

EARLY AND LATE STAGES in the history of the San Andreas fault are depicted. Twenty-five million years ago (*left*) Baja California presumably nestled against mainland Mexico. The first section of oceanic rise between the Murray fracture zone and the Pioneer fracture zone has just collided with the continent. Trench deposits are uplifted and become part of the Coast Ranges of California. The block containing the present San Francisco area (*dark color*) is about to start its long northward journey. A block immediately to the east (*light color*) becomes attached to the Pacific plate and eventually is jammed against the San Bernardino Mountains. Three million years ago (*middle*) the Gulf of California has started to open. As the peninsula moves away from mainland

Mexico a series of rifts appear, fill with magma and are offset by numerous fractures. Baja California may have been torn off in one piece or in slivers. Southwest California and Baja California today (*right*) continue to slide northwest against the North American mainland. The illustration shows major fault systems and offshore fracture zones. On the basis of unique rock formations geologists infer that the Los Angeles area has moved northwest about 130 miles (*D'* to *D*) in the past 20 million years or less. Other studies indicate that the Palo Alto region has been carried about 200 miles (*C'* to *C*). Coastal rocks to the north of San Francisco have been displaced at least 300 miles (*A'* to *A*) and perhaps as much as 650 miles (*A''* to *A*) in the past 30 million years.

and a fracture zone [see bottom illustration on page 876]. The interaction of oceanic and continental plates or of two continental plates is apparently much more complicated, and this is one reason the new concepts were developed by study of the ocean bottom rather than continental geology.

The boundary between two oceanic plates can be a deep oceanic trench, an oceanic rise or a strike-slip fault depending on whether the plates are approaching, receding from or moving past each other. The forces involved are respectively compressional, tensional and shearing. When a thick continental plate is involved, compression can also result in high upthrust and folded mountain ranges. The Himalayas resulted from the collision of the subcontinent of India with Asia. I shall show that the transverse ranges in California were formed in a similar way. Tension can result in a wide zone of crustal thinning, normal faulting and volcanism; it can also create a fairly narrow rift of the kind found in the Gulf of California and the Red Sea [see top illustration on page 876].

The interaction of western North America with the Pacific plate has led to large horizontal motions along the San Andreas fault, to concentrated rifting as in the Salton Sea trough and the Gulf of California, to diffuse rifting and normal faulting accompanied by volcanism in the basins and range province of California, Nevada, Utah and Arizona, to large vertical uplift by overthrusting as in the transverse ranges north and west of Los Angeles, to the generation of large batholiths such as the Sierra Nevada and to the incorporation of deep-sea trench material on the edge of the continent. Ultimately the geology, tectonics and seismicity of California can be related to the collision of North America with the Pacific plate.

Most of the Pacific Ocean is bounded by trenches and island arcs. Trenches border Japan, Alaska, Central America, South America and New Zealand. Island arcs are represented by the Aleutians, the Kuriles, the Marianas, New Guinea, the Tongas and Fiji. The arcs are themselves bordered by trenches. All these areas are characterized by andesitic volcanism and deep-focus earthquakes. Western North America is lacking a trench and has only shallow earthquakes, but the geology indicates that there was once a trench off the West Coast, and in fact there was once a rise. The present absence of a rise and a trench, the absence of deep-focus earthquakes and the existence of uplifted deep-sea sediments are all related.

Tracing back the history of the interaction of the Pacific plate with the North American plate, one is forced to conclude that the northern part and the southern part of the San Andreas fault originated at different times and in different ways. The northern part was evidently formed about 30 million years ago when a portion of rise between the Mendocino fracture zone and the Murray fracture zone approached an offshore trench bordering the southern part of North America. At that time the west coast of North America resembled the present Pacific coast of South America: there was a deep trench offshore, high mountains paralleled the coastline and large underthrusting earthquakes were associated with the downgoing lithosphere.

Origin of the Fault

As the rise approached the continent both the geometry and the dynamics of interaction changed [see illustrations on pages 8 and 9]. Depending on the spreading rate of the new crust generated at the rise and the rate at which the rise itself approaches the continent, the relative motion between the rise and the continental plate will decrease, stop or reverse when the rise hits the trench. The forces keeping the trench in existence will therefore decrease, stop or reverse, leading to uplift of the sedimentary material that has been deposited in the trench. In classical geologic terms these are known as eugeosynclinal deposits. Although they have been exposed to only moderate temperatures, they have been subjected to great pressures, both hydrostatic (owing to their depth of burial) and directional (owing to the horizontal compressive forces between the impinging plates). Eugeosynclinal sediments are therefore strongly deformed and become even more so as they are contorted and sheared during uplift. Much of the western edge of California and Baja California is underlain by this material, called the Franciscan formation. The formation is physically attached to the Pacific plate and is therefore moving northwest with respect to the rest of North America. The present boundary is the northern part of the San Andreas fault. Today this section of the San Andreas system extends from Cape Mendocino, north of San Francisco, to somewhere south and east of Santa Barbara, near the beginning of the great bend of the San Andreas fault, where the San Andreas and the Garlock faults intersect.

Meanwhile, 30 million years ago, another section of the rise south of the Murray fracture zone was still offshore, together with an active trench. Baja California was still attached to the mainland of Mexico and the Gulf of California had not yet opened up. The southern part of the San Andreas fault had not yet been formed.

The abrupt change in the direction of the coastline south of the tip of Baja California suggests that here the rise approached the continent more end on than broadside. A sliver of existing continent was welded onto the Pacific plate and rifted away from Mexico, thus forming Baja California and the Gulf of California. Thereafter Baja California participated in the northwesterly motion of the Pacific plate, with the result that the Gulf of California widened progressively with time.

About five million years were required for northern California, which had broken off from Baja California, to be carried about 200 miles to the northwest. At the end of that time the Gulf of California and the Salton Sea trough had not yet opened. The faults that delineate the major geologic blocks in southern California had not yet been activated. The block bearing the San Gabriel fault, now north of San Fernando, occupied the future Salton Sea trough. The transverse ranges will eventually be formed from the Santa Barbara, San Gabriel and San Bernardino blocks by strong compression from the south when Baja California breaks loose from mainland Mexico. This also opens up the Gulf of California and the Salton Sea trough.

As northern California is being carried away from Baja California by the Pacific plate another segment of oceanic rise south of the Murray fracture zone approaches the southern half of Baja California, where the situation described above is repeated except that the rise crest encounters a sharp bend in the coastline and the trench hits just south of the tip of the peninsula. Now instead of approaching the continent more or less broadside the rise approaches the continent end on. Mainland Mexico is still decoupled from that part of the Pacific plate to the west of the rise by the rise and the trench. Baja California, however, is now coupled to the northwestward-moving Pacific plate and Baja California is torn away from the mainland. This happened between four and six million years ago. Magma from the upper mantle wells up into the rift, forming a new rise that works its way north into the widening gulf. Alternatively, the entire peninsula of Baja California could have broken off from the mainland at the

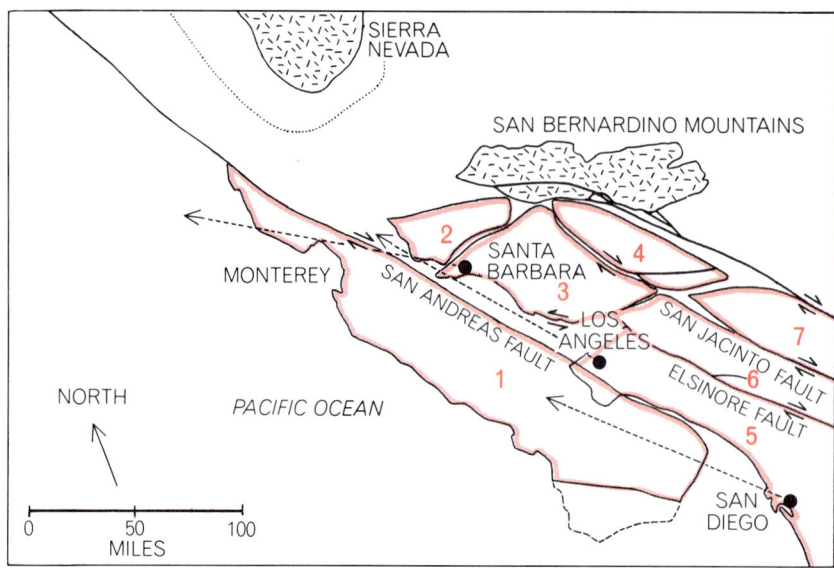

SEQUENCE OF SIMPLIFIED VIEWS shows the movement of major blocks in southern California over the past 12 million years. In the first view (*above*) the Gulf of California has not yet appreciably opened but the block carrying the Coast Ranges (*1*) has started to move rapidly northwest with activation of northern portion of San Andreas fault. Dots show origin and arrows show displacement of San Diego, Los Angeles and Santa Barbara.

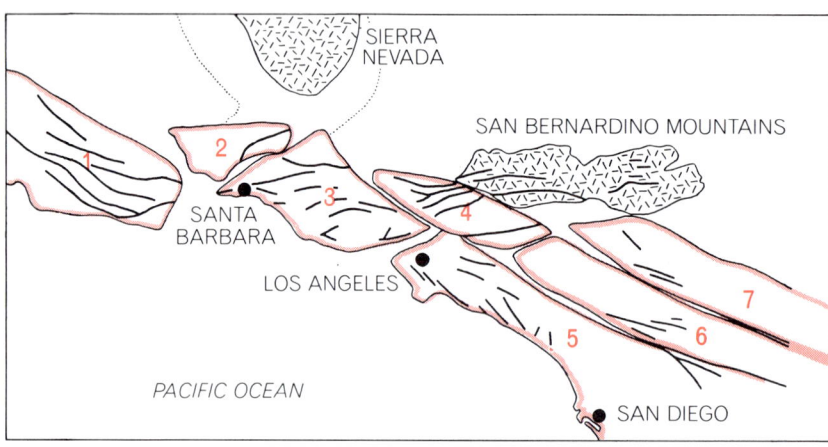

TWO MILLION YEARS AFTER ACTIVATION of the southern portion of the San Andreas fault four blocks (*2, 3, 4, 7*) have been forced against the deep roots of the Sierra Nevada and San Bernardino Mountains. Compressive forces create the transverse ranges. Meanwhile the block carrying the Coast Ranges (*1*) has been carried far to the northwest.

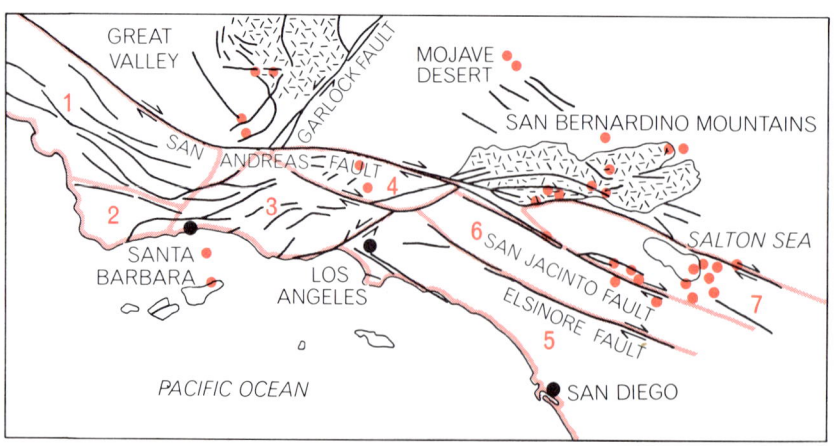

GEOLOGY OF SOUTHERN CALIFORNIA TODAY is dominated by compressive forces operating in the big bend of the San Andreas fault, which connects the southern and northern parts of the system. Colored dots show the location of earthquakes in the recent past.

same time. As the peninsula, including parts of southern California, moves north it collides with parts of the continent that are still attached to the main North American plate. This results in compression, overthrusting and shearing and the eventual formation of the transverse ranges.

The southern part of the San Andreas fault system was therefore formed by the rifting off of a piece of continent. Today it represents the boundary between two parts of the continental plate that are moving with respect to each other. This part of the San Andreas fault was formed well east, or inland, of the southward projection of the northern San Andreas.

The northerly march of southern California and Baja California seems to have been blocked when the moving plate encountered the thick continental crust to the north, particularly the massive granitic San Bernardino mountain range, which includes the 11,485-foot San Gorgonio Mountain. Since large and high mountain ranges have deep roots, the crust in this region is probably much thicker than normal, perhaps as thick as 50 kilometers. Earthquakes in this region are all shallower than 20 kilometers, which may be the thickness of the sliding plates. The blocks veer westward and are strongly overthrust as they attempt to get around the obstacle; this movement generates the big bend in the San Andreas fault system. The deflected blocks eventually join up with the northern California block.

Earthquake Country

From a social and economic point of view earthquakes are one of the most important manifestations of plate interaction. From a scientific point of view they supply a third dimension to the study of faults and the nature of the interactions between crustal blocks, including the stresses involved and the nature of the motions.

Seismologists at the University of California at Berkeley and at the Cal Tech Seismological Laboratory have been keeping track of earthquake activity in California for more than 40 years. Both groups have installed arrays of seismometers that telemeter seismic data to their laboratories for processing and dissemination to the appropriate public agencies. During the 36-year period 1934 through 1969 there were more than 7,300 earthquakes with a Richter magnitude of 4 or greater in southern California and adjacent regions [*see illustration on page 884*]. Many thousands more earthquakes of smaller magnitude are routinely located and reported in the seismological bulletins. Although damage depends on local geological conditions and the nature of the earthquake, a rough rule of thumb is that a nearby earthquake of magnitude 3.5 or greater can cause structural damage. The average annual number of earthquakes of magnitude 3 or greater in southern California recorded since 1934 is 210; the number in any one year has varied from a low of 97 to a high of 391. The strongest earthquake in this period was the Kern County event of magnitude 7.7 in 1952. The aftershocks of that event increased the total number of events for several years thereafter.

In general the larger the earthquake, the greater the displacement across a fault and the greater the length of fault that breaks. The great earthquakes of 1906 and 1857 respectively caused large displacements across the northern and central parts of the San Andreas fault and relieved the accumulated strain in these areas. The accumulation of strain in southern California is relieved mainly by slip on a series of parallel faults and by overthrusting on faults at an angle to the main San Andreas system; that is what happened in the Kern County and San Fernando earthquakes. The unique east-west-trending transverse ranges were formed in this way. In the process deep-seated ancient rocks were uplifted and exposed by erosion.

Another seismically active area associated with major faults is south of the Mojave Desert near San Bernardino, where the faults show a sudden change in direction. The central part of the Mojave Desert is also moderately active. This is consistent with the idea that the sliding lithosphere is diverted by the San Bernardino Mountains. Faults and evidence of relatively recent volcanic events abound in the area. The northern part of Baja California is also quite active. An interesting feature of seismicity maps of southern California is the alignment of earthquakes in zones that trend roughly northeast-southwest, approximately at right angles to the major trend of the San Andreas system.

The map on the next page shows that the San Andreas fault itself has played only a small role in the seismicity of southern California over the past 30-odd years. One must not forget, however, that the great earthquake of 1857 probably broke the San Andreas fault for about 100 miles northwest and southeast from the epicenter. That epicenter is thought to have been near Fort Tejon, which is close to the projected intersection of the Garlock and San Andreas

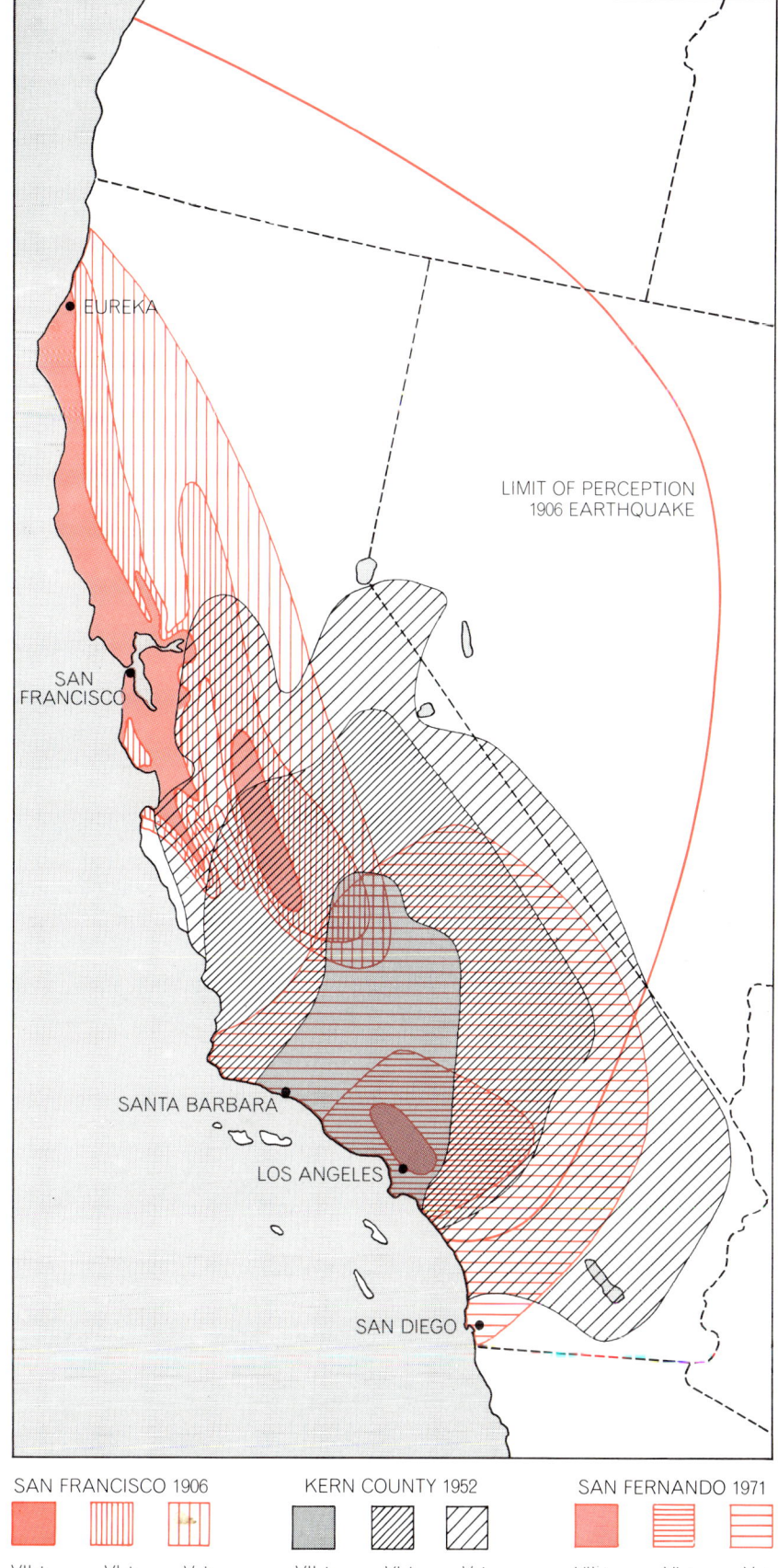

ISOSEISMAL CONTOUR MAP shows the pattern and intensity of ground-shaking produced by the 1906 San Francisco earthquake of magnitude 8.3, the 1952 Kern County earthquake of magnitude 7.7 and the 1971 San Fernando earthquake of magnitude 6.6. The Roman numerals indicate levels of perceived intensities as defined by the modified Mercalli scale. A short description of each level in the scale appears in the text (*page 885*).

faults; the actual location of the epicenter is uncertain by hundreds of miles because there were no seismic instruments in those days. Since that time this part of southern California has been remarkably quiet and seems to be locked, generating neither earthquakes nor creep. Activity along the San Andreas fault picks up near Coalinga, which is about midway between Bakersfield and San Francisco. Alignments of earthquakes are apparent along the San Jacinto and Imperial faults in the Salton Sea trough near the Mexican border. Although these faults lie west of the main San Andreas fault, they are part of the San Andreas system. The White Wolf fault, which is northwest of and parallel to the Garlock fault, has also been quite active, particularly after

the Kern County earthquake, which occurred on this fault. The White Wolf fault lines up with the Santa Barbara Channel area, which has similarly been quite active.

One way to quantify the seismicity of southern California is to count the number of earthquakes per year per 1,000 square kilometers and compare this figure for the world as a whole. For example, southern California averages one earthquake of magnitude 3 or greater per year per 1,000 square kilometers. Thus within the entire region there are about 200 such earthquakes per year. The rate for earthquakes of magnitude 6.6, the size of the San Fernando earthquake, is about one every five or six years. The actual rates, however, vary

considerably from year to year and depend somewhat on the time interval of the sample. The number of earthquakes decreases rapidly with size, and the average recurrence interval is not well established for the larger earthquakes. Southern California is about 10 times more active seismically than the world as a whole, which is simply to say that California is earthquake country.

Although certain areas in southern California are relatively free of earthquakes, none is immune from their effects. One of the largest quiet areas is the western part of the Mojave Desert wedge. This is surprising because the region is bounded on the northwest and southwest by areas that are obviously under large compression, as is shown by the upthrust mountains in the transverse and Tehachapi ranges and the large overthrust earthquakes that occurred in Kern County and San Fernando. It appears that the region is being protected from the northwesterly march of the southern California–Baja California block by the San Bernardino batholith and may represent a stagnation area in the lee of the mountains. Only a small number of earthquakes are centered near San Diego, although the larger earthquakes in northern Baja California and in the mountains between San Diego and the Salton Sea are felt in San Diego. The Great Central Valley north of Bakersfield and the eastern part of the Sierra Nevada are fairly inactive, as is a large area north of Santa Barbara in the Coast Ranges.

Magnitude and Intensity

It is somewhat deceptive to plot earthquakes as small points on a map. The points represent the epicenter: the point on the surface above the initial break. Once the break is started it can continue, if the earthquake is a major one, for hundreds of miles. Earthquakes of the thrust type, which result from a failure in compression, typically first break many miles below the surface; the surface break and maximum damage can be 10 miles or more from the epicenter. The distance over which strong shocks were felt during three large California earthquakes in this century (1906, 1952 and 1971) can be represented by plotting isoseismals: lines of equal intensity [see illustration on preceding page]. The shape of the pattern varies with the type of earthquake and with the nature of the local geology; structures on deep sedimentary basins or on uncompacted fill get a more intense shaking than structures on bedrock. The isoseismals of the

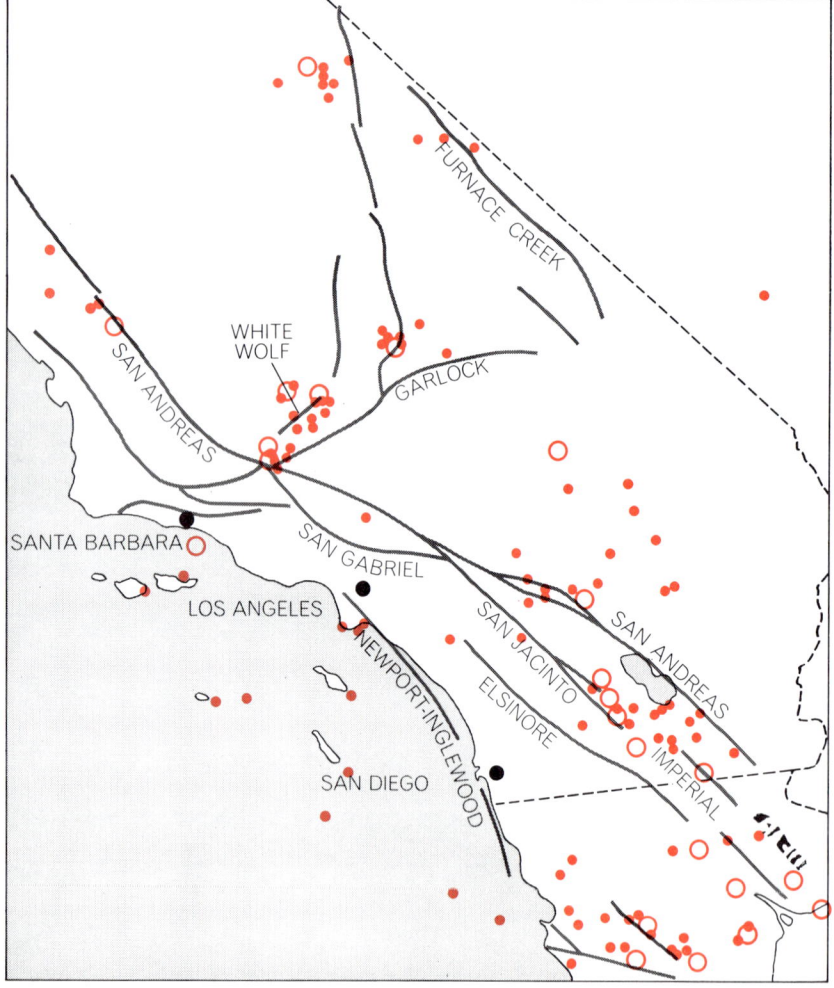

THIRTY-SIX-YEAR EARTHQUAKE RECORD shows the epicenters of all events of magnitude 5 or greater recorded in southern California and in the northern part of Baja California from 1934 through 1969. The epicenter is the point on the earth's surface above the initial break. Dots show earthquakes between 5 and 5.9 in magnitude. Open circles indicate earthquakes of magnitude 6 or greater. The hypocenter, the point of the initial break in the earth's crust, is often many miles below the surface in thrust-type earthquakes, a type frequently observed in this region. In the 36-year period southern California and adjacent regions experienced more than 7,300 earthquakes with a magnitude of 4 or more. Earthquakes are about 10 times more frequent in this area than they are in the world as a whole.

San Francisco earthquake are long and narrow, both because of the orientation of the fault and the length of the faulting and because of the northwest-southeast trend of the valleys. The orientation of the valleys in turn is controlled by the orientation of the San Andreas fault.

The public and the news media are confused about the various measures of the size of an earthquake. There are many parameters associated with an earthquake; they are usually regarded as fault parameters. They include the length, depth and orientation of the fault, the direction of motion, the rupture velocity, the radiated energy, the causal stresses and their orientation, the stress drop (which is related to the strength or the friction along the fault), the energy spectrum, the amount of offset or displacement and the time history of the motion. Most of these parameters can be estimated from seismic records, even from signals recorded several thousand kilometers from the earthquake. To obtain high precision, however, one needs records from many well-distributed seismic stations together with field observations at the site of the earthquake.

The magnitude on the Richter scale is a number assigned to an earthquake from instrumental readings of the amplitude of the seismic waves on a standard seismometer, the Wood-Anderson torsion seismometer. The amplitude must be suitably corrected for spreading and attenuation in the earth, and for instrumental response if a non-standard instrument is used. The magnitude is closely related to the energy of the earthquake, the single most important quantity by which earthquakes can be ranked one against another. If all the corrections are adequately made, a seismologist anywhere in the world will assign the same magnitude. In practice, because of the complicated radiation pattern of earthquakes and because of the distortion of the waves traveling through the earth, the initial magnitude assigned by various observatories may differ slightly. The magnitude scale is logarithmic and is open-ended at both ends. It is not a scale with a maximum value of 10, as is often reported in the press, and negative magnitudes are routinely measured by seismologists working on microearthquakes.

The intensity scale was developed for engineering purposes and is a qualitative measure of the intensity of ground vibration and structural damage. These qualitative assessments are assigned Roman numerals from I to XII. Unlike the magnitude of an earthquake, the inten-

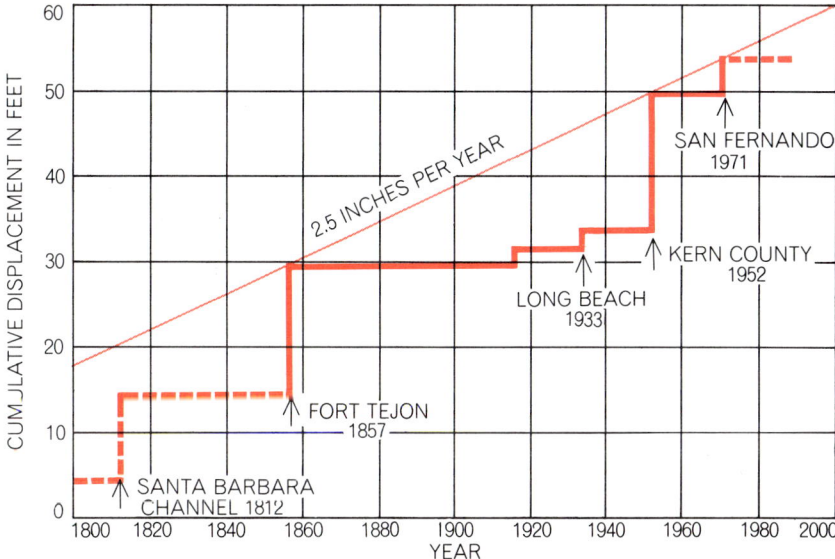

CUMULATIVE DISPLACEMENTS directly related to earthquakes indicate that southern California west of the San Andreas fault system is sliding northwestward at an average rate of 2½ inches per year. Major earthquakes relieve stresses that have built up over decades.

sity varies with distance and depends on the nature of the local ground. In general alluvial valleys, soft sediments and areas of uncompacted fill will magnify ground-shaking and will register higher intensities than adjacent areas on solid rock.

The intensity scale in common usage today is the Modified Mercalli Intensity Scale. The following characterizations of intensity, abridged from longer descriptions, indicate the kind of observations on which the Mercalli scale is based:

I. Not felt except by a very few under special circumstances. Birds and animals are uneasy; trees sway; doors and chandeliers may swing slowly.

II. Felt only by a few persons at rest, particularly on the upper floors of buildings.

III. Felt indoors, but many people do not recognize as an earthquake. Vibrations like the passing of light trucks. Duration of the shaking can be estimated.

IV. Windows, dishes and doors rattle. Walls make creaking sounds. Sensation like the passing of heavy trucks. Felt indoors by many, outdoors by few.

V. Felt by nearly everyone; many awakened. Small unstable objects are displaced or upset; plaster may crack.

VI. Felt by all; many are frightened and run outdoors. Some heavy furniture is moved; books are knocked off shelves and pictures off walls. Small church and school bells ring. Occasional damage to chimneys, otherwise structural damage is slight.

VII. Most people run outdoors. Difficult to stand up. Noticed by drivers of automobiles. Damage is negligible in

buildings of good design and construction, slight to moderate in well-built ordinary structures, considerable in poorly built or badly designed structures. Waves on ponds and pools.

Intensity VII corresponds to the general experience within five or 10 miles of the surface faults associated with the San Fernando earthquake of last February. The following intensity levels were experienced in a small area of the northern San Fernando Valley and would be widely experienced in more severe earthquakes.

VIII. Steering of automobiles affected. Frame houses move on foundations if not bolted down; loose panel walls are thrown out. Some masonry walls fall. Chimneys twist or fall. Damage is slight in specially designed structures, great in poorly constructed buildings. Heavy furniture is overturned.

IX. General panic. Damage is considerable in specially designed structures; partial collapse of substantial buildings. Serious damage to reservoirs and underground pipes. Conspicuous cracks in the ground.

X. Most masonry and frame structures are destroyed with their foundations. Some well-built wooden structures are destroyed. Rails are bent slightly. Large landslides.

XI. Few, if any, masonry structures remain standing. Bridges are destroyed. Broad fissures in the ground. Rails are bent severely.

XII. Damage is nearly total. Objects are thrown into the air.

It is clear that the Mercalli intensity scale is people-oriented; anyone can estimate the intensity from his own experience during an earthquake. The National Oceanic and Atmospheric Administration compiles information on intensities by mailing out questionnaires to a sample of the population living in an area that has experienced a sizable earthquake.

In order to obtain more exact information about the ground motions involved in earthquakes engineers have developed strong-motion accelerometers that automatically trigger and start to record when shaken severely. Most of these instruments are installed in the seismic areas of the U.S., with a particularly heavy concentration in and around Los Angeles. The instruments are expensive and must be located very close to an earthquake to provide useful data. More than 250 of the instruments were triggered during the San Fernando earthquake, and a wealth of engineering data will be provided by these records.

A strong-motion instrument records ground acceleration as a function of time. Accelerations are commonly reported as fractions of a g, the acceleration due to gravity at the earth's surface.

One g is roughly 10 meters per second per second. In designing a building to withstand moderate earthquakes, engineers are concerned chiefly with the maximum accelerations, the period or frequency of shaking and the duration of shaking. Buildings in earthquake-hazard regions with stringent building codes are usually designed to withstand at least .1 g of acceleration; this corresponds to an intensity of about VII on the Mercalli scale.

Although there is no direct correlation between intensity and magnitude, the zone of destruction increases as the magnitude increases for shallow-focus earthquakes. In general the larger the magnitude of an earthquake, the longer the fault length, the larger the displacement across the fault and the longer the duration of shaking. The longer fault length alone accounts for much of the increased area of destruction. For example, the San Francisco earthquake of 1906 had an intensity of VII or greater out to a distance of 500 miles from the epicenter, and this may not have been the largest California earthquake in historic times. The San Francisco earth-

quake had a magnitude of 8.3. The 1952 Kern County earthquake (magnitude 7.7) had an intensity of VII or greater out to 50 miles. The recent San Fernando earthquake (magnitude 6.6) damaged older structures out to 25 miles. An earthquake of magnitude 5.5, the Parkfield earthquake of 1966, produced comparable damage to a distance of 10 miles.

The February Earthquake

The San Fernando earthquake occurred in the San Gabriel Mountains just north of the San Fernando Valley, a densely populated northern suburb of Los Angeles. The San Gabriel Mountains are part of the structural province of the transverse ranges: the band of east-west-trending mountains, valleys and faults that is characterized by strong and geologically recent tectonic deformation. Geologists recognize that the region is one of recent crustal shortening caused by north-south compression. The mountains, produced by buckling and thrusting, are one result of this crustal shortening. They have been thrusting over the valleys to the south for at least five million years along fault planes that dip to the north or northeast.

Although many faults are known to have been active in this area in the past several thousand years, the San Fernando earthquake produced the first historic example of surface faulting. The San Gabriel Mountains rise abruptly some 5,000 feet above the San Fernando Valley and the Los Angeles basin to the south. During the earthquake of February 9 a wedge-shaped prism of the crystalline basement rock comprising the San Gabriel Mountains was thrust over the San Fernando Valley to the southwest, thereby raising the elevation of a section of the San Gabriel Mountains and sliding it slightly to the west. The displacement is consistent with the motions that have been occurring for millions of years, as one can infer from geologic offsets and uplifts. It also agrees with the general picture presented here, namely that the transverse ranges were formed by the collision of the southern and Baja California block with the central and northern California block, and with the concept that the southern California block is being diverted to the west by the massive San Bernardino batholith. One can infer that the thickening of the crust involved in the overthrusting and uplift of the San Gabriel Mountains made this region an additional obstacle to the northwesterly march of

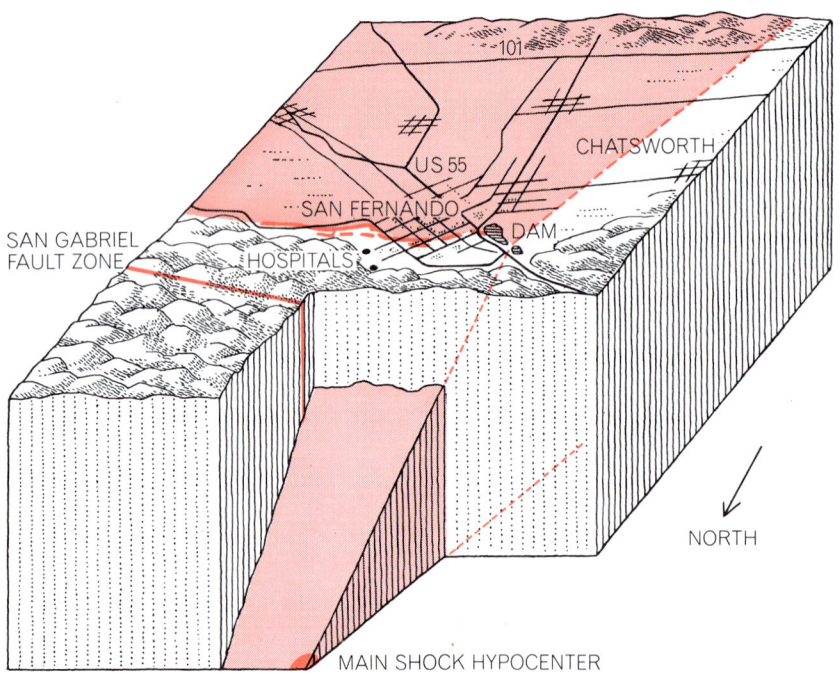

HYPOCENTER OF SAN FERNANDO EARTHQUAKE (*dark color*) of last February was 13 kilometers deep and 12 kilometers north of the area where the principal ground-shaking occurred. The earthquake collapsed sections of two hospitals in the San Fernando Valley, taking 64 lives, and so seriously weakened the earthen wall of the Van Norman Dam at the northern end of the San Fernando Valley that 80,000 people living below the dam had to leave their homes until the water level in the reservoir could be lowered. Total damage caused by the earthquake is estimated at $500 million to $1 billion. This three-dimensional view is based on a drawing prepared by two of the author's colleagues, Bernard Minster and Thomas Jordan, who worked with information supplied by geologists and geophysicists of the California Institute of Technology. The view is looking toward Los Angeles.

the southern California block. If it did, this would lend additional support to the notion that the plates in California are only 15 to 20 kilometers thick. An intriguing possibility is that the upper part of the crust is sliding with relatively little friction on a layer of rock rich in the mineral serpentine.

The hypocenter, or point of initial rupture, of the San Fernando earthquake was at a depth of 13 kilometers under the San Gabriel Mountains. The fault motion was propagated to the surface along a fault inclined northward at an angle of 45 degrees and broke the surface near the cities of San Fernando and Sylmar, at the boundary between the crystalline rocks of the mountains and the sediments of the valley [see illustration on facing page]. Two heavily damaged hospitals were between the epicenter and the surface break and were therefore on the upthrust, or elevated, block. The hundreds of aftershocks following the earthquake covered an area of approximately 300 square kilometers; the total volume of rock lifted up was about 2,500 cubic kilometers.

Even though the elevation difference between the peaks of the San Gabriel Mountains, such as Mount Wilson and Mount Baldy, and the floors of the adjacent valleys is impressive, it does not represent the total uplift. Erosion removes material from the mountains and deposits it in the valleys. The total amount of differential vertical motion probably exceeds two and a half miles, and horizontal displacements in the transverse ranges probably exceed 25 miles. Many thousands of earthquakes of the San Fernando type must have occurred in the area over the past several million years.

Seismic surveillance of the region with instruments dates back only four decades. In this period the northern San Fernando Valley was less active seismically than many other parts of the greater Los Angeles area, although it was comparable to the average for all southern California. On the basis of the seismic data there was no reason to believe the San Fernando area was any more or less likely than any other region of recent mountain-building in southern California to experience a large earthquake. On the other hand, the thrusting and bending associated with the geologic processes in the region, and the tilting that was associated with the earthquake and its aftershocks, suggest that a dense network of tiltmeters could provide a warning of the next large earthquake here.

The Author

DON L. ANDERSON is professor of geophysics at the California Institute of Technology and director of the Cal Tech Seismological Laboratory. After receiving his bachelor's degree in geology and geophysics at Rensselaer Polytechnic Institute in 1955, he spent a year with an oil company, followed by two years of geophysical research with the Air Force Cambridge Research Center and the Arctic Institute of North America. He received his Ph.D. in geophysics and mathematics from Cal Tech in 1962. His current research activities include seismological investigations of the earth's interior, the physics of the interiors of the moon and terrestrial planets, volcanology, theoretical seismology and studies pertaining to the evolution of the earth and the solar system. In addition he is principal investigator of the Viking seismology experiment that will be landed on Mars in 1976.

Bibliography

RELATIONSHIP BETWEEN SEISMICITY AND GEOLOGIC STRUCTURE IN THE SOUTHERN CALIFORNIA REGION. C. R. Allen, P. St. Amand, C. F. Richter and J. M. Nordquist in Bulletin of the Seismological Society of America, Vol. 55, No. 4, pages 753–797; August, 1965.

SPREADING OF THE OCEAN FLOOR: NEW EVIDENCE. F. J. Vine in Science, Vol. 154, No. 3755, pages 1405–1415; December 16, 1966.

PROCEEDINGS OF CONFERENCE ON GEOLOGIC PROBLEMS OF SAN ANDREAS FAULT SYSTEM. Edited by William R. Dickinson and Arthur Grantz in Stanford University Publication: Geological Sciences, Vol. XI. School of Earth Sciences, 1968.

IMPLICATIONS OF PLATE TECTONICS FOR THE CENOZOIC TECTONIC EVOLUTION OF WESTERN NORTH AMERICA. Tanya Atwater in Geological Society of America Bulletin, Vol. 81, No. 12, pages 3513–3535; December, 1970.

SCIENTIFIC
AMERICAN December 1971, Vol. 225, No. 6, pp. 80–88

OFFPRINT **897**

THE ROTATION OF THE EARTH

by D. E. Smylie and L. Mansinha

An analysis of recent measurements indicates that one of the main
nonuniformities of the earth's rotation—its tendency to wobble
gently about its rotation axis—may be excited by major earthquakes.

The fact that our planet rotates has far-reaching implications. Besides providing the daily rhythm of our lives the phenomenon of rotation makes the earth a noninertial frame of reference, requiring the introduction of the apparent centrifugal and Coriolis forces to earth-based mechanics. The centrifugal force in turn gives rise to the principal feature of the earth's shape: its departure from sphericity, or the fact that it is flattened at the poles and bulges at the Equator. The Coriolis force, on the other hand, plays an important role in determining the wind patterns in the atmosphere, the currents in the ocean and the flow of material in the earth's outer core. The influence of the Coriolis force on the motions of the outer core is thought to be critical to the operation of the self-exciting dynamo that generates the main magnetic field of the earth; its effect in turn may explain the near-alignment of the earth's magnetic axis and its rotation axis.

It is hardly surprising, therefore, that the problem of describing the rotation of the earth accurately has intrigued physicists for more than two centuries. The realization that even the theory of the rotation of a rigid body is not a simple subject goes back to the great 18th-century mathematician Leonard Euler, whose investigations in this area form a significant part of classical physics.

The earth of course is not a rigid body but consists of several parts that are either liquid or deformably solid, and as a consequence its rotation is exceedingly complex. The nonuniformity of the earth's rotation manifests itself in several ways; for example, the rotation exhibits variations in speed (corresponding to changes in the length of the day), variations in the direction of the axis of rotation in space (precession) and variations

in the orientation of the rotation axis within the earth (wobble). Recent measurements have shown that the continuous excitation required to maintain the natural wobble of the earth about its rotation axis may be supplied by major earthquakes.

It was Euler who first demonstrated that a rigid body, in addition to being capable of a stable, uniform rotation about either its axis of greatest rotational inertia or its axis of least rotational inertia, can wobble freely when the spin axis is displaced from either of these two principal axes. (Rotational inertia is measured by the sum of the products of all the mass elements of the body with the squares of their distances from the rotation axis. The resulting quantity is called the moment of inertia. The dynamical behavior of a rigid body in rotation can be completely characterized by the moments of inertia about three mutually perpendicular principal axes through its center of mass.)

If the moments of inertia about two of the principal axes are equal, and the rotation axis is slightly displaced from the third axis, the resulting motion can be represented as two circular cones rolling on each other without slip [see illustration on opposite page]. The line of contact is the instantaneous spin axis, the axis of the small cone is the angular-momentum axis and the axis of the large cone is the symmetry axis. These three axes lie in the same plane. To an observer in space the small cone would appear fixed and the large one would roll uniformly around it, giving the body a wobble. An observer riding the body (provided that he has some spatial reference objects) would see the axis of rotation uniformly describe the surface of the large cone within the body as though

it were fixed and the small cone rolled on its interior.

When the Euler theory is applied to the earth, it predicts that the latitude of a given point on the earth's surface will vary regularly with a period of 304 days. Since the latitude of an observer is the complement of the angle between the rotation axis and the local vertical, the conical motion of the spin axis through the body of the earth results in a periodic variation in latitude. The period of the latitude variation in days is almost equal to the ratio of the angle of the large "earth-fixed" cone to the angle of the small "space-fixed" cone. (To be precise, the period in sidereal days, or days measured by rotation with respect to the stars, is equal to the ratio of the angles minus one.) This ratio is equal in turn to the ratio of the polar moment of inertia to its excess over the equatorial moment of inertia, which is known from measurements of the rate of precession of the spin axis in space to be 306 to one.

A succession of prominent scholars, among them Friedrich Wilhelm Bessel of Germany, James Clerk Maxwell of Britain and Simon Newcomb of the U.S. searched the latitude records for a variation with the Euler period. A premature announcement of the confirmation of the existence of the Euler wobble was made by Lord Kelvin in his 1876 presidential address to the British Association for the Advancement of Science. His conclusion was based on the amplitude of .05 second of arc, which Newcomb had obtained in his analysis of latitudes measured at Washington, D.C.

The reality of the variation of latitude was proved beyond doubt by a special series of observations organized by the International Geodetic Association and the U.S. Coast and Geodetic Survey in 1891. Simultaneous measurements were made at a number of locations in Europe and the U.S., and at Waikiki in Hawaii. Waikiki is roughly 180 degrees removed in longitude from the European stations and its latitude variation showed the expected opposite sense.

In the same year S. C. Chandler, an actuary and amateur astronomer living in Cambridge, Mass., reported that he had found the latitude variation to contain two periodic components with periods of 428 days and one year. There was no evidence of the 304-day period that Euler's rigid body theory had predicted and that Kelvin had suggested Newcomb had found. Chandler's 428-day variation, now called the Chandler wobble, was quickly ascribed by Newcomb to a free wobble with a period lengthened from 304 days by the response of a deformable earth to rotational forces.

When the spin axis shifts within the body of the earth, the action of the centrifugal forces is modified, resulting in an adjustment of the symmetry axis through 30 percent of the angular distance toward the new position of the rotation axis. In the absence of disturbances the Chandler wobble can still be visualized as a cone-on-cone motion. The symmetry axis is no longer the axis of the large earth-fixed cone, however, but rather describes a cone coaxial with it.

The realization that the lengthened Chandler period was caused by the response of a deformable earth to rotational forces not only led to one of the first estimates of the overall rigidity of the earth but also presented a long-standing puzzle of geophysics: What is

EARTH'S WOBBLE can, in the absence of other disturbances, be visualized by an earth-bound observer in terms of a small cone rolling inside a large cone without slip, the line of contact being the instantaneous spin axis. The axis of the small cone is the angular-momentum axis. In a rigid earth the axis of the large cone would be the symmetry axis. Since the real earth is deformable, the symmetry axis is in fact shifted 30 percent of the angular distance toward the rotation axis. Deformation in response to the varying rotational forces increases the period of the wobble from 304 days to 435 days. The ratio of the angles of the two cones is almost equal to the period of the wobble in days. Viewed from space, the large "earth-fixed" cone rolls once a day around the small "space-fixed" cone. The angles of the cones are grossly exaggerated in the illustration; actually the large cone subtends an angle of only about .3 second of arc at the earth's center. All four axes always lie in roughly the same plane.

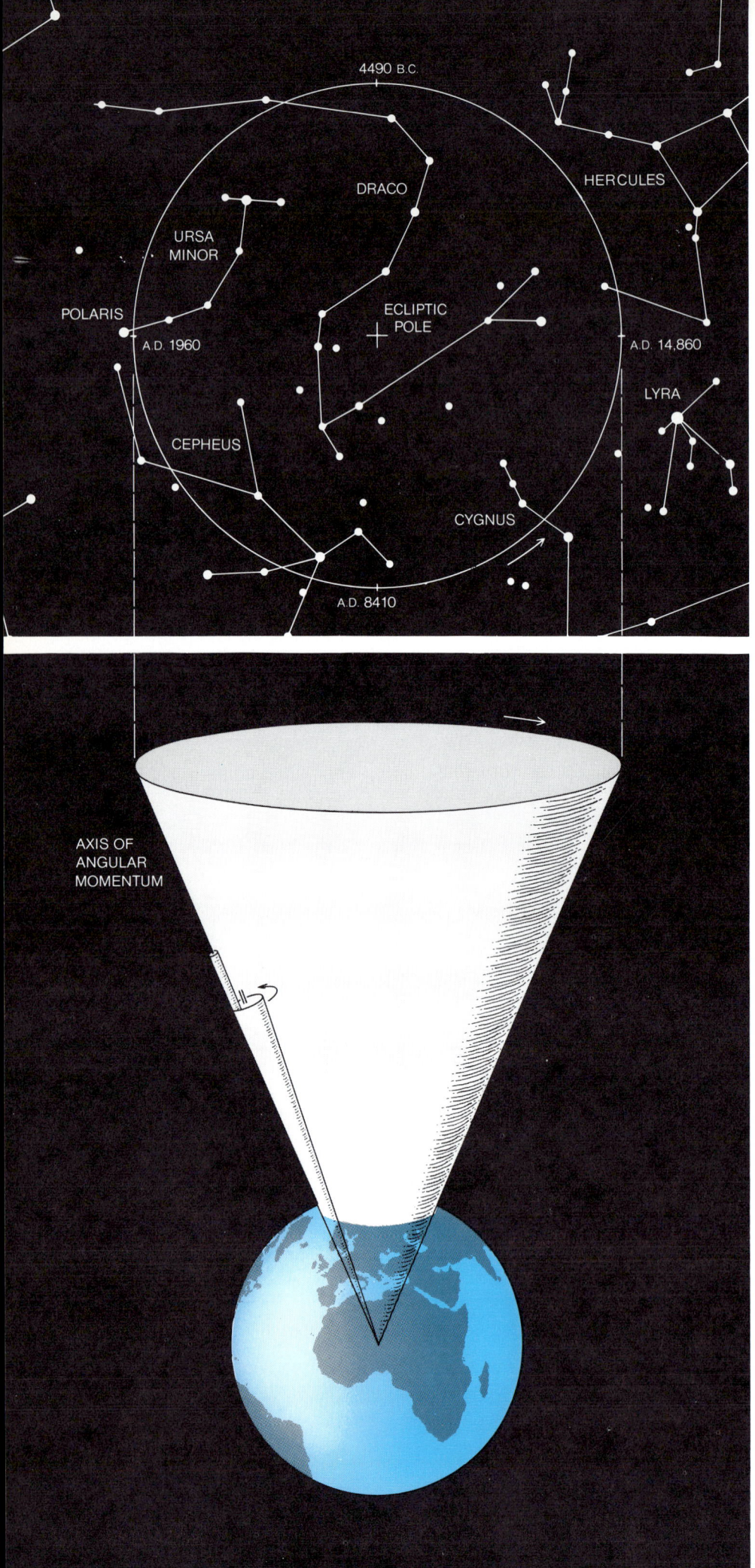

the Chandler wobble's source of excitation? Deformation implies the dissipation of energy by inelastic processes, so that the wobble must be more or less continuously excited.

By the time of Chandler's discovery the precession of the earth's spin axis in space was well understood, both observationally and theoretically. The phenomenon was known to the ancients and Isaac Newton had given an account of it in terms of his gravitational theory. The plane of the earth's Equator is inclined to the plane of the sun's apparent orbit (the plane of the ecliptic) by 23.5 degrees, and the plane of the moon's orbit is slightly inclined to the plane of the ecliptic (by five degrees). The gravitational attractions of the sun and the moon, acting together on the earth's equatorial bulge, produce a torque in the sense required to bring the equatorial plane into coincidence with the ecliptic plane. Because of the earth's rotation this does not happen. Instead the additional angular momentum adds vectorially to the angular momentum associated with the rotation; the resultant is a vector unchanged in magnitude but altered in spatial orientation. The effect of the gravitational attraction of the moon is about twice that of the sun. As a consequence the celestial pole moves almost uniformly along a circle on the celestial sphere with a period of 25,800 years. (The celestial pole is the point where the rotation axis pierces the celestial sphere, a globe of indefinite radius on which the stars can be regarded as being fixed. The celestial pole is now near the star Polaris; 5,000 years ago it was near Alpha Draconis; 5,000 years hence it will be near Alpha Cephei.)

PRECESSION of the earth's rotation axis cuts out a cone in space subtending an angle of 47 degrees at the earth's center (*bottom*). The projection of this motion on the celestial sphere (*top*) describes a circle with a period of 25,800 years. On this time scale Polaris serves only temporarily as the pole-star; the position of the pole at other epochs is indicated by the dates adjacent to the circle. Superimposed on the precession of the earth's rotation axis in space are small irregularities called nutations (*not shown*), the principal one being an elliptical motion with semiaxes of nine and seven seconds of arc and a period of 18.6 years, which is caused by changes in the position of the moon's orbit. All these motions are unrelated to the free wobble of the earth depicted on the preceding page. In addition the geographic and rotation axes both move daily about the axis of angular momentum.

The precession of the earth's axis of rotation is subject to small superimposed irregularities called nutations. The principal nutation is an elliptical motion of the celestial pole with semiaxes of nine and seven seconds of arc and a period of 18.6 years. This irregularity results from a variation of the same period in the orientation of the plane of the moon's orbit with respect to the plane of the ecliptic. The line of intersection of these two planes (called the line of nodes) regresses with an 18.6-year period under the joint attractions of the sun and the earth. Whereas the steady precession can be accounted for satisfactorily by regarding the earth as being rigid, Sir Harold Jeffreys of the University of Cambridge and R. O. Vincente of the University of Lisbon have shown that the presence of the liquid core has a detectable effect on the amplitude of the nutation.

The interest aroused by the verification of the reality of a latitude variation and by Chandler's discovery of the lengthened Euler period led to the establishment of the International Latitude Service. By the turn of the century it was operating observing stations at six locations, all at 39 degrees eight minutes north latitude. The stations at Mizusawa in Japan and Ukiah, Calif., have operated continuously ever since. Observations at Carloforte, on a small island off Sardinia, were interrupted briefly during World War II. The Cincinnati observatory ceased operations in 1916. The Gaithersburg, Md., station was closed for financial reasons from 1915 through 1931, and the station at Tschardjui in the U.S.S.R. had to be moved to Kitab because of a change in the course of the Amu Darya, a large river nearby. Mizusawa is the present headquarters of the International Polar Motion Service, successor to the International Latitude Service.

The development of atomic clocks by late 1955 led the international time service, the Bureau International de l'Heure in Paris, to begin the compilation and reduction of latitude measurements from a group of observatories, including the five International Latitude Service stations. At the millisecond level of timekeeping that atomic clocks allow, time variations and latitude variations are inextricably bound together. For example, with the pole of rotation at P, an observer at O measures his latitude as the angle OCM, where C is the earth's center and the meridian circle through O meets the Equator at M [see illus-

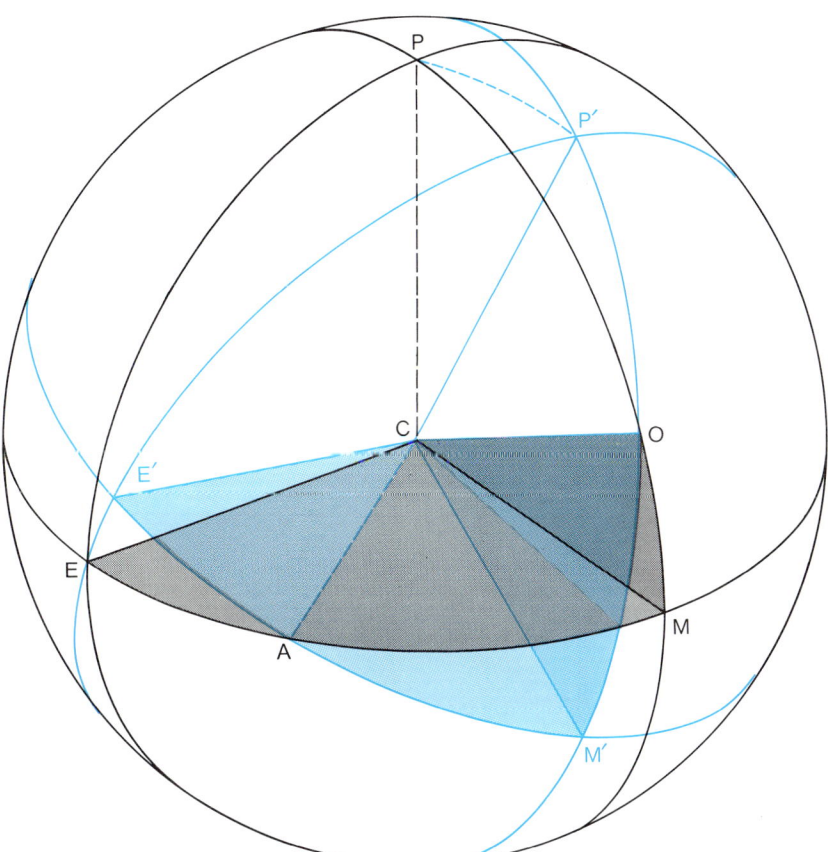

POLAR MOTION changes both the latitude observed at a particular point on the earth's surface and the sidereal time at that point. (Sidereal time is time measured by the rotation of the earth with respect to the stars.) If the pole moves from point P to point P' on the earth's surface, for example, the Equator pivots about an axis passing through the surface point A and the center of the earth (C). The latitude observed at point O changes from the angle OCM to the angle OCM'. Sidereal time prior to the movement of the pole is given by the angle ECM and after by the angle E'CM'. Before the polar shift E is the point on the earth's Equator directly below the vernal equinox on the celestial sphere (that is, the point on the celestial sphere where the sun crosses the celestial equator on its springtime journey northward), whereas E' is the same point on the Equator after the polar shift.

tration above]. Sidereal time at O is given by the angle ECM, where E is the point on the Equator directly below the vernal equinox on the celestial sphere (the point on the celestial sphere where the sun crosses the celestial equator in its springtime journey northward). If the pole of rotation now shifts to the new position P' on the earth's surface but the celestial pole remains fixed, both the latitude of O and its sidereal time are altered. To earthbound observers the Equator appears to pivot on an axis passing through point A and the center of the earth. The latitude at O is now the angle OCM' measured along the new meridian circle through O to the point M' where it intersects the new Equator. Sidereal time at O is now given by the angle E'CM', where E' is the point on the Equator directly below the vernal equinox on the celestial sphere.

One of the simplest and most accurate instruments for making observations of time and latitude is the photographic zenith tube [see illustration on next page]. Light from stars near the zenith (the projection of the local vertical on the celestial sphere) passes through a lens of long focal length. It is reflected from a mercury surface and brought to a focus on a photographic plate. Four successive exposures are made. During each exposure the plate is driven west to east to keep the star images stationary. The lens head and plate are rotated 180 degrees between each exposure. Exposures begin at precisely timed equal intervals.

The result is a pattern of star images set on the vertexes of parallelograms. Sidereal time at the transit of a star across the local meridian is equal to the star's right ascension. The local latitude is equal to the star's declination minus half of the distance between its north and south images on the photographic plate.

Right ascension and declination are the angular coordinates of a star on the celestial sphere and are available from star catalogues, established over long periods of careful observation. Right ascension is the angle measured from the vernal equinox eastward along the celestial equator to the point where the great circle from the celestial pole through the star's position intersects the celestial equator. The declination is the elevation angle of the star above the celestial equator.

Simultaneous compilation and reduction of time and latitude measurements at a group of stations allow that part of the variation of sidereal time due to polar motion to be separated from that part due to variations in the earth's speed of rotation. Information on the speed of rotation has also been obtained from recorded astronomical observations, principally on the discrepancy in longitude of the positions of the sun and the moon from the predictions of gravitational theory made under the assumption of a uniform rotation rate. The effects of tidal interactions can be removed from such records. When the nontidal variations in

the length of the day back as far as 1800 are represented in the form of a smooth curve, irregular fluctuations appear with a time scale of approximately 10 years and an amplitude of a few milliseconds [see top illustration on opposite page].

The late E. H. Vestine of the Carnegie Institution of Washington first presented evidence to suggest that these irregular fluctuations in the length of the day are correlated with changes in the rate of drift of the earth's main magnetic field, which is thought to be carried westward with the highly conducting outer liquid core of the earth. The transfer of angular momentum between the

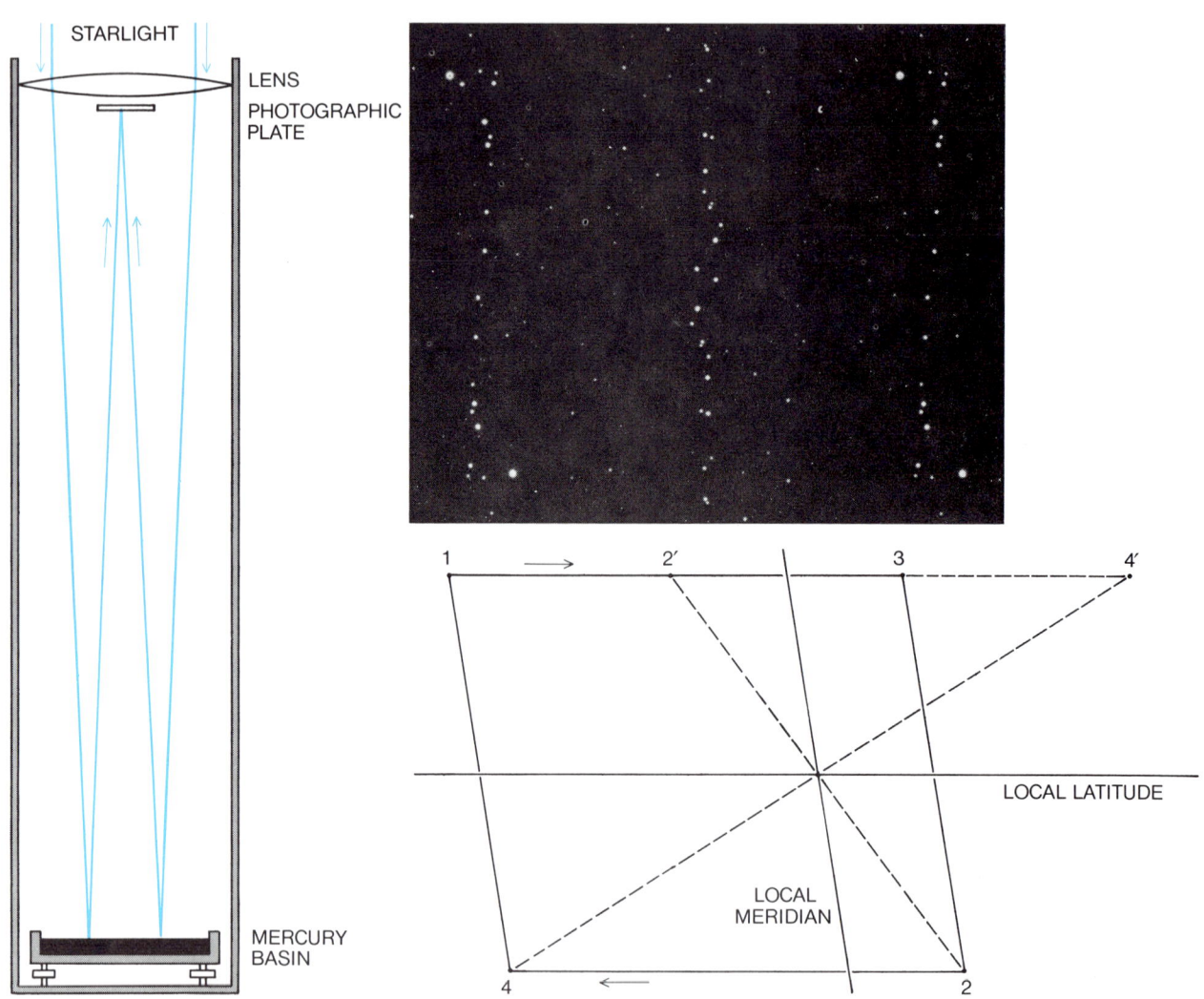

PHOTOGRAPHIC ZENITH TUBE (left) is used for making observations of latitude and sidereal time. Light from stars near the zenith (the projection of the local vertical on the celestial sphere) passes through a lens of long focal length and is reflected from a mercury surface onto a photographic plate in the focal plane. During exposure the plate is driven west to east to keep the star images stationary. Four exposures are made at precisely timed equal intervals. Between each exposure the lens and the plate are rotated together by 180 degrees. Latitude and time are deduced from the resulting pattern of star images (upper right). The portion of the exposed photographic plate shown was made with a photographic zenith tube located at Ottawa by R. W. Tanner of the Canada Department of Energy, Mines and Resources. One millimeter on the photograph corresponds to an angular distance of about 10 seconds of arc. Rotating the plate by 180 degrees between exposures results in the star images appearing at the vertexes of a parallelogram (lower right). The first image is at position 1. The second image would appear at 2' but actually appears at 2 because both the plate and the lens have been rotated about a vertical axis by 180 degrees. Similarly, the fourth exposure results in an image at position 4 instead of 4'. Arrows indicate the directions of the plate's motion during the exposures. From the known position of the star on the celestial sphere, the instantaneous latitude and sidereal time can be deduced from measurements on the photographic plate.

outer core and the mantle through electromagnetic coupling would explain the correlation. Electromagnetic coupling has been shown by M. G. Rochester of the Memorial University of Newfoundland to be just barely adequate to accomplish the requisite transfer.

Since 1955 the greater resolution afforded by atomic timekeeping has provided much more detail on such variations in the speed of the earth's rotation. Superimposed on the 10-year fluctuations are annual and semiannual variations [see middle illustration at right]. The earth runs slow in the spring and early summer of the Northern Hemisphere and fast in the fall. The annual variation amounts to an amplitude of about half a millisecond in the length of the day and the semiannual variation amounts to about a third of a millisecond.

Walter H. Munk and his co-workers at the University of California at San Diego have shown that most of the annual variation can be accounted for by seasonal changes in the winds of the atmosphere. These also contribute to the semiannual variation along with ocean tides. The question of the origins of the annual and semiannual variations is by no means closed. A long list of effects have been considered, including seasonal changes in vegetation, snow load, air mass, ground water and atmospheric loading of the oceans.

Much greater difficulty is encountered in trying to find an explanation for the "sudden events" that are occasionally observed [see bottom illustration at right]. These amount to defects in the earth's timekeeping of several milliseconds established in periods as short as one week. A very large impulsive torque that acts first in one direction and then in the opposite direction is required. S. K. Runcorn of the University of Newcastle upon Tyne has suggested that large toroidal magnetic fields carried suddenly to the surface of the earth's core by hydromagnetic waves burst like sunspots and provide the necessary impulsive torque on the mantle. Runcorn has spent several years recording signals on abandoned undersea telegraph cables in an attempt to detect the electric fields that would be associated with strong toroidal-magnetic-field changes at the surface of the core. Encouraging results have been obtained, but the recordings are difficult to interpret with certainty.

Owing to the continuous operation of the International Latitude Service since its formation, the polar-motion rec-

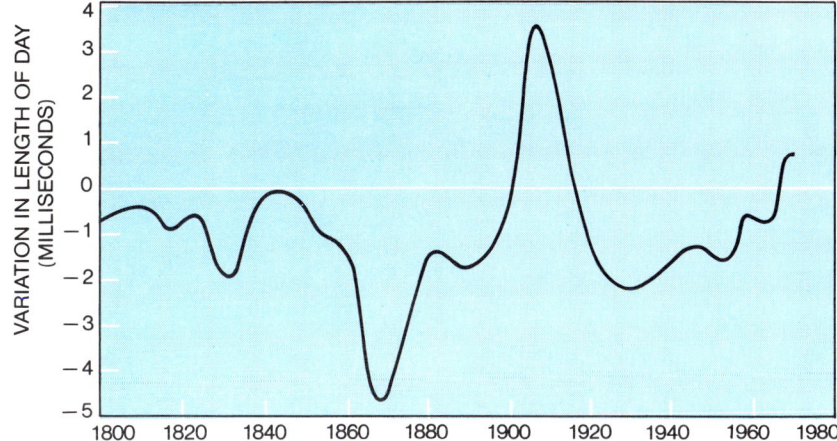

FLUCTUATIONS IN LENGTH OF DAY attributable to nontidal forces are represented on this graph for the period from 1800 to 1970. Before the development of atomic clocks in 1955 the variation in the speed of the earth's rotation could only be detected on a long time scale by measuring the discrepancy between the observed positions of the moon and the planets and those predicted by gravitational theory. As the graph reveals, the fluctuations occur on a time scale of about a decade. The explanation of these irregular fluctuations may lie in the electromagnetic coupling of the earth's mantle to its fluid outer core.

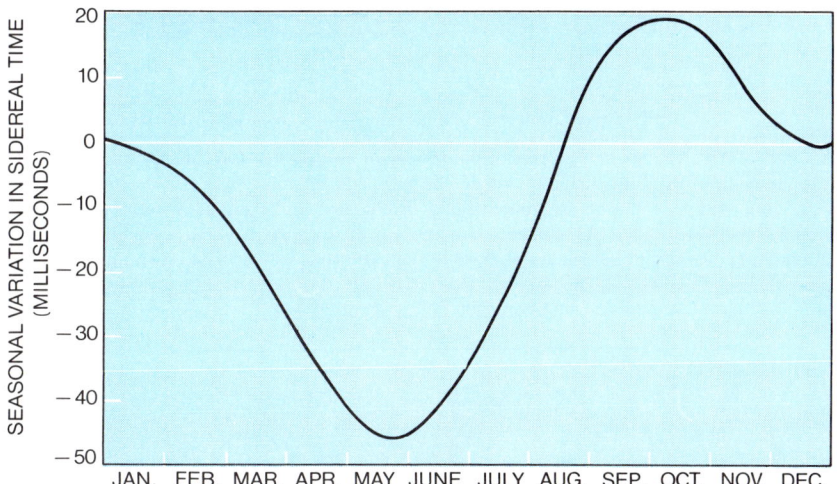

SEASONAL AND SEMISEASONAL VARIATION in the speed of the earth's rotation is charted in this graph in terms of the variation in sidereal time at a given location during the year 1970. A linear drift associated with the decade fluctuations shown at top has been removed. The earth runs slow in the spring and early summer of the Northern Hemisphere, fast in the fall. Atmospheric winds and ocean tides account for most of this variation.

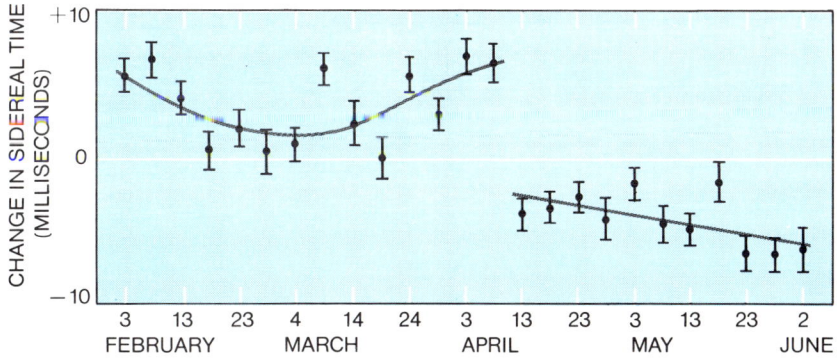

"SUDDEN EVENT" in the earth's rotation appears in this graph of the variation of sidereal time at a given location. Changes of up to 10 milliseconds occur occasionally in sidereal time. No entirely satisfactory explanation has yet been given for such changes. The graph was prepared by B. Guinot of the Bureau International de l'Heure in Paris.

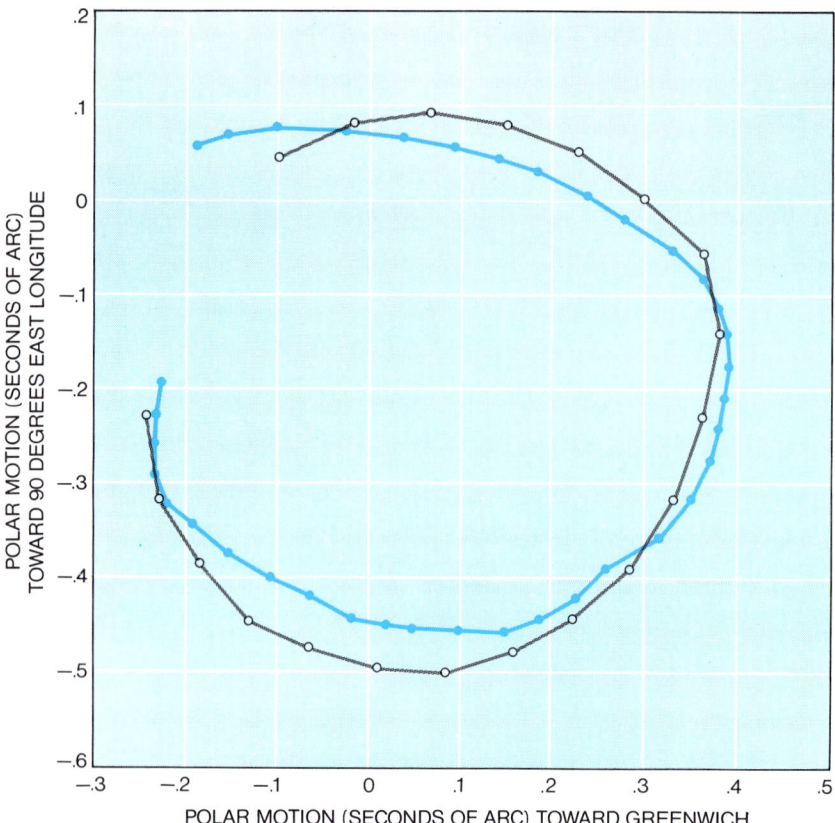

POLE PATH FOR 1957 is represented here by open circles, denoting measurements made at the stations of the International Latitude Service, and by colored dots, showing the positions computed by the Bureau International de l'Heure in Paris. Both paths are highly smoothed versions of the observations. The motion of the pole is counterclockwise and is expressed in seconds of arc; at the surface of the earth one second of arc is approximately equal to 100 feet. The origin is near the mean North Pole for the period 1900–1905.

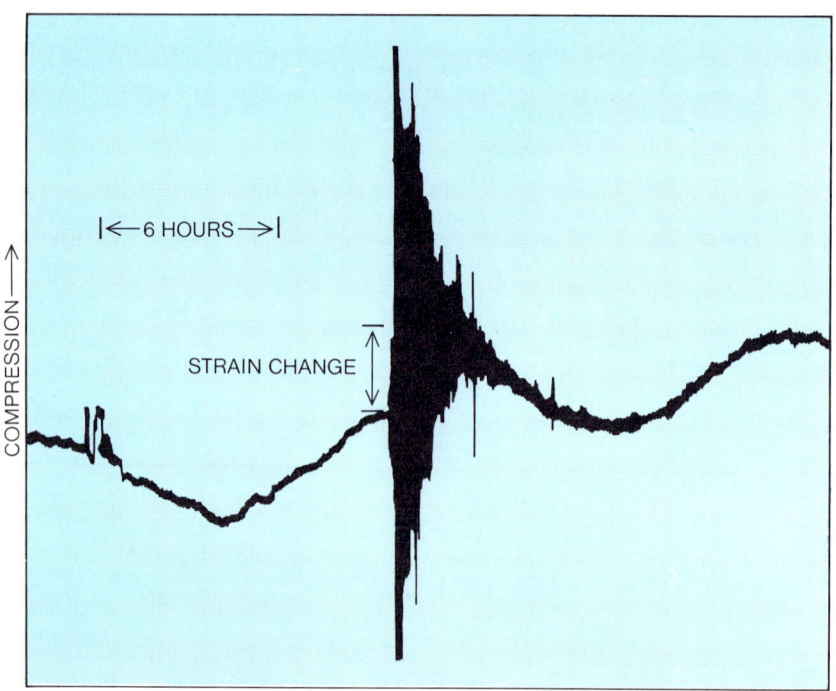

ABRUPT CHANGE IN STRAIN caused by the great Alaskan earthquake on March 27, 1964, was recorded by a strainmeter at Kipapa in Hawaii by L. Blayney of the Seismological Laboratory of the California Institute of Technology. The sinusoidal fluctuations correspond to tidal motions in the solid earth. Such distant strain measurements indicated that the displacements caused by major earthquakes were more extensive than had been thought.

ord is now more than 70 years long. The production of a pole-path record from independent observations, and lately the simultaneous reduction of both time and latitude measurements by the Bureau International de l'Heure, have provided an opportunity to compare results. Disagreements as large as .1 second of arc show up [*see top illustration at left*].

The origin used in expressing polar-motion measurements is near the mean pole position for the period 1900–1905. Evidently the mean pole position has shifted since then by as much as .3 second of arc. This low-frequency motion, called the secular polar shift, could result from a permanent rearrangement of mass, such as might be caused by earthquakes.

The long record of latitude measurements accumulated since Chandler's discovery has allowed considerable improvement in the resolution of the frequency spectrum of the polar motion. The annual motion appears almost as a line feature in the spectrum. The Chandler wobble has a broadened peak centered on a period of 435 days. On the assumption that the excitation of the Chandler wobble is random, the breadth of the resonance gives an indication of the damping. A decay time of 20 years has been estimated in this way. Therefore left to itself the Chandler motion would decay to half its initial amplitude in about 14 years. Since it has been observed to exist for more than 80 years, a source of energy to regenerate the wobble is implied.

From an analysis of a long series of barometric pressure records Munk and E. S. M. Hassan of the University of California at San Diego have shown that the annual component of the polar motion is due mainly to the seasonal variation in the mass distribution of the atmosphere. This finding confirms a suggestion made originally in 1901 by Rudolf Spitaler of the University of Prague. The shift in air mass is principally associated with the extremely high pressure that prevails over Siberia in the winter.

Because the Chandler resonance is not far removed from an annual period, it was thought for some time that seasonal effects might be the source of its excitation. The seasonal variation in atmospheric mass distribution was particularly suspect. As the record of polar motion grew in length, however, it became apparent that the annual motion was distinct from the Chandler resonance. Munk and Hassan have calculated that the contribution of the atmosphere to the excitation of the Chandler wobble is between 10 and 100 times less energetic than the observed motion requires.

A connection between earthquakes and polar motion was made by the seismologist John Milne as long ago as 1906. Quantitative considerations, based on the assumption that the displacements caused by earthquakes were largely local, appeared to rule out the possibility that they could be the source of the excitation. This was the prevailing view until 1965, when Frank Press of the Massachusetts Institute of Technology used both the theoretical predictions of elasticity theory and the recordings of distant strain measurements to argue that the displacements were much more extensive than had been thought.

The displacements given by elasticity theory had been worked out earlier by J. A. Steketee, M. A. Chinnery and Rochester at the University of Toronto and applied to the interpretation of measured offsets near earthquake faults. Working at the University of Western Ontario and the University of British Columbia, we applied this theory to calculate the Chandler-wobble excitation that could be expected from earthquakes. We found that between a sixth and a third of the observed level could be accounted for. This was several thousand times greater than earlier estimates based on the assumption that the displacements were entirely local.

We have since improved the theoretical calculations to take into account self-gravitation, the fact that the earth has a liquid core and the variation in the elastic properties, density and gravity of the earth's interior. There is now substantial agreement between the computed and the observed levels of excitation.

A test of the theory has been made on the polar-motion observations. Earthquakes are sudden events, and the displacements they cause are expected to be established suddenly. Mathematically such changes in time are represented by "step functions." A redistribution of mass in the earth, as might be caused by a step-function displacement of material by an earthquake, results in a polar motion that consists of two successive circular arcs [see top illustration at right]. Before the earthquake the pole traverses the first arc at a uniform angular rate. At the time of the earthquake the pole begins to traverse a second circular arc at the same uniform angular rate. There is only a slight immediate change in the path of the pole; the principal discontinuity is a second-order one in the curvature of the path.

With pole positions uniformly spaced in time the angle subtended at the cen-

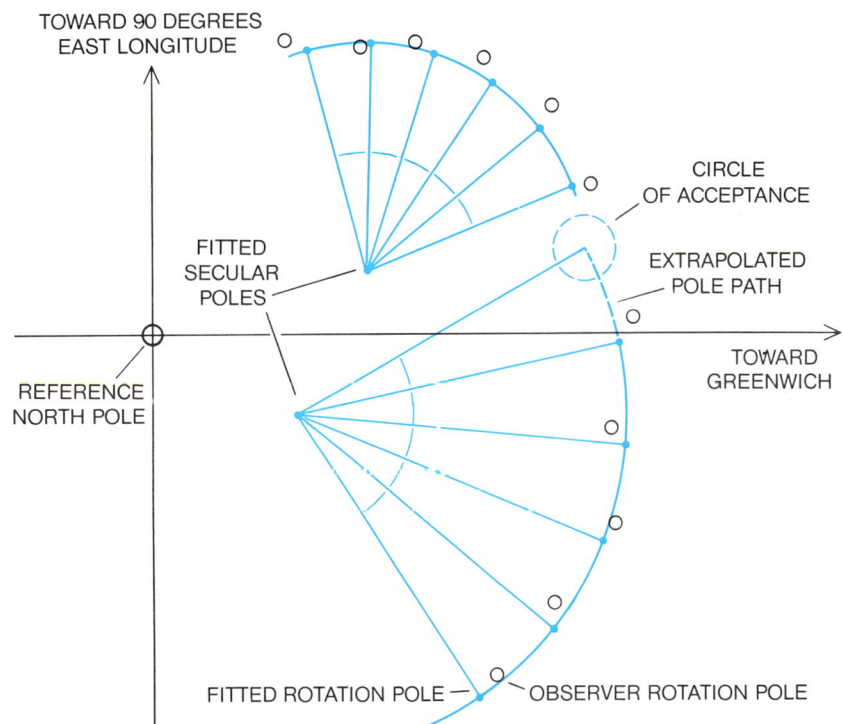

METHOD USED to analyze polar-motion data to detect signals of the form expected from earthquake excitation is demonstrated here. The angle subtended between data points equally spaced in time is a constant determined by the period of the earth's wobble. A circular arc is fitted to the data by the least-squares technique until the measured pole position lies outside a small "circle of acceptance" drawn around the predicted pole position. The arc is then broken and a new arc is fitted to this and subsequent pole positions.

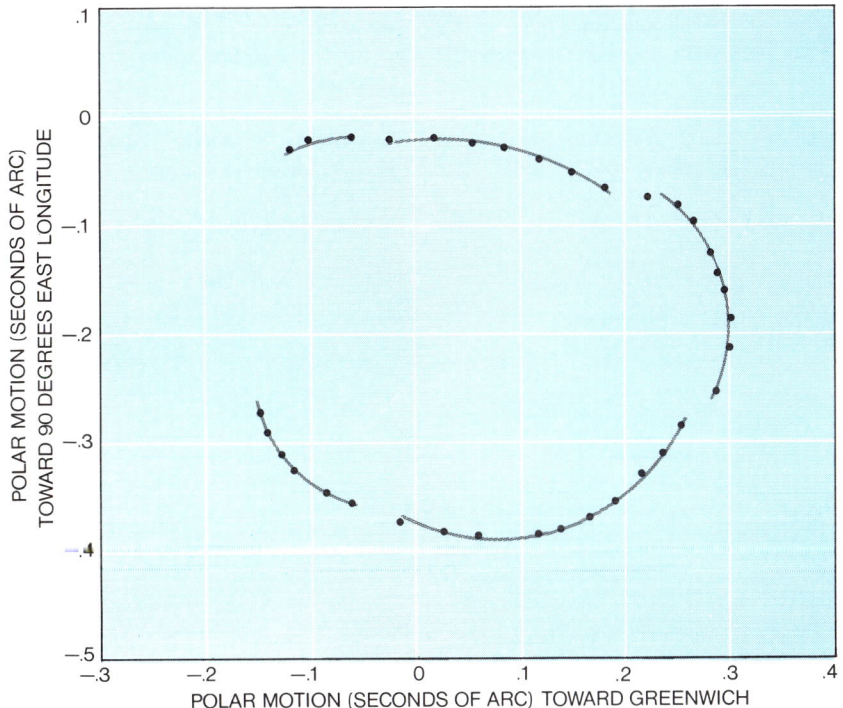

CIRCULAR ARCS were fitted to the 1957 pole-path record of the Bureau International de l'Heure by the method shown above after the effect of the pole's annual motion was removed. An unusual correlation of earthquakes with a magnitude greater than 7.5 with breaks in the arcs was found for the 11-year period analyzed (from 1957 to 1968). Most of the large earthquakes in this period took place before 1964. Low seismic activity (in terms of the incidence of major earthquakes) during the past seven years has not allowed a reliable test of this observation to be made using more recent measurements of polar motion.

ter between two successive positions on a circular arc is known from the Chandler period. When the subtended angle is fixed, a circular arc can be fitted uniquely to two pole positions. A prediction of the next location can be made by extending the arc fitted to the first two positions through the subtended angle. Because of the presence of noise in the polar-motion measurements it is necessary to allow some tolerance in comparing the predicted position with the measured one. A small circle of acceptance is drawn around the predicted position. If the measured pole position lies outside the circle, the fitted arc is broken and a new one is fitted to this pole position and succeeding ones. If the measured pole position lies inside the circle of acceptance, the arc is refitted to this position and the preceding ones by the mathematical procedure of least squares, and a prediction of the next location is made.

In this way a series of breaks in the pole path was computed after the annual motion had been removed. The locations of these breaks in the pole path were then compared with the times of occurrence of major earthquakes. They were found to be highly correlated. The probability of obtaining an equal or better correlation on a random basis is about .1 percent. Of great interest is the unexpected result that the breaks in the pole path often *precede* the correlated earthquake. It is not yet clear whether this premonition is real or results from the smoothing procedures used in reducing the observations to obtain pole positions. The questions raised by the lengthened period of the Chandler wobble and its damping time have been perhaps less puzzling than the problem of its excitation, but completely satisfactory geophysical answers have yet to be given.

The study of the propagation of seismic waves inside the earth has greatly improved our knowledge of the interior's physical properties since the time of Newcomb. We can say with some confidence that the elasticity of the mantle lengthens the period of the Chandler wobble by 120 days. Because the outer core of the earth and the oceans are fluid, their effects are much more difficult to calculate. Neglecting flow (an omission that must surely be unjustified), the oceans are calculated to lengthen the period by 40 days and the outer core to shorten it by 30 days.

The inelastic dissipation of seismic waves in the earth's mantle has been extensively studied. Although the periods of seismic waves are all less than one hour and the results should be extrapolated to periods as long as the period of the Chandler wobble with caution, the damping of the wobble appears to be much too severe to attribute it to the inelasticity of the mantle. Both viscous and electromagnetic dissipation in the outer core seem inadequate except under the most extreme assumptions. Ocean-bottom friction has been dismissed and reinstated. No definite conclusions can be drawn as yet.

Even less certain is the role the gravitational torques of the moon and the sun play in events in the earth's interior. The core of the earth is considerably less oblate than the outer layers of the mantle and the tidal torque on it is correspondingly smaller. There is no evidence in surface measurements of the main magnetic field, however, of the violent magnetic disturbances that would result from a large differential motion. It is assumed that the core and the mantle are somehow coupled on the 25,800-year time scale of the earth's precession.

The nature of this coupling is critical to understanding the dissipation of tidal energy. It is known that the earth is slowing down in its speed of rotation because of this dissipation. The length of the day increases at the rate of about two milliseconds per century as a consequence. Dissipation in shallow seas was long suspect but now appears inadequate. The effect of the core cannot be accurately estimated without a knowledge of the coupling mechanism.

There has been no sudden jump in the quality of observations of the polar motion comparable to the one that followed the development of atomic clocks in the measurement of time. Several new methods of measurement currently being developed promise greatly increased accuracy. Already Doppler tracking of U.S. Navy navigation satellites has produced pole paths comparable to those of the Bureau International de l'Heure and the International Latitude Service. The techniques of very-long-base-line interferometry, used in radio astronomy, also promise to provide a method of accurately determining the earth's orientation in space. Laser-ranging of satellites is currently being tested in an international geodetic program.

Much remains to be learned about the earth's rotation. Challenging theoretical problems that draw on nearly every aspect of geophysics remain to be solved. As our theoretical understanding increases, demands on measurement become ever more stringent and tax the ingenuity of experimentalists.

The Authors

D. E. SMYLIE and L. MANSINHA teach respectively at the University of British Columbia and the University of Western Ontario. Smylie holds a bachelor's degree in engineering physics from Queen's University at Kingston, a master's degree in mathematics from the University of Toronto and a doctorate in earth physics from the same institution. Mansinha was born in India and received all his education there except for his Ph.D., which he obtained from the University of British Columbia in 1963.

Bibliography

THE ROTATION OF THE EARTH: A GEOPHYSICAL DISCUSSION. Walter H. Munk and Gordon J. F. Macdonald. Cambridge University Press, 1960.

THE EARTH-MOON SYSTEM. Edited by B. G. Marsden and A. G. W. Cameron. Plenum Press, 1966.

PHYSICS OF THE EARTH. Frank D. Stacey. John Wiley & Sons, Inc., 1969.

THE EARTH: ITS ORIGIN, HISTORY AND PHYSICAL CONSTITUTION. Sir Harold Jeffreys. Cambridge University Press, 1970.

EARTHQUAKE DISPLACEMENT FIELDS AND THE ROTATION OF THE EARTH. Edited by L. Mansinha, D. E. Smylie and A. E. Beck. Springer-Verlag New York, 1970.

SCIENTIFIC
AMERICAN January 1972, Vol. 226, No. 1, pp. 70–77

OFFPRINT 898

GEOTHERMAL POWER

by Joseph Barnea

The pressure on energy resources has generated new interest in the earth's heat. The emphasis is on exploring for new geothermal areas and developing new ways to extract work from steam and hot water.

An old source of power for man's work has begun to attract new interest. Natural underground reservoirs of steam and hot water are now being tapped on a significant scale, and it will come as a surprise to many people to learn that the harnessing of this geothermal energy has already reached an aggregate capacity of a million kilowatts in plants around the world. At the present rate of development it is likely that by the end of this decade the production of electric power from steam fields will be quadrupled.

The heat of many geothermal reservoirs comes from a large body of molten rock that has been pushed up into the earth's crust from great depths by geologic forces. This dome of magma heats the rocks in the crust near the surface, which in turn heats the water in fissured or porous rocks to a temperature of perhaps 500 degrees Fahrenheit. Being at depths of as much as six miles, the water is under high pressure and is therefore liquid. Where the hot water can escape through a fissure it begins to boil, and part of it flashes off as steam. The geothermal energy can be tapped by a well driven into the fissure or down to the porous layer.

Interest in this source of energy has quickened in the past few years. Recent explorations have revealed that the resource is larger and more extensive than had been supposed. A generation ago the hot springs and steam fields that had long been known in a few localities around the world were believed to be merely local freaks of nature. There is evidence now that reservoirs of steam and hot water are actually widespread in the earth's crust. Signs of their presence have been detected on most of the continents and on a number of islands. It seems possible that such fields will also be found under the seas. Some of the explored fields are known to hold large quantities of energy. A single steam field in northern California, the Geysers field, is estimated to have a potential capacity of three million kilowatts, and surveys that have been made in the Imperial Valley of southern California have indicated a potential of 20 million kilowatts in that area.

The incentive for undertaking a major effort to tap geothermal fields has been heightened by projects showing that in addition to electric power they can yield other useful products. The geothermal steam or hot water can be applied to desalting seawater, to heating houses, greenhouses and swimming pools and to providing nonelectrical energy for refrigeration and air conditioning. Moreover, the hot water itself is a source of extractable minerals and can serve to provide potable water. These additional dividends increase the economic attractiveness of investment for the exploitation of this great earth resource, which up to now has served man mainly as a resort attraction (in the form of health spas) and as a somewhat esoteric and certainly minor source of power supporting small generating plants at a few sites around the globe.

Hot springs, where water from heated strata flows naturally to the surface, have of course been known and used since ancient times. The Romans developed these watering places for medical and recreational purposes all around the Mediterranean and to the outskirts of their empire as far as Bath in the British Isles; there were also medical spas in ancient Japan and elsewhere in the Far East. Hot springs still flourish as health resorts today in Japan, in France and other centers in continental Europe, in Africa and in many other places outside the Anglo-Saxon world.

The realization that the steam in the crust might be tapped for power came at the beginning of this century. In 1904 the first electricity plant so powered was built and plugged into a steam field in northern Italy now known as the Larderello field. Over the following decades there was a slow and tentative growth of interest in geothermal energy. More plants were built on the Larderello field, and other small-scale projects for the use of natural steam or hot water for power, industrial purposes and heating were developed in Japan, Hungary, the U.S.S.R., Iceland, New Zealand and elsewhere. In the U.S. the first geothermal power plant, of 12,500-kilowatt capacity, was commissioned in 1960 on the Geysers field, which is by far the largest field yet discovered in the world.

To those investigators who early recognized the potentialities of geothermal energy the development of this resource has seemed agonizingly slow. There are several reasons why things have not gone more rapidly. Judging from the surface indications (the comparative rarity of hot springs or steam holes) the geothermal energy that might be available appeared to be highly localized and minor in amount. The explorations and the discovery of fields have been limited to those that show surface signs, because little information has been available on geological indications that might signal the presence of hidden fields. In the past such fields could be found only by speculative drilling, and it seemed that the expense of drilling would be justified only for fields located at a shallow depth. The paucity of research and information on the geothermal resources in the crust, the lack of guides for exploration and the shortage of trained specialists and

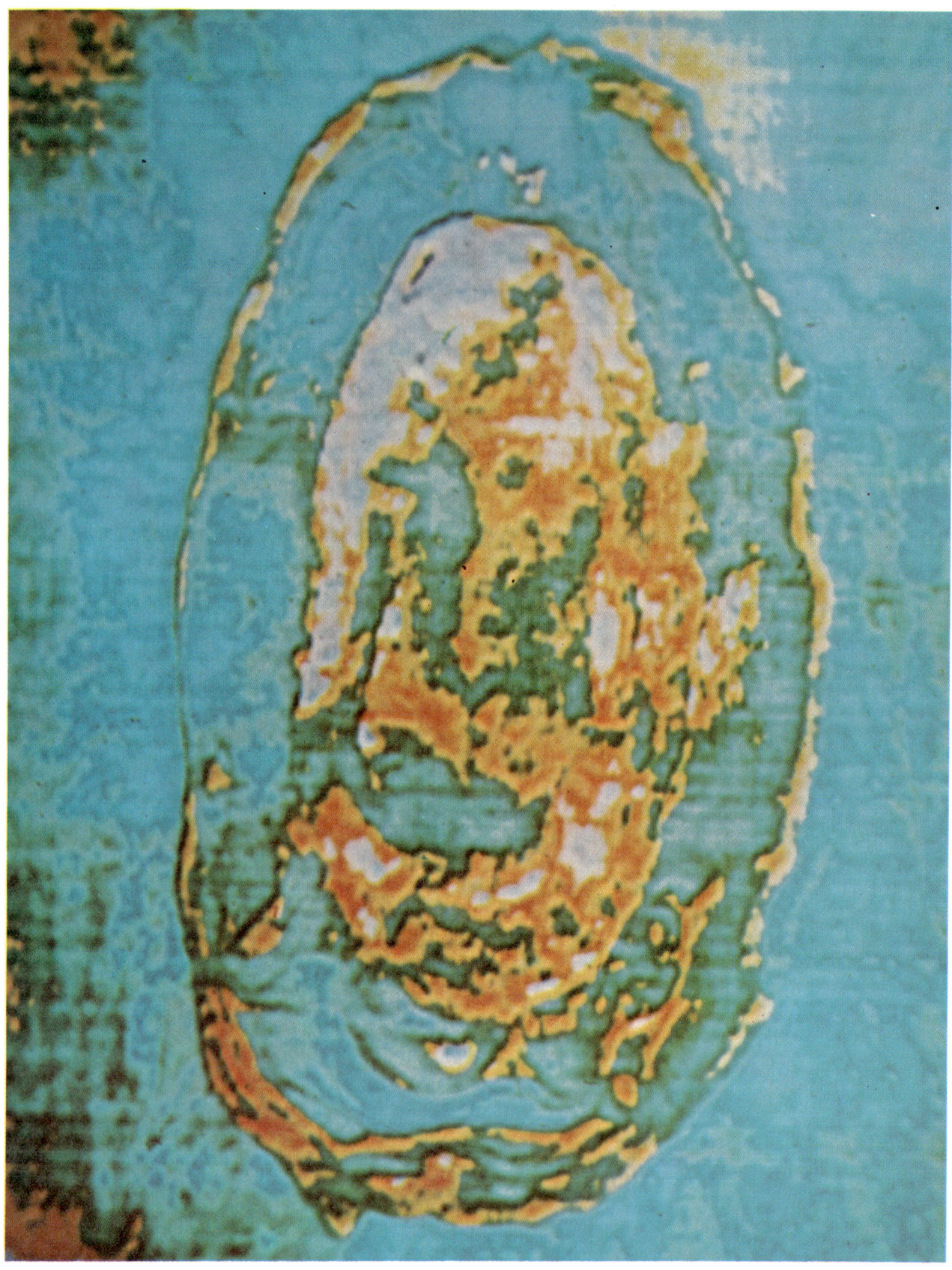

GEOTHERMAL SOURCE associated with a volcanic crater in the Rift Valley in Ethiopia was explored by means of infrared photography in a project carried out by the United Nations. The explorations were made by airplane using black-and-white infrared film, which shows hotter areas as white and cooler areas as progressively darker shades of gray. Promising photographs, such as the one shown here, were processed through a densitometer, which converts the density of the photograph in terms of ground temperature and applies a predetermined color scheme to indicate the differences. In this photograph the hottest areas are orange and the coolest ones are blue. The technique provided the first measurement of the extent of this geothermal source and range of temperatures in area.

technicians in this field have in the past combined to retard progress.

Nevertheless, the enterprise is now moving forward at an accelerating pace. It has been given impetus in the U.S. by the Geothermal Steam Act, adopted by Congress in December, 1970, which establishes the development of U.S. geothermal resources as a national goal. What is now needed is a worldwide expansion of efforts in research, exploration and the training of experts for this work.

The sources of usable geothermal energy in the earth fall into three classes: dry steam fields, wet steam fields and fields of lesser heat content consisting of water at temperatures below the boiling point (at atmospheric pressure). Each type of geothermal energy has its special uses and also capabilities for a variety of applications.

The dry steam fields are filled mainly with steam itself, under pressure and at relatively high temperatures. This steam is usable directly for the production of electric power. It can be piped right to the turbine and therefore simplifies the requirements for plant equipment; the investment in plant may be as low as $100 per kilowatt. In order to minimize piping costs the plant must be located close to the steam wells; moreover, since the steam emerges from the field at low pressure and large amounts of steam must be handled, the effective size of the turbines is limited. This means that the plant cannot be very large. The upper limit at present is about 55 megawatts. Power generators of this magnitude, each fed by 10 to 15 steam wells, are now being installed at the Geysers field in California.

The steam from a dry field can be put to uses other than power production. The water condensed from the steam after it has given up its energy can provide a supply of fresh water. In locations near the ocean or a saltwater lake the steam could be employed as a heating medium in distillation plants for producing potable water by subatmospheric boiling, the steam in this case being provided without any cost for fuel.

So far the existence of five important dry steam fields has been established: the Larderello field in Italy, the Geysers field in California, the Valle Caldera field in New Mexico and two fields in Japan. In the absence of systematic exploration it is not yet possible to estimate how many other such fields may lie hidden in the earth's crust.

On the basis of discoveries made to date, it seems that wet steam fields may

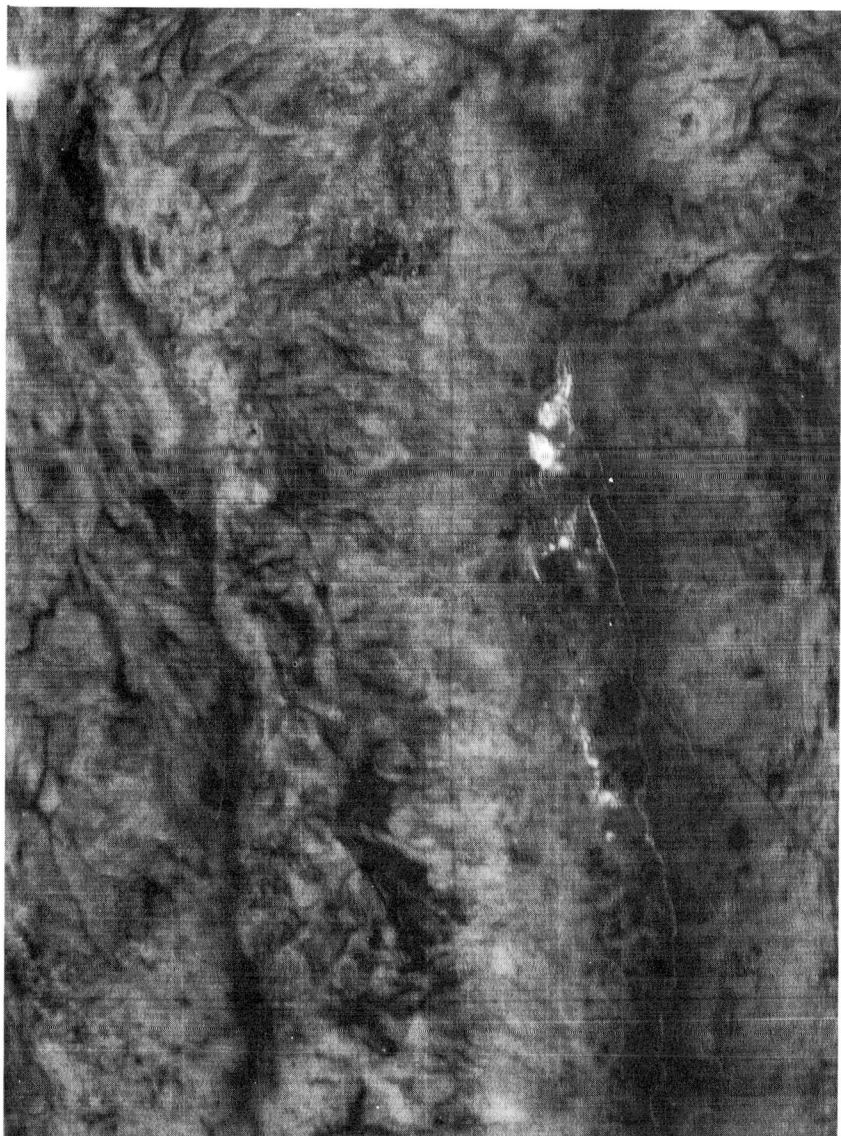

INFRARED VIEW of the steam field at The Geysers was obtained by the U.S. Geological Survey. The aerial photograph was made before dawn in order to minimize the effect of the sun on the temperature of the ground. Light areas at right center are geothermal areas.

be 20 times more abundant than dry steam fields. The wet field is filled with hot water (above its boiling point at atmospheric pressure) that does not become steam until the pressure is released by drilling into the field. The superheated water in the wet field, typically at temperatures ranging from 180 to 370 degrees Celsius (about 350 to 700 degrees Fahrenheit), flashes into a mixture of steam and water as it comes to the surface. About 10 to 20 percent of the discharge, by weight, is steam; the rest is hot water. The steam can be used for power production; the hot water has a multitude of potential uses.

The pioneering stages of the harnessing of geothermal energy have been marked by concentration on a single application of the yield from the wells. At

the Wairakei wet steam field in New Zealand, for example, the steam fraction is fed to a power-generating plant and the hot water is discarded into a river. In this respect geothermal energy has paralleled the history of the discovery of petroleum, which at first was used only for kerosene lamps. Now geothermal development is entering a more sophisticated stage through the analysis of its components and their combination into multipurpose projects.

There are already installations in which the steam of a wet steam field is devoted to the production of power and some of the hot geothermal water is distilled, without the addition of any more heat, to make fresh water. (Distillation is possible because the pressure in the flash-distillation plant is kept below at-

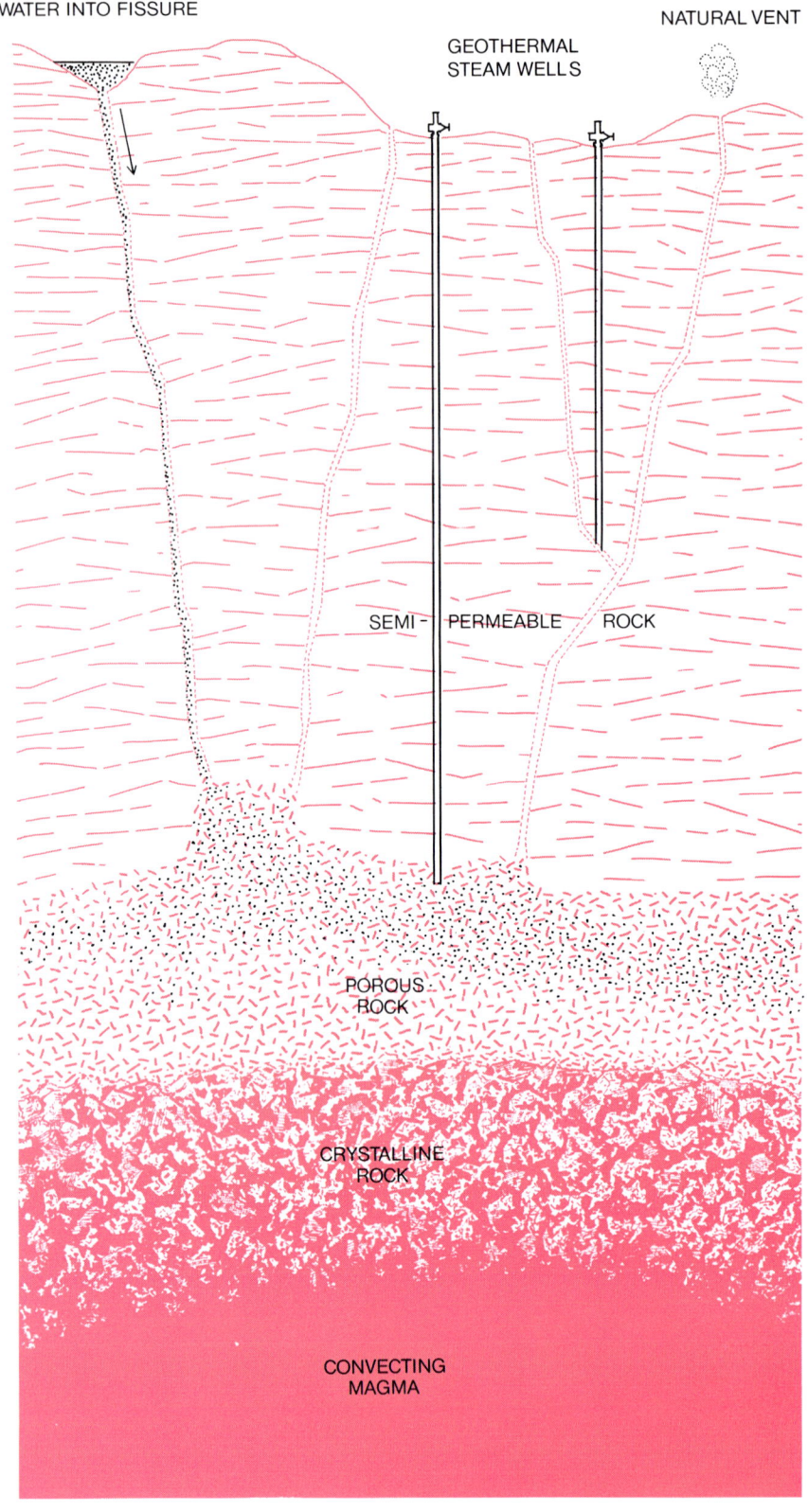

WATER INTO FISSURE

GEOTHERMAL
STEAM WELLS

NATURAL VENT

SEMI - PERMEABLE   ROCK

POROUS
ROCK

CRYSTALLINE
ROCK

CONVECTING
MAGMA

GEOLOGICAL SETTING of a geothermal energy source is portrayed. The heat comes from magma, or molten rock, that has been pushed up into the earth's crust. By convection of the magma the heat moves through crystalline rock to a layer of porous rock containing water that has percolated down from the ground, sometimes to great depths. Over the porous rock is relatively impermeable rock that serves as a cap to contain the heat. Being deep in the ground, the water is under high pressure and is therefore liquid, although its temperature may be some 500 degrees Fahrenheit. It expands and rises in a natural vent; as the pressure drops, water begins to boil and produce steam. A well can tap the vent or the porous layer.

mospheric pressure by exhaust pumps.) A further step is planned at a wet steam field recently discovered at El Tatio in Chile. There the Chilean government in cooperation with the United Nations is investigating the development of a facility that will generate three products [*see illustration on page 904*]. The steam will first be used to produce electricity. Hot water produced from the steam will go through a desalination plant, producing fresh water, and the effluent from the hot-water feed will be concentrated in a mineral-rich brine from which valuable minerals will be extracted in evaporation ponds.

Following the accidental discovery of mineral-rich geothermal brines in southern California and in the Red Sea the UN began a systematic search for mineral brines as part of its program of geothermal investigations. Two discoveries of potential economic importance have been made so far, one in Ethiopia and one in Chile.

At Kawerau in New Zealand a paper and pulp company is using the hot water from a wet steam field for heating in industrial processes. In Iceland the hot water from such fields has long been applied to industrial uses and household and district heating. In Japan the applications include uses in experimental fish-farming projects, cleaning, cooking, soil-heating and bathing. Househeating with hot-well water is being developed on a large scale in several countries, notably Japan, the U.S.S.R. and Hungary (where the cost of such heating is reported to be only a fourth of that with fuel-burning systems). In the U.S. househeating from hot wells is being applied on a small scale in Boise, Idaho, and Klamath Falls, Ore.

The use of geothermal water in air conditioning is based on a process that employs water as the refrigerant and a solution of lithium bromide as a low-temperature absorbent fluid. As in other refrigerating systems the refrigerant is vaporized, thereby extracting heat from the surroundings. Then, however, the refrigerant is taken up by the absorbent. External heat (in this case supplied by geothermal water) drives the refrigerant off the absorbent as a gas; the gas is condensed to liquid, which returns to the evaporator to begin the cycle again. Two Russian investigators of applications of geothermal energy, A. N. Tikhonov and I. M. Dvorov, recently reported that a machine of this kind, used in a system providing refrigeration in summer and heat in winter, is being mass-produced

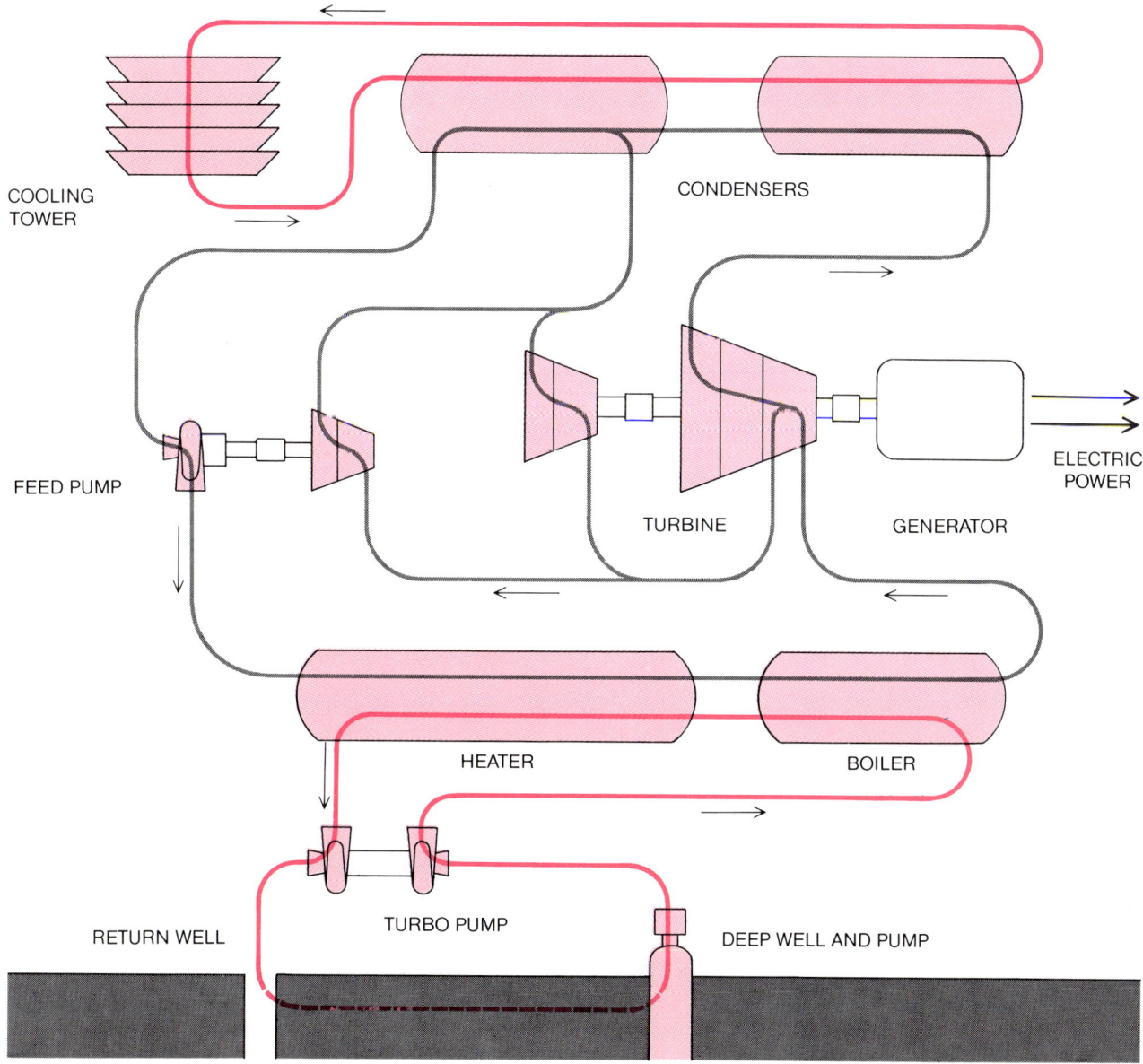

POWER PLANT using geothermal hot water instead of steam has been designed with a low-boiling-point heat absorbent such as Freon or isobutane as the driving fluid. Geothermal water is pumped through a heat-exchange system (*bottom*), where the absorbent takes up the heat. The absorbent evaporates and drives the system of turbines connected to the generator. Absorbent next goes to the condensers, where it is condensed into liquid again by water from the cooling tower and returned to heat exchanger for a new cycle.

in the U.S.S.R. Such a system has also been installed in a hotel in New Zealand, which reports that the energy cost, using geothermal water, is only a tenth of that for a system using electrically operated compressors.

The third type of geothermal field, called a low-temperature field, has only recently begun to receive attention. Fields in this class generally consist of large bodies of water in the range of 50 to 82 degrees C. (about 120 to 180 degrees F.). They are found in sedimentary deposits, notably in Hungary, where the field was discovered accidentally while drilling for petroleum was in progress. The hot water from this type of field

is most efficiently used for heating: in houses, greenhouses, mines in cold climates and industrial plants. The use of such water from low-temperature fields in the U.S.S.R. is reported to have represented a saving of about 15 million tons of fuel in 1970.

The new Geothermal Steam Act of the U.S. stresses the multipurpose approach in the development of geothermal energy resources. To this end it will be necessary to plan on a comprehensive scale, treating the problem with an approach like that for the development of an entire river basin. This means that we shall need planners who are acquainted with all the technologies and economic

considerations involved, from exploration to the numerous possible applications.

Much study has already been given to the costs of exploitation of geothermal energy for various purposes. Since a number of special factors are involved in this new technology, standards for estimating costs have not yet been developed; however, the UN, in response to a proposal made at the Symposium on the Development and Utilization of Geothermal Resources, which was held in Pisa in 1970, is expected to appoint a committee of experts to formulate uniform costing procedures, so that costs in various situations and various countries can be compared.

Some of the costs are already well known from experience. Drilling a steam well costs from $50 to $150 per meter, depending on conditions, so that the drilling cost of a field 1,000 meters deep will be between $50,000 and $150,000 for one well. There are also ready answers on the costs of piping, valves and the various items of equipment for a power plant. The cost of operation for delivery of the heat from a steam field to the plant is likewise well established; with proper management this cost is only about one to three cents per million British thermal units.

What, then, are the special costs? The most important ones are related to the question of the life expectancy of the available heat supply in a field or a given well. This obviously is difficult to estimate. There are reasons to believe, however, that with proper management a geothermal field will last for many years, particularly if it is recharged by ground water or by artificial injection of gas or geothermal effluent water. At the present·stage of development I believe the lifetime of a typical field can prudently be assumed to be about 30 years for purposes of estimating the amortization of the investment used in developing it. To the initial investment we must add a special cost having to do with maintenance: the wells have to be cleaned regularly and sometimes even redrilled because of the precipitation of chemicals from the steam or hot water.

The experience thus far gained furnishes us with approximate cost figures for the various applications of geother-

mal energy. In a single-purpose installation producing only electric power at base load the cost of the power produced is between three and six mills per kilowatt-hour, including full amortization of all the investments over a reasonable period. In desalination plants the cost would probably be in the range of 20 to 50 cents per 1,000 gallons of freshwater yield—far below the costs of other desalination systems. For househeating, air conditioning and similar purposes the use of geothermal energy makes possible savings of up to 90 percent or more, as we have already noted. A hotel in the city of Rotorua in New Zealand reports that the operating cost of a heating and air-conditioning system based on the use of geothermal energy in a lithium bromide absorption installation is only 12 cents per million kilocalories, as against $2.40 per million kilocalories for an oil-burning system involving approximately the same investment in equipment.

These estimates are calculated from the experience of single-purpose facilities. With the development of multipurpose plants the dividends made possible by extraction of all the benefits in the crude outflow from the geothermal field (like the extraction of the various products from crude petroleum) should reduce the cost of the individual applications.

Not the least of the attractions of geothermal energy is that it can be used at little or no cost to the environment. Unlike fossil or fissionable fuels, it does not pollute the biosphere with combustion products or radiation; unlike hydropower systems, it does not flood fertile

lands or generate stresses that may lead to earthquakes. It does present two hazards. The steam and hot water from many fields contain small amounts of boron and other chemicals, which can be harmful when discharged into streams. Trials at the Geysers field and at a geothermal field in El Salvador have shown, however, that the contaminated effluent can be injected back into the field without reducing production from the wells. There is reason to believe this problem will not be difficult to control. The other hazard is that the land may subside where large amounts of water are withdrawn from geothermal reservoirs. Some subsidence has occurred at the Wairakei field, which has been depleted of 70 million tons of water per year and as a result has changed in part from a wet to a dry steam field. This problem too can be controlled, by limiting withdrawals from the field to a safe rate and by recharging it with water, as is now done to prevent subsidence in petroleum fields.

The UN is taking an active interest in geothermal energy. In cooperation with the government of Italy it conducted the symposium on geothermal energy in Pisa, where it was demonstrated that geothermal applications are marked by considerable international interest and collaboration. Although the scale of this research is severely limited in funding, a variety of imaginative ideas are being explored.

One project is concerned with the possibility of producing electricity in low-temperature fields. The heat from the geothermal water is used in a heat exchanger to boil a secondary fluid with a

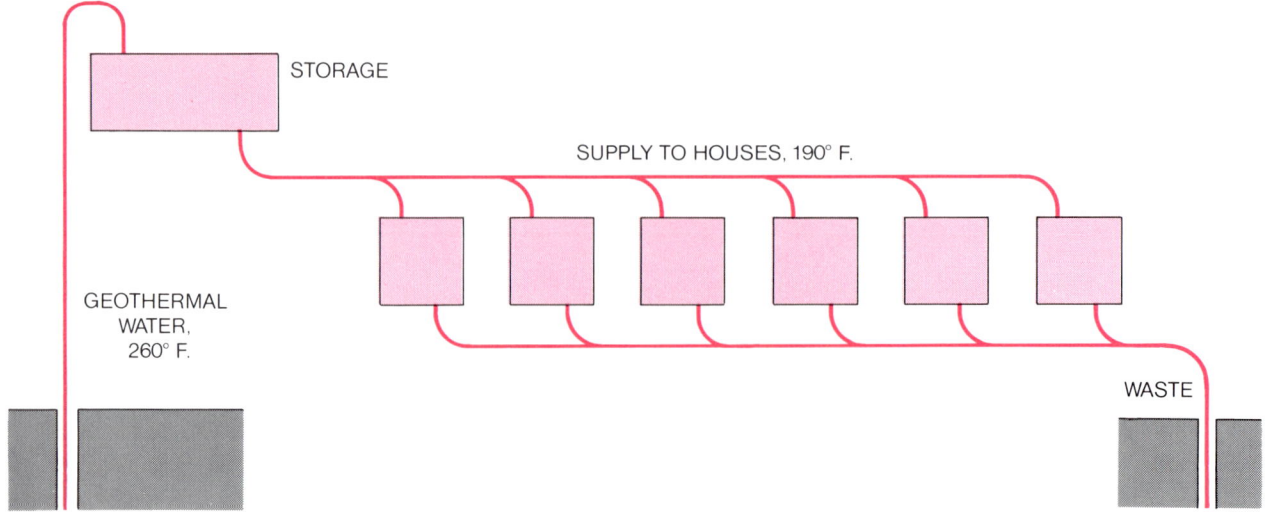

HEATING OF HOUSES and other buildings is done in a few places by a scheme such as the one shown here. Geothermal water is pumped to a storage tank, from which it flows to the buildings. Such systems are in use or being developed in several countries.

STEAM WELLS tap geothermal energy for the production of power at a plant operated by the Pacific Gas and Electric Company at The Geysers, about 90 miles north of San Francisco. Since 1960 the company has brought plant capacity at the site to 192,000 kilowatts.

LARDERELLO GEOTHERMAL FIELD in Italy has been used for generating electric power since 1904. It now has a capacity of 380,-000 kilowatts. The chimney-like structures at left are hyperbolic cooling towers that are associated with the power plant at the site.

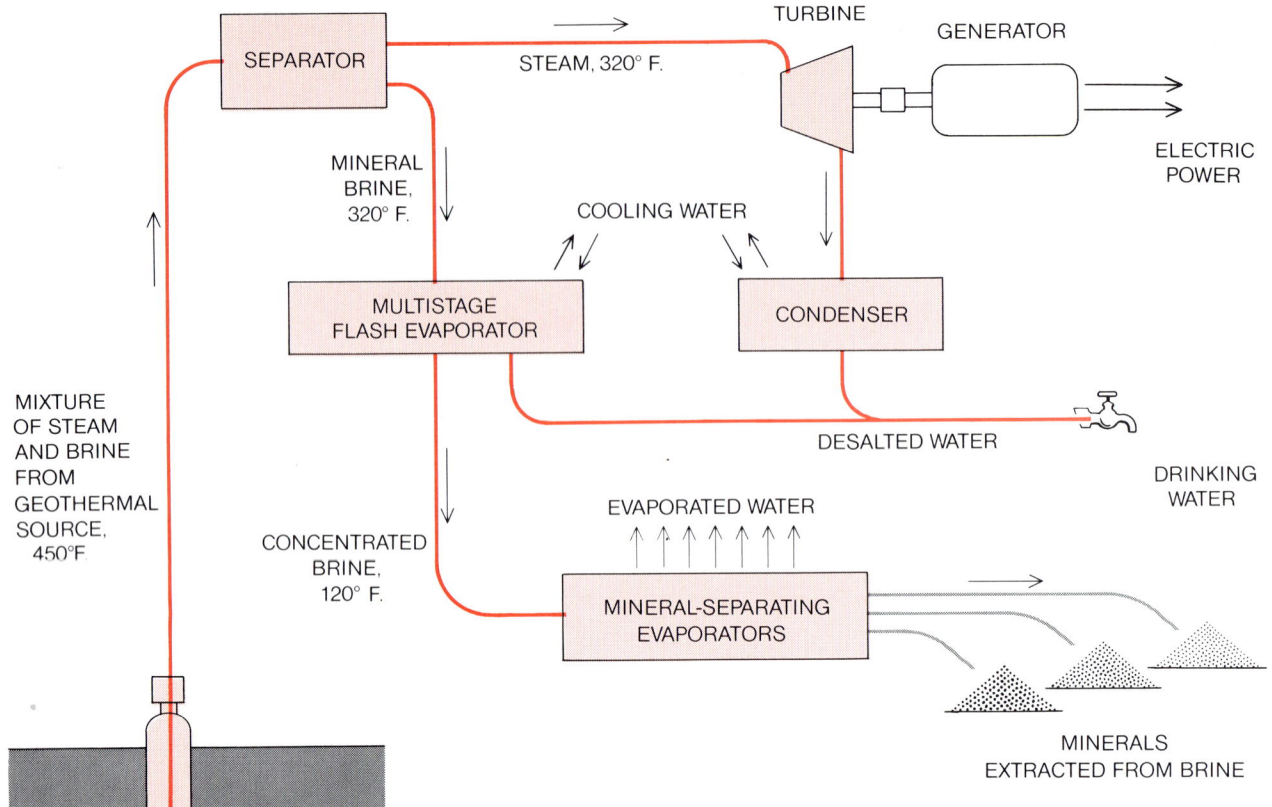

MULTIPURPOSE DEVELOPMENT based on geothermal energy is being designed by the UN and the government of Chile for a geothermal field recently discovered in Chile. In this case the geothermal source produces a mixture of steam and mineral-rich brine. The steam and brine are separated, and the steam drives a turbine to produce electric power while the brine is put through an evaporator that concentrates it, thereby producing desalted water. The concentrated brine goes to a separator that extracts the minerals.

low boiling point, which then drives the power turbine. Such a plant, installed on a field providing water at 81 degrees C., is in operation at Kamchatka in the U.S.S.R.; it uses Freon as the secondary fluid. Similar small plants have recently been built in Japan.

Among all the research needs the paramount one is the development of techniques of exploration to search the earth for geothermal reservoirs, hidden as well as visible. This will call for extensive geological, geochemical and geophysical studies and testing. (The UN recently resorted to infrared surveying in a large-scale search from the air for possible geothermal sites in Ethiopia and Kenya.) From the standpoint of geology, interest naturally focuses on areas underlain by rocks of high porosity, since these are likely to hold large quantities of water. From the standpoint of utility and benefit, one hopes to find geothermal reservoirs in arid areas where underground water itself, as well as energy and minerals, would be a boon to the region.

From surface indications alone it appears there are belts of geothermal reservoirs along the western side of the

Americas from Alaska all the way down to Chile, in the Middle East (Turkey) and East Africa throughout the African Rift Valley and in the Far East along the "Circle of Fire" of volcanic activity that surrounds the Pacific Ocean. In Turkey two-thirds of the country is believed to have geothermal potential, and there are good prospects for this resource in almost all the countries around the Mediterranean. The many spas of hot waters throughout Europe suggest that geothermal reservoirs should be widespread on that continent. Recent discoveries by drilling in Europe and elsewhere also indicate that a potential exists in many regions that had not previously been considered for exploration. The U.S. may have similar possibilities: drillers came on geothermal reservoirs in Louisiana and Texas recently during deep drilling for petroleum.

Inexpensive power and heating would be very helpful to many developing countries. Some of them, notably in Central America, are rich in geothermal resources—indeed, Central America has much more of this potential energy than it could use itself. Large-scale explora-

tion and development of its abundant geothermal fields would be very worthwhile, however, particularly for the region's economy, if the power potential were fully developed and marketed in the U.S. by way of long-distance transmission lines.

As new information becomes available the magnitude of geothermal energy resources is beginning to be appreciated. On the basis of a reconnaissance, which included airborne infrared scanning over a large area, carried out by the government of Ethiopia and the UN it has been estimated that a part of the Afar region in Ethiopia may have an exploitable geothermal potential sufficient to meet the present need for electric power for the whole of Africa. There are in addition other areas in Ethiopia that are believed to have a geothermal potential of similar magnitude.

At this stage it is impossible to estimate the magnitude of the exploitable resources of geothermal energy that lie hidden under our feet in the earth's crust. The world's energy needs and exciting recent discoveries, however, certainly warrant a great effort of exploration for this ready-made store of energy.

The Author

JOSEPH BARNEA is director of the Resources and Transport Division of the United Nations. After completing work at a technically oriented high school in Germany he studied economics in Germany, Britain and Switzerland, where he obtained his Ph.D. He worked in Palestine and then Israel, specializing in natural resources of the Middle East, and joined the UN in 1951. He writes of his interest in "new concepts of exploring for natural resources" and in "interdisciplinary and multipurpose" approaches to both exploration and development.

Bibliography

PROCEEDINGS OF THE UNITED NATIONS CONFERENCE ON NEW SOURCES OF ENERGY. VOL. II, GEOTHERMAL ENERGY: I. United Nations, 1964.

PROCEEDINGS OF THE UNITED NATIONS CONFERENCE ON NEW SOURCES OF ENERGY. VOL. III, GEOTHERMAL ENERGY: II. United Nations, 1964.

THE ECONOMIC POTENTIAL OF GEOTHERMAL RESOURCES IN CALIFORNIA. Geothermal Resources Board, State of California, January, 1971.

GEOTHERMAL RESOURCE INVESTIGATIONS, IMPERIAL VALLEY OF CALIFORNIA—STATUS REPORT. United States Department of the Interior, Bureau of Reclamation, April, 1971.

UNITED NATIONS SYMPOSIUM ON THE DEVELOPMENT AND UTILIZATION OF GEOTHERMAL RESOURCES (A SUMMARY REPORT). United Nations ST/TAO/Ser. C/126, 1971.

THE EXPLOITATION OF GEOTHERMAL RESOURCES IN THE PRESENT AND FUTURE. Gunnar Bodvarsson. United Nations, in press.

SCIENTIFIC
AMERICAN March 1972, Vol. 226, No. 3, pp. 30–38

OFFPRINT **899**

GEOSYNCLINES, MOUNTAINS
AND CONTINENT-BUILDING

by Robert S. Dietz

A geosyncline is a huge deposit of sedimentary rock that forms at
the edge of a continent. When it is compressed, it buckles up into
a mountain range. It also enables a continent to grow by accretion.

A geosyncline is a long prism of sedimentary rock laid down on a subsiding region of the earth's crust. It has long been recognized that geosynclines are fundamental geologic units. Furthermore, it has been a dictum of geology that they eventually evolve into mountains consisting of folded sedimentary strata. The laying down of such sediments and their subsequent folding constitute a basic geologic cycle that requires a few hundred million years. Until recently the original nature of geosynclines has been inferred only by studying folded mountains. It was commonly believed that there are no nascent (unfolded) geosynclines in the world today, but this would defy another geologic dictum: that the present is the key to the past.

In recent years the study of marine geology has been revolutionized by the concept of plate tectonics, which holds that the earth's crust is divided into a mosaic of about eight rigid but shifting plates in which the continents are embedded and drift along as passive passengers. With this concept the evolution of ocean basins has been rather clearly resolved. The question arises: Must plate tectonics stay at sea, or is it also the prime mover of the geosyncline mountain-building cycle? In other words, can it account for the collapse of geosynclines and the growth of continents? I am among those who believe it can. Some notable advocates of this new concept of continental evolution are John Dewey and John M. Bird of the State University of New York at Albany, Andrew Mitchell and Harold Reading of the University of Oxford and William R. Dickinson of Stanford University.

When one examines the structure of ancient folded mountains, one finds that

the classic geosyncline is divided into a couplet: two adjacent and parallel structures consisting of a eugeosyncline (true geosyncline) and a miogeosyncline (lesser geosyncline), often shortened to eugeocline and miogeocline. Now that the ocean floor is becoming better known, one need not look far to find an example of the geosynclinal couplet in process of formation. A probable example of a "living" eugeocline is the continental rise that lies seaward of the continental slope off the eastern U.S. Landward of the rise and capping the continental shelf is a wedge of sediments that becomes progressively thicker as it extends toward the shelf edge. This wedge seems to be a living miogeocline.

In dimensions and in the overall character of its rocks and stratigraphy the modern continental-rise prism closely matches typical ancient eugeoclines. It parallels the Atlantic seaboard for 2,000 kilometers, forming an apron 250 kilometers wide from the continental slope to the abyssal plain [see illustrations on pages 910–11]. Seismic studies reveal that the rise is the top of a huge planoconvex lens of sediments whose maximum depth is about 10 kilometers. The sediments are turbidites, deposited by the muddy suspensions known as tur-

bidity currents. Such suspensions periodically cascade down submarine canyons and pour across the continental rise, depositing sedimentary fans that eventually coalesce into an apron. Turbidites consist of thin graded beds of poorly sorted particles of silt and sand in which coarse material is at the base and finer material is at the top. The gradation in particle size reflects the differential rate of settling from a single injection of muddy sand. Interlayered with the graded beds are fine clays (pelagites) that slowly settle from the overlying water as a "gentle rain" between major influxes of turbidity currents.

Collapsed eugeoclines in ancient folded mountains are similarly composed of thick and repetitive sequences of turbidites; these strata are usually termed flysch or graywacke. Mixed with the graywackes are thin limestones, ironstones and cherts formed from the skeletons of radiolarians, indicating that the sediments were deposited in deep water. True fossils are sparse, but many eugeoclinal sequences of the lower Paleozoic era contain graptolites: extinct plantlike animals that settled down from the surface.

Close examination of the graded beds also reveals what are called sole mark-

COLLISION OF CONTINENTS is depicted in this view of the Zagros Mountains in Iran along the Persian Gulf taken from the spacecraft *Gemini 12* in November, 1966. The mountains are uplifted folds of sedimentary strata, originally deposited as a geosyncline, whose cores have been exposed by erosion. The foldbelt has apparently been thrown up by the collision of the Arabian block, rotating counterclockwise, with the Eurasian block, rotating clockwise. Since the Arabian block is part of the African block, the folding represents the collision between Africa and Eurasia. The Zagros Mountains and the shallow Persian Gulf are both part of the Arabian block that extends to the Red Sea. The suture between the Arabian block and the Eurasian block is marked by a major thrust fault that passes through the upper right corner of the photograph just beyond the mountain chains.

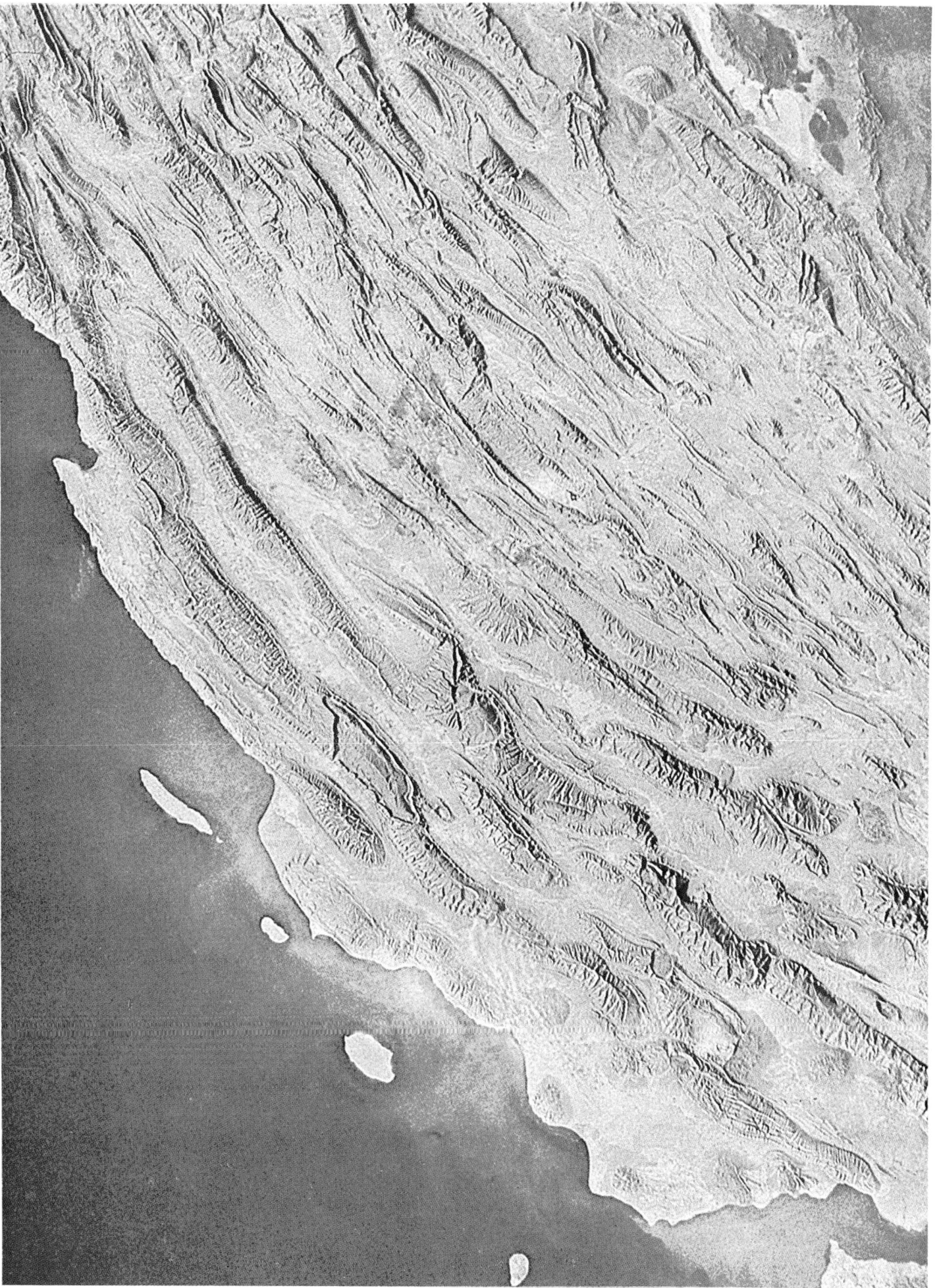

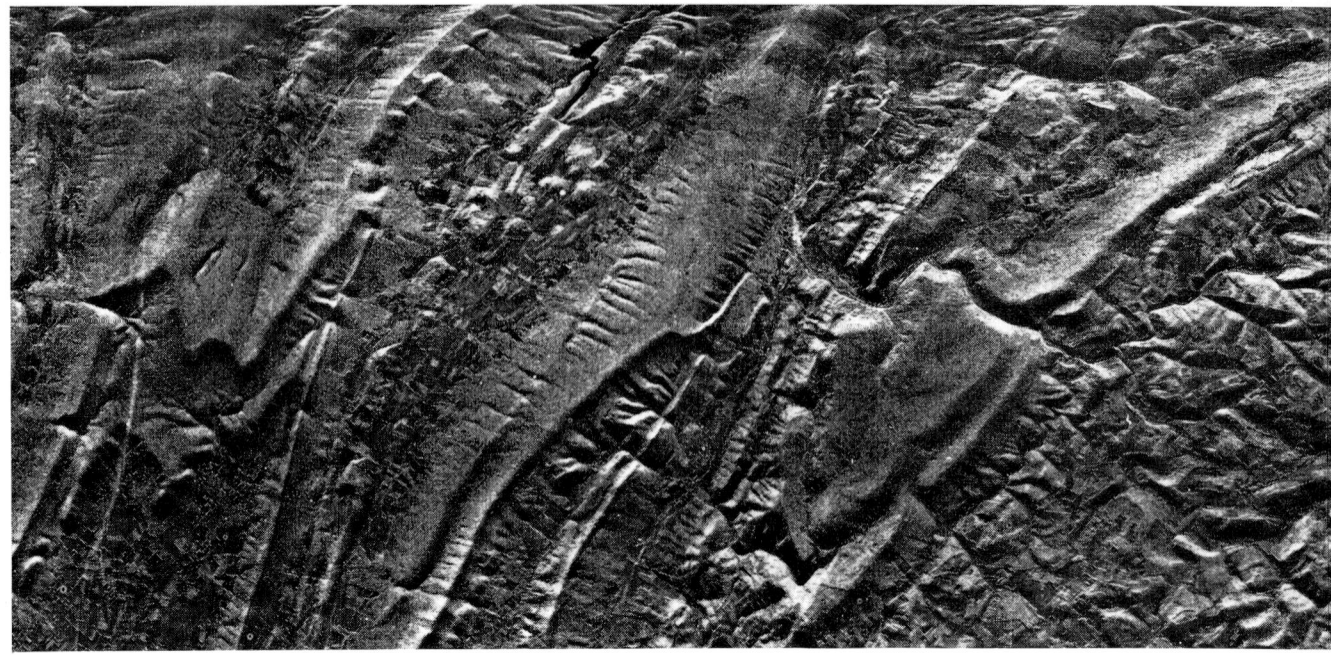

FOLDED APPALACHIAN MOUNTAINS in western Pennsylvania are depicted in this image produced by side-looking radar. The picture covers a region 25 miles long parallel to the Maryland border, centered approximately at 78 degrees 45 minutes west longitude. The picture is printed with north at the bottom so that the landscape appears to be lighted from the top. (The illuminant, of course, is the radar beam transmitted from an airplane.) If the picture is inverted, features that are actually elevated appear depressed and vice versa. This image and the sequence of three views at the bottom of these two pages were made by the National Aeronautics

ings or "flysch figures," for example ripple marks of a kind that could have been produced by turbidity currents. There can be little doubt that most of these sequences are the uplifted and eroded remnants of former continental-rise prisms. The crystalline Appalachians, which are that part of the Appalachians lying seaward of the Blue Ridge Mountains and equivalent ranges to the north and south, bear the clear imprint of being a collapsed continental-rise prism

laid down in the early Paleozoic some 450 to 600 million years ago. The original prism has been much altered by intrusions and metamorphism.

The sedimentary wedge that underlies the coastal plain and continental shelf along the Atlantic seaboard appears to be an actively growing miogeocline. The wedge thickens as it progresses seaward, attaining a total thickness of between three and five kilometers along the shelf edge. Laid down

on a basement of Paleozoic rocks, the wedge is composed of well-sorted shallow-water sediments deposited during the past 150 million years under conditions much like those of today. The stratified beds exhibit characteristics indicating they were deposited across the continental shelf in alluvial plains, in lagoons, along shorelines and offshore. Taking into account expected changes in the pattern of sedimentation over geologic time, the present Atlantic marine

APPALACHIAN FOLDBELT north of Harrisburg, Pa., is an extension of the foldbelt shown at the top of the page. In these side-looking radar views north is at the right. The three pictures cover a distance of 75 miles from just south of Mechanicsburg to the vicinity of a town called Jersey Shore on the West Branch of Susque-hanna River. The Susquehanna River itself appears in the first frame at the left. The folded Appalachians were probably created in a late compressional stage of the collision between Africa and North America more than 450 million years ago, which caused "rugfolds" in the strata of sedimentary rock that formed part of a

and Space Administration in collaboration with the Remote Sensing Laboratory of the University of Kansas. The *K*-band radar system that produced the images was built by the Westinghouse Electric Corporation.

deposits closely resemble the ancient miogeoclinal foldbelts of the Paleozoic era and earlier. For example, the modern sedimentary wedge is much like the one found in the folded Appalachians of Pennsylvania. Both wedges are characterized by "thickening out," signifying that they grow steadily thicker toward the east before they abruptly terminate.

If the foregoing analysis is correct, one must conclude that geosynclines are actively forming along many continental margins today: eugeoclines at the base of the continental slope and miogeoclines capping the continental shelves. It remains to be shown, however, that the crustal shifting associated with plate tectonics can convert these sedimentary prisms into the mountainous foldbelts that make up the fabric of the continents, mostly as ancient eroded mountain roots rather than as modern mountain belts. In order to examine this possibility we must first summarize some of the basic concepts of plate tectonics.

The approximately eight rigid but shifting plates into which the earth is currently divided are thought to be about 100 kilometers thick. Most of the plates support at least one massive continental plateau, often referred to as a craton. We can visualize the ideal plate as being rectangular, although only the plate supporting the Indian craton approaches this simple shape. Along one edge of a crustal plate there is a subduction zone, usually marked by a trench, where the plate dives steeply into the earth's mantle, attaining a depth as great as 700 kilometers before being fully absorbed into the mantle. On the opposite side of the plate from the subduction zone is a mid-ocean rift, or pull-apart zone. As the rift opens, the gap is quickly healed from below by the inflow of liquid basalt and quasi-solid mantle rock. The other two opposed sides of the plate, connecting the rifts to the trenches, are shears called transform faults.

Hence three types of plate boundary are possible: divergent junctures (the mid-ocean rifts where new ocean crust is created), shear junctures (the transform faults where the plates slip laterally past one another, so that crust is conserved) and convergent junctures (trenches where two plates collide, with one being subducted and consumed). Only the last of the three, the convergent juncture, can help to explain how the sedimentary prism of a submarine geosyncline might be collapsed into a folded range of mountains. As the plate carrying a prism collides with a plate carrying a continental craton one would expect the prism to be compressed into folds. Thrusting and crustal thickening would follow, assisted by isostatic forces that act to keep adjacent crustal masses in balance. Such forces would cause the collapsed prism to be uplifted. The entire process would be accompanied by the generation and intrusion of magma, together with extensive metamorphism of the crustal rocks.

A grand theme of plate tectonics is that ocean basins are not fixed in size or shape; they are either opening or closing. Today the Atlantic Ocean is opening and the Pacific Ocean is closing. The drifting of the continents is another theme; every continent must have a leading edge and a trailing edge. For the past 200 million years the Pacific coast of North America has been the leading edge and the Atlantic coast the trailing edge. The trailing margin is tectonically stable, and since the continental divide is near the mountainous Pacific rim, most of the sediments are ultimately dumped into the Atlantic Ocean, including the Gulf of Mexico. Therefore it is primarily along a trailing edge that the great geosynclinal prisms are deposited.

Consider, however, what would hap-

Paleozoic geosyncline (*see illustration on page 912*). After folding the region was eroded to a level plain and then uplifted. The modern mountains were subsequently etched out according to the hardness of the various strata. Thus the ridges are composed mainly of dense sandstone and can be either synclines (troughs) or anticlines (arches). The V-shaped chevrons in the first frame are synclines that plunge to the northeast. The Susquehanna River established its course when the entire region was reduced to a level surface, so that its course has been superimposed on the folded structure, thereby cutting directly across the folds and creating water gaps.

pen if, with the changing patterns of plate motions, a subduction zone (a trench) were created along a former trailing edge, forming a new plate boundary. The Atlantic would be transformed into a closing ocean with its geosynclinal prisms riding toward the trench. The continental margin and the trench would eventually collide, collapsing the eugeocline into a contorted mountainous foldbelt and also folding the miogeocline to a much lesser extent. Before that happened the continental margin would encounter and incorporate an island arc, similar to the island arcs found along the perimeter of the western Pacific. These arcs are created by tectonic and magmatic activity triggered by the plunging crustal plate. It is also quite possible that the Atlantic Ocean would close entirely, causing North Africa to collide with eastern North America. The collision of India with the underbelly of Asia, throwing up the Himalaya rampart, would be a present-day analogy. One can imagine many possible scenarios, depending on the geometry of plate boundaries and other variables.

The creation of a eugeoclinal foldbelt is of course considerably more than simply the accordion-like collapse of a continental-rise prism. The foldbelt is sheared into thrust faults and the landward edge of the eugeocline is common-

ly thrust onto the adjacent miogeocline. The descending crustal plate is not entirely consumed within the earth's hot mantle, with the result that low-density magmas buoyantly rise and invade the eugeocline. This leads to intrusions of granodiorite (a granite-quartz rock) and the growth of volcanic mountains consisting of andesite (the rock characteristic of the Andes). This lava is highly explosive because it is charged with water sweated out of the descending plate. Magma is not generated from the plunging lithosphere until it has reached a considerable depth. As a result the eugeocline can be subdivided into two parallel geologic belts. Toward the sea one finds sedimentary rock transformed at high pressure and low temperature; farther inland the sedimentary rock has been altered predominantly at low pressure and high temperature by the numerous intrusions of magma. From the new marginal mountain range, delta and river deposits sweep back across the continent, covering the miogeocline with a suite of continental shales and conglomerates collectively called molasse.

The concept that the geosynclinal cycle is controlled by plate tectonics provides some new answers to old questions about geosynclines. For example, is mountain-building periodic and world-

wide or is it random in space and time? The answer must be both yes and no. On the one hand, the crustal plates are highly intermeshed; the drift of any one plate has global repercussions, giving rise to synchronous mountain-building. Any brief interval of rapid plate motion would also be one of widespread mountain-building. On the other hand, the rate of plate convergence is highly dependent on the latitudinal distance from the relative pole of rotation of that plate and on the particular geometry of the plate boundary.

A law of plate tectonics states that sea-floor spreading (injection of new ocean crust) proceeds at right angles to a rift; the crustal plate, however, may be subducted into a trench at any angle. The rate of subduction and the attendant distorting of the crust therefore vary greatly from place to place, as can be observed on the perimeter of the Pacific today. Thus it would seem that although mountain-building over the span of geologic time may reach crescendos, it must also be continuous and random.

The plate-tectonic version of the geosynclinal cycle predicts that miogeoclines are ensialic, or laid down on continental crust (sial), whereas eugeoclines are ensimatic, or deposited on oceanic crust (sima). This differs from the earlier view that all geosynclines are ensialic,

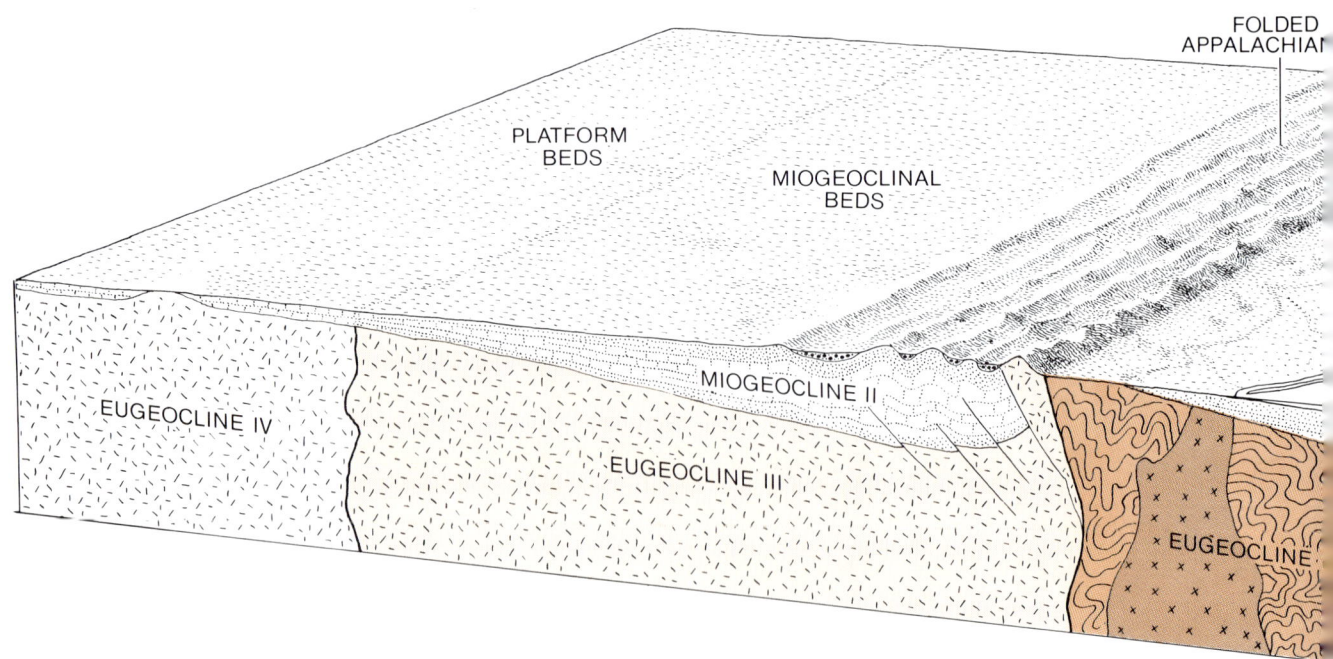

SUCCESSION OF EUGEOCLINES underlies nearly all North America below the relatively undisturbed cover beds. These contorted and intruded prisms constitute the fundamental fabric of continents, known as the basement complex. A new geosynclinal couplet is being deposited today. It consists of a miogeocline (lesser geosyncline) of shallow water beds that caps the coastal plain and continental shelf paralleled by a eugeocline (true geosyncline) that is formed at the base of the continental slope by detritus

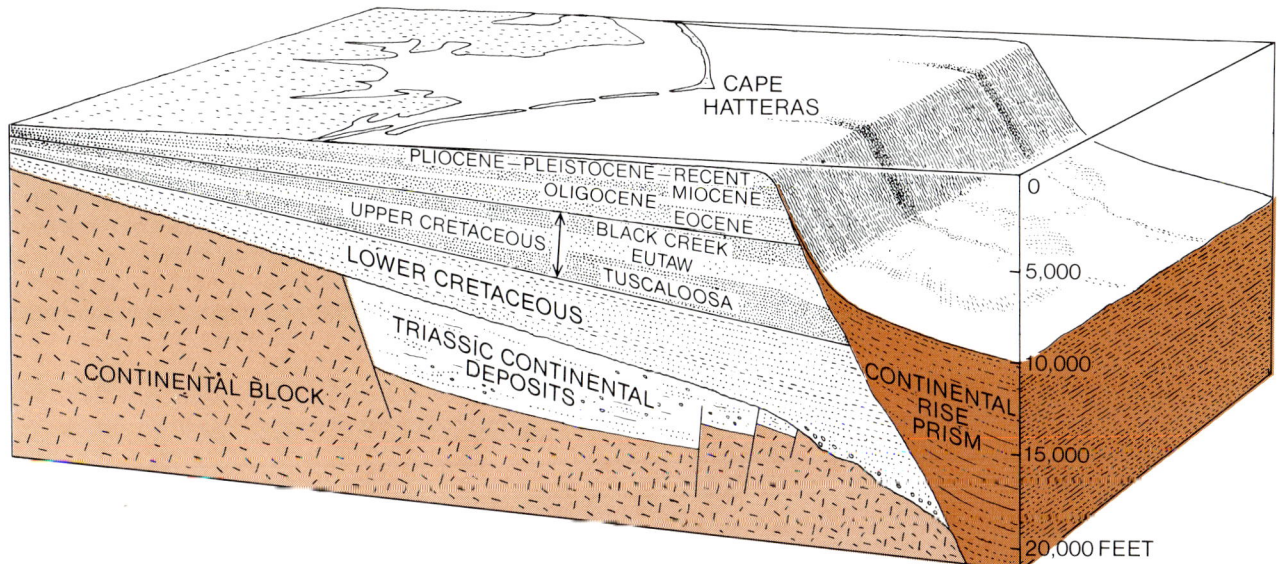

LIVING GEOSYNCLINAL COUPLET off the Atlantic coast of the U.S. consists of a miogeocline, strata laid down on the shallow continental shelf during the past 150 million years, and a eugeocline prism (*dark color*), consisting of thin beds of sand and mud deposited by turbidity currents flowing down the continental slope. The material in the Triassic basin represents continental deposits laid down before the foundering of the continental margin under tension 190 million years ago, prior to opening of Atlantic Ocean.

which is certainly incorrect. Early investigators observed that a granitic basement is invariably present under miogeoclines and evidently reasoned that a similar sialic basement, although it was unseen, must also be present under eugeoclines. A collapsed eugeocline is as thick as the continental plate, about 35 kilometers, so that its basement is beyond the depth of even the deepest boreholes. We can infer that the ultimate basement is simatic, however, by observing detached fragments that are caught up in the contorted mélange of the eugeoclinal pile. These fragments include samples of oceanic crust (for example radiolarian cherts and sodium-altered lavas) and upper-mantle rocks (for example serpentinites and peridotites).

The ensimatic location of eugeoclines can also account for their tectonic style. They are tightly folded, faulted, tumbled and dynamically metamorphosed into an almost unmappable mélange. This contorted state is understandable, since the ocean floor is shearing under the eugeocline and thrusting the sedimentary pile against the continental slope. Extensive tectonic thickening and interleaving must occur before the pile will rise to mountainous altitudes. On the other hand, the miogeocline beds are protected by the stable continental slab, so that they are simply thrown into a series of loose, open, ruglike folds.

It now seems amusing to recall that 19th-century geologists, using a wrinkled apple as an analogy, interpreted folded mountains to mean that the earth was shrinking. Today it seems clear that eugeoclines are deposited at the edge of a continent on oceanic crust seaward of the continental slope, so that folded mountains really show that the continents are growing larger through marginal accretion. Mountain-building is therefore evidence of an even more fundamental geologic process: the growth of continents. The continents grow not as a layer cake but as a craton that is divided vertically into zones with an old nucleus and young margins.

An important aspect of geosynclines requiring explanation is that they are laid down on foundations that are continuously subsiding. This aspect is par-

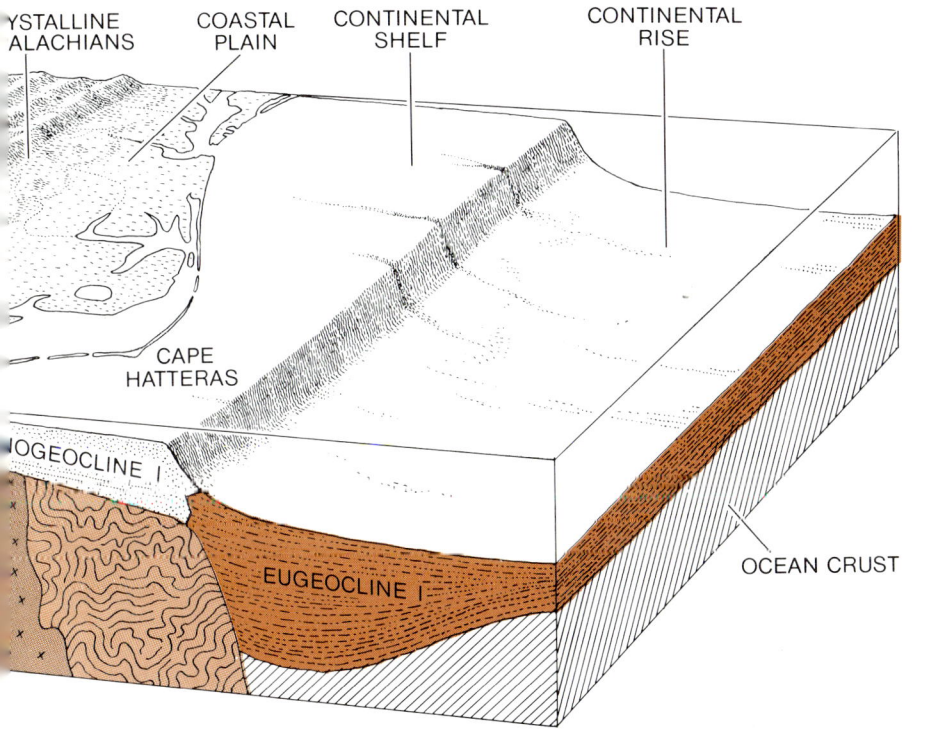

washed over the shelf edge. If at some future time the sea floor were to thrust against the continent, the modern eugeocline (*I*) would collapse into a new foldbelt like the earlier ones. The hypothetical mechanism that creates foldbelts is shown on page 913. This diagram and others are based on drawings by the author's colleague John C. Holden.

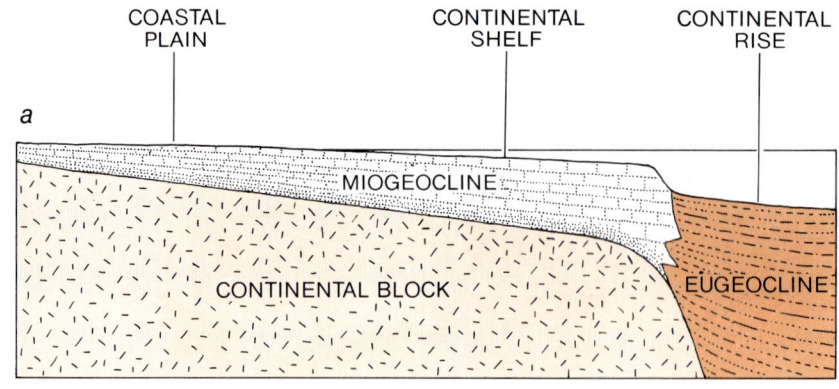

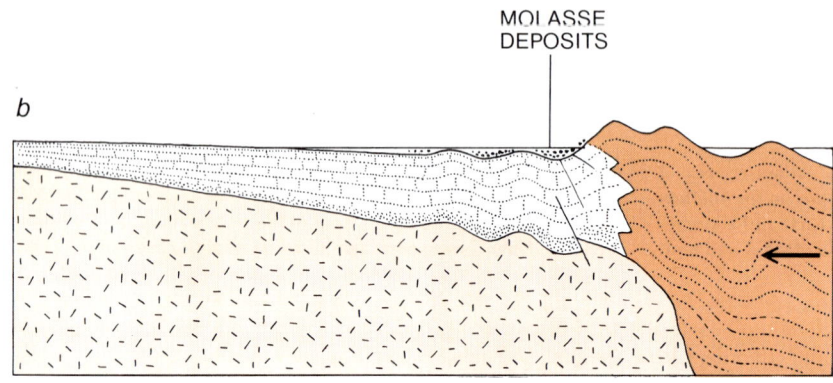

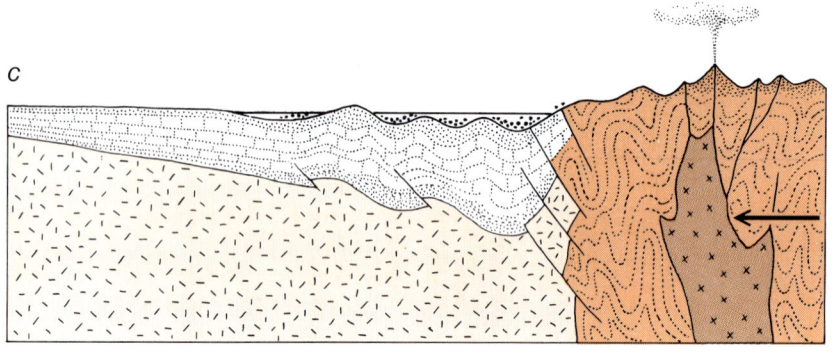

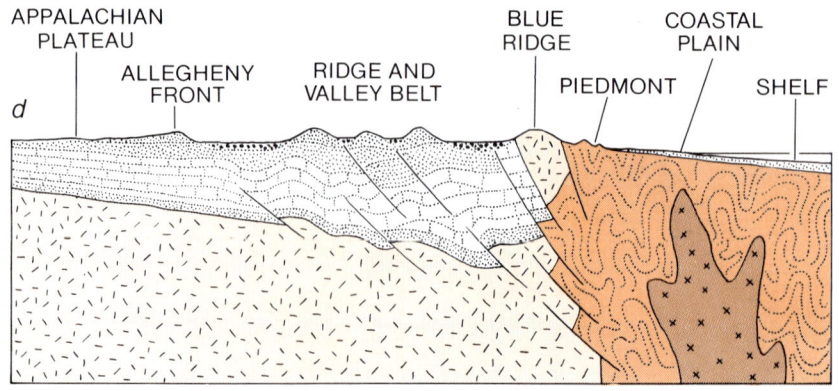

CRUMPLING OF EARLIER GEOSYNCLINAL COUPLET, apparently laid down in late Precambrian and early Paleozoic time more than 450 million years ago, produced the Appalachian foldbelt. The four-part sequence shows how the miogeocline, or western part of the geosynclinal couplet, was folded into the series of ridges between the Blue Ridge line and the Allegheny front. The eugeocline, altered by heat, pressure and volcanism, formed a lofty range of mountains, now almost completely eroded, east of the Blue Ridge line.

ticularly evident in miogeoclines, which can attain a total thickness at their seaward edge of five kilometers even though they are entirely composed of beds deposited in shallow water. This phenomenon is nicely accounted for by plate tectonics: the margins of rift oceans inherently have, as one geologist has expressed it, a "certain sinking feeling."

Let us take as an example the Atlantic Ocean between the U.S. and the bulge of Africa. This new ocean basin was created about 180 million years ago by the insertion of a spreading rift that split North America away from Africa [see illustration on opposite page]. Attendant swelling of the mantle arched the continents upward along the rift line by about two kilometers. Erosion then beveled the raised edges, thinning the margins of the two continental plates.

A modern example of crustal arching associated with incipient rifting of the crust can be observed in the high dorsal of Africa from Ethiopia southward. The Red Sea provides a more advanced stage of a newly opening ocean basin. Along the flanks of this linear trough crustal arching has stripped away young rocks, exposing "windows" of Precambrian basement.

With the insertion of new oceanic crust by sea-floor spreading, the ocean grew ever wider. In the process the continental edges subsided, as is demonstrated by the sloping flanks of the mid-Atlantic ridge today. When the ocean was smaller, the continental edges had to ride down a similar slope. Eventually the inflated mantle under the ridge reverts to normal mantle, but this takes 100 million years or more. Therefore as a geosyncline is laid down on the trailing edges of a drifting continent it slowly subsides for reasons external to the sedimentary deposit itself.

Additional subsidence, however, is caused by the steadily growing mass of the sedimentary apron, which must be isostatically compensated because the earth's crust is not sufficiently strong to sustain the load. For every three meters of sediment deposited the crust sinks about two meters. This crustal failure, however, is spread over a large geographic area, so that the growth and subsidence of a huge continental-rise prism causes a sympathetic downward flexing of the adjacent continental margin. As the continental shelf slowly tilts, wedges of shallow-water sediments are deposited.

Over the millenniums, as the shoreline transgresses and regresses repeatedly across the shelf, a large composite

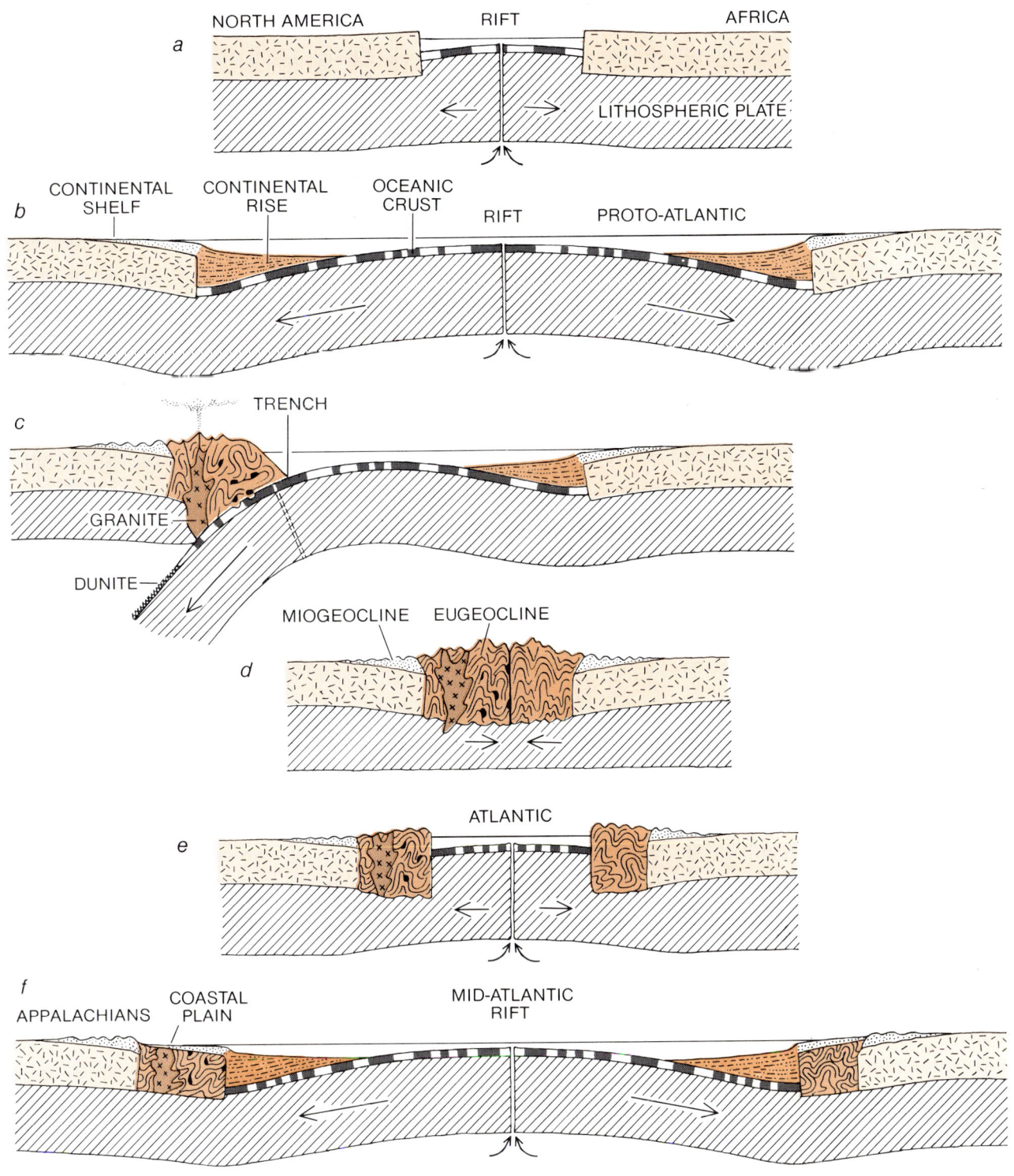

MECHANISM OF CRUMPLING that produced the Appalachian foldbelt is depicted on the hypothesis that the Atlantic Ocean has opened, closed and reopened. In the late Precambrian (*a*), North America and Africa are split apart by a spreading rift, which inserts a new ocean basin. By the process of sea-floor spreading (*b*) the ancestral Atlantic Ocean opens. New oceanic crust is created as the plates on each side move apart. As the crust cools, its direction of magnetization takes the sign of the earth's magnetic field; the field periodically reverses, and the reversals are represented by the striped pattern. On the margin of each continent sediments produce the geosynclinal couplet: miogeocline on the continental shelf, eugeocline on the ocean floor itself. The ancestral Atlantic now begins to close (*c*). The lithosphere breaks, forming a new plate boundary, and a trench is produced as the lithosphere de-

scends into the earth's mantle and is resorbed. The consequent underthrusting collapses the eugeocline, creating the ancient Appalachians. The eugeocline is intruded with ascending magmas that create plutons of granite and volcanic mountains of andesite. The proto-Atlantic is now fully closed (*d*). The opposing continental masses, each carrying a geosyncline couplet, are sutured together, leaving only a transform fault (*vertical black line*). The shear contains squeezed-up pods of ultramafic mantle rock. Sediments eroded from the mountain foldbelt create deltas and fluvial deposits collectively called molasse. North America and Africa were apparently joined in this way between 350 million and 225 million years ago. About 180 million years ago (*e*) the present Atlantic reopened near the old suture line. Today (*f*) the central North Atlantic is opening at the rate of three centimeters per year, creating new geosynclines.

megawedge of shallow-water sediments caps the shelf. The abundant supply of sediment usually ensures that the top of the prism is maintained close to sea level. Excess detritus bypasses the shelf, is temporarily dumped on the continental slope and is then carried onto the continental rise by turbidity currents. The shelf edge and the continental-rise prism comprise a couplet within which there is constant interplay.

Like the sedimentary wedge under the Atlantic coastal plain today, the early Paleozoic Appalachian miogeocline thickens in the seaward direction. The abrupt termination of this miogeocline was long a mystery to early geologists who mapped it. They suggested that a missing seaward limb had been thrust upward and completely eroded away or that it had foundered into an ancient oceanic basin. This hypothetical land mass of Appalachia was the geological equivalent of the legendary Atlantis. The wedgelike structure of the existing continental-plain prism provides a satisfactory solution to the puzzle: the hypothesized seaward limb never existed. We now see that the thickening out of sedimentary deposits at the shelf edge is a normal mode of sedimentation. One way in which this may happen is that reefs of

coral and algae build up along the margin of the continental shelf, creating a carbonate dam behind which other shelf sediments accumulate.

The mechanism of building continents by the peripheral accretion of collapsed continental rises seems also to ensure that the sedimentary deposits become dry land. (We take for granted that continents are above sea level, but it should be remembered that the mid-ocean ridge system, which approaches the continents in importance as a topographic feature, almost never rises above the sea surface.) The sedimentary apron gradually thickens until it approaches the height of the continental slope (about five kilometers), but upward growth ceases once the slope is completed and sea level is attained.

As we have seen, however, isostasy is at work, causing the oceanic crust to subside under the sedimentary load. The result is that a fully developed sedimentary prism attains an overall thickness of about 15 kilometers. When the prism is subsequently collapsed into a eugeosynclinal foldbelt, it becomes thicker still. The attendant metamorphism and granitic intrusion (which increases the total mass of rock) give rise to a monolithic structure that is more than 35 kilometers thick, thicker than a continental plate.

Thus new foldbelts not only rise above sea level but also throw up rugged mountain ranges.

The hypothesis that geosynclines are deposited along a continental margin and then crushed against the continent as a result of plate tectonics seems to explain satisfactorily how geosynclines are transformed into folded mountains. The close relation between eugeoclines and foldbelts is not one of cause and effect but a simple consequence of location: geosynclines are laid down along continental margins and such margins are the locus of interaction between continents and subduction zones.

In spite of the vast span of geologic time and the rigors of erosion, the continents remain in a good state of health. We can predict that they always will be: detritus lost to the oceans is eventually carried back to the continents and collapsed into accretionary belts that also incorporate new igneous rock. Although the earthquakes that punctuate mountain-building are sometimes disastrous to man's culture, they are acts of continental construction. The great flood—the complete inundation of the erosionally leveled continents—will always threaten but will never come to pass.

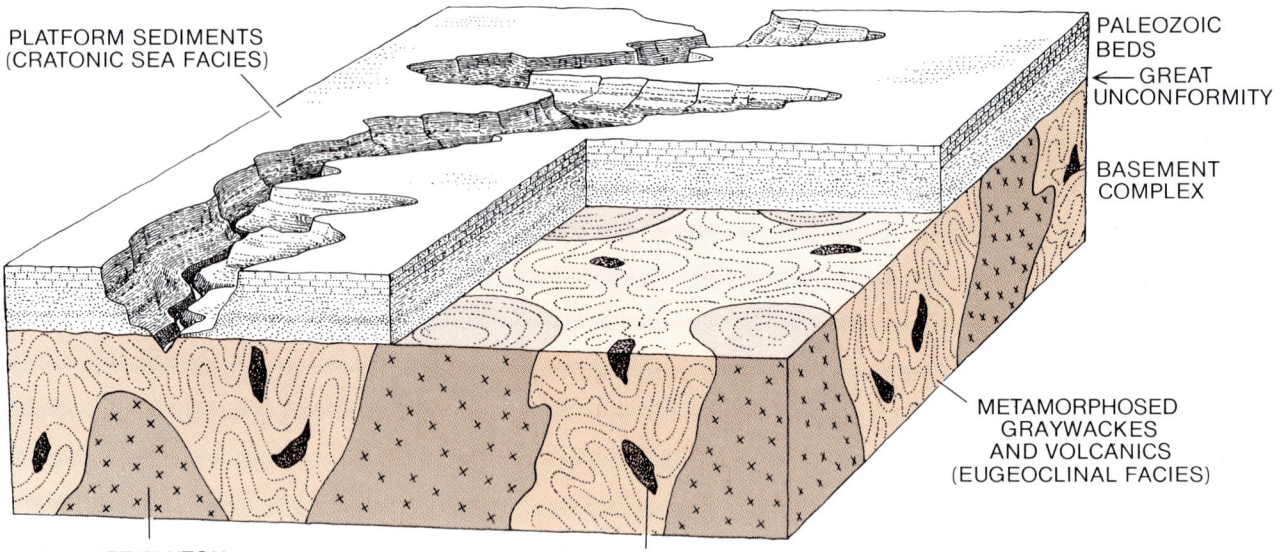

PLATFORM SEDIMENTS (CRATONIC SEA FACIES)

PALEOZOIC BEDS

← GREAT UNCONFORMITY

BASEMENT COMPLEX

METAMORPHOSED GRAYWACKES AND VOLCANICS (EUGEOCLINAL FACIES)

GRANITE PLUTON

ULTRAMAFIC PODS

DEEP FABRIC OF CONTINENTS, the basement complex, is the fundamental rock unit of the continental plateaus, or cratons. This complex is usually obscured from view by miogeoclinal beds or by the coating of shallow-sea deposits that have invaded the continents from time to time. (Hudson Bay is a modern example of such an invading shallow sea.) Long a puzzle to geologists, the basement is composed of eugeoclinal foldbelts that have undergone intensive folding, metamorphism and intrusion. Geologists once thought that these "roots of mountains" indicated that the earth had contracted while cooling. The folds were likened to the skin of a dried apple. The present interpretation is that the eugeoclinal facies was laid down on the ocean floor and subsequently was crumpled against the continental margin, building up an onion-like vertically zoned

craton. On this view the continents have grown larger rather than smaller with time. Moreover, the basement complex need not be Archean (composed of the oldest rocks) as formerly supposed, because the high degree of metamorphism does not necessarily indicate great antiquity and repeated mountain-building events. Instead it reflects the intensity of the collapsing process; once accreted, the foldbelt is not usually mobilized again. It has long been known that the granites of the basement complex are always younger than the metamorphosed sediments they intrude. On the other hand, the included pods of ultramafic rocks are always older, since they are detached fragments of the oceanic foundation on which the eugeocline was deposited. If geologists ever find "the original crust of the earth," it will be one of these pods within the oldest foldbelts.

The Author

ROBERT S. DIETZ is a marine geologist with the National Oceanic and Atmospheric Administration, working at the Atlantic Oceanographic and Meteorological Laboratories in Miami. His degrees are from the University of Illinois, where he received his Ph.D. in 1941. He has written extensively on geosynclines, plate tectonics, sea-floor spreading, continental drift, marine mineral resources and deep-research vehicles. "As a departure from investigating the ocean floor," he writes, "I occasionally take trips to various parts of the world to study geologic scars of ancient meteoritic or cometary impacts. These are not circular holes in the ground but usually complex disrupted domes revealing evidence of intense shock." For his research in geotectonics Dietz recently received the Walter H. Bucher medal of the American Geophysical Union and the gold medal of the U.S. Department of Commerce.

Bibliography

NORTH AMERICAN GEOSYNCLINES: THE GEOLOGICAL SOCIETY OF AMERICA, MEMOIR 48. Marshall Kay. Geological Society of America, 1951.

COLLAPSING CONTINENTAL RISES: AN ACTUALISTIC CONCEPT OF GEOSYNCLINES AND MOUNTAIN BUILDING. Robert S. Dietz in *The Journal of Geology*, Vol. 71, No. 3, pages 314–333; May, 1963.

MIOGEOCLINES (MIOGEOSYNCLINES) IN SPACE AND TIME. Robert S. Dietz and John C. Holden in *The Journal of Geology*, Vol. 74, No. 5, Part 1, pages 566-583; September, 1966.

CONTINENTAL MARGINS, GEOSYNCLINES, AND OCEAN FLOOR SPREADING. Andrew H. Mitchell and Harold G. Reading in *The Journal of Geology*, Vol. 77, No. 6, pages 629–646; November, 1969.

MOUNTAIN BELTS AND THE NEW GLOBAL TECTONICS. John F. Dewey and John M. Bird in *Journal of Geophysical Research*, Vol. 75, No. 14, pages 2625–2647; May 10, 1970.